Lecture Notes in Statistics 109

Edited by P. Bickel, P. Diggle, S. Fienberg, K. Krickeberg, I. Olkin, N. Wermuth, S. Zeger

AF698069

Springer
New York
Berlin
Heidelberg
Barcelona
Budapest
Hong Kong
London
Milan
Paris
Santa Clara
Singapore
Tokyo

Helmut Rieder
(Editor)

Nova Southeastern
University Libraries
3301 College Avenue
Ft. Lauderdale, FL 33314

Robust Statistics, Data Analysis, and Computer Intensive Methods

In Honor of Peter Huber's 60th Birthday

Springer

Helmut Rieder
Universität Bayreuth
Lehrstuhl Mathematik VII
D-95440 Bayreuth
Universitätstrasse 30
Gebäude NM II
Germany

Library of Congress Cataloging-in-Publication Data Available
Printed on acid-free paper.

© 1996 Springer-Verlag New York, Inc.
All rights reserved. This work may not be translated or copied in whole or in part without the written permission of the publisher (Springer-Verlag New York, Inc., 175 Fifth Avenue, New York, NY 10010, USA), except for brief excerpts in connection with reviews or scholarly analysis. Use in connection with any form of information storage and retrieval, electronic adaptation, computer software, or by similar or dissimilar methodology now known or hereafter developed is forbidden. The use of general descriptive names, trade names, trademarks, etc., in this publication, even if the former are not especially identified, is not to be taken as a sign that such names, as understood by the Trade Marks and Merchandise Marks Act, may accordingly be used freely by anyone.

Camera ready copy provided by the author.
Printed and bound by Braun-Brumfield, Ann Arbor, MI.
Printed in the United States of America.

9 8 7 6 5 4 3 2 1

ISBN 0-387-94660-8 Springer-Verlag New York Berlin Heidelberg

Table of Contents

Peter J. Huber

born in Wohlen/Aargau, Switzerland	25 March 1934

Education

Ph. D. ETH Zürich (Mathematics)	1961
Diploma ETH Zürich (Mathematics)	1958

Teaching and Research Positions

Universität Bayreuth	
Lehrstuhl für Mathematische Statistik	1992–present
Massachusetts Institute of Technology	
Full Professor (Applied Mathematics)	1988–1992
Harvard University	
Full Professor (Statistics)	1978–1988
Eidg. Technische Hochschule Zürich	
Full Professor (Mathematical Statistics)	1964–1978
Harvard University	
Visiting Professor	1977–1978
Yale University	
Visiting Professor	1971–1972
Princeton University	
Visiting Professor	1970–1971
Cornell University	
Visiting Professor	1966–1967
Visiting Professor	1963–1964
University of California, Berkeley	
MSRI, Visiting Professor	1982
Miller Fellow	1962–1963
Research Fellowship (Schweiz. Nationalfonds)	1961–1962

Memberships

International Statistical Institute
Institute of Mathematical Statistics (Fellow)
American Mathematical Society
Association for Computing Machinery
Bernoulli Society
American Association for the Advancement of Science (Fellow)
National Computer Graphics Association
American Academy of Arts and Sciences (Fellow)

Professional Activities

Board Member, Institute for Mathematics and its Applications	1984–1987
Council Member, Institute of Mathematical Statistics	1983–1986
Member, Panel on Computers in Statistics Research (ONR "Eddy Panel")	1986
Member, Panel on Future Directions in Computational Mathematics (NSF, "Rheinboldt Panel")	1985
Associate Editor—Zeitschrift für Wahrscheinlichkeitstheorie	1972–1987
Associate Editor—Annals of Statistics	
Chairman of several conferences (IMS/AMS Summer School, Oberwolfach, DMV–Seminar, ...)	

Honors

Dr. h. c. , Université de Neuchâtel	1994
Humboldt Award	1988
Wald Lecturer, IMS Annual Meeting	1972
NSF Senior Foreign Scientist Fellowship, Princeton	1970–1971
Silver Medal of the Eidgenössische Technische Hochschule	1961

List of Publications

[1] Zu einem Mathematischen Keilschrifttext (VAT 8512). *Isis* **46** (1955) 104–106.

[2] Bemerkungen über Mathematische Keilschrifttexte. *Enseignement mathematique* **3** (1957) 19–27.

[3] Zur täglichen Bewegung des Jupiter nach Babylonischen Texten. *Z. Assyriol.* N. F. **18** (1957) 265–303.

[4] Über den Nullpunkt der Babylonischen Ekliptik. *Centaurus* **5** (1958) 192–208.

[5] Homotopical cohomology and Čech cohomology. *Math. Ann.* **144** (1961) 73–76.

[6] Homotopy theory in general categories. *Math. Ann.* **144** (1961) 361–385.

[7] Standard constructions in abelian categories. *Math. Ann.* **146** (1962) 321–325.

[8] A remark on a paper of Trawinski and David. *Ann. Math. Stat.* **34** (1963) 92–94.

[9] (with H. Bühlmann). Pairwise comparison and ranking in tournaments. *Ann. Math. Stat.* **34** (1963) 501–510.

[10] Pairwise comparison and ranking: Optimum properties of the row sum procedure. *Ann. Math. Stat.* **34** (1963) 511–520.

[11] Robust estimation of a location parameter. *Ann. Math. Stat.* **35** (1964) 73–101.

[12] A robust version of the probability ratio test. *Ann. Math. Stat.* **36** (1965) 1753–1758.

[13] Neuere Fortschritte in der Theorie der Bayesschen Statistik. *Mitteil. d. Vereinig. schweiz. Versicherungsmath.* **65** (1965) 163–171.

[14] Strict efficiency excludes superefficiency (abstract). *Ann. Math. Stat.* **37** (1966) 1425.

[15] The behavior of maximum likelihood estimates under nonstandard conditions. *Proc. 5th Berkeley Symp. Math. Stat. Prob.*, Vol. I (1967) 221–233.

[16] Robust confidence limits. *Z. Wahrsch. verw. Gebiete* **10** (1968) 269–278.

[17] *Théorie de l' Inférence Statistique Robuste.* Les Presses de l' Université de Montréal (1969).

[18] Robust estimation. *Mathematical Centre Tracts* **27** (1968) 1–25, Amsterdam.

[19] Studentizing robust estimates. In *Nonparametric Techniques in Statistical Inference* (M. L. Puri, ed.), (1970) 453–463, Bloomington.

[20] (with B. Kleiner, H. Fluhler, G. Dumermuth). Spectrum analysis of the electroencephalogram. *Computer Programs in Biomedicine.* Amsterdam, Vol. I (1970) 183–197.

[21] (with B. Kleiner, T. Gasser, G. Dumermuth). Statistical methods for investigating phase relations in stationary stochastic processes. *IEEE Transactions on Audio and Electroacoustics* Vol. AU-19 (1971) 78–86.

[22] Robust statistics: A review (The A. Wald lecture). *Ann. Math. Stat.* **43** (1972) 1041–1067.

[23] (with Andrews, Bickel, Hampel, Rogers, Tukey). *Robust Estimates of Location: Survey and Advances.* Princeton University Press, Princeton, New Jersey (1972).

[24] (with V. Strassen). Minimax tests and the Neyman–Pearson lemma for capacities. *Ann. Statist.* **1** (1973) 251–263, **2** (1974) 223–224.

[25] Robust regression: Asymptotics, conjectures and Monte Carlo (Wald memorial lecture). *Ann. Statist.* **1** (1973) 799–821.

[26] The use of choquet capacities in statistics (1973). *Bull. I.S.I.* Vol. XLV (1973) 181–188.

[27] Fisher information and spline interpolation. *Ann. Statist.* **2** (1974) 1029–1033.

[28] (with R. Dutter). Numerical solution of robust regression problems. COMPSTAT 1974, *Proc. in Computational Statistics* (G. Bruckman, ed.), Physica-Verlag, Vienna (1974) 165–172.

[29] Some mathematical problems arising in robust statistics. *Proc. Int. Congress of Mathematicians*, Vancouver (1974) 821–824.

[30] *Einführung in die Wahrscheinlichkeit und Statistik* (nach einer Vorlesung). Verlag der Fachvereine an der ETH Zürich (1975).

[31] Robustness and designs. In *A Survey of Statistical Design and Linear Models* (J. N. Srivastava, ed.), North–Holland Publishing Company (1975) 287–301.

[32] Application vs. abstraction: The selling out of mathematical statistics? In *Proc. Conf. Directions for Mathematical Statistics* (S. G. Ghurye, ed.), Applied Probability Trust (1975) 84–89.

[33] Kapazitäten statt Wahrscheinlichkeiten? Gedanken zur Grundlegung der Statistik. *Jber. Deutsch. Math.-Verein.* **78** H.2 (1976) 81–92.

[34] Robust methods of estimation of regression coefficients. *Math. Operationsforsch. Statist. Ser. Statistics* **8** No. 1 (1977) 41–53.

[35] Early cuneiform evidence for the planet Venus. In *Scientists Confront Velikovsky* (D. Goldsmith, ed.), Cornell University Press (1977) 117–144.

[36] Robust covariances. In *Statistical Decision Theory and Related Topics II* (S. S. Gupta and D. S. Moore, eds.), Academic Press (1977) 165–191.

[37] *Robust Statistical Procedures.* Society for Industrial and Applied Mathematics, Philadelphia (1977).

[38] (with A. Aaboe). A text concerning subdivision of the synodic motion of Venus from Babylon: BM 37151. In *Ancient Near Eastern Studies in Memory of J. J. Finkelstein*, Connecticut Acad. Arts & Sciences, Mem. **19** (1977) 1–4.

[39] (with Gasser, Largo, Prader and Stützle). Analysis of the adolescent growth spurt using smoothing spline functions. *Ann. Human Biology* **5** (1978) 421–434.

[40] Robust smoothing. In *Robustness in Statistics* (R. L. Launer and G. N. Wilkinson, eds.), Academic Press (1979) 33–47.

[41] Robustness, computing and data analysis. In *Interactive Statistics* (D. McNeil, ed.), North-Holland (1979) 85–90.

[42] (with Gasser, Largo, Molinari, Prader and Stützle). Shape-invariant modelling. *Ann. Human Biology* **7** (1980) 507–528.

[43] *Robust Statistics.* Wiley, New York (1981), 308 pp.

[44] (with R. Dutter). Numerical methods for the nonlinear robust regression problem. *J. Statist. Comput. Simul.* (1981) 79–113.

[45] (with D. Donoho and H. M. Thoma). The use of kinematics displays to represent high dimensional data. In *Computer Science and Statistics: Proc. 13th Symp. Interface* (W. F. Eddy, ed.), Springer-Verlag (1981) 274–278.

[46] *Astronomical Dating of Babylon II and UR III*. Undena Press, Malibu, CA (1982).

[47] (with D. Donoho). The notion of breakdown point. In *Festschrift in Honor of Erich Lehmann* (Bickel, Doksum and Hodges, eds.), Wadsworth (1982) 157–184.

[48] (with D. Donoho, E. Ramos and H. M. Thoma). Kinematic display of multivariate data. *Proc. 3rd Ann. Conf. NCGA* (1982) 393–400.

[49] Discussion of Burguete et al., "On unification of the asymptotic theory of nonlinear economic models". *Econometric Reviews* **1** (1982) 191–192.

[50] Current issues in robust statistics. In *Proc. Symp. Recent Advances in Statistics*. Academic Press, New York (1983) 183–196.

[51] Statistical graphics: History and overview. *Proc. 4th Ann. Conf. NCGA* (1983) 667–676.

[52] Minimax aspects of bounded influence regression (with discussions). *J. Amer. Statist. Assoc.* **78** (1983) 66–80.

[53] Kinematic displays. In *Encyclopedia of Statistical Sciences* (Kotz and Johnson, eds.), **4**, John Wiley (1983) 381–384.

[54] Finite sample breakdown of M- and P-estimators. *Ann. Statist.* **12** (1984) 119–126.

[55] Data analysis: In search of an identity. *Proc. Berkeley Conf. in Honor of Jerzy Neyman and Jack Kiefer*, Vol I (Lucien M. LeCam and Richard A. Olshen, eds.), Wadsworth (1985) 65–78.

[56] Projection pursuit. *Ann. Statist.* **13** (1985) 435–475. Invited Discussion Paper.

[57] Environments for supporting statistical strategy. In *Artificial Intelligence* (W. A. Gale, ed.), Addison Wesley, Reading, MA (1986) 285–294.

[58] Data analysis implications for command language design. In *Foundations for Human-Computer Communication* (R. Hopper and I. A. Newman, eds.), Elsevier, North-Holland (1986) 488–504.

[59] Discussion of R. A. Thisted, "Computing environments for data analysis". *Statistical Science* **1** (1986) 273.

[60] Dating by lunar eclipse omina. In *From Ancient Omens to Statistical Mechanics* (J. L. Berggren and B. R. Goldstein, eds.), Copenhagen (1987) 3–13.

[61] Experiences with 3-d scatter plots. *J. Amer. Statist. Assoc.* **82** (1987) 448–453.

[62] Astronomical evidence for the long and against the middle and short chronologies. In *High, Middle or Low? Acts of an Intl. Coll. on Absolute Chronology, Gothenburg, 20–22. August 1987* (P. Åström, ed.), Studies in Mediterranean Archaeology and Literature, P. Åströms Förlag, Gothenburg, Vol. 56, 5–17.

[63] The place of the L_1-norm in robust estimation. *Comput. Statist. and Data Analysis* **5** (1987) 255–262.

[64] Discussion of Becker et al., "Dynamic graphics for data analysis". *Statistical Science* **2** (1987) 385–386.

[65] Discussion of J. Hodges, "Uncertainty, policy analysis and statistics". *Statistical Science* **2** (1987) 279–281.

[66] Spurious structure? Or: Improved bounds for the Kolmogorov index. Technical Report CICS-P-77 (1988).

[67] (with W. Vach). The design of a "Lab assistant" for data analysis. Technical Report (1988).

[68] (with E. H. Huber-Buser). ISP: Why a Command Language? In *Fortschritte der Statistik-Software I* (F. Faulbaum und H.-M. Uehlinger, eds.), Gustav Fischer Verlag, Stuttgart (1988) 349–360.

[69] *Projection Pursuit and Exploratory Data Analysis.* Technical Report, M.I.T. Dept. of Mathematics (1990).

[70] Between robustness and diagnostics. In *Directions in Robust Statistics and Diagnostics* (W. Stahel and S. Weisberg, eds.), IMA Volumes in Mathematics and Its Applications (1991) **33** 121–130.

[71] Goals versus algorithms. In *Data Analysis and Statistical Inference* (S. Schach and G. Trenkler, eds.), (1992) 567–579.

[72] Issues in computational data analysis. In *Comput. Statist.* (Y. Dodge and J. Whittaker, eds.), (1992) Vol. 2, 3–13.

[73] Projection pursuit and robustness. In *New Directions in Statistical Data Analysis and Robustness.* Birkhäuser Verlag, (1993) 139–146.

[74] Huge data sets. In *Proc. Comput. Statist.* (R. Dutter and W. Grossmann, eds.), (1994) 3–13.

[75] Languages for statistics and data analysis. In *Comput. Statist.* (P. Dirschedl and R. Ostermann, eds.), (1994), 53–80.

[76] Zufall oder Unordnung? In *Festschrift für Kurt Weichselberger* (H. Rinne, B. Rüger and H. Strecker, eds.), Physica-Verlag, Vienna (1995) 45–59.

[77] Robustness: Where are we now? (1995) *Student*, to appear.

[78] Speculations on the Path of Statistics. Tukey–Festschrift, to appear.

[79] Massive Data Sets Workshop: The Morning After. To appear.

Preface

To celebrate Peter Huber's 60th birthday in 1994, our university had invited for a festive occasion in the afternoon of Thursday, June 9. The invitation to honour this outstanding personality was followed by about fifty colleagues and former students from, mainly, all over the world. Others, who could not attend, sent their congratulations by mail and e-mail (P. Bickel: "... It's hard to imagine that Peter turned 60...").

After a welcome address by Adalbert Kerber (dean), the following lectures were delivered.

Volker Strassen (Konstanz):	*Almost Sure Primes and Cryptography —an Introduction*
Frank Hampel (Zürich):	*On the Philosophical Foundations of Statistics*[1]
Andreas Buja (Murray Hill):	*Projections and Sections— High-Dimensional Graphics for Data Analysis.*

The distinguished speakers lauded Peter Huber a hard and fair mathematician, a cooperative and stimulating colleague, and an inspiring and helpful teacher.

The Festkolloquium was surrounded with a musical program by the University's Brass Ensemble.[2]

The subsequent Workshop "Robust Statistics, Data Analysis and Computer Intensive Methods" in Schloss Thurnau, Friday until Sunday, June 9–12, was organized about the areas in statistics that Peter Huber himself has markedly shaped. In the time since the conference, most of the contributions could be edited for this volume—a late birthday present—that may give a new impetus to further research in these fields.

Helmut Rieder
Bayreuth, September 1995

Acknowledgements Deutsche Forschungsgemeinschaft granted essential financial support for the Workshop. The Festkolloquium was supported by the University of Bayreuth, in particular, by the University-Club and my colleagues of the Mathematics Department. Mrs. B. Daniel, Mrs. A. Geißer, T. Gneiting, W. Olbricht, E. Tabakis and M. Schlather helped me with the organization. B. Benzing prepared the files for, and P. Ruckdeschel assisted me in, the final editing. I thank them all.

[1] this volume

[2] Also, Strassen's talk contained a musical quote from Richard Wagner's "Valkyrie".

Bootstrap Variable–Selection and Confidence Sets

Rudolf Beran
University of California, Berkeley

Abstract

This paper analyzes estimation by bootstrap variable–selection in a simple Gaussian model where the dimension of the unknown parameter may exceed that of the data. A naive use of the bootstrap in this problem produces risk estimators for candidate variable–selections that have a strong upward bias. Resampling from a less overfitted model removes the bias and leads to bootstrap variable–selections that minimize risk asymptotically. A related bootstrap technique generates confidence sets that are centered at the best bootstrap variable–selection and have two further properties: the asymptotic coverage probability for the unknown parameter is as desired and the confidence set is geometrically smaller than a classical competitor. The results suggest a possible approach to confidence sets in other inverse problems where a regularization technique is used.

Key words and phrases: Curve estimation, bootstrap failure, risk estimate, shrink bootstrap, coverage probability.
AMS 1991 subject classifications: Primary 62G07; secondary 62G09.

1 Introduction

Certain statistical estimation problems, such as curve estimation, signal recovery, or image reconstruction, share two distinctive features: the dimension of the parameter space exceeds that of the data and each component of the unknown parameter may be important. In such problems, ordinary least squares or maximum likelihood estimation typically overfits the model. One general approach to estimation in such problems has three stages: First, devise a promising class of candidate estimators, such as penalized maximum likelihood estimators corresponding to a family of penalty functions or Bayes estimators generated by a family of prior distributions. This step is sometimes called using a regularization technique. Second, estimate the risk of each candidate estimator. Third, use the candidate estimator with smallest estimated risk.

Largely unresolved to date is the question of constructing accurate confidence sets based on such adaptive,. regularized estimators. Even obtaining reliable estimators of risk can be difficult. This paper treats both matters in the following problem, which is relatively simple to analyze explicitly, yet sufficiently general to indicate potential directions for other problems that involve a regularization technique. Suppose that X_n is an observation on a discretized signal ξ that is measured with error at n time points. The errors are independent, identically distributed, Gaussian random variables with means zero. Thus, X_n is a random vector whose distribution is $\mathcal{N}(\xi_n, \sigma_n^2 I_n)$. Both ξ_n and σ_n^2 are unknown. The problem is to estimate the signal ξ_n.

The integrated squared error of an estimator $\hat{\xi}_n$ is

$$L_n(\hat{\xi}_n, \xi_n) = n^{-1}|\hat{\xi}_n - \xi_n|^2, \tag{1.1}$$

where $|\cdot|$ is Euclidean norm. Under this loss, Stein (1956) showed that X_n, the maximum likelihood or least squares estimator of ξ_n, is inadmissible for $n \geq 3$. Better estimators for ξ_n include the James–Stein (1961) estimator, locally smoothed estimators such as the kernel variety treated by Rice (1984), and variable–selection estimators, to be described in the next paragraph. Each of these improved estimators accepts some bias in return for a greater reduction in variance.

A variable–selection approach to estimating ξ_n consists of three steps: first, transform X_n orthogonally to $X_n' = OX_n$; second, replace selected components of X_n' with zero; and third, apply the inverse rotation O^{-1} to the outcome of step two. The vector generated by such a process will be called a *variable–selection* estimator of ξ_n.

How shall we choose the orthogonal matrix O? Ideally, the components of the rotated mean vector $O\xi_n$ would be either very large or very small relative to measurement error. The nature of the experiment that generated X_n may suggest that O be a finite Fourier transform, or an analysis of variance transform, or an orthogonal polynomial transform, or a wavelet transform. Important though it is, we will not deal further, in this paper, with the choice of O.

Having rotated X_n, how shall we choose which components of X_n' to zero out? Thereafter, how shall we construct, around the variable–selection estimator, an accurate confidence set for ξ_n? A plausible answer is to compare candidate variable–selections through their bootstrap risks and then bootstrap the empirically best candidate estimator to obtain a confidence set for ξ_n. Efron and Tibshirani (1993, Chapter 17) discussed simple bootstrap estimators of mean squared prediction error. However, Freedman et al. (1988) and Breiman (1992) showed that simple bootstrap estimators of mean-squared prediction error can be untrustworthy for variable–selection.

This paper treats variable–selection for estimation rather than prediction and allows the dimension of the unknown parameter to increase with sample size n. The second point is very important. A stronger model assumption used by Speed and Yu (1993) and others—that the dimension of the parameter space is fixed for all n—restricts the possible bias induced by candidate variable–selections. In such restricted models, variable–selection by C_p does not choose well. On the other hand, C_p can be asymptotically correct when the dimension of the parameter space increases quickly with n and the selection class is not too large (cf. Section 2). Rice (1984, Section 3) and Speed and Yu (1993, Section 4) discuss other instances and aspects of this phenomenon.

Section 2 of this paper proves for our estimation problem that naive bootstrapping-resampling from a $\mathcal{N}(X_n, \hat{\sigma}_n^2 I_n)$ model, where $\hat{\sigma}_n^2$ estimates σ_n^2, yields upwardly biased risk estimators for candidate variable–selections. However, resampling from a $\mathcal{N}(\tilde{\xi}_n, \hat{\sigma}_n^2)$ distribution, where $\tilde{\xi}_n$ is obtained by suitably shrinking some of the components of X_n' toward zero, corrects the bias and generates a good bootstrap variable–selection estimator $\hat{\xi}_{n,B}$ for ξ_n. Using a related shrinkage bootstrap, Section 3 then constructs confidence sets centered at $\hat{\xi}_{n,B}$ that have correct asymptotic coverage probability for ξ_n and small geometrical error. Here as well, two plausible but naive bootstrap algorithms give wrong answers.

2 Bootstrap Selection Estimators

This section proposes bootstrap selection estimators for ξ_n and analyzes their asymptotic losses (which equal the asymptotic risks). The choice of bootstrap algorithm proves critical to the success of bootstrap selection. Naive bootstrapping does not work.

The signal vector $X_n = (X_{n,1}, \ldots, X_{n,n})'$ has a $\mathcal{N}(\xi_n, \sigma_n^2 I_n)$ distribution on $\mathbb{R}^n$. For brevity, write $\theta_n = (\xi_n, \sigma_n^2)$ and let $P_{\theta,n}$ denote the above normal distribution. Because the estimation problem is invariant under rotation of the coordinate system, we will simplify notation by assuming, without any loss of generality, that the orthogonal matrix O is the identity matrix. Then, the variable selection is done directly on the components of X_n. Consider candidate estimators for ξ_n that have the form

$$\hat{\xi}_n(A) = (a_{n,1}(A)X_{n,1}, \ldots, a_{n,n}(A)X_{n,n})', \tag{2.1}$$

where A ranges over subsets of $[0,1]$ and $a_{n,i}(A) = 1$ if $i/(n+1) \in A$ and vanishes otherwise. The goal is to choose A, on the basis of the data X_n, so as to minimize, at least asymptotically, the loss of the corresponding candidate estimator $\hat{\xi}_n(A)$.

Success of this formulation of variable selection appears to require restrictions on the possible values of A. In the paper, we assume that A is the union of m ordered closed intervals:

$$A = \bigcup_{i=1}^{m} [t_{2i-1}, t_{2i}], \tag{2.2}$$

where $0 \le t_1 \le \ldots \le t_{2m} \le 1$ and m is fixed. The pseudo-distance between two such sets A and B is defined to be

$$d(A, B) = \mu(A \triangle B), \tag{2.3}$$

where μ is Lebesgue measure. After forming equivalence classes, the collection $\mathcal{S}(m)$ of all subsets having the form (2.2) is a compact metric space under d.

Let $\mathcal{A}$ be a compact subset of $\mathcal{S}(m)$, possibly $\mathcal{S}(m)$ itself, that contains the unit interval $[0,1]$ as an element. Consider the candidate estimators $\hat{\xi}_n(A)$ that are generated as A ranges over $\mathcal{A}$. Since $[0,1]$ is an element of $\mathcal{A}$, the unbiased estimator X_n is among these candidate estimators. Let A^c be the complement of A in $[0,1]$. The quadratic loss of $\hat{\xi}_n(A)$ is then

$$\begin{aligned} L_n(\hat{\xi}_n(A), \xi_n) &= n^{-1}|\hat{\xi}_n(A) - \xi_n|^2 \\ &= n^{-1} \sum_{i/(n+1)\in A} (X_{n,i} - \xi_{n,i})^2 + \nu_n(A^c), \end{aligned} \tag{2.4}$$

where ν_n is the non-negative measure defined by

$$\nu_n(A) = n^{-1} \sum_{i/(n+1)\in A} \xi_{n,i}^2. \tag{2.5}$$

Estimators of this loss or of the associated risk are naturally phrased in terms of the discrete uniform measure

$$\mu_n(A) = n^{-1} \sum_{i/(n+1)\in A} 1 \tag{2.6}$$

and the empirical measure

$$\hat{\lambda}_n(A) = n^{-1} \sum_{i/(n+1)\in A} X_{n,i}^2. \tag{2.7}$$

Consider the following two bootstrap risk estimators:

Naive bootstrap. Suppose $\hat{\sigma}_n^2$ is a consistent estimator of σ_n^2, such as the variance estimators to be discussed in Section 3. Let X_n^* be a random vector such that the conditional distribution of X_n^* given X_n is $\mathcal{N}(X_n, \hat{\sigma}_n^2 I_n)$. Let $\xi_n^*(A)$ denote the recalculation from X_n^* of the candidate estimator $\hat{\xi}_n(A)$. Let E_* denote expectation with respect to the conditional distribution of X_n^* given X_n. The *naive bootstrap* risk estimator produced by the scheme is

$$\begin{aligned} \hat{R}_{n,N}(A, \hat{\sigma}_n^2) &= E_* L_n(\xi_n^*(A), X_n) \\ &= E_*\Big[n^{-1} \sum_{i/(n+1)\in A} (X_{n,i}^* - X_{n,i})^2 + n^{-1} \sum_{i/(n+1)\in A^c} X_{n,i}^2\Big] \\ &= \hat{\sigma}_n^2 \mu_n(A) + \hat{\lambda}_n(A^c). \end{aligned} \tag{2.8}$$

Unfortunately, if ν_n converges weakly to ν and σ_n^2 converges to σ^2 as n increases, then $\hat{R}_{n,N}(A)$ converges in probability to

$$\bar{\rho}(A) = \sigma^2 + \nu(A^c). \tag{2.9}$$

The actual asymptotic loss or risk of $\hat{\xi}_n(A)$ is

$$\rho(A) = \sigma^2 \mu(A) + \nu(A^c), \tag{2.10}$$

where μ is Lebesgue measure. Theorem 2.1 below gives details. The upward asymptotic bias in $\hat{R}_{n,N}(A, \hat{\sigma}_n^2)$ renders it useless for selection among the candidate estimators.

Shrink bootstrap. Let $[\cdot]_+$ denote the positive-part function. The modified estimator

$$\hat{R}_{n,B}(A, \hat{\sigma}_n^2) = \hat{\sigma}_n^2 \mu_n(A) + [\hat{\lambda}_n(A^c) - \hat{\sigma}_n^2 \mu_n(A^c)]_+ \tag{2.11}$$

corrects the asymptotic bias in $\hat{R}_{n,N}$ and converges in probability to $\rho(A)$, the correct asymptotic loss of $\hat{\xi}_n(A)$; see Theorem 2.1. Moreover, the risk estimator $\hat{R}_{n,B}(A, \hat{\sigma}_n^2)$ can also be viewed as a bootstrap estimator:

Let

$$\hat{s}_n(A^c) = [1 - \hat{\sigma}_n^2 \mu_n(A^c)/\hat{\lambda}_n(A^c)]_+ \tag{2.12}$$

and define $\tilde{\xi}_n(A) = (\tilde{\xi}_{n,1}(A), \ldots, \tilde{\xi}_{n,n}(A))'$ by

$$\tilde{\xi}_{n,i}(A) = \begin{cases} X_{n,i} & \text{if } i/(n+1) \in A \\ \hat{s}_n^{1/2}(A^c) X_{n,i} & \text{if } i/(n+1) \in A^c \end{cases}. \tag{2.13}$$

Let X_n^* now be a random vector such that the conditional distribution of X_n^* given X_n is $\mathcal{N}(\tilde{\xi}_n(A), \hat{\sigma}_n^2 I_n)$. As before, let $\xi_n^*(A)$ denote the recalculation from X_n^* of the candidate estimator $\hat{\xi}_n(A)$. Now the bootstrap risk is

$$\begin{aligned} E_* L_n(\xi_n^*(A), \tilde{\xi}_n) &= E_*\Big[n^{-1} \sum_{i/(n+1) \in A} (X_{n,i}^* - X_{n,i}^2) \\ &\qquad + n^{-1} \sum_{i/(n+1) \in A^c} \hat{s}_n(A^c) X_{n,i}^2\Big] \\ &= \hat{\sigma}_n^2 \mu_n(A) + \hat{s}_n(A^c) \hat{\lambda}_n(A^c) \qquad (2.14) \\ &= \hat{R}_{n,B}(A, \hat{\sigma}_n^2). \end{aligned}$$

The *shrink bootstrap* method just described has two notable features: It depends on the candidate set A and it shrinks some, but not all, of the components of X_n towards the origin. In defining $\tilde{\xi}_n(A)$, we could shrink as well the components of X_n for which $i/(n+1) \in A$ without changing the final evaluation in (2.14). In this sense, $\hat{R}_{n,B}(A, \hat{\sigma}_n^2)$ is the bootstrap risk generated by a family of shrink bootstrap algorithms. The shrinkage factor in (2.13) corrects the overfitting of ξ_n that occurs in the naive bootstrap.

The idea of bootstrap variable selection is to choose the candidate estimator whose estimated loss is smallest. Thus, let $\hat{A}_{n,B}$ be any set in $\mathcal{A}$ such that

$$\hat{R}_{n,B}(\hat{A}_{n,B}, \hat{\sigma}_n^2) = \min_{A \in \mathcal{A}} \hat{R}_{n,B}(A, \hat{\sigma}_n^2). \tag{2.15}$$

The minimum is achieved because, for each n, $\hat{R}_{n,B}(\cdot, \hat{\sigma}_n^2)$ has a finite number of possible values. We will call

$$\hat{\xi}_{n,B} = \hat{\xi}_n(\hat{A}_{n,B}) \tag{2.16}$$

a *bootstrap selection* estimator generated by $\{\hat{\xi}_n(A) : A \in \mathcal{A}\}$, the candidate estimators.

Let $\|\cdot\|_{\mathcal{A}}$ denote supremum norm taken over all sets $A \in \mathcal{A}$. To study the locally uniform convergences of $\hat{R}_{n,N}$ and $\hat{R}_{n,B}$, we introduce two conditions:

C1. $\mathcal{A}$ is a compact subset of $\mathcal{S}(m)$, in the metric d, that contains $[0,1]$ as an element. The sequence $\{\theta_n = (\xi_n, \sigma_n^2) : n \geq 1\}$ is such that

$$\lim_{n\to\infty} \|\nu_n - \nu\|_{\mathcal{A}} = 0, \qquad \lim_{n\to\infty} \sigma_n^2 = \sigma^2 \tag{2.17}$$

for some bounded, d-continuous, non-negative measure ν on $\mathcal{A}$ and some finite positive σ^2.

C2. For every sequence $\{\theta_n : n \geq 1\}$ that satisfies condition C1,

$$\operatorname*{plim}_{n\to\infty} \hat{\sigma}_n^2 = \sigma^2. \tag{2.18}$$

Here plim stands for the limit in $P_{\theta_n,n}$-probability.

Theorem 2.1 *Suppose that conditions C1 and C2 hold. Then, for every* $\epsilon > 0$,

$$\operatorname*{plim}_{n\to\infty} \|\hat{R}_{n,N}(\cdot, \hat{\sigma}_n^2) - \bar{\rho}\|_{\mathcal{A}} = 0 \tag{2.19}$$

where $\bar{\rho}$ *is defined in (2.9). By contrast,*

$$\operatorname*{plim}_{n\to\infty} \|L_n(\hat{\xi}_n(\cdot), \xi_n) - \rho\|_{\mathcal{A}} = 0$$

$$\operatorname*{plim}_{n\to\infty} \|\hat{R}_{n,B}(\cdot, \hat{\sigma}_n^2) - \rho\|_{\mathcal{A}} = 0 \tag{2.20}$$

where ρ *is defined in (2.10). Consequently,*

$$\operatorname*{plim}_{n\to\infty} L_n(\hat{\xi}_{n,B}, \xi_n) = \min_{A\in\mathcal{A}} \rho(A). \tag{2.21}$$

If the limiting loss ρ *has a unique minimizer* $A_0 \in \mathcal{A}$, *then*

$$\operatorname*{plim}_{n\to\infty} d(\hat{A}_{n,B}, A_0) = 0. \tag{2.22}$$

The theorem proof is in Section 4. By equations (2.20) and (2.21), the limiting loss of $\hat{\xi}_{n,B}$ coincides with the limiting loss of the unrealizable candidate estimator that minimizes loss over all selections A in $\mathcal{A}$. In this sense, the bootstrap selection estimator $\hat{\xi}_{n,B}$ is asymptotically optimal. Because $[0,1]$ is an element of $\mathcal{A}$,

$$\min_{A\in\mathcal{A}} \rho(A) \leq \rho([0,1]) = \sigma^2 \tag{2.23}$$

with equality only in special circumstances ($\nu = \sigma^2\mu$ or $\mu(A_0) = 1$, $\nu(A_0^c) = 0$; for example). Thus, $\hat{\xi}_{n,B}$ asymptotically dominates the unbiased estimator X_n.

An alternative to the shrink bootstrap risk estimator $\hat{R}_{n,B}$ replaces the positive-part function in (2.11) with the identity function. The result is the risk or loss estimator

$$\begin{aligned}\hat{R}_{n,C}(A, \hat{\sigma}_n^2) &= \hat{\sigma}_n^2[\mu_n(A) - \mu_n(A^c)] + \hat{\lambda}_n(A^c) \\ &= \hat{\lambda}_n(A^c) + \hat{\sigma}_n^2\,[\,2\mu_n(A) - 1\,]. \end{aligned} \tag{2.24}$$

Unlike $\hat{R}_{n,B}$, this risk estimator can assume negative values.

Let $\hat{A}_{n,C}$ be any value of $A \in \mathcal{A}$ that minimizes $\hat{R}_{n,C}(A, \hat{\sigma}_n^2)$. We will call $\hat{\xi}_{n,C} = \hat{\xi}_n(\hat{A}_{n,C})$ a C_p*-estimator* generated by the candidate estimators $\{\hat{\xi}_n(A) : A \in \mathcal{A}\}$. This terminology recognizes the analogy between (2.24) and risk estimators discussed by Mallows (1973) in a different context. Conclusions (2.20), (2.21) and (2.22) remain valid when $\hat{R}_{n,B}$, $\hat{A}_{n,B}$,and $\hat{\xi}_{n,B}$ are replaced by $\hat{R}_{n,C}$, $\hat{A}_{n,C}$, and $\hat{\xi}_{n,C}$, respectively.

Other variable selection criteria, such as Akaike's (1974) AIC, Shibata's (1981) method, and several competitors discussed by Rice (1984, Section 3), Speed and Yu (1993, Section 4) might also be used to choose A. Under Conditions C1 and C2, these methods do not minimize asymptotic loss in the sense of (2.21).

3 Bootstrap Confidence Sets

A confidence ball for ξ_n, centered at an estimator $\hat{\xi}_n$ and having radius $\hat{d}_n$, is

$$C_n(\hat{\xi}_n, \hat{d}_n) = \{t \in R^k : |\hat{\xi}_n - t| \le \hat{d}_n\}. \tag{3.1}$$

This section studies confidence balls centered at the bootstrap selection estimator $\hat{\xi}_{n,B}$. The first goal is to devise a bootstrap radius $\hat{d}_{n,B}$ such that the coverage probability $P_{\theta,n}[C_n(\hat{\xi}_{n,B}, \hat{d}_{n,B}) \ni \xi_n]$ converges to α as n increases. The second goal is to determine the *geometric loss* of $C_n(\hat{\xi}_n, \hat{d}_n)$ for various choices of $(\hat{\xi}_n, \hat{d}_n)$:

$$\begin{aligned} GL_n(C_n, \xi_n) &= n^{-1/2} \sup_{t \in C_n} |t - \xi_n| \\ &= n^{-1/2}|\hat{\xi}_n - \xi| + n^{-1/2}\hat{d}_n. \end{aligned} \tag{3.2}$$

Geometric loss measures the error of $C_n(\hat{\xi}_n, \hat{d}_n)$ as a set-valued estimator of ξ_n. It has a projection-pursuit interpretation that stems from the well-known identity $|x| = \sup\{u'x \colon |u| = 1\}$.

Both the definition of $\hat{\xi}_{n,B}$ and the construction of confidence balls centered at $\hat{\xi}_{n,B}$ require a good estimator of σ_n^2. One possibility, used in Rice (1984), is

$$\hat{\sigma}_{n,1}^2 = [2(n-1)]^{-1} \sum_{i=2}^{n} (X_{n,i} - X_{n,i-1})^2. \tag{3.3}$$

The consistency or asymptotic normality of $\hat{\sigma}^2_{n,1}$ requires that the first-order squared differences $\{(\xi_{n,i} - \xi_{n,i-1})^2\}$ be sufficiently small, in a sense that Condition D1 below makes precise.

A second estimator of σ^2_n works under the assumption that ξ_n lies in a subspace of dimension $n' < n$. Suppose that n' is the integer part of cn, where c is a fraction strictly between 0 and 1. By making an appropriate orthogonal transformation, assume without loss of generality that $X_n = (X_{n'}, Y_{n-n'})$, where $X_{n'}$ has a $\mathcal{N}(\xi_{n'}, \sigma^2_n I_{n'})$ distribution in n' dimensions, $Y_{n-n'}$ has a $\mathcal{N}(0, \sigma^2_n I_{n-n'})$ distribution in $n - n'$ dimensions, and $X_{n'}$, $Y_{n-n'}$ are independent. In this canonical formulation, a bootstrap selection estimator of $\xi_{n'}$ can be formed from $X_{n'}$ and the variance estimator

$$\hat{\sigma}^2_{n,2} = (n - n')^{-1} |Y_{n-n'}|^2. \tag{3.4}$$

The distribution of $(n-n')\hat{\sigma}^2_{n,2}/\sigma^2_n$ is chi-squared with $n-n'$ degrees of freedom.

The essential features of $\hat{\sigma}^2_{n,1}$ and $\hat{\sigma}^2_{n,2}$ are expressed in the following two assumptions:

D1. The variance estimator $\hat{\sigma}^2_{n,1}$ is defined by (3.3). The sequence of mean vectors $\{\xi_n : n \geq 1\}$ satisfies

$$\lim_{n\to\infty} n^{-1/2} \sum_{i=2}^{n} (\xi_{n,i} - \xi_{n,i-1})^2 = 0. \tag{3.5}$$

D2. The variance estimator $\hat{\sigma}^2_{n,2}$ and X_n are independent random variables. The distribution of $b_n \hat{\sigma}^2_{n,2}/\sigma^2_n$ is chi-squared, where $\{b_n : n \geq 1\}$ is a sequence of constants such that $\lim_{n\to\infty} b_n/n = b < \infty$.

Under D1 and A1, the asymptotic distribution of $n^{1/2}(\hat{\sigma}^2_{n,1} - \sigma^2_n)$ is $\mathcal{N}(0, 3\sigma^4)$, as in Gasser et al. (1986) or by the reasoning in Section 4. Under D2 and A1, the asymptotic distribution of $n^{-1/2}(\hat{\sigma}^2_{n,2} - \sigma^2_n)$ is $\mathcal{N}(0, 2b^{-1}\sigma^4)$.

To construct confidence balls, we begin by finding the asymptotic distribution of

$$D_n(\xi_n, X_n, \hat{\sigma}^2_{n,j}) = n^{1/2}[L_n(\hat{\xi}_{n,B}, \xi_n) - \hat{R}_{n,C}(\hat{A}_{n,B}, \hat{\sigma}^2_{n,j})]. \tag{3.6}$$

The quantity D_n compares the loss of $\hat{\xi}_{n,B}$ with the simple estimator (2.24) of its risk. On the one hand, the asymptotic distribution of D_n turns out to be normal with mean zero (Theorem 3.2 below). On the other hand, referring $D_n(\xi_n, X_n, \hat{\sigma}^2_{n,j})$ to the α-th quantile of its bootstrap distribution generates a confidence ball centered at $\hat{\xi}_{n,B}$ that has asymptotic coverage probability α for ξ_n (Theorem 3.3 below). There is no apparent advantage to replacing $\hat{R}_{n,C}$ in (3.6) with the more complex bootstrap risk estimator $\hat{R}_{n,B}$.

In the remainder of the paper, the notation Dj stands for either condition D1 or D2, according to the value of j.

Theorem 3.1 *Suppose that Conditions C1 and Dj hold and that the limiting loss* ρ *has a unique minimum at* $A_0 \in \mathcal{A}$. *Then*

$$\mathcal{L}[D_n(\xi_n, X_n, \hat{\sigma}^2_{n,j})|P_{\theta_n,n}] \implies \mathcal{N}(0, \tau^2_j(\nu, \sigma^2, A_0)), \tag{3.7}$$

where

$$\tau^2_1(\nu, \sigma^2, A_0) = 2\sigma^4 + \sigma^4[2\mu(A_0) - 1]^2 + 4\sigma^2\nu(A^c_0) \tag{3.8}$$

and

$$\tau^2_2(\nu, \sigma^2, A_0) = 2\sigma^4 + 2b^{-1}\sigma^4[2\mu(A_0) - 1]^2 + 4\sigma^2\nu(A^c_0). \tag{3.9}$$

This result is proved in Section 4. The same asymptotic distributions hold if the bootstrap selection estimator in the definition of $D_n(\xi_n, X_n, \hat{\sigma}^2_{n,j})$ is replaced by the C_p-estimator $\hat{\xi}_{n,C}$. Moreover, comparing the proof of Theorem 3.1 with its counterpart for the C_p-estimator establishes the following asymptotic equivalence in loss:

$$\operatorname*{plim}_{n\to\infty} n^{1/2}|L_n(\hat{\xi}_{n,B}, \xi_n) - L_n(\hat{\xi}_{n,C}, \xi_n)| = 0. \tag{3.10}$$

To successfully bootstrap the sampling distribution of $D_n(\xi_n, X_n, \hat{\sigma}^2_{n,j})$ requires an algorithm that recognizes both the data-based selection $\hat{A}_{n,B}$ and the structure of the variance estimator $\hat{\sigma}^2_{n,j}$. For every $A \in \mathcal{A}$, let

$$\tilde{D}_n(\xi_n, A, X_n, \hat{\sigma}^2_{n,j}) = n^{1/2}[L_n(\hat{\xi}_n(A), \xi_n) - \hat{R}_{n,C}(A, \hat{\sigma}^2_{n,j})]. \tag{3.11}$$

We consider two cases:

Bootstrapping $D_n(\xi_n, X_n, \hat{\sigma}^2_{n,1})$. Let

$$\hat{s}_n = [1 - \hat{\sigma}^2_{n,1}\mu_n(\hat{A}^c_{n,B})/\hat{\lambda}_n(\hat{A}^c_{n,B})]_+ \tag{3.12}$$

and define $\tilde{\xi}_n = (\tilde{\xi}_{n,1}, \ldots, \tilde{\xi}_{n,n})$ by

$$\tilde{\xi}_{n,i} = \begin{cases} X_{n,i} & \text{if } i/(n+1) \in \hat{A}_{n,B} \\ \hat{s}^{1/2}_n X_{n,i} & \text{if } i/(n+1) \in \hat{A}^c_{n,B} \end{cases}. \tag{3.13}$$

Let $E^*_n = (E^*_{n,1}, \ldots, E^*_{n,n})$ be a random vector such that the conditional distribution of E^*_n given X_n is $\mathcal{N}(0, \hat{\sigma}^2_{n,1}I_n)$. Define $X^*_n = (X^*_{n,1}, \ldots, X^*_{n,n})'$ and $\sigma^{*2}_{n,1}$ by

$$X^*_n = \tilde{\xi}_n + E^*_n \tag{3.14}$$

$$\sigma^{*2}_{n,1} = \hat{\sigma}^2_{n,1} + [2(n-1)]^{-1}\sum_{i=2}^{n}(E^*_{n,i} - E^*_{n,i-1})^2. \tag{3.15}$$

The partial bootstrap estimator of $\mathcal{L}[D_n(\xi_n, X_n, \hat{\sigma}^2_{n,1})|P_{\theta_n,n}]$ is then

$$\hat{H}_{n,B,1} = \mathcal{L}[D_n(\tilde{\xi}_n, \hat{A}_{n,B}, X^*_n, \sigma^{*2}_{n,1})|X_n]. \tag{3.16}$$

Bootstrapping $D_n(\xi_n, X_n, \hat{\sigma}^2_{n,2})$. Redefine $\hat{s}_n$ and E^*_n above by replacing $\hat{\sigma}^2_{n,1}$ with $\hat{\sigma}^2_{n,2}$. Define $\tilde{\xi}_n$ by (3.13) and X^*_n as in (3.14). Let $\sigma^{*2}_{n,2}$ be a random variable such that $\mathcal{L}[b_n\sigma^{*2}_{n,2}/\hat{\sigma}^2_{n,2}|X_n]$ is chi-squared with b_n degrees of freedom and such that $\sigma^{*2}_{n,2}$, E^*_n are conditionally independent, given X_n. The actual construction of $\sigma^{*2}_{n,2}$ will normally require a separate bootstrap scheme. The partial bootstrap estimator of $\mathcal{L}[D_n(\xi_n, X_n, \hat{\sigma}^2_{n,2})|P_{\theta_n,n}]$ is then

$$\hat{H}_{n,B,2} = \mathcal{L}[D_n(\tilde{\xi}_n, \hat{A}_{n,B}, X^*_n, \hat{\sigma}^{*2}_{n,2})|X_n] \tag{3.17}$$

Theorem 3.2 *Suppose that Conditions C1 and Dj hold and that the limiting loss ρ has a unique minimum at $A_0 \in \mathcal{A}$. Then*

$$\hat{H}_{n,B,j} \Longrightarrow \mathcal{N}(0, \tau^2_j(\nu, \sigma^2, A_0)) \tag{3.18}$$

in $P_{\theta_n,n}$-probability, where $\tau^2_j(\nu, \sigma^2, A_0)$ is defined by (3.8) and (3.9).

Both bootstrap algorithms in Theorem 3.2 shrink toward zero these components of X_n that are not selected by $\hat{\xi}_{n,B}$. The construction (3.13) of $\tilde{\xi}_n$ is critical for the weak convergence (3.18). If we took instead $\tilde{\xi}_n = X_n$, overfitting ξ_n, then the asymptotic variance in (3.18) would become $\tau^2_j(\nu + \sigma^2\mu, \sigma^2, A_0)$. If we used $\tilde{\xi}_n = \hat{\xi}_{n,B}$, underfitting ξ_n for bootstrap purposes, the asymptotic variance in (3.18) would become $\tau^2_j(0, \sigma^2, A_0)$. These conclusions follow by the method used to prove Theorem 3.2. Thus, neither of these alternative bootstrap algorithms yield consistent estimators of the sampling distribution of $D_n(\xi_n, X_n, \hat{\sigma}^2_{n,j})$.

For $0 < \alpha < 1$, let $\hat{H}^{-1}_{n,B,j}(\alpha)$ be the αth quantile of the bootstrap distribution $\hat{H}_{n,B,j}$ defined in (3.16) or (3.17). Under Condition Dj, define the *bootstrap selection confidence set* for ξ_n to be $C_{n,B,j} = C_n(\hat{\xi}_{n,B}, \hat{d}_{n,B,j})$, where

$$\hat{d}_{n,B,j} = [n\hat{R}_{n,C}(\hat{A}_{n,B}, \hat{\sigma}^2_{n,j}) + n^{1/2}\hat{H}^{-1}_{n,B,j}(\alpha)]^{1/2}_{+}. \tag{3.19}$$

The following theorem justifies this confidence set centered at $\hat{\xi}_{n,B}$.

Theorem 3.3 *Suppose that Conditions C1 and Dj hold and that the limiting risk ρ has a unique minimizer $A_0 \in \mathcal{A}$ such that $\rho(A_0) > 0$. Then*

$$\lim_{n\to\infty} P_{\theta_n,n}(C_{n,B,j} \ni \xi_n) = \alpha \tag{3.20}$$

and

$$\operatorname*{plim}_{n\to\infty} GL_n(C_{n,B,j}, \xi_n) = 2\rho^{1/2}(A_0). \tag{3.21}$$

If $\rho(A_0) = 0$ then

$$\liminf_{n\to\infty} P_{\theta_n,n}(C_{n,B,j} \ni \xi_n) \geq \alpha. \tag{3.22}$$

Remark 3.4 Only when $\mu(A_0) = \nu(A^c_0) = 0$, the exceptional case $\rho(A_0) = 0$ arises. This occurs when all but an asymptotically vanishing fraction of the components of ξ_n are zero. ////

Remark 3.5 A more familiar confidence set for ξ_n in the normal model is $C_{n,F} = C_n(X_n, \hat{\sigma}_n \chi_n^{-1/2}(\alpha))$, where $\chi_n^{-1/2}(\alpha)$ is the square root of the αth quantile of the chi-squared distribution with n degrees of freedom. Under Conditions C1 and C2, we have

$$\lim_{n\to\infty} P_{\theta_n,n}(C_{n,F} \ni \xi_n) = \alpha$$
$$\operatorname*{plim}_{n\to\infty} GL_n(C_{n,F}, \xi_n) = 2\sigma, \tag{3.23}$$

the second convergence relying on (3.2) and the normal approximation to the chi-squared distribution. It follows from (2.23), (3.21) and (3.23) that, at asymptotic coverage probability α, the bootstrap-selection confidence balls $C_{n,B,j}$ are both asymptotically smaller than the confidence ball $C_{n,F}$. ////

Remark 3.6 As an alternative to bootstrapping, the asymptotic variances in Theorem 3.1 may be estimated consistently from the sample, using $\hat{\sigma}^2_{n,j}$ for σ^2 and $[\hat{\lambda}_n(\hat{A}^c_{n,B}) - \hat{\sigma}^2_{n,j}\mu(\hat{A}^c_{n,B})]_+$ for $\nu(A^c_0)$. Equation (4.4) below justifies the second of these estimators. The estimated normal limit distributions then yield critical values and confidence sets for which an analog of Theorem 3.3 holds. ////

4 Proofs

The theorem proofs rely on ideas from Beran (1994) augmented by bootstrap considerations. Let $E_{n,i} = X_{n,i} - \xi_{n,i}$ and, for every set $A \in \mathcal{A}$, define

$$W_{n,1}(A, \theta_n) = n^{-1/2} \sum_{i/(n+1)\in A} (E^2_{n,i} - \sigma^2_n)$$
$$W_{n,2}(A, \theta_n) = n^{-1/2} \sum_{i/(n+1)\in A} \xi_{n,i} E_{n,i}. \tag{4.1}$$

Let $D(\mathcal{A})$ denote the set of all bounded functions having at most jump discontinuities on the compact set $\mathcal{A}$. Metrize $D(\mathcal{A})$ by supremum norm $\|\cdot\|_{\mathcal{A}}$. The σ-algebra is that generated by open balls. Under Condition C1, the two processes $W_{n,j}(\theta_n) = \{W_{n,j}(A, \theta_n) : A \in \mathcal{A}\}$ are random elements of $D(\mathcal{A})$.

Let $B_j = \{B_j(A) : A \in \mathcal{A}\}$ be two independent Gaussian processes on $\mathcal{A}$ with mean zero and covariance structure

$$\operatorname{Cov}[B_1(A), B_1(A')] = \mu(A \cap A')$$
$$\operatorname{Cov}[B_2(A), B_2(A')] = \nu(A \cap A') \tag{4.2}$$

where μ is Lebesgue measure and ν is the bounded non-negative measure defined in Condition C1. Both processes are random elements of $D(\mathcal{A})$ that have d-continuous sample paths.

Lemma 4.1 *Suppose that Condition C1 holds. Then the bivariate processes* $\{(W_{n,1}(\theta_n), W_{n,2}(\theta_n))\}$ *converge weakly as random elements of* $D(\mathcal{A}) \times D(\mathcal{A})$ *to the process* $(2^{1/2}\sigma^2 B_1, \sigma B_2)$.

Convergence of the finite-dimensional distributions is straightforward. For tightness, see LeCam (1983, Lemma 4) or Alexander and Pyke (1986, Section 4).

PROOF OF THEOREM 2.1 The definitions (2.5), (2.7) and (4.1) entail

$$\hat{\lambda}_n(A) = \nu_n(A) + \hat{\sigma}_n^2 \mu_n(A) + n^{-1/2} W_{n,1}(A, \theta_n) + 2n^{-1/2} W_{n,2}(A, \theta_n). \quad (4.3)$$

Consequently, by Lemma 4.1,

$$\operatorname*{plim}_{n \to \infty} \|\hat{\lambda}_n(\cdot) - [\nu(\cdot) + \sigma^2 \mu(\cdot)]\|_{\mathcal{A}} = 0. \quad (4.4)$$

Then (2.19) and the second convergence in (2.20) follow from (4.4), Condition C2, and the definitions (2.8), (2.11) of $\hat{R}_{n,N}$ and $\hat{R}_{n,B}$.

On the other hand, by (2.4) and (4.1),

$$L_n(\hat{\xi}_n(A), \xi_n) = \nu_n(A^c) + \sigma_n^2 \mu_n(A) + n^{-1/2} W_{n,1}(A, \theta_n). \quad (4.5)$$

The first convergence in (2.20) follows from Lemma 4.1.

Definition (2.16) of $\hat{\xi}_{n,B}$, (2.20), and the triangle inequality imply

$$\operatorname*{plim}_{n \to \infty} \hat{R}_{n,B}(\hat{A}_{n,B}, \hat{\sigma}_n^2) = \min_{A \in \mathcal{A}} \rho(A) \quad (4.6)$$

and

$$\operatorname*{plim}_{n \to \infty} [\hat{R}_{n,B}(\hat{A}_{n,B}, \hat{\sigma}_n^2) - L_n(\hat{\xi}_{n,B}, \xi_n)] = 0. \quad (4.7)$$

Conclusion (2.21) thus follows.

Limit (4.6) and the second limit in (2.20) imply that

$$\operatorname*{plim}_{n \to \infty} \rho(\hat{A}_{n,B}) = \rho(A_0), \quad (4.8)$$

where A_0 is the unique minimizer of ρ over $\mathcal{A}$. Suppose that (2.22) does not hold. By considering a subsequence, we may assume without loss of generality that convergence (4.8) occurs almost surely while

$$P_{\theta_n,n}[d(\hat{A}_{n,B}, A_0) > \epsilon] \geq \delta \quad (4.9)$$

for some positive ϵ and δ. Because ρ is d-continuous on the compact $\mathcal{A}$ and A_0 uniquely minimizes ρ over $\mathcal{A}$, the almost sure version of (4.8) implies that $d(\hat{A}_{n,B}, A_0)$ tends to 0 with probability one. This contradicts (4.8), thereby proving (2.22). ////

PROOF OF THEOREM 3.1 As above, the definitions (2.4) and (2.24) of L_n and of $\hat{R}_{n,C}$ respectively entail that

$$L_n(\hat{\xi}_{n,B}, \xi_n) = \nu(\hat{A}_{n,B}^c) + \sigma_n^2 \mu_n(\hat{A}_{n,B}) + n^{-1/2} W_{n,1}(\hat{A}_{n,B}, \theta_n) \quad (4.10)$$

and

$$\begin{aligned}\hat{R}_{n,C}(\hat{A}_{n,B}, \hat{\sigma}^2_{n,j}) = \nu_n(\hat{A}^c_{n,B}) + \sigma^2_n \mu_n(\hat{A}^c_{n,B}) \\ + n^{-1/2} W_{n,1}(\hat{A}^c_{n,B}, \theta_n) + 2n^{-1/2} W_{n,2}(\hat{A}^c_{n,B}, \theta_n) \\ + \hat{\sigma}^2_{n,j}[\mu_n(\hat{A}_{n,B}) - \mu_n(\hat{A}^c_{n,B})].\end{aligned} \tag{4.11}$$

Consequently,

$$\begin{aligned}D_n(\xi_n, X_n, \hat{\sigma}^2_{n,j}) &= n^{1/2}[L_n(\hat{\xi}_{n,B}, \xi_n) - \hat{R}_{n,C}(\hat{A}_{n,B}, \hat{\sigma}^2_{n,j})] \\ &= W_{n,1}(\hat{A}_{n,B}, \theta_n) - W_{n,1}(\hat{A}^c_{n,B}, \theta_n) \\ &\quad - 2W_{n,2}(\hat{A}^c_{n,B}) \\ &\quad - n^{-1/2}(\hat{\sigma}^2_{n,j} - \sigma^2_n)[2\mu_n(\hat{A}_{n,B}) - 1].\end{aligned} \tag{4.12}$$

Under Condition D1,

$$\begin{aligned}\hat{\sigma}^2_{n,1} &= 2^{-1}[(n-1)^{-1}\sum_{i=2}^{n}(E^2_{n,i} + E^2_{n,i-1})] \\ &\quad + (n-1)^{-1}\sum_{i=2}^{n} E_{n,i}E_{n,i-1} + o_p(n^{-1/2}) \\ &= \sigma^2_n + 2^{-1}\{W_{n,1}([2/(n+1), 1], \theta_n) \\ &\quad + W_{n,1}([0, (n-1)/(n+1)], \theta_n)\} \\ &\quad + (n-1)^{-1}\sum_{i=2}^{n} E_{n,i}E_{n,i-1} + o_p(n^{-1/2}).\end{aligned} \tag{4.13}$$

The argument for Lemma 4.1 and the martingale central limit theorem, applied to the quadratic term in the last line of (4.13), which is uncorrelated with $W_{n,1}$ and $W_{n,2}$, imply that

$$n^{1/2}(\hat{\sigma}^2_{n,1} - \sigma^2_n) \implies 2^{1/2}\sigma^2 B_1([0,1]) + \sigma^2 Z, \tag{4.14}$$

where Z is a $\mathcal{N}(0,1)$ random variable such that B_1, B_2 and Z are independent. Moreover,

$$D_n(\xi_n, X_n, \hat{\sigma}^2_{n,1}) = S_n(E_n, \xi_n, \sigma^2_n) + o_p(1), \tag{4.15}$$

where $S_n(E_n, \xi_n, \sigma^2_n)$ is defined by substituting (4.13) into (4.12) and dropping the remainder term.

The d-continuity of μ and ν together with the convergence (2.22) imply that $\text{plim}_{n\to\infty}\mu(\hat{A}_{n,B}) = \mu(A_0)$ and that $\text{plim}_{n\to\infty}\nu(\hat{A}_{n,B}) = \nu(A_0)$. The foregoing considerations yield

$$\begin{aligned}S_n(E_n, \xi_n, \sigma^2_n) \implies 2^{1/2}\sigma^2 B_1(A_0) - 2^{1/2}B_1(A^c_0) - 2\sigma B_2(A^c_0) \\ - [2\mu(A_0) - 1][2^{1/2}B_1([0,1]) + Z]\sigma^2.\end{aligned} \tag{4.16}$$

For $j = 1$, the weak convergence (3.7) follows from (4.2), (4.15) and (4.16). We will use (4.16) again to prove Theorem 3.2.

Under Condition D2, $\hat{\sigma}^2_{n,2}$ is independent of X_n and

$$n^{1/2}(\hat{\sigma}^2_{n,2} - \sigma^2_n) \implies 2^{1/2} b^{-1/2} \sigma^2 Z, \tag{4.17}$$

where Z again is a $\mathcal{N}(0,1)$ random variable such that B_1, B_2 and Z are independent. Because of (4.12), (2.22) and Lemma 4.1,

$$\begin{aligned} D_n(\xi_n, X_n, \hat{\sigma}^2_{n,2}) \implies & 2^{1/2}\sigma^2 B_1(A_0) - 2^{1/2} B_1(A_0^c) - 2\sigma B_2(A_0^c) \\ & - [2\mu(A_0) - 1] 2^{1/2} b^{-1/2} \sigma^2 Z, \end{aligned} \tag{4.18}$$

which establishes (3.7) for $j = 2$. ////

Proof of Theorem 3.2 Suppose that $\{\nu_n\}$, $\{\sigma^2_n\}$, and $\{A_n \in \mathcal{A}\}$ are such that

$$\lim_{n\to\infty} \|\nu_n - \tilde{\nu}\|^2, \qquad \lim_{n\to\infty} \sigma^2_n = \sigma^2, \qquad \lim_{n\to\infty} d(A_n, A_0) = 0 \tag{4.19}$$

and $\tilde{\nu}$ is d-continuous. By the reasoning for Theorem 3.1,

$$\tilde{D}_n(\xi_n, A_n, X_n, \hat{\sigma}^2_{n,1}) = \tilde{S}_n(E_n, A_n, \xi_n, \sigma^2_n) + o_p(1), \tag{4.20}$$

where $\tilde{S}_n(E_n, A_n, \xi_n, \sigma^2_n)$ is obtained from the definition of $S_n(E_n, \xi_n, \sigma^2_n)$ by replacing $\hat{A}_{n,B}$ with A_n. As in Theorem 3.1,

$$\mathcal{L}[\tilde{S}_n(E_n, A_n, \xi_n, \sigma^2_n) | P_{\theta_n,n}] \implies \mathcal{N}(0, \tau_1^2(\tilde{\nu}, \sigma^2, A_0)) \tag{4.21}$$

and

$$\mathcal{L}[\tilde{D}_n(\xi_n, A_n, X_n, \hat{\sigma}^2_{n,2} | P_{\theta_n,n}] \implies \mathcal{N}(0, \tau_2^2(\tilde{\nu}, \sigma^2, A_0)). \tag{4.22}$$

Next, consider the empirical measure

$$\tilde{\nu}_n(A) = n^{-1} \sum_{i/(n+1)\in A} \tilde{\xi}^2_{n,i} = \hat{\lambda}_n(A \cap \hat{A}_{n,B}) + \hat{s}_n \hat{\lambda}_n(A \cap \hat{A}^c_{n,B}). \tag{4.23}$$

Under either Condition D1 or D2, it follows from (4.4) and (2.22) that

$$\operatorname*{plim}_{n\to\infty} \|\tilde{\nu}_n - \tilde{\nu}\|_{\mathcal{A}} = 0, \qquad \operatorname*{plim}_{n\to\infty} \hat{\sigma}^2_{n,j} = \sigma^2, \tag{4.24}$$

where $\tilde{\nu}$ is the measure on $\mathcal{A}$ defined by

$$\begin{aligned} \tilde{\nu}(A) = & \nu(A \cap A_0) + \sigma^2 \mu(A \cap A_0) \\ & + s[\nu(A \cap A_0^c) + \sigma^2 \mu(A \cap A_0^c)] \end{aligned} \tag{4.25}$$

$$s = \nu(A_0^c)/[\nu(A_0^c) + \sigma^2 \mu(A_0^c)]. \tag{4.26}$$

For the case $j = 1$, (3.14) and the reasoning for (4.20) entail that

$$D_n(\tilde{\xi}_n, \hat{A}_{n,B}, X_n^*, \hat{\sigma}^{*2}_{n,1}) = \tilde{S}_n(E_n^*, \hat{A}_{n,B}, \tilde{\xi}_n, \hat{\sigma}^{*2}_{n,1}). \tag{4.27}$$

Consequently, by (4.24), (2.22) and (4.21),

$$\hat{H}_{n,B,1} \implies \mathcal{N}(0, \tau_1^2(\tilde{\nu}, \sigma^2, A_0)) \tag{4.28}$$

in $P_{\theta_n,n}$-probability. This limit law agrees with (3.18) because $\tilde{\nu}(A_0^c)$ from (4.25) equals $\nu(A_0^c)$ in (3.8).

For the case $j = 2$, (4.24), (2.22) and (4.22) yield

$$\hat{H}_{n,B,2} \implies \mathcal{N}(0, \tau_2^2(\tilde{\nu}, \sigma^2, A_0)), \tag{4.29}$$

which agrees with (3.18) as $\tilde{\nu}(A_0^c)$ from (4.25) again equals $\nu(A_0^c)$ in (3.9). ////

PROOF OF THEOREM 3.3 This result follows from Theorems 3.1 and 3.2. The argument parallels the proof of Theorem 3.2 in Beran (1994). ////

Acknowledgements

This paper was written while I was a guest of the Statistics group in the Institut für Angewandte Mathematik at Universität Heidelberg. The research was supported in part by National Science Foundation grant DMS 9224868. Comments by Lutz Dümbgen were especially helpful.

References

[1] Akaike, H. (1974).: A new look at statistical model identification. *IEEE Trans. Auto. Cont.* **19** 716–723.

[2] Alexander, K.S. and Pyke, R. (1986): A uniform central limit theorem for set-indexed partial-sum processes with finite variance. *Ann. Prob.* **14** 582–597.

[3] Beran, R. (1994): *Confidence sets centered at C_p-estimators.* Preprint.

[4] Breiman, L. (1992): The little bootstrap and other methods for dimensionality selection in regression: X-fixed prediction error. *J. Amer. Statist. Assoc.* **87** 738–754.

[5] Efron, B. and Tibshirani, R.J. (1993): *An Introduction to the Bootstrap.* Chapman & Hall, New York.

[6] Freedman, D.A., Navidi, W., and Peters, S.C. (1988): On the impact of variable selection in fitting regression equations. *On Model Uncertainty and its Statistical Implications* (T.K. Dijkstra, ed.), 1–16. Springer-Verlag, Berlin.

[7] Gasser, T., Sroka, L., and Jennen-Steinmetz, C. (1986): Residual variance and residual pattern in nonlinear regression. *Biometrika* **73** 625–633.

[8] James, W. and Stein, C. (1961): Estimation with quadratic loss. *Proc. Fourth Berkeley Symp. Math. Statist. Prob.*, Vol. 1, 361–380. Univ. of California Press, Berkeley.

[9] LeCam, L. (1983): A remark on empirical measures. *Festschrift for Erich Lehmann* (P.J. Bickel, K. Doksum and J.L. Hodges, eds.), 305–327. Wadsworth, Belmont, California.

[10] Mallows, C. (1973): Some comments on C_p. *Technometrics* **15** 661–675.

[11] Rice, J. (1984): Bandwidth choice for nonparametric regression. *Ann. Statist.* **12** 1215–1230.

[12] Shibata, R. (1981): An optimal selection of regression variables. *Biometrika* **68** 45–54.

[13] Speed, T.P. and Yu, B. (1993): Model selection and prediction: Normal regression. *Ann. Inst. Statist. Math.* **45** 35–54.

[14] Stein, C. (1956): Inadmissibility of the usual estimator for the mean of a multivariate normal distribution. *Proc. Third Berkeley Symp. Math. Statist. Prob.* **1** 197–206. Univ. of California Press, Berkeley.

Robust Estimation in the Logistic Regression Model

Ana M. Bianco
Universidad de Buenos Aires

Víctor J. Yohai *
Universidad de San Andrés and Universidad de Buenos Aires.

Abstract

A new class of robust and Fisher-consistent M-estimates for the logistic regression models is introduced. We show that these estimates are consistent and asymptotically normal. Their robustness is studied through the computation of asymptotic bias curves under point-mass contamination for the case when the covariates follow a multivariate normal distribution. We illustrate the behavior of these estimates with two data sets. Finally, we mention some possible extensions of these M-estimates for a multinomial response.

Key words and phrases: logistic regression model, M-estimates, robust estimates.
AMS 1991 subject classifications: Primary 62F35, 62J12; secondary 62F12.

1 Introduction

Let us consider a generalized linear model with a Bernoulli dependent variable Y, (i. e., Y takes values 0 and 1), q covariates $X_1, \dots, X_q$ and a link function P. Then, if $\mathbf{X} = (X_1, \dots, X_q)'$, for any $\mathbf{x} \in \mathbb{R}^q$, we have

$$\Pr(Y = 1|\mathbf{X} = \mathbf{x}) = P(\mathbf{x}'\boldsymbol{\beta}_0), \tag{1.1}$$

for some $\boldsymbol{\beta}_0 \in \mathbb{R}^q$. We will assume that P is a continuous and strictly increasing distribution function. For example, P may be the logistic function

$$P(t) = \frac{\exp(t)}{1 + \exp(t)}, \tag{1.2}$$

or the probit function Φ corresponding to the standard normal distribution function.

* This research was partially done while V. J. Yohai was visiting the Department of Statistics of the University of Washington supported by ONR Contract No. CN00014–91–J–1074. The authors thank Professor Raymond J. Carroll for providing the food stamp data set.

Consider a sample $\mathbf{Z}_1 = (\mathbf{X}_1', Y_1), \ldots, \mathbf{Z}_n = (\mathbf{X}_n', Y_n)$ of n independent observations, where $\mathbf{X}_i = (X_{i1}, \ldots, X_{iq})'$ for $i = 1, \ldots, n$. The $\mathbf{X}_i$'s may be fixed or random. The maximum likelihood (ML) estimate which is obtained by maximizing

$$\sum_{i=1}^{n} [Y_i \ln(P(\mathbf{X}_i'\boldsymbol{\beta})) + (1 - Y_i) \ln(1 - P(\mathbf{X}_i'\boldsymbol{\beta}))] \tag{1.3}$$

gives an asymptotically efficient procedure for estimating $\boldsymbol{\beta}_0$. Alternatively, consider the deviances $D_i(\boldsymbol{\beta}) = Y_i(-\ln(P(\mathbf{X}_i'\boldsymbol{\beta}))) + (1-Y_i)(-\ln(1-P(\mathbf{X}_i'\boldsymbol{\beta})))$; then the ML-estimate may be also defined by the minimization of

$$\sum_{i=1}^{n} D_i(\boldsymbol{\beta}). \tag{1.4}$$

However, the ML-estimate is extremely sensitive to the presence of anomalous data in the sample. Pregibon (1981,1982) has exemplified this sensitivity. Pregibon (1981), Cook and Weisberg (1982) and Johnson (1985) have proposed procedures to identify observations which are influential for estimating $\boldsymbol{\beta}_0$.

Pregibon (1982) defined robust estimates by modifying the ML-estimate goal function given in (1.4). He proposed to minimize

$$\sum_{i=1}^{n} \rho(D_i(\boldsymbol{\beta})), \tag{1.5}$$

where ρ is a suitable monotone loss function which increases slower than the identity. He suggested to use as ρ a Huber type function in the family given by

$$\rho(t) = \begin{cases} t & \text{if } t \leq c \\ 2(tc)^{1/2} - c & \text{if } t > c. \end{cases} \tag{1.6}$$

However, the estimators defined by the minimization of (1.5) are not consistent. In fact, it is simple to show that they are asymptotically biased when the model (1.1) holds.

Copas (1988) found an approximate expression for the bias of these estimates, and he showed that it may be non-negligible. Moreover, if the ρ function is given in the family (1.6), the resulting estimate is not robust against outliers with high leverage covariates.

Stefanski, Carroll and Ruppert (1986) proposed a class of robust estimates for generalized linear models. They obtained Hampel-optimal bounded influence estimators: estimators which minimize some functional of the asymptotic covariance matrix subject to a bound of some norm of the influence curve. Künsch, Stefanski and Carroll (1989) found similar optimal estimators but with the additional constraint of conditional unbiasness. These two classes of bounded influence estimates require the estimation of a robust covariance matrix of $\mathbf{X}$. Carroll and Pederson (1993) studied the bias of these estimators for small uncontaminates samples. They found that there are estimates in the Mallows subclass which have a bias comparable with the ML-estimator.

Copas (1988) has proposed a robust estimation procedure which is based on a logistic regression model which contains the outlier generation mechanism. The estimation procedure is maximum likelihood applied to this model.

In this paper we propose a class of M-estimators which can be thought as a Fisher-consistent version of the estimators given by (1.5). These estimators are defined by the minimization of

$$\sum_{i=1}^{n}[\rho(D_i(\boldsymbol{\beta}))+G(P(\mathbf{X}_i'\boldsymbol{\beta}))+G(1-P(\mathbf{X}_i'\boldsymbol{\beta}))], \tag{1.7}$$

where ρ is a bounded, differentiable and nondecreasing function and

$$G(t)=\int_0^t \psi(\text{-}\ln u)\,du, \tag{1.8}$$

where $\psi(t)=\rho'(t)$. As we will see in Section 2 this choice of $G(t)$ makes the M-estimate Fisher-consistent.

The fact that ρ is bounded makes these estimates qualitatively robust in Hampel's sense; that is, a small fraction of arbitrary outliers in the sample has a small effect on the estimate. An advantage of these estimates is that they do not need a robust estimate of the scatter matrix of the $\mathbf{X}$'s. The estimates defined by the minimization of (1.7) are a natural generalization of the M-estimates with bounded ρ function used for regression. Since these M-estimates for regression have good robustness properties (e.g., a high breakdown point—see Yohai (1987)), it seems reasonable to expect also a robust behavior of our proposal.

In particular, we study a family of ρ functions given by

$$\rho_c(t)=\begin{cases} t-t^2/2c & \text{if } t\le c\\ c/2 & \text{if } t>c,\end{cases} \tag{1.9}$$

where c is a positive number. Observe that the ρ function corresponding to the ML-estimate is just the identity, which is an unbounded function. One way to robustify the identity function is by truncating it for values larger than a constant c. The correction term $t^2/2c$ in (1.9) makes the truncation process smoother. It is clear that when c tends to ∞ the corresponding ρ_c function converges to the identity and therefore the corresponding M-estimate approaches the ML-estimate. We use this family mainly because of the simplicity of the corresponding G functions. However, similar results may be obtained with other ρ-functions using different truncation schemes.

In Section 2 we prove the Fisher-consistency and we state the asymptotic properties of the M-estimators which are proved in the Appendix. In Section 3, we compute asymptotic biasses of the M-estimates under outlier contamination.

These asymptotic biasses are compared with those of the most robust estimate in the family of conditional unbiased bounded influenced (CUBI) estimators proposed by Künsch et al. (1989) In Section 4, we illustrate the performance of the M-estimators defined by the minimization of (1.7) with some examples. In Section 5 we refer to possible extensions of these M-estimates for the case of a multinomial response vector. Section 6 is an Appendix with some proofs.

2 Consistency and Asymptotic Normality

In order to obtain the asymptotic results, we will require the following assumptions on the loss ρ and the link function P:

Assumptions on ρ:

A1. $\rho(t)$ is bounded with bounded and continuous derivative ψ

A2. $\psi(t) \geq 0$ and there exist some $c > -\ln 1/2$ such that $\psi(t) > 0$ for all $0 < t < c$

A3. $\rho(0) = 0$

A4. ψ is differentiable, except for at most a finite numbers of points

A5. $\psi(-\ln p)$ is Lipschitz on p

Assumptions on P**:**

B1. $P(t)$ is a strictly increasing function.

B2. $\lim_{t\to\infty} P(t) = 1$.

B3. $\lim_{t\to-\infty} P(t) = 0$.

B4. $P(t)$ is continuous.

B5. $P(t)$ is twice differentiable.

The following Lemma is necessary to prove the Fisher-consistency of the proposed estimators.

Lemma 2.1 *Let* Y *be a Bernoulli variable such that* $\Pr(Y = 1) = p_0$ *and assume that* ρ *satisfies A1 and A2 then*

$$L(p_0, p) = \mathrm{E}_{p_0}(Y\rho(-\ln p) + G(p) + (1-Y)\,\rho(-\ln(1-p)) + G(1-p)) \quad (2.1)$$

has a unique minimum at $p = p_0$.

PROOF It is enough to prove that

$$\frac{\partial L}{\partial p}(p_0, p) < 0 \quad \text{for all} \quad p < p_0 \quad (2.2)$$

and

$$\frac{\partial L}{\partial p}(p_0, p) > 0 \quad \text{for all} \quad p > p_0. \quad (2.3)$$

Since

$$L(p_0, p) = p_0\rho(\text{-}\ln p) + G(p) + (1-p_0)\rho(\text{-}\ln(1-p)) + G(1-p),$$

we have

$$\begin{aligned}\frac{\partial L}{\partial p}(p_0, p) &= -\frac{p_0}{p}\ \psi(\text{-}\ln p) + \psi(\text{-}\ln p) + \frac{1-p_0}{1-p}\psi(\text{-}\ln(1-p)) - \psi(\text{-}\ln(1-p)) \\ &= (p - p_0)\left(\frac{\psi(\text{-}\ln p)}{p} + \frac{\psi(\text{-}\ln(1-p))}{1-p}\right). \end{aligned} \tag{2.4}$$

Using A1 and A2 we get

$$\frac{\psi(\text{-}\ln p)}{p} + \frac{\psi(\text{-}\ln(1-p))}{1-p} > 0. \tag{2.5}$$

(2.4) and (2.5) imply (2.2) and (2.3). ////

The next theorem shows the Fisher-consistency of the estimates defined by the minimization of (1.7).

Theorem 2.2 *Assume that $\rho(t)$ verifies A1–A3 and P verifies B1–B4. Suppose that for all $\boldsymbol{\alpha} \in \mathbb{R}^q$, $\boldsymbol{\alpha} \neq 0$, we have* $\Pr(\mathbf{X}'\boldsymbol{\alpha} = 0) < 1$. *Then for all $\boldsymbol{\beta} \in \mathbb{R}^q$, $\boldsymbol{\beta} \neq \boldsymbol{\beta}_0$ we have*

$$\mathrm{E}_{\boldsymbol{\beta}_0}(\phi(Y, \mathbf{X}, \boldsymbol{\beta}_0)) < \mathrm{E}_{\boldsymbol{\beta}_0}(\phi(Y, \mathbf{X}, \boldsymbol{\beta})),$$

where ϕ is defined by

$$\phi(Y, \mathbf{X}, \boldsymbol{\beta}) = \rho(D(\boldsymbol{\beta})) + G(P(\mathbf{X}'\boldsymbol{\beta})) + G(1 - P(\mathbf{X}'\boldsymbol{\beta})). \tag{2.6}$$

PROOF Taking conditional expectations with respect to $\mathbf{X}$, it is immediate that

$$\begin{aligned}\mathrm{E}_{\boldsymbol{\beta}_0}(\phi(Y, \mathbf{X}, \boldsymbol{\beta}_0)) = \mathrm{E}(&P(\mathbf{X}'\boldsymbol{\beta}_0)\rho(-\ln(P(\mathbf{X}'\boldsymbol{\beta}))) \\ &+ (1 - P(\mathbf{X}'\boldsymbol{\beta}_0))\rho(-\ln(1 - P(\mathbf{X}'\boldsymbol{\beta}))) \\ &+ G(P(\mathbf{X}'\boldsymbol{\beta})) + G(1 - P(\mathbf{X}'\boldsymbol{\beta}))) \\ = \mathrm{E}(&L(P(\mathbf{X}'\boldsymbol{\beta}_0), P(\mathbf{X}'\boldsymbol{\beta}))),\end{aligned}$$

where L is defined in (2.1). Using the Lemma and the assumptions on P, it follows that $L(P(\mathbf{X}'\boldsymbol{\beta}_0), P(\mathbf{X}'\boldsymbol{\beta}))$ reaches its minimum when $\mathbf{X}'\boldsymbol{\beta} = \mathbf{X}'\boldsymbol{\beta}_0$. Hence, since $\Pr(\mathbf{X}'\boldsymbol{\beta} = \mathbf{X}'\boldsymbol{\beta}_0) < 1$ for all $\boldsymbol{\beta} \neq \boldsymbol{\beta}_0$, the result follows. ////

We will assume that $\mathbf{Z}_1 = (\mathbf{X}_1', Y_1), \ldots, \mathbf{Z}_n = (\mathbf{X}_n', Y_n)$ are independent and identically distributed with

$$\Pr(Y_i = 1 | \mathbf{X}_i = \mathbf{x}) = P(\mathbf{x}'\boldsymbol{\beta}_0).$$

We also will need to assume

C1. $\Pr(\mathbf{X}'\boldsymbol{\alpha} = 0) = 0\ \forall \boldsymbol{\alpha} \in \mathbb{R}^q\ \boldsymbol{\alpha} \neq \mathbf{0}$

The following theorem gives the consistency of the M-estimates.

Theorem 2.3 *Assume A1 to A3, B1 to B4 and C1. Then the estimate defined by minimizing (1.7) converges almost surely to the true parameter* $\boldsymbol{\beta}_0$.

The proof, whose details can be found in the Appendix, follows very closely Lemma 1 and Theorem 1 of Huber (1967). The difference is that Huber's assumption A5 of Lemma 1 does not hold. This assumption is used by Huber to show that the estimate stays ultimately in a compact set almost surely. This fact is proved separately in our proof.

Define the following q-dimensional vector and $q \times q$ matrix:

$$\Psi(Y, \mathbf{X}, \boldsymbol{\beta}) = \frac{\partial \phi}{\partial \boldsymbol{\beta}}(Y, \mathbf{X}, \boldsymbol{\beta})$$

$$J(Y, \mathbf{X}, \boldsymbol{\beta}) = \frac{\partial^2 \phi}{\partial \boldsymbol{\beta}^2}(Y, \mathbf{X}, \boldsymbol{\beta}) = \frac{\partial \Psi}{\partial \boldsymbol{\beta}}(Y, \mathbf{X}, \boldsymbol{\beta}).$$

The next theorem states the asymptotic normality of the M-estimates for the logistic model. A similar result holds for a general link function P.

Theorem 2.4 *Assume* $\mathrm{E}(\|X\|^3) < \infty$, *where* $\|\ \|$ *denotes Euclidean norm, A1–A5, P satisfies (1.2) and C1. Let* $\widehat{\boldsymbol{\beta}}_n$ *be the estimate defined by the minimization of (1.7) converges in distribution to a q-dimensional normal distribution with* $\mathbf{0}$ *mean and covariance matrix* $C(\boldsymbol{\beta}_0)$ *given by:*

$$C(\boldsymbol{\beta}_0) = \mathrm{E}(J(Y, \mathbf{X}, \boldsymbol{\beta}_0))^{-1}\, \mathrm{E}(\Psi(Y, \mathbf{X}, \boldsymbol{\beta}_0)\Psi(Y, \mathbf{X}, \boldsymbol{\beta}_0)')(\mathrm{E}(J(Y, \mathbf{X}, \boldsymbol{\beta}_0))^{-1})'.$$

The proof is based on Theorem 3 of Huber (1967); for details see the Appendix.

3 Asymptotic Biasses under Contamination

Let G_0 be the distribution of $\mathbf{X}$ under the nominal model and F_0 the corresponding joint distribution of $(\mathbf{X}', Y)$ given by model (1.1). For fixed $\boldsymbol{\xi} \in \mathbb{R}^q$ and $\eta \in \mathbb{R}$, let $F_{\epsilon, \boldsymbol{\xi}, \eta} = (1-\epsilon)F_0 + \epsilon \delta_{(\boldsymbol{\xi}', \eta)}$ be the contaminated distribution, where $\delta_{(\boldsymbol{\xi}', \eta)}$ denotes the point-mass distribution on $(\boldsymbol{\xi}', \eta)$. Then, under $F_{\epsilon, \boldsymbol{\xi}, \eta}$ we will observe with probability $(1-\epsilon)$ the vector $(\mathbf{X}', Y)$ coming from a population with the central distribution F_0, and with probability ϵ the outlier point $(\boldsymbol{\xi}', \eta)$.

Let $\mathbf{T}$ be an $\mathbb{R}^q$-valued functional defined on a "large" subset of the space of distribution functions F on $\mathbb{R}^{q+1}$. This subset is assumed to include all empirical distributions F_n corresponding to a sample $(\mathbf{X}_1', Y_1), \ldots, (\mathbf{X}_n', Y_n)$ of size n from F. Then, $\mathbf{T}_n = \mathbf{T}(F_n)$ is an estimate of $\boldsymbol{\beta}_0$. We assume that $\mathbf{T}$ is continuous under the topology of convergence in distribution. Therefore, $\mathbf{T}(F)$ is the asymptotic value of the estimate when the sample is drawn under the distribution F.

We measure the robustness of an estimate by the mean asymptotic bias produced by a contamination of size ϵ at the point $(\boldsymbol{\xi}', \eta)$ at the conditional probability of $Y=1$ given $\mathbf{X}$. More precisely, this bias is defined by

$$B(\mathbf{T}, F_0, \epsilon, \boldsymbol{\xi}, \eta) = \mathrm{E}_{G_0}(|P(\mathbf{x}'\boldsymbol{\beta}_0) - P(\mathbf{x}'\mathbf{T}(F_{\epsilon,\boldsymbol{\xi},\eta}))|).$$

We have computed the mean asymptotic bias curves $B(\mathbf{T}, F_0, \epsilon, \boldsymbol{\xi}, \eta)$ for the M-estimates based on ρ_c with $c = -\log(.03)$, which will be denoted as $\mathrm{M}_{0.03}$. These curves are compared with those of the most robust CUBI-estimator proposed by Künsch et al. (1989) and with the ML-estimate. The value of ϵ was taken equal to 0.05.

These computations were done for four different models corresponding to one covariate and the intercept; that is, $\mathbf{X} = (1, X)'$, where X has distribution $\mathcal{N}(0,1)$. These four models were chosen so that $\Pr(Y=1) = 0.5$, 0.4, 0.3 and 0.2. The corresponding parameters β_{01} and β_{02} are shown in Table 1.

$\Pr(Y=1)$	β_{01}	β_{02}
0.5	0	4.36
0.4	-1.16	4.20
0.3	-2.16	3.71
0.2	-2.80	2.82

Table 1: Parameter values of the selected models

We consider only the most influential point-mass contaminations, that is at points (ξ, η) with $\eta = 1$ if $P(\beta_{01} + \beta_{02}\xi) < 0.5$ and $\eta = 0$ in other case.

The asymptotic mean bias curves are shown in Figure 1. The solid line corresponds to the estimate $\mathrm{M}_{0.03}$, the dashed line to the CUBI-estimate and the dotted line to the ML-estimate. The maximum values of the mean asymptotic bias curves corresponding to the estimator $\mathrm{M}_{0.03}$ and to the CUBI estimator are shown in Table 2.

	$\Pr(Y=1)$			
Estimator	0.5	0.4	0.3	0.2
CUBI	0.091	0.091	0.089	0.083
$\mathrm{M}_{0.03}$	0.071	0.075	0.085	0.109

Table 2: Maximum Mean Asymptotic Biasses for $\epsilon = 0.05$

From Figure 1 we observe that the ML-estimate is extremely sensitive to outlier contamination. On the other hand the $\mathrm{M}_{0.03}$ and CUBI estimators have a much more stable behavior. For moderate outliers these two estimators have a

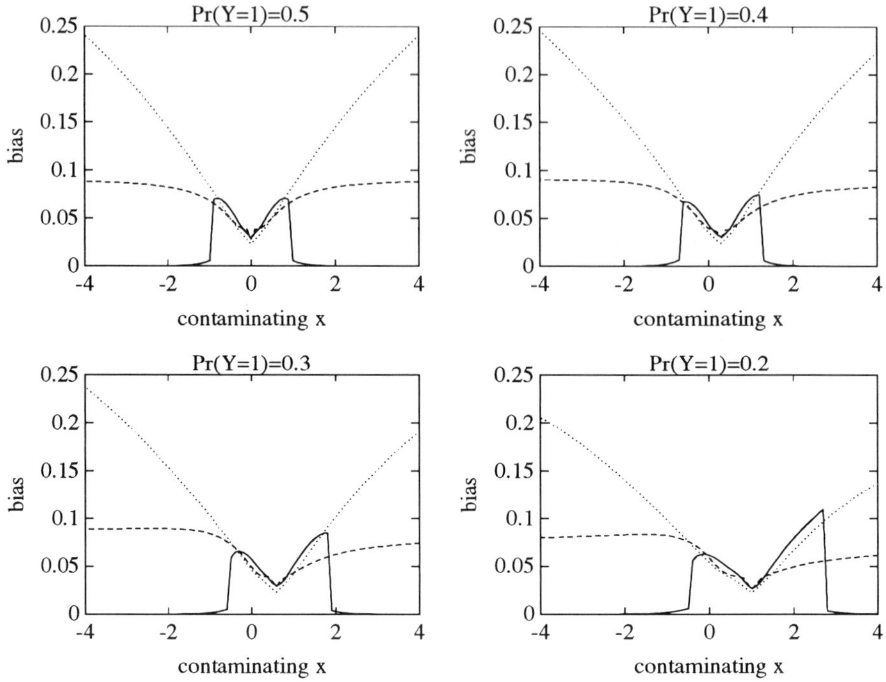

Figure 1: Mean Asymptotic Bias Curves

similar robustness performance, but when the absolute value of the contaminating ξ becomes large, the estimator $\mathrm{M}_{0.03}$ has a clear advantage, since it rejects very wild outliers. According to the values of the maximum asymptotic biasses in Table 2, the estimator $\mathrm{M}_{0.03}$ is more robust for $\Pr(Y=1)$ close to 0.5 and the CUBI estimator when $\Pr(Y=1)$ is close to 0. By symmetry the case of $\Pr(Y=1)$ close to 1 should behave similarly to the latter case.

4 Examples

In this section we illustrate the behavior of the M-estimator proposed in the paper using the logistic model for two different data sets.

In both examples we compute an initial estimate by a stepwise procedure as follows. The procedure begins by fitting models with one covariate each time and the minimization of (1.7) is carried over using a starting grid followed by the minimization method of Golden Section Search. We select the one-variable model which minimizes (1.7). Then we leave fixed the coefficient corresponding to the first selected variable and fit a two covariate model as in the one variable case. We continue in this way until all variables are entered. To get the final estimate we used the Downhill Simplex Method due to Nelder and Mead (1965).

4.1 Example: Leukemia Data

These data have been considered by Cook and Weisberg (1982, Ch. 5, p. 193) and Johnson (1985) to illustrate the identification of influential observations. The data consist of 33 leukemia patients. The response variable is 1 when the patient survives more than 52 weeks. Two covariates are considered:

WBC : White blood cell count
AG : Presence or absence of certain morphologic characteristic in the white cells

The model also includes an intercept.

Cook and Weisberg (1982) detected that there was one observation (# 15) corresponding to a patient with WBC = 100.000 who survived for a long period of time. This observation was very influential when the maximum likelihood estimator was used. They also noticed that removing this observation a much better overall fit was obtained, and that the survival fitted probabilities of those observations corresponding to patients with small values of WBC increased significatively.

In Table 3 we give the estimated coefficients corresponding to (i) the maximum likelihood estimator with the complete sample (ML_{33}), (ii) the maximum likelihood estimator after removing the influential observation (ML_{32}) and (iii) the robust M-estimate ($\mathrm{M}_{0.03}$) corresponding to the family ρ_c with $c = -\log(.03)$. In Figure 2 we show the fitted probabilities corresponding to these three estimates. The solid line corresponds to the $\mathrm{M}_{0.03}$ estimate, the dashed line to the ML_{32} estimate and the dotted line to the ML_{33} estimate.

Estimator	MLE_{33}	MLE_{32}	$\mathrm{M}_{0.03}$
Intercept	-1.307	0.212	0.104
WBC	-0.00003	-0.0002	-0.0002
AG	2.261	2.558	2.392

Table 3: Cook Data: Estimated Coefficients

From Table 2 it can be observed the similarity between the coefficients fitted with the $\mathrm{M}_{0.03}$ and the ML_{32} estimates. A similar comment holds for the fitted probabilities of Figure 2.

In Figure 3 we plot the residuals $Y - P(\boldsymbol{\beta}'\mathbf{x})$ for the $\mathrm{M}_{0.03}$ (with asterisks) and ML_{33} (with circles) estimates. From this plot it is clear that the $\mathrm{M}_{0.03}$ estimate gives a better fit for most of the points.

4.2 Example: Food Stamp Data

The following example was taken from Stefanski et al. (1986) and Künsch et al. (1989). The response variable indicates participation on the Food Stamp

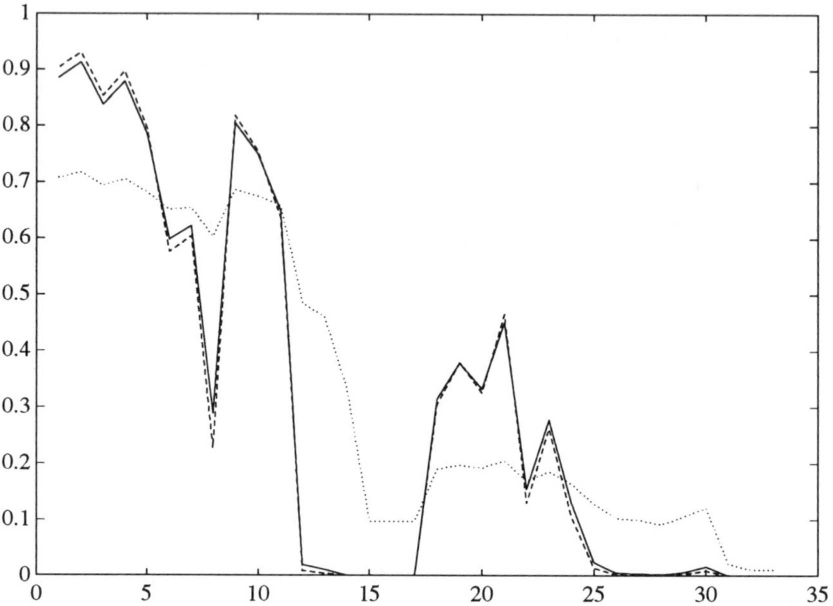

Figure 2: Cook Data: Fitted Probabilities

Program and the covariates include log(monthly income + 1) (LMI) and two dicothomus variables: tenancy (TEN) and supplemental income (SI). The model also has an intercept. The data consist of 150 observations.

In Table 4 we display the estimates corresponding to different robust procedures and to the ML-estimate. The robust procedures considered are: $\mathrm{M}_{0.03}$ and the CUBI-estimator based on Hampel's ψ function which appears in Table 2 of Künsch et al. (1989).

Estimator	MLE	MLE_{147}	CUBI	$\mathrm{M}_{0.03}$
Intercept	0.926	7.552	6	7.330
TEN	-1.850	-2.268	-1.8	-1.894
SI	0.896	0.863	0.7	0.691
LMI	-0.333	-1.461	-1.18	-1.419

Table 4: Food-stamp Data: Estimated Coefficients

The comparison between the probabilities fitted using the estimate $\mathrm{M}_{0.03}$ and the observed y's, reveals observations 5 and 66 as clear outliers, and observa-

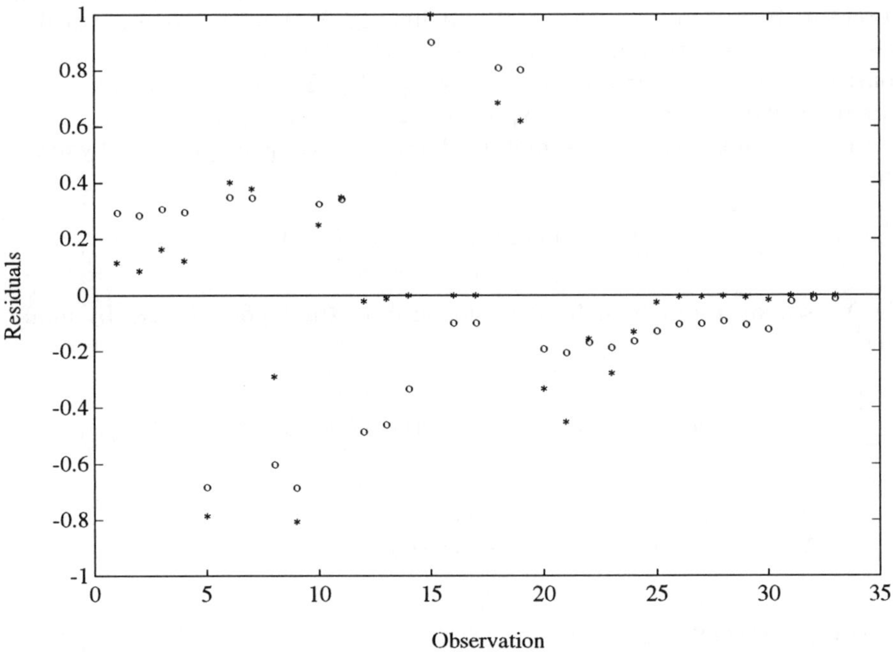

Figure 3: Cook Data: Residuals

tion 137 as a possible outlier. In fact, their y values are 0, 1 and 1, and their fitted probabilities are 0.9993 and 0.0097 and 0.0329 respectively. In Table 4 we also show the ML-estimate computed after removing these three observations (ML_{147}), which is very similar to $\mathrm{M}_{0.03}$ and to the CUBI-estimator based on Hampel's ψ-function given by Künsch et al. (1989).

5 Extensions

The class of M-estimators proposed in this paper can be extended in a natural way to multinomial regression models. Let $\mathbf{Y} = (Y_1, \ldots, Y_k)'$ be the response vector where $Y_i = 0$ or 1 and $\sum_{i=1}^k Y_i = 1$ and let $\mathbf{X} = (X_1, \ldots, X_q)'$ be the vector of covariates. Assume that for any $\mathbf{x} \in \mathbb{R}^q$ and $j = 1, \ldots, k-1$

$$\Pr(Y_j = 1|\mathbf{X} = \mathbf{x}) = \frac{P(\mathbf{x}'\boldsymbol{\beta}_j)}{1 + \sum_{h=1}^{k-1} P(\mathbf{x}'\boldsymbol{\beta}_h)} = P_j^*(\mathbf{x}, \boldsymbol{\beta}_1, \ldots, \boldsymbol{\beta}_{k-1}),$$

while

$$\Pr(Y_k = 1|\mathbf{X} = \mathbf{x}) = \frac{1}{1 + \sum_{h=1}^{k-1} P(\mathbf{x}'\boldsymbol{\beta}_h)} = P_k^*(\mathbf{x}, \boldsymbol{\beta}_1, \ldots, \boldsymbol{\beta}_{k-1}),$$

where $\boldsymbol{\beta}_j \in \mathbb{R}^q$, $j = 1, \ldots, k-1$ and P is a continuous, strictly increasing function satisfying $\lim_{t\to-\infty} P(t) = 0$ and $\lim_{t\to\infty} P(t) = \infty$. As a particular case we have the logistic multinomial model with $P(t) = e^t$.

Consider a sample $\mathbf{Z}_1 = (\mathbf{X}_1', \mathbf{Y}_1), \ldots, \mathbf{Z}_n = (\mathbf{X}_n', \mathbf{Y}_n)$ of n independent observations, where $\mathbf{X}_i' = (X_{i1}, \ldots, X_{iq})$ and $\mathbf{Y}_i' = (Y_{i1}, \ldots, Y_{ik})$, $i = 1, \ldots, n$. Then the maximum likelihood estimators of $\boldsymbol{\beta}_1, \ldots, \boldsymbol{\beta}_{k-1}$ are obtained by maximizing

$$\sum_{i=1}^{n}\sum_{j=1}^{k} Y_{ij} \ln(P_j^*(\mathbf{x}_i, \boldsymbol{\beta}_1, \ldots, \boldsymbol{\beta}_{k-1})).$$

The M-estimates may be defined analogously to the binomial case, by minimizing

$$\sum_{i=1}^{n}\sum_{j=1}^{k}[Y_{ij}\rho(-\ln(P_j^*(\mathbf{x}_i, \boldsymbol{\beta}_1, \ldots, \boldsymbol{\beta}_{k-1}))) + G(P_j^*(\mathbf{x}_i, \boldsymbol{\beta}_1, \ldots \boldsymbol{\beta}_{k-1}))],$$

where G is defined in (1.8). A result similar to the one given in Section 2 implies that these M-estimates are also Fisher-consistent.

6 Appendix

Some auxiliary results will be needed to prove the consistency of M-estimates.

We will assume that P verifies B1 to B4.

Lemma 6.1 *Let $\phi(y, \mathbf{x}, \boldsymbol{\beta})$ be defined as in (2.6). Assume that $\rho(t)$ satisfies conditions A1 to A3. Then*

(i) $\phi(y, \mathbf{x}, \boldsymbol{\beta}) > 0$.

(ii) There exists a constant $K > 0$ such that $\phi(y, \mathbf{x}, \boldsymbol{\beta}) < K$.

PROOF It is immediate from the assumptions. ////

The next Lemma gives the Fisher-consistency of the M-estimators.

Lemma 6.2 *If $\rho(t)$ satisfies A1 to A3 and $\mathbf{X}$ satisfies condition C1, then for any $\mathbf{t} \in \mathbb{R}^q$ with $||\mathbf{t}|| = 1$ we have:*

$$\mathrm{E}_{\beta_o}\big(\lim_{\lambda\to\infty} \phi(Y, \mathbf{X}, \lambda\mathbf{t})\big) > \mathrm{E}_{\beta_o}(\phi(Y, \mathbf{X}, \boldsymbol{\beta}_0)) \quad \textit{for all } \boldsymbol{\beta} \neq \boldsymbol{\beta}_0.$$

PROOF Since ϕ is bounded, the Dominated Convergence Theorem implies

$$\begin{aligned}
\mathrm{E}_{\beta_o}\big(\lim_{\lambda\to\infty} \phi(Y, \mathbf{X}, \lambda\mathbf{t})\big) &= \lim_{\lambda\to\infty} \mathrm{E}_{\beta_o}(\phi(Y, \mathbf{X}, \lambda\mathbf{t})) \\
&= \mathrm{E}_{\beta_o}\big(\lim_{\lambda\to\infty} \mathrm{E}(\phi(Y, \mathbf{X}, \lambda\mathbf{t})|\mathbf{X})\big).
\end{aligned}$$

We also have

$$\mathrm{E}(\phi(Y, \mathbf{X}, \lambda \mathbf{t})|\mathbf{X}) = L(P(\mathbf{X}'\boldsymbol{\beta}_0), P(\lambda \mathbf{X}'\mathbf{t})),$$

where $L(p_0, p)$ is defined in (2.1). Using the facts that (i) $L(p_0, p)$ takes its minimum value at $p = p_0$, (ii) $\lim_{\lambda\to\infty} P(\lambda\mathbf{X}'\mathbf{t}) = 0$ or 1 with probability one and (iii) L is continuous, we get

$$\begin{aligned}\lim_{\lambda\to\infty} \mathrm{E}(\phi(Y, \mathbf{X}, \lambda \mathbf{t})|\mathbf{X}) &= L(P(\mathbf{X}'\boldsymbol{\beta}_0), \lim_{\lambda\to\infty} P(\lambda \mathbf{X}'\mathbf{t})) \\ &> L(P(\mathbf{X}'\boldsymbol{\beta}_0), P(\mathbf{X}'\boldsymbol{\beta}_0)) = \mathrm{E}(\phi(Y, \mathbf{X}, \boldsymbol{\beta}_0)|\mathbf{X}).\end{aligned}$$

Then by C1, we obtain

$$\mathrm{E}_{\beta_o}\big(\lim_{\lambda\to\infty} \phi(Y, \mathbf{X}, \lambda\mathbf{t})\big) > \mathrm{E}_{\beta_o}(\phi(Y, X, \boldsymbol{\beta}_0)).$$

This ends the proof. ////

Define $B(\boldsymbol{\alpha}, \varepsilon) = \{\boldsymbol{\beta} \in \mathbb{R}^q : \|\boldsymbol{\beta} - \boldsymbol{\alpha}\| < \varepsilon\}$. Then the following Lemma holds.

Lemma 6.3 *Assume that* $\rho(t)$ *satisfies A1 to A3. Let* $\mathbf{X} \in \mathbb{R}^q$ *be a random vector such that* $\Pr(\mathbf{X}'\mathbf{s} = 0) = 0$ *for every* $\mathbf{s} \in \mathbb{R}^q$. *Then:*

(i) for all $\boldsymbol{\alpha} \in \mathbb{R}^q$,

$$\lim_{\varepsilon\to 0} \mathrm{E}_{\beta_o}\Big(\inf_{\boldsymbol{\beta}\in B(\boldsymbol{\alpha},\varepsilon)} \phi(Y, \mathbf{X}, \boldsymbol{\beta}) - \phi(Y, \mathbf{X}, \boldsymbol{\alpha})\Big) = 0.$$

(ii) For every $\mathbf{t} \in \mathbb{R}^q$ *such that* $\|\mathbf{t}\| = 1$ *there exists some* $\varepsilon > 0$ *such that*

$$\mathrm{E}_{\beta_o}\Big(\liminf_{\lambda\to\infty} \inf_{\mathbf{s}\in B(\mathbf{t},\varepsilon)} \phi(Y, \mathbf{X}, \lambda\mathbf{s})\Big) > \mathrm{E}_{\beta_o}(\phi(Y, \mathbf{X}, \boldsymbol{\beta}_0)).$$

PROOF

(i) Since ϕ is continuous and bounded (i) is a consequence of the Dominated Convergence Theorem.

(ii) Since $\Pr(\mathbf{X}'\mathbf{s} = 0) = 0$ for all $\mathbf{s} \in \mathbb{R}^q$, we have that given $\mathbf{t} \in \mathbb{R}^q$ such that $\|\mathbf{t}\| = 1$ and $\delta > 0$, there exist positive numbers ε and M such that:

$$\Pr(|\mathbf{X}'\mathbf{t}| > \varepsilon \text{ and } \|\mathbf{X}\| < M) > 1 - \delta.$$

Let $C = \{\mathbf{x} \in \mathbb{R}^q : |\mathbf{x}'\mathbf{t}| > \varepsilon \text{ and } \|\mathbf{x}\| < M\}$ and take $\eta < \varepsilon/2M$. Then , it is easy to show that for any $\mathbf{x} \in C$ and $\mathbf{s} \in B(\mathbf{t}, \eta)$ we have $\mathrm{sg}(\mathbf{x}'\mathbf{s}) = \mathrm{sg}(\mathbf{x}'\mathbf{t})$ and $|\mathbf{x}'\mathbf{s}| > \varepsilon/2$, where sg denotes the sign function. Therefore, it is easy to show that for any $\mathbf{x} \in C$ we have

$$\liminf_{\lambda\to\infty} \inf_{\mathbf{s}\in B(\mathbf{t},\eta)} \phi(y, \mathbf{x}, \lambda\mathbf{s}) = \lim_{\lambda\to\infty} \phi(y, \mathbf{x}, \lambda\mathbf{t}). \tag{6.1}$$

Let $D = \mathrm{E}_{\beta_o}(\lim_{\lambda\to\infty}\phi(Y,\mathbf{X},\lambda\mathbf{t})) - \mathrm{E}_{\beta_o}(\phi(Y,\mathbf{X},\boldsymbol{\beta}_0))$, which is positive by Theorem 2.2, and let $K = \sup\phi(y,\mathbf{x},\boldsymbol{\beta})$. Take now $\delta < D/K$ and denote by $\mathrm{I}_A(t)$ the indicator function of the set A. Then using the fact that ϕ is positive and (6.1) we have

$$\begin{aligned} &\mathrm{E}_{\beta_o}\left(\liminf_{\lambda\to\infty}\inf_{\mathbf{s}\in B(\mathbf{t},\varepsilon)}\phi(Y,\mathbf{X},\lambda\mathbf{s})\right)\\ &\geq \mathrm{E}_{\beta_o}\left(\mathrm{I}_C(\mathbf{X})\liminf_{\lambda\to\infty}\inf_{\mathbf{s}\in B(\mathbf{t},\varepsilon)}\phi(Y,\mathbf{X},\lambda\mathbf{s})\right)\\ &= \mathrm{E}_{\beta_o}(\mathrm{I}_C(\mathbf{X})\lim_{\lambda\to\infty}\phi(Y,\mathbf{X},\lambda\mathbf{t}))\\ &> \mathrm{E}_{\beta_o}(\phi(Y,\mathbf{X},\boldsymbol{\beta}_0)). \end{aligned}$$

which is the result. ////

PROOF OF THEOREM 2.3 First we will prove that the sequence $\widehat{\boldsymbol{\beta}}_n$ ultimately stays in a compact set C.

Let $S = \{\mathbf{t}\in\mathbb{R}^q : \|\mathbf{t}\| = 1\}$. Then, if $\mathbf{t}\in S$, by Lemma 6.3 (ii), there exist some positive numbers ε_t, η_t and M_t such that:

$$\mathrm{E}_{\beta_o}\left(\inf_{\lambda>M_t}\inf_{\mathbf{s}\in B(\mathbf{t},\varepsilon_t)}\phi(Y,\mathbf{X},\lambda\mathbf{s})\right) > \mathrm{E}_{\beta_o}(\phi(Y,\mathbf{X},\boldsymbol{\beta}_0)) + 3\eta_t.$$

The Strong Law of Large Numbers implies that, almost surely, there exists some n_t such that for all $n\geq n_t$,

$$\begin{aligned} \inf_{\lambda>M_t}\inf_{\mathbf{s}\in B(\mathbf{t},\varepsilon_t)}\frac{1}{n}\sum_{i=1}^n\phi(Y_i,\mathbf{X}_i,\lambda\mathbf{s}) &\geq \frac{1}{n}\sum_{i=1}^n\inf_{\lambda>M_t}\inf_{\mathbf{s}\in B(\mathbf{t},\varepsilon_t)}\phi(Y_i,\mathbf{X}_i,\lambda\mathbf{s})\\ &> \mathrm{E}_{\beta_o}(\phi(Y,\mathbf{X},\boldsymbol{\beta}_0)) + 2\eta_t. \end{aligned}$$

Let $B_{t_i}, 1\leq i\leq N$ be a finite subcovering of S. Then, for sufficiently large M and n_1^* and sufficiently small ε and η, if $n\geq n_1^*$ we have that

$$\inf_{\lambda>M}\inf_{\mathbf{s}\in B(\mathbf{t},\epsilon)}\frac{1}{n}\sum_{i=1}^n\phi(Y_i,\mathbf{X}_i,\lambda\mathbf{s}) > \mathrm{E}_{\beta_o}(\phi(Y,\mathbf{X},\boldsymbol{\beta}_0)) + 2\eta. \tag{6.2}$$

On the other hand, by the Strong Law of Large Numbers there exists n_2^* such that if $n\geq n_2^*$ then

$$\frac{1}{n}\sum_{i=1}^n\phi(Y_i,\mathbf{X}_i,\boldsymbol{\beta}_0) < \mathrm{E}_{\beta_o}(\phi(Y,\mathbf{X},\boldsymbol{\beta}_0)) + \eta. \tag{6.3}$$

Let $n^* = \max(n_1^*, n_2^*)$ and $C = \{\boldsymbol{\beta}\in\mathbb{R}^q : \|\boldsymbol{\beta}\|\leq M\}$. Then (6.2) and (6.3) imply that:

$$\frac{1}{n}\sum_{i=1}^n\phi(Y_i,\mathbf{X}_i,\boldsymbol{\beta}_0) < \inf_{\boldsymbol{\beta}\notin C}\frac{1}{n}\sum_{i=1}^n\phi(Y_i,\mathbf{X}_i,\boldsymbol{\beta})$$

and

$$P(\exists n^* : n \geq n^* \widehat{\boldsymbol{\beta}}_n \in C) = 1).$$

Let U be an open neighborhood of $\boldsymbol{\beta}_0$. Since $\mathrm{E}_{\beta_0}(\phi(Y, \mathbf{X}, \boldsymbol{\beta}))$ is continuous in $\boldsymbol{\beta}$, it attains its infimum in the compact $C - U$. Then, by the Lemma 6.1, there exists $\eta > 0$ such that

$$\inf_{\beta \in C-U} \mathrm{E}_{\beta_0}(\phi(Y, \mathbf{X}, \boldsymbol{\beta})) > \mathrm{E}_{\beta_0}(\phi(Y, \mathbf{X}, \boldsymbol{\beta}_0)) + 4\eta.$$

Part (i) of Lemma 6.2 implies that for any $\boldsymbol{\alpha} \in C - U$ there exists some ε_α such that:

$$\begin{aligned}\mathrm{E}_{\beta_0}\Big(\inf_{\beta \in B(\alpha, \varepsilon_\alpha)} \phi(Y, \mathbf{X}, \boldsymbol{\beta})\Big) &> \mathrm{E}_{\beta_0}(\phi(Y, \mathbf{X}, \boldsymbol{\alpha})) - \eta \\ &> \mathrm{E}_{\beta_0}(\phi(Y, \mathbf{X}, \boldsymbol{\beta}_0)) + 3\eta.\end{aligned}$$

Let $B(\boldsymbol{\alpha}_1, \varepsilon_{\alpha_1}), \ldots, B(\boldsymbol{\alpha}_N, \varepsilon_{\alpha_N})$ be a finite subcovering of the compact set $C - U$. Using the Strong Law of Large Numbers, we can almost surely find n_j, $j = 1, \ldots, N$ such that, if $n \geq n_j$ for $j = 1, \ldots, n$, then

$$\begin{aligned}\inf_{\beta \in B(\alpha_j, \varepsilon_{\alpha_j})} \frac{1}{n} \sum_{i=1}^{n} \phi(Y_i, \mathbf{X}_i, \boldsymbol{\beta}) &\geq \frac{1}{n} \sum_{i=1}^{n} \inf_{\beta \in B(\alpha_j, \varepsilon_{\alpha_j})} \phi(Y_i, \mathbf{X}_i, \boldsymbol{\beta}) \\ &> \mathrm{E}_{\beta_0}\Big(\inf_{\beta \in B(\alpha_j, \varepsilon_{\alpha_j})} \phi(Y, \mathbf{X}, \boldsymbol{\beta})\Big) - \eta \\ &> \mathrm{E}_{\beta_0}(\phi(Y, \mathbf{X}, \boldsymbol{\beta}_0)) + 2\eta. \qquad (6.4)\end{aligned}$$

On the other hand, the Strong Law of Large Numbers implies that we can almost surely find n_0 such that $n \geq n_0$ implies:

$$\inf_{\beta \in U} \frac{1}{n} \sum_{i=1}^{n} \phi(Y_i, \mathbf{X}_i, \boldsymbol{\beta}) \leq \frac{1}{n} \sum_{i=1}^{n} \phi(Y_i, \mathbf{X}_i, \boldsymbol{\beta}_0) < \mathrm{E}_{\beta_0}(\phi(Y, \mathbf{X}, \boldsymbol{\beta})) + \eta. \qquad (6.5)$$

Let $n_1^* = \max(n_0, n_1, \ldots, n_N)$, then (6.4) and (6.5) imply that, for all $n \geq n_1^*$,

$$\begin{aligned}\inf_{\beta \in C-U} \frac{1}{n} \sum_{i=1}^{n} \phi(Y_i, \mathbf{X}_i, \boldsymbol{\beta}) &> \mathrm{E}_{\beta_0}(\phi(Y_i, \mathbf{X}_i, \boldsymbol{\beta}_0)) + 2\eta \\ &> \inf_{\beta \in U} \frac{1}{n} \sum_{i=1}^{n} \phi(Y_i, \mathbf{X}_i, \boldsymbol{\beta}) + \eta \qquad (6.6)\end{aligned}$$

Finally, (6.5) and (6.6) imply that $\widehat{\boldsymbol{\beta}}_n \in U$ for all $n \geq \max = (n^*, n_1^*)$. ////

PROOF OF THEOREM 2.4 Let $\lambda(\boldsymbol{\beta}) = \mathrm{E}(\Psi(Y, \mathbf{X}, \boldsymbol{\beta}))$ and

$$U(y, \mathbf{x}, \boldsymbol{\beta}, d) = \sup_{\|\boldsymbol{\tau} - \boldsymbol{\beta}\| \leq d} \|\Psi(y, \mathbf{x}, \boldsymbol{\beta}) - \Psi(y, x, \boldsymbol{\tau})\|.$$

According to Theorem 3 of Huber (1967) and its corollary, it suffices to show that there exist positive numbers a, b, c, d_0 such that

(i) $\|\lambda(\boldsymbol{\beta})\| \geq a\|\boldsymbol{\beta} - \boldsymbol{\beta}_0\|$ if $\|\boldsymbol{\beta} - \boldsymbol{\beta}_0\| \leq d_0$

(ii) $\mathrm{E}(U(Y, \mathbf{X}, \boldsymbol{\beta}, d)) \leq b \cdot d$

(iii) $\mathrm{E}(U^2(Y, \mathbf{X}, \boldsymbol{\beta}, d)) \leq c \cdot d$ and that $\mathrm{E}(J(Y, \mathbf{X}, \boldsymbol{\beta}_0))$ is nonsingular.

First, it will be shown that $\Sigma = \mathrm{E}(J(Y, \mathbf{X}, \boldsymbol{\beta}_0))$ is nonsingular. It is easy to show that $\Sigma = \mathrm{E}(V)$ where V is the random matrix $V = d(\mathbf{X})\mathbf{X}\mathbf{X}'$ with

$$\begin{aligned} d(\mathbf{X}) &= P(\mathbf{X}'\boldsymbol{\beta}_0)(1 - P(\mathbf{X}'\boldsymbol{\beta}_0))[\psi(-\ln(P(\mathbf{X}'\boldsymbol{\beta}_0)))(1 - P(\mathbf{X}'\boldsymbol{\beta}_0)) \\ &\quad + \psi(-\ln(1 - P(\mathbf{X}'\boldsymbol{\beta}_0)))P(\mathbf{X}'\boldsymbol{\beta}_0)]. \end{aligned}$$

Then for all $\mathbf{a} \in \mathbb{R}^q, \mathbf{a} \neq 0, \mathbf{a}'\Sigma\mathbf{a} = \mathrm{E}(d(\mathbf{X})(\mathbf{a}'\mathbf{X})^2)$. Since, by C1, $\mathbf{a}'\mathbf{X} \neq 0$ almost surely and, by A2, $d(\mathbf{X}) > 0$, we obtain $\mathbf{a}'\Sigma\mathbf{a} > 0$.

Now we will prove that $\lambda(\boldsymbol{\beta})$ is differentiable at $\boldsymbol{\beta} = \boldsymbol{\beta}_0$ and

$$\left.\frac{\partial \lambda}{\partial \boldsymbol{\beta}}\right|_{\boldsymbol{\beta} = \boldsymbol{\beta}_0} = \Sigma. \tag{6.7}$$

Let $q(y, p) = (y - p)[\psi(-\ln p)(1 - p) + \psi(-\ln(1 - p))p]$ for $y = 0$ or 1 and $p \in (0, 1)$. Condition (A5) implies that there exists a constant M such that

$$|q(y, p) - q(y, p')| < M|p - p'|.$$

Then

$$\begin{aligned} \frac{\partial \lambda_i}{\partial \beta_j}(\boldsymbol{\beta}) &= \lim_{h \to 0} \frac{\lambda_i(\boldsymbol{\beta} + \mathbf{e}_j h) - \lambda_i(\boldsymbol{\beta})}{h} \\ &= \lim_{h \to 0} \int x_i \frac{[q(y, P(\mathbf{x}'(\boldsymbol{\beta} + \mathbf{e}_j h))) - q(y, P(\mathbf{x}'\boldsymbol{\beta}))]}{h} \, dP_{\mathbf{X}Y}, \end{aligned} \tag{6.8}$$

where $\mathbf{e}_1, \ldots, \mathbf{e}_q$ denotes the canonical basis in $\mathbb{R}^q$ and $P_{\mathbf{X}Y}$ is the underlying distribution of $(\mathbf{X}, Y)$. Since, with probability one, the integrand in (6.8) converges to $\partial \Psi_i(y, \mathbf{x}, \boldsymbol{\beta})/\partial \beta_j$, and $q(y, p)$ is Lipschitz in p, the Dominated Convergence Theorem implies that we may interchange the order of differentiation and integration and this proves (6.7). In order to prove the differentiability of $\lambda(\boldsymbol{\beta})$ at $\boldsymbol{\beta} = \boldsymbol{\beta}_0$, it is enough to show that $\partial \lambda_i(\boldsymbol{\beta})/\partial \beta_j$ is continuous at $\boldsymbol{\beta} = \boldsymbol{\beta}_0$. Since

$$\begin{aligned} \frac{\partial \lambda_i}{\partial \beta_j}(\boldsymbol{\beta}) &= \int x_i x_j \{P(\mathbf{x}'\boldsymbol{\beta})(1 - P(\mathbf{x}'\boldsymbol{\beta}))[\psi(-\ln(P(\mathbf{x}'\boldsymbol{\beta})))(1 - P(\mathbf{x}'\boldsymbol{\beta})) \\ &\quad + \psi(-\ln(1 - P(\mathbf{x}'\boldsymbol{\beta})))P(\mathbf{x}'\boldsymbol{\beta})] \\ &\quad + (P(\mathbf{x}'\boldsymbol{\beta}) - Y)[-\psi'(-\ln(P(\mathbf{x}'\boldsymbol{\beta}))(1 - P(\mathbf{x}'\boldsymbol{\beta}))^2 \\ &\quad - \psi(-\ln(P(\mathbf{x}'\boldsymbol{\beta}))P(\mathbf{x}'\boldsymbol{\beta})(1 - P(\mathbf{x}'\boldsymbol{\beta})) \\ &\quad + \psi'(-\ln(1 - P(\mathbf{x}'\boldsymbol{\beta})))(P(\mathbf{x}'\boldsymbol{\beta}))^2 \\ &\quad + \psi(-\ln(1 - P(\mathbf{x}'\boldsymbol{\beta})))P(\mathbf{x}'\boldsymbol{\beta})(1 - P(\mathbf{x}'\boldsymbol{\beta}))]\} dP_{\mathbf{X}Y} \\ &= \int \frac{\partial \Psi_i}{\partial \beta_j}(y, \mathbf{x}, \boldsymbol{\beta}) dP_{\mathbf{X}Y}, \end{aligned} \tag{6.9}$$

this follows immediately from the Dominated Convergence Theorem and the fact that the integrand in (6.9) is bounded and almost surely continuous.

The differentiability of λ and (6.7) imply $\|\lambda(\boldsymbol{\beta})-\Sigma(\boldsymbol{\beta}_0-\boldsymbol{\beta})\| = \mathrm{o}(\|\boldsymbol{\beta}-\boldsymbol{\beta}_0\|)$. Then

$$\begin{aligned} \|\lambda(\boldsymbol{\beta})\| &\geq \|\Sigma(\boldsymbol{\beta}_0 - \boldsymbol{\beta})\| - o(\|\boldsymbol{\beta} - \boldsymbol{\beta}_0\|) \\ &\geq \|\Sigma^{-1}\|^{-1}\|(\boldsymbol{\beta} - \boldsymbol{\beta}_0)\| - o(\|\boldsymbol{\beta} - \boldsymbol{\beta}_0\|), \end{aligned} \tag{6.10}$$

where, if A is a $q \times q$ matrix, $\|A\|$ is the norm of maximum eigenvalue. There exists $d_0 > 0$ such that if $\|\boldsymbol{\beta} - \boldsymbol{\beta}_0\| < d_0$ then

$$\mathrm{o}(\|\boldsymbol{\beta} - \boldsymbol{\beta}_0\|) < 1/2\|\Sigma^{-1}\|^{-1}\|\boldsymbol{\beta} - \boldsymbol{\beta}_0\|. \tag{6.11}$$

Then, if we take $2a = \|\Sigma^{-1}\|^{-1}$, from (6.10) and (6.11) we obtain (i).

In order to prove (ii) and (iii), we will bound $U(Y, \mathbf{X}, \boldsymbol{\beta}, d)$. According to (6.8) we have

$$\begin{aligned} U(Y, \mathbf{X}, \boldsymbol{\beta}, d) &= \sup_{\|\boldsymbol{\tau} - \boldsymbol{\beta}\| \leq d} \|\mathbf{x}q(y, P(\mathbf{x}'\boldsymbol{\beta})) - \mathbf{x}q(y, P(\mathbf{x}'\boldsymbol{\tau}))\| \\ &\leq \sup_{\|\boldsymbol{\tau} - \boldsymbol{\beta}\| \leq d} M\|\mathbf{x}\||P(\mathbf{x}'\boldsymbol{\beta}) - P(\mathbf{x}'\boldsymbol{\tau})|. \end{aligned} \tag{6.12}$$

Since $P(t)$ is also Lipschitz in t with constant 1, we have

$$U(y, \mathbf{x}, \boldsymbol{\beta}, d) \leq M\|\mathbf{x}\|^2 d. \tag{6.13}$$

Then, choosing $b = \mathrm{E}(M\|\mathbf{X}\|^2)$, (ii) follows. Since $|P| < 1$ by (6.12) and (6.13), we get

$$U^2(y, \mathbf{x}, \boldsymbol{\beta}, d) \leq 2M^2\|\mathbf{x}\|^3 d.$$

Then (iii) holds with $c = 2M^2\,\mathrm{E}(\|\mathbf{X}\|^3)$. ////

References

[1] Carroll, R.J. and Pederson, S. (1993): On robustness in the logistic regression model. *Biometrika* **55** 693–706.

[2] Cook, R.D. and Weisberg, S. (1982): *Residuals and Influence in Regression.* Chapman & Hall, London.

[3] Copas, J.B. (1988): Binary regression models for contaminated data. *J. Roy. Statist. Soc. B* **50** 225–265.

[4] Hampel, F.R. (1971): A general definition of qualitative robustness. *Ann. Math. Statist.* **42** 1887–1896.

[5] Huber, P.J. (1967): The behavior of maximum likelihood estimates under non-standard conditions. *Proc. Fifth Berkeley Symp. Math. Statist. Prob.* **1** 221–233. Univ. of California Press, Berkeley.

[6] Johnson, W. (1985): Influence measures for logistic regression: another point of view. *Biometrika* **72** 59–66.

[7] Künsch, H.R., Stefanski, L.A. and Carroll, R.J. (1989): Conditionally unbiased bounded influence estimation in general regression models, with applications to generalized linear models. *J. Amer. Statist. Assoc.* **84** 460–466.

[8] Martin, R.D., Yohai, V.J. and Zamar, R.H. (1989): Min-max bias robust regression. *Ann. Statist.* **17** 1608–1630.

[9] Nelder, J.A. and Mead, R. (1965): *Computer Journal* **7** 308.

[10] Pregibon, D. (1981): Logistic regression diagnostics. *Ann. Statist.* **9** 705–724.

[11] Pregibon, D. (1982): Resistent fits for some commonly used logistic models with medical applications. *Biometrics* **38** 485–498.

[12] Stefanski, L.A., Carroll, R.J. and Ruppert D. (1986): Optimally bounded score functions for generalized linear models with applications to logistic regression. *Biometrika* **73** 413–425.

[13] Yohai, V.J. (1987): High breakdown point and high efficiency robust estimates for regression. *Ann. Statist.* **15** 692–656.

The m out of n Bootstrap and Goodness of Fit Tests with Double Censored Data

Peter J. Bickel
University of California, Berkeley

Jian-Jian Ren
University of Nebraska, Lincoln

Abstract

This paper considers the use of the m out of n bootstrap (Bickel, Götze, and van Zwet, 1994) in setting critical values for Cramér–von Mises goodness of fit tests with doubly censored data. We show that, as might be expected, the usual n out of n nonparametric bootstrap fails to estimate the null distribution of the test statistic. We show that if the m out of n bootstrap with $m \to \infty$, $m = o(n)$ is used to set the critical value of the test, the proposed testing procedure is asymptotically level α, has the correct asymptotic power function for $\sqrt{n}$ alternatives and is asymptotically consistent.

Key words and phrases: Bootstrap, Cramér–von Mises statistics, doubly censored data, goodness of fit tests.
AMS 1991 subject classifications: Primary 62G10; secondary 62E20.

1 Introduction

It is logically clear but not always evident or appreciated that the usual nonparametric bootstrap (the n out of n bootstrap) should fail when one tries to estimate the distribution of test statistics under a semiparametric (restricted nonparametric) hypothesis and ignores the restrictions imposed by the hypothesis. For example, Freedman (1981) points out that in setting confidence intervals on the usual slope estimate for regression through the origin, one must resample not the residuals but the residuals centered at their mean. If one considers setting confidence bands as the dual of hypothesis testing, a moment's thought will show that not centering the residuals is tantamount to not imposing the model requirement that the expectation of the error is zero.

For more recent examples, see Härdle and Mammen (1993), Mammen (1992), Bickel, Götze and van Zwet (1994). The same phenomenon occurs when one considers hypothesis testing problems using a censored or incomplete sample. Imposing hypothesis restrictions for censored data can be very complicated given the unknown distributions of the censoring variables. In the situation we consider here, it is impossible to estimate the censoring distribution! Specifically, we consider Cramér–von Mises goodness of fit tests with doubly censored data. The problem is introduced in Section 2, where we show that the usual n out of n nonparametric bootstrap fails to estimate the null distribution of the test statistic. In Section 3, we propose that one uses the m out of n bootstrap to set the critical value of the test and show that the proposed testing procedure is asymptotically consistent and has correct power against $\sqrt{n}$ alternatives with the proofs deferred to Section 5. For the exponential distribution family, Section 4 presents a simulation result which compares the power functions using the true critical value and the one by the m out of n bootstrap.

One should note that the testing procedure proposed here includes the goodness of fit tests for proportional hazards model with right censored data (Wei (1984)) as a special case, and that the proposed procedure easily applies to many other models with censored data or with uncensored data. An alternative to the m out of n bootstrap method in the model of this paper, the Fredholm Integral Equation (FIE) method, has been constructed by Ren (1993). The m out of n bootstrap method, although less powerful to second order, has the advantage of easy implementation and generalization to essentially any semiparametric test.

In collaboration with Götze, van Zwet and others we are in the process of studying this broad applicability and the critical issue of choice of m. In the example we study in Section 4, $m = \sqrt{n}$ is chosen arbitrarily and evidently performs reasonably well.

We close the basic heuristics of this method which are evident in the classic problem of testing $H : \mu = 0$ vs $K : \mu > 0$ when $X_1, \ldots, X_n$ are iid $\mathcal{N}(\mu, 1)$. The correct test is to reject if $\sqrt{n}\,\bar{X}_n \geq z(1-\alpha)$, where $\bar{X}_n = \frac{1}{n}\sum_{i=1}^n X_i$ and $z(1-\alpha)$ is the $1-\alpha$ quantile of $\mathcal{N}(0,1)$. Suppose that we instead use the bootstrap critical value of $\sqrt{n}\,\bar{X}_n$. That is call $\hat{c}_{m,n}$ the $1-\alpha$ quantile of the distribution of $\sqrt{m}\,\bar{X}^*_m$, where $X^*_1, \ldots, X^*_m$ are a sample of size m with replacement from $X_1, \ldots, X_n$ and reject if $\sqrt{n}\,\bar{X}_n \geq \hat{c}_{m,n}$. It is easy to see that if $m \to \infty$ with $m = o(n)$, then $\hat{c}_{m,n} = z_{1-\alpha} + o_p(1)$ not only for $\mu = 0$ but also for $\mu = \frac{\Delta}{\sqrt{n}}$, $\Delta > 0$. Thus, the local power function of our test is the same as that of the z test. Furthermore, if $\mu > 0$ is fixed, then $\hat{c}_{m,n} = O_p(\sqrt{m})$ and $P_\mu\{\sqrt{n}\,\bar{X}_n \geq \hat{c}_{m,n}\} \to 1$ since $m = o(n)$. This is the kind of behavior we exhibit in our more complicated context.

2 Goodness of Fit Tests with Doubly Censored Data

In medical follow-up studies and in reliability studies, the data available to be analyzed are often incomplete due to various reasons. Recently, some more complicated types of censored data than right censored data, such as doubly censored data, interval censored data, truncated data, etc., have started to catch the attention of statisticians, as these data occur in important clinical trials.

For instance, doubly censored data are presented in a recent study of the age-dependent growth rate of primary breast cancer (Peer et al. (1993)). Other examples of doubly censored data encountered in practice have been given by Gehan (1965), Mantel (1967), Peto (1973), Turnbull (1974), and others. In statistical analysis, assumptions on the underlying lifetime distributions are often made for various reasons. To check if these assumptions are correct, one would naturally wish to conduct a goodness of fit test. In this study, we are interested in the goodness of fit test with doubly censored data.

Let $X_1, X_2, \ldots, X_n$ be independent observations on a nonnegative r.v. X with a continuous d.f. F. If it is wished to test the null hypothesis

$$H_0 : F = F_0, \tag{2.1}$$

where F_0 is a specified d.f., then the Cramér–von Mises test statistics are given by

$$T_n = n \int_0^{\infty} [F_n(x) - F_0(x)]^2 dF_0(x), \tag{2.2}$$

where F_n is the empirical d.f. based on $X_1, X_2, \ldots, X_n$. In this study, one observes not $\{X_i\}$ but a doubly censored sample:

$$W_i = \begin{cases} X_i & \text{if } Z_i \le X_i \le Y_i, \quad \delta_i = 1 \\ Y_i & \text{if } X_i > Y_i, \quad \delta_i = 2 \\ Z_i & \text{if } X_i < Z_i, \quad \delta_i = 3 \end{cases} \tag{2.3}$$

where (X_i, Y_i, Z_i), $i = 1, 2, \ldots, n$, are independent observations on (X, Y, Z) for nonnegative variables X, Y and Z, and the r.v.'s Y_i and Z_i are called *right* and *left censoring variables*, respectively. This means that X_i is observable whenever X_i lies in the interval $[Z_i, Y_i]$, and we know whether $X_i < Z_i$ or $X_i > Y_i$ and observe the value of Z_i or Y_i accordingly. The usual assumptions on this model are that X_i and (Y_i, Z_i) are independent and $P\{Y > Z\} = 1$ (see Chang and Yang (1987), Chang (1990), Gu and Zhang (1993)). Right censored data, that is, $Z \equiv -\infty$ in (2.3), is a special case of the doubly censored data (2.3). The problem considered here is to test the goodness of fit of F based on (W_i, δ_i).

Let (W_i, δ_i) be distributed as (W, δ), and denote

$$Q_j(t) = P\{W \le t, \delta = j\}, \quad j = 1, 2, 3,$$

$$S_X(t) = P\{X > t\}, \quad S_Y(t) = P\{Y > t\}, \quad S_Z(t) = P\{Z > t\}.$$

By Turnbull (1974), Tsai and Crowley (1985), the *Nonparametric Maximum Likelihood Estimator* (NPMLE) $\hat{S}_n$ of S_X is one of the solutions of the following equation:

$$\hat{S}_n(t) = 1 - Q^{(n)}(t) + \int_{u \le t} \frac{\hat{S}_n(t)}{\hat{S}_n(u)} dQ_2^{(n)}(u) - \int_{u > t} \frac{1 - \hat{S}_n(t)}{1 - \hat{S}_n(u)} dQ_3^{(n)}(u), \tag{2.4}$$

where

$$Q_j^{(n)}(t) = \frac{1}{n}\sum_{i=1}^{n} I\{W_i \le t, \delta_i = j\}, \qquad j = 1, 2, 3,$$

$$Q^{(n)}(t) = \sum_{j=1}^{3} Q_j^{(n)}(t),$$

with $\int_{u\le t} = 0$ if $S_X^{(n)}(t) = 0$ and $\int_{u>t} = 0$ if $S_X^{(n)}(t) = 1$ (The solutions of (2.4) are not unique, see Gu and Zhang (1993) for examples, but all solutions are asymptotically equivalent). This equation (2.4) is obtained by generalizing the idea of the self-consistent estimator for right censored data originally given by Efron (1967, (7.4) on page 840). One may see Efron (1967) for an intuitive explanation for such an equation. The NPMLE $\hat{S}_n$ may be found numerically by the method suggested by Mykland and Ren (1994), which is based on the EM algorithm.

Let U denote the uniform distribution function on $[0,1]$, and let

$$\hat{F}_n = 1 - \hat{S}_n \quad \text{and} \quad \hat{U}_n = \hat{F}_n \circ F_0^{-1}. \tag{2.5}$$

$\hat{F}_n$ is the NPMLE of F. For the test (2.1), one would naturally want to replace F_n in (2.2) by $\hat{F}_n$; that is, to use

$$T_n = n\int_0^{\infty} [\hat{F}_n(x) - F_0(x)]^2 dF_0(x) = n\int_0^1 [\hat{U}_n - U]^2 dU \tag{2.6}$$

as the test statistics. For the rest of the paper, we will always refer T_n to (2.6). If $U_0 = F \circ F_0^{-1}$, under some regularity conditions Gu and Zhang (1993) showed that

$$\sqrt{n}\,[\hat{U}_n - U_0] \text{ weakly converges to } G_F, \qquad \text{as } n \to \infty \tag{2.7}$$

where G_F is a Gaussian process with mean 0 and a covariance function γ. The covariance function γ is determined by S_X, S_Y and S_Z and its specific formulation is given by Ren (1993). Hence, it is easy to see that the null distribution of T_n is given by

$$T_n \xrightarrow{\mathcal{D}} \int_0^1 G_{F_0}^2 dU = \sum_{j=1}^{\infty} \lambda_j Z_j^2, \qquad n \to \infty \tag{2.8}$$

where Z_j are independent normal random variables with mean 0 and variance 1, and λ_j are the eigenvalues for the following eigenvalue problem:

$$\int_0^1 \gamma(s,t)\phi(t)dt = \lambda\phi(s), \qquad \phi \in L^2[0,\beta]. \tag{2.9}$$

One should note that even under H_0, the limiting covariance function γ of the NPMLE $\hat{F}_n$ depends on unknown survival functions S_Y and S_Z. To set the critical value of the test, one needs to estimate the unknown null distribution given by (2.8). As expected, the usual nonparametric n out of n bootstrap fails in this case.

To see the reason for this failure and justify our other arguments, we need to extend the central limit theorem for the bootstrapped empirical process of Giné and Zinn (1990) to the doubly censored case considered in this study. We state the results as below with a proof sketched in Section 5 under the assumption (A1)–(A6) outlined in Chang (1990). Under these conditions (A1)–(A6), Chang (1990) established (2.7) on $[0,T]$ for any $T>0$.

Let m be the bootstrap sample size satisfying $m \to \infty$ as $n \to \infty$, and let $\hat{F}_m^*$ be the NPMLE based on the bootstrap sample. Then

Proposition 2.1 *Suppose that* $\sqrt{n}\,[\hat{F}_n - F\,]$ *weakly converges to a Gaussian process* Z *as* $n \to \infty$. *Then, with probability* 1, $\sqrt{m}\,[\hat{F}_m^* - \hat{F}_n]$ *weakly converges to* Z *as* $n \to \infty$.

Immediately, we have that with probability 1,

$$\sqrt{m}\,[\hat{U}_m^* - \hat{U}_n] \text{ weakly converges to } G_F, \tag{2.10}$$

where $\hat{U}_m^* = \hat{F}_m^* \circ F_0^{-1}$. Let T_m^* be given by (2.6) based on the bootstrap sample. Hence, for $m=n$, i.e., when the n out of n bootstrap is used, we have

$$\sqrt{n}\,[\hat{U}_n^* - U\,] = \sqrt{n}\,[\hat{U}_n^* - \hat{U}_n] + \sqrt{n}\,[\hat{U}_n - U\,] = G_n^* + G_n, \tag{2.11}$$

and thus

$$T_n^* = \int_0^1 [G_n^*]^2 dU + 2\int_0^1 G_n^* G_n dU + \int_0^1 [G_n]^2 dU. \tag{2.12}$$

Let $\boldsymbol{O}_n = \{(W_i, \delta_i);\ 1 \le i \le n\}$, then from (2.7) and (2.10), we know that under H_0,

$$P\{T_n^* \le t|\,\boldsymbol{O}_n\} \longrightarrow P\Big\{\int_0^1 [G_{F_0}^* + G_{F_0}]^2 dU \le t|G_{F_0}\Big\}, \quad \text{as } n \to \infty,$$

where $G_{F_0}^*$ and G_{F_0} are two independent centered Gaussian processes with the same covariance function. As stated in the introduction, we need to "bootstrap" taking the restrictions of H_0 into account. For the simpler right censored case, one may draw a sample $X_1^*, \ldots, X_n^*$ from F_0 and a sample $Y_1^*, \ldots, Y_n^*$ from the estimator of S_Y (see Chang and Yang (1987) for the formulation of such an estimator) to obtain the bootstrap sample. It is well known that the n out of n bootstrap method estimates the null distribution consistently. However, this method does not easily generalize to the doubly censored case. The reason is that in the doubly censored case, one would need to obtain a sample of $(Y_1^*, Z_1^*), \ldots, (Y_n^*, Z_n^*)$ from an estimator which estimates the joint distribution of (Y,Z), because Y and Z are usually not independent random variables.

From (2.11) and (2.12), we see that in a general way, the problem can be solved by resampling fewer observations, i.e., using $m = o(n)$ with $m \to \infty$ as the bootstrap sample size. More specifically, with $m = \mathrm{o}(n)$, by (2.7), we have that (2.11) becomes

$$\begin{aligned} \sqrt{m}\,|\hat{U}^*_m - U\,] &= \sqrt{m}\,[\hat{U}^*_m - \hat{U}_n] + \sqrt{m}\,[\hat{U}_n - U\,] \\ &= \sqrt{m}\,[\hat{U}^*_m - \hat{U}_n] + \mathrm{o}_p(1) = G^*_m + \mathrm{o}_p(1). \end{aligned} \tag{2.13}$$

Hence, from (2.10), we have that (2.12) becomes

$$T^*_m = \int_0^1 [G^*_m]^2 dU + \mathrm{o}_p(1), \tag{2.14}$$

which along with (2.10) implies that under H_0,

$$P\{T^*_m \leq t|\,\boldsymbol{O}_n\} \quad \longrightarrow \quad P\Big\{\int_0^1 [G_{F_0}]^2 dU \leq t\Big\}, \qquad \text{as } n \to \infty.$$

Therefore, the m out of n bootstrap is consistent under H_0. Based on this, we propose the m out of n bootstrap tests for the test (2.1) in the next section.

3 The m out of n Bootstrap

For the test (2.1) with a doubly censored sample (2.3), we propose that one uses $m = \mathrm{o}(n)$ as the bootstrap sample size, and use C^*_α as the critical value of the test, where for $0 < \alpha < 1$ and $\boldsymbol{O}_n = \{(W_i, \delta_i)\,;\ 1 \leq i \leq n\}$, C^*_α is given by

$$P\{T^*_m \geq C^*_\alpha|\,\boldsymbol{O}_n\} = \alpha. \tag{3.1}$$

This is called the m *out of* n *bootstrap* method. In the following, we investigate the asymptotic properties of this test procedure

Let the limiting null distribution of T_n be distributed as

$$T = \int_0^1 G^2_{F_0} dU \tag{3.2}$$

and let C^0_α be the true critical value given by

$$P\{T \geq C^0_\alpha\} = \alpha. \tag{3.3}$$

Proposition 3.1 *Let $\|\cdot\|$ denote the uniform norm and let $m = \mathrm{o}(n)$ with $m \to \infty$ as $n \to \infty$. Then, when Proposition 2.1 holds,*

(i) if $\|F - F_0\| = \mathrm{O}(1/\sqrt{n})$, we have

$$C^*_\alpha \stackrel{P}{\longrightarrow} C^0_\alpha, \qquad \textit{as } n \to \infty; \tag{3.4}$$

(ii) if $F = F_1 \neq F_0$ *, we have*

$$C_\alpha^* \xrightarrow{P} \infty, \qquad \text{as } n \to \infty; \tag{3.5}$$

with

$$C_\alpha^*/T_n \xrightarrow{P} 0, \qquad \text{as } n \to \infty. \tag{3.6}$$

The proof of Proposition 3.1 is given in Section 5.

From (3.4), (2.7) and (3.3), we have

$$P\{T_n \geq C_\alpha^* | H_0\} \longrightarrow \alpha, \quad \text{as } n \to \infty. \tag{3.7}$$

Consider the contiguous alternatives

$$H_n : F = F_n \neq F_0, \tag{3.8}$$

where for a bounded function Δ, $\sqrt{n}[F_n - F_0] = \Delta_n \to \Delta$, as $n \to \infty$. Denoting $U_n = F_n \circ F_0^{-1}$, we have

$$\sqrt{n}[\hat{U}_n - U] = \sqrt{n}[\hat{U}_n - U_n] + \Delta_n \circ F_0^{-1}.$$

Note that under the condition of Corollary 1 of Gu and Zhang (1993) and under H_0, $\sqrt{n}[\hat{U}_n - U]$ is, in probability, equivalent to a continuous operator of the empirical processes $\sqrt{n}[Q_j^{(n)} - Q_j]$, $j = 1, 2, 3$ (see the proof of Theorem 2 by Gu and Zhang, 1993). Hence, by contiguity, we have that for $\Delta_0 = \Delta \circ F_0^{-1}$,

$$T_n \xrightarrow{\mathcal{D}} T_{\Delta_0} = \int_0^1 [G_{F_0} + \Delta_0]^2 dU = \sum_{j=1}^{\infty} \lambda_j Z_{\Delta j}^2, \qquad \text{as } n \to \infty \tag{3.9}$$

where λ_j's are given by (2.8) and $Z_{\Delta j}$ are independent normal r.v.'s with non-zero mean and variance 1. Hence, from (3.4) we have

$$P\{T_n \geq C_\alpha^* | H_n\} \longrightarrow P\{T_{\Delta_0} \geq C_\alpha^0\}, \qquad \text{as } n \to \infty \tag{3.10}$$

where $P\{T_{\Delta_0} \geq C_\alpha^0\}$ is the true asymptotic power of the test (2.1) under the contiguous alternatives (3.8). Moreover, under the fixed alternative

$$H_1 : F = F_1 \ (\neq F_0) \tag{3.11}$$

we know that (3.6) implies

$$P\{T_n \geq C_\alpha^* | H_1\} \longrightarrow 1, \qquad \text{as } n \to \infty. \tag{3.12}$$

To summarize, (3.7), (3.10) and (3.12) give the following theorem on the asymptotic properties of the proposed m out of n bootstrap test procedure.

Theorem 3.2 *Let* $m = o(n)$ *with* $m \to \infty$ *as* $n \to \infty$. *In the test (2.1) with data (2.3), if* T_n *given by (2.6) is used as the test statistic and if* C_α^* *given by (3.1) is used as the critical value of the test, then, when Proposition 2.1 holds,*

(i) *for* $0 < \alpha < 1$, $\lim_{n\to\infty} P\{T_n \geq C^*_\alpha | H_0\} = \alpha$;

(ii) *for contiguous alternatives* H_n *in (3.8)*, $\lim_{n\to\infty} P\{T_n \geq C^*_\alpha | H_n\} = P\{T_{\Delta_0} \geq C^0_\alpha\}$;

(iii) *for the fixed alternative* H_1 *in (3.11)*, $\lim_{n\to\infty} P\{T_n \geq C^*_\alpha | H_1\} = 1$.

4 A Small Simulation Study

In this section, we consider the goodness of fit test (2.1) with $F_0 = \text{Exp}(1)$ against the exponential distributions $\text{Exp}(\mu)$, where $\text{Exp}(\mu)$ denotes the exponential distribution with mean μ. We denote the power functions of the test with the true critical value C^0_α given by (3.3) and the m out of n bootstrap critical value C^*_α given by (3.1) as

$$P_0(\mu) = P\{T_n \geq C^0_\alpha | \text{Exp}(\mu)\} \tag{4.1}$$

and

$$P_b(\mu) = P\{T_n \geq C^*_\alpha | \text{Exp}(\mu)\}, \tag{4.2}$$

respectively, where T_n is given by (2.6). For the Fredholm Integral Equation (FIE) method proposed by Ren (1993), we denote the power function as

$$P_f(\mu) = P\{T_n \geq C^f_\alpha | \text{Exp}(\mu)\}, \tag{4.3}$$

where C^f_α is the critical value estimated by FIE such that $P\{T \geq C^f_\alpha\} = \alpha$. The details on the implementation of the FIE method may be found in Ren and Ledder (1995).

In Figure 1, we compare the curves of P_0, P_b and P_f, which are obtained by the simulation with sample size $n = 200$, $m = \sqrt{n}$ and $\alpha = 0.05$. Because the use of the EM algorithm in (2.4) is very time consuming to conduct the simulation study for doubly censored data, in Figure 1 we consider the right censored case with the right censoring variable Y from $\text{Exp}(3)$. One may note that for the right censored case, we have $S_Z \equiv 0$ and the solution $\hat{S}_n$ of (2.4) is the Kaplan–Meier estimator (Chang and Yang (1987)). In our study, the true critical value C^0_α is obtained from the Monte Carlo method. All simulation results are based on 300 runs, and for each run, the percentiles of P_0, P_b, P_f, C^0_α, C^*_α and C^f_α are obtained from 400 replications of the procedures.

From Figure 1, we easily see that the simulation results are consistent with our asymptotic results in Section 3. As μ gets further away from 1, the power of the test using the m out of n bootstrap critical value C^*_α, although smaller than the true one by C^0_α, goes to 1. As μ gets closer to 1, the power of the test using C^*_α goes to α. We also note that the curve P_f by FIE is closer to P_0 than P_b by bootstrap tests. However, the FIE method method is constructed through solving the Fredholm integral equation of the second kind, and the programming of FIE is no trivial matter. Also, FIE does not easily extend to other hypothesis testing problems.

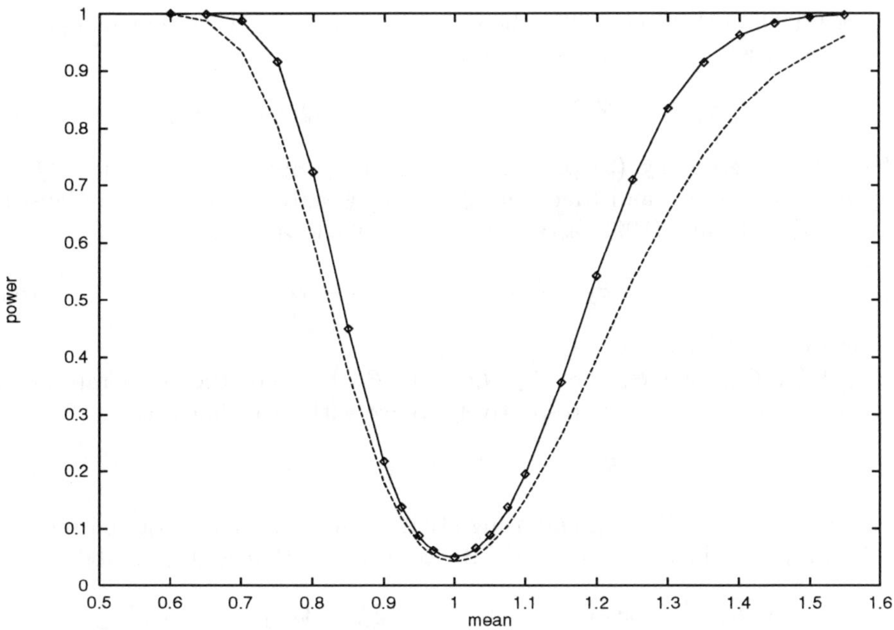

Figure 1: P_0 : — ; P_b : - - - ; P_f : ◇◇◇ ; $X \sim \mathrm{Exp}(\mu)$, $Y \sim \mathrm{Exp}(3)$

A better choice of m and extrapolation as discussed in Bickel, Götze and van Zwet (1994) should also improve the power behavior of the m out of n bootstrap.

5 Proofs

PROOF OF PROPOSITION 2.1 The proposition may be shown under weaker conditions. For simplicity, in this paper we will sketch the proof under the assumptions A1 through A6 outlined in Chang (1990).

Under these assumptions, Chang (1990) showed that for any $T > 0$,

$$\sqrt{n}\,[\boldsymbol{S}_n - \boldsymbol{S}] \text{ weakly converges to } \boldsymbol{Z}, \quad \text{as } n \to \infty \tag{5.1}$$

where $\boldsymbol{Z}$ is a Gaussian process on $[0, T]$,

$$\boldsymbol{S}_n = (S_X^{(n)}, S_Y^{(n)}, S_Z^{(n)}), \quad \boldsymbol{S} = (S_X, S_Y, S_Z), \quad S_X^{(n)} = \hat{S}_n, \tag{5.2}$$

and $S_Y^{(n)}$, $S_Z^{(n)}$ are the estimators of S_Y, S_Z, respectively, given by Chang and Yang (1987). Let

$$\boldsymbol{Q}_n = (Q_1^{(n)}, Q_2^{(n)}, Q_3^{(n)}) \quad \text{and} \quad \boldsymbol{Q} = (Q_1, Q_2, Q_3). \tag{5.3}$$

Chang (1990, (14) of page 395) showed that there exist a continuous operator L_1 and a linear operator L_2 such that

$$\sqrt{n}\,[\boldsymbol{S}_n - \boldsymbol{S}] = L_1 \circ L_2(\sqrt{n}\,[\boldsymbol{Q}_n - \boldsymbol{Q}]) + L_1(\boldsymbol{\theta}_n), \tag{5.4}$$

where $\boldsymbol{\theta}_n$ is given by (9) of Chang (1990, page 394). Noting that $\boldsymbol{Q}_n$ is the empirical process and that $\sqrt{n}\,[\boldsymbol{Q}_n - \boldsymbol{Q}]$ weakly converges to a Gaussian process $\boldsymbol{Z}_Q$, Chang (1990) established (5.1) by showing that

$$\boldsymbol{\theta}_n \overset{\text{a.s.}}{\longrightarrow} 0, \qquad \text{as } n \to \infty \tag{5.5}$$

uniformly on $[0,T]$.

Let $\boldsymbol{S}^*_m$, $\boldsymbol{Q}^*_m$ and $\boldsymbol{\theta}^*_m$ be $\boldsymbol{S}_n$, $\boldsymbol{Q}_n$ and $\boldsymbol{\theta}_n$ based on the bootstrap sample from (W_i, δ_i), $i = 1, \ldots, n$, respectively. Since with probability 1,

$$\boldsymbol{S}_n \longrightarrow \boldsymbol{S}, \qquad \text{as } n \to \infty, \tag{5.6}$$

uniformly on $[0,T]$ (Chang and Yang (1987)), the conditions required for (5.4) hold with probability 1, as $n \to \infty$. Hence, we have that with probability 1,

$$\sqrt{m}\,[\boldsymbol{S}^*_m - \boldsymbol{S}_n] = L_1 \circ L_2(\sqrt{m}\,[\boldsymbol{Q}^*_m - \boldsymbol{Q}_n]) + L_1(\boldsymbol{\theta}^*_m). \tag{5.7}$$

From Giné and Zinn (1990), we know that with probability 1,

$$\sqrt{m}\,[\boldsymbol{Q}^*_m - \boldsymbol{Q}_n] \text{ weakly converges to } \boldsymbol{Z}_Q, \qquad \text{as } n \to \infty. \tag{5.8}$$

Since L_1 is a continuous operator and L_2 is a linear operator, by Theorem 5.1 of Billingsley (1968), (5.8) implies that, as $n \to \infty$, with probability 1, $L_1 \circ L_2(\sqrt{m}\,[\boldsymbol{Q}^*_m - \boldsymbol{Q}_n])$ weakly converges to the Gaussian process $\boldsymbol{Z} = L_1 \circ L_2(\boldsymbol{Z}_Q)$. Therefore, by (5.7), it suffices to show that,

$$\boldsymbol{\theta}^*_m \overset{\text{a.s.}}{\longrightarrow} 0, \qquad \text{as } n \to \infty \tag{5.9}$$

uniformly on $[0,T]$.

The key to (5.9) is that one needs to show that with probability 1,

$$\|\boldsymbol{S}^*_m - \boldsymbol{S}_n\| \overset{\text{a.s.}}{\longrightarrow} 0, \qquad \text{as } n \to \infty. \tag{5.10}$$

The rest of the proof just follows line by line of Lemma 3.1 through Lemma 3.3 in Chang (1990).

Denoting the equation (2.7) of Chang and Yang (1987) as $\boldsymbol{E}(\boldsymbol{Q}_n, \boldsymbol{S}_n) = \boldsymbol{0}$, $\boldsymbol{S}^*_m$ is the solution of $\boldsymbol{E}(\boldsymbol{Q}^*_m, \boldsymbol{S}^*_m) = \boldsymbol{0}$. Since $\boldsymbol{Q}_n$ is the empirical process, from (5.6) we know that with probability 1, $\boldsymbol{E}(\boldsymbol{Q}_n, \boldsymbol{S}_n)$ has limit $\boldsymbol{E}(\boldsymbol{Q}, \boldsymbol{S})$ as $n \to \infty$. Since each component of $\boldsymbol{S}^*_m$ is uniformly bounded and nonincreasing, by Helly's Theorem, for each fixed t, $\boldsymbol{S}^*_m$ has a convergent subsequence, say with limit $\boldsymbol{S}^*_0$. This limit $\boldsymbol{S}^*_0$ should satisfy the equation $\boldsymbol{E}(\boldsymbol{Q}, \boldsymbol{S}) = \boldsymbol{0}$. From the identifiability for the solution of $\boldsymbol{E}(\boldsymbol{Q}, \boldsymbol{S}) = \boldsymbol{0}$ studied by Chang and Yang (1987), we know that $\boldsymbol{S}^*_0$ must be the solution of $\boldsymbol{E}(\boldsymbol{Q}, \boldsymbol{S}) = \boldsymbol{0}$, i.e., $\boldsymbol{S}_0^* = \boldsymbol{S}$. Since each component of $\boldsymbol{S}_m^*$ is uniformly bounded and non-increasing and since each component of $\boldsymbol{S}$ is continuous, we have that with probability 1,

$$\| \boldsymbol{S}_m^* - \boldsymbol{S} \| \stackrel{\text{a.s.}}{\longrightarrow} 0, \qquad \text{as } n \to \infty. \tag{5.11}$$

Hence, (5.10) follows from (5.6) and (5.11). ////

Proof of Proposition 3.1 (i) Note that

$$\sqrt{m}\,[\hat{U}_m^* - U\,] = \sqrt{m}\,[\hat{U}_m^* - \hat{U}_n] + \sqrt{m}\,[\hat{U}_n - U_0] + \sqrt{m}\,[U_0 - U\,]. \tag{5.12}$$

Hence, by (2.7) and the assumptions: $\|F - F_0\| = \mathrm{O}(1/\sqrt{n}\,)$ and $m = \mathrm{o}(n)$ with $m \to \infty$ as $n \to \infty$, we have

$$\sqrt{m}\,[\hat{U}_m^* - U\,] = \sqrt{m}\,[\hat{U}_m^* - \hat{U}_n] + \mathrm{o}_p(1), \quad \text{as } n \to \infty \tag{5.13}$$

where $\mathrm{o}_p(1)$ uniformly converges to 0 in probability as $n \to \infty$. From (2.10) and (3.2), we know that with probability 1,

$$\sup_{-\infty<x<\infty} |P\{\hat{T}_m^* \le x |\, \boldsymbol{O}_n\} - P\{T \le x\}| \longrightarrow 0, \qquad \text{as } n \to \infty \tag{5.14}$$

where

$$\hat{T}_m^* = m \int_0^1 [\hat{U}_m^* - \hat{U}_n]^2 dU. \tag{5.15}$$

Hence, from (6.13) and (6.14), we have that

$$\sup_{-\infty<x<\infty} |P\{T_m^* \le x |\, \boldsymbol{O}_n\} - P\{T \le x\}| \stackrel{P}{\longrightarrow} 0, \qquad \text{as } n \to \infty. \tag{5.16}$$

For C_α^* given by (3.1) and C_α^0 by (3.3), we know that (5.16) implies

$$|P\{T_m^* \ge C_\alpha^* |\, \boldsymbol{O}_n\} - P\{T \ge C_\alpha^*\}| = |\alpha - P\{T \ge C_\alpha^*\}|$$

$$= |P\{T \ge C_\alpha^0\} - P\{T \ge C_\alpha^*\}| \stackrel{P}{\longrightarrow} 0, \quad \text{as } n \to \infty.$$

Therefore, (3.4) follows from the continuity of the d.f. of T.
(ii) When $F = F_1 \ne F_0$, from (5.12), (2.7), (2.10) and the assumption that $m = \mathrm{o}(n)$, we have

$$\begin{aligned} m[\hat{U}_m^* - U\,]^2 &= m[\hat{U}_m^* - \hat{U}_n]^2 + m[U_0 - U\,]^2 + \\ &\quad + 2m[\hat{U}_m^* - \hat{U}_n][\,U_0 - U\,] + \mathrm{o}_p(1) \end{aligned} \tag{5.17}$$

and thus by (5.15) and (2.10),

$$T_m^* = \hat{T}_m^* + m \int_0^1 [\,U_0 - U\,]^2 dU + 2m \int_0^1 [\hat{U}_m^* - \hat{U}_n][\,U_0 - U\,]dU + \mathrm{o}_p(1) \tag{5.18}$$

$$= O_p(1) + m \int_0^1 [U_0 - U]^2 dU + O_p(\sqrt{m})$$

Noting that (3.1) implies

$$P\{T_m^* < C_\alpha^* | \boldsymbol{O}_n\} = 1 - \alpha, \tag{5.19}$$

hence, (3.5) follows from (5.18) and

$$m \int_0^1 [U_0 - U]^2 dU = m \int_0^\infty [F_1 - F_0]^2 dF_0 \longrightarrow \infty, \qquad \text{as } n \to \infty. \tag{5.20}$$

To show (3.6), one just needs to notice that by (2.7),

$$T_n = O_p(1) + O_p(\sqrt{n}) + n \int_0^1 [U_0 - U]^2 dU,$$

where $O_p(1)$ is bounded in probability, and that by (3.1),

$$\alpha = P\{T_m^* \geq C_\alpha^* | \boldsymbol{O}_n\} = P\{T_m^*/T_n \geq C_\alpha^*/T_n | \boldsymbol{O}_n\}.$$

Therefore, (3.6) follows from (5.18), (5.20) and $T_m^*/T_n \to 0$ as $n \to \infty$. ////

References

[1] Bickel, P.J., Götze, F. and Van Zwet, W.R. (1994): *Resampling Fewer than n Observations: Gains, Losses and Remedies for Losses*. Preprint.

[2] Billingsley, P. (1968): *Convergence of Probability Measures*. John Wiley and Sons, New York.

[3] Chang, M.N. and Yang, G.L. (1987): Strong consistency of a nonparametric estimator of the survival function with doubly censored data. *Ann. Statist.* **15** 1536–1547.

[4] Chang, M.N. (1990): Weak convergence of a self-consistent estimator of the survival function with doubly censored data. *Ann. Statist.* **18** 391–404.

[5] Efron, B. (1967): The two sample problem with censored data. *Proc. Fifth Berkeley Symp. Math. Statist. Prob.* **4** 831–853.

[6] Freedman, D.F. (1981): Bootstrapping regression models. *Ann. Statist.* **12** 1218–1228.

[7] Gehan, E.A. (1965): A generalized two-sample Wilcoxon test for doubly censored data. *Biometrika* **52** 650–653.

[8] Giné, E. and Zinn, J. (1990): Bootstrapping general empirical measures. *Ann. Prob.* **18** 852–869.

[9] Gu, M.G. and Zhang, C.H. (1993): Asymptotic properties of self-consistent estimators based on doubly censored data. *Ann. Statist.* **21**, 611–624.

[10] Härdle, W. and Mammen, E. (1993): Comparing nonparametric versus parametric regression fits. *Ann. Statist.* **21** 1926–1947.

[11] Mammen, E. (1992): *When Does Bootstrap Work?* Springer Verlag, New York.

[12] Mykland, P.A. and Ren, J. (1994): Self-Consistent and Maximum Likelihood Estimation for Doubly Censored Data. Submitted to *Ann. Statist.* (under revision).

[13] Peer, P.G., Van Dijck, J.A., Hendriks, J.H., Holland, R. and Verbeek, A.L. (1993): Age-dependent growth rate of primary breast cancer. *Cancer* **71**(11), 3547–3551.

[14] Peto, R. (1973): Experimental survival curves for interval-censored data. *Appl. Statist.* **22** 86–91.

[15] Ren, J. (1993): Generalized Cramér-von Mises tests of goodness of fit with censored data. *Ann. Inst. Statist. Math.*, to appear.

[16] Ren, J. and Ledder, G. (1995): On Cramér-von Mises Tests of Goodness of Fit with Censored Data. *Communications in Statistics*, to appear.

[17] Tsai, W.Y. and Crowley, J. (1985): A large sample study of generalized maximum likelihood estimators from incomplete data via self-consistency. *Ann. Statist.* **13** 1317–1334.

[18] Turnbull, B.W. (1974): Nonparametric estimation of a survivorship function with doubly censored data. *J. Amer. Statist. Assoc.* **69** 169–173.

[19] Wei, L.J. (1984): Testing goodness of fit for the proportional hazards model with censored observations. *J. Amer. Statist. Assoc.* **79**, no. 387, 649–652.

What Criterion for a Power Algorithm?

Andreas Buja

AT&T Bell Laboratories, New Jersey

Abstract

If Breiman and Friedman's ACE algorithm is used with smoothing splines as building blocks, the question arises what criterion is being optimized. The obvious guess turns out to be wrong. The problem is that the building blocks are no longer orthogonal projections in the usual inner product. We describe the non-standard inner product that does yield orthogonal projections, and we indicate what the resulting "penalized R^2"-criterion is.

Key words and phrases: ACE, power algorithms, curve fitting, smoothing splines, penalty-based regularization, response transformation.
AMS 1991 subject classifications: Primary 62G07; secondary 62H25.

1 The ACE Criterion

We start with a description of the ACE method. Later, we modify it by introducing penalized least squares smoothers as building blocks.

ACE is defined as a method for estimating nonlinear transformations of response and predictors in regression problems. ACE minimizes the fraction of unexplained variance after regressing a transformed response on transformed predictors. Denoting the transformation of the response Y by $\theta = \theta(Y)$ and the transformations of the predictors X_j $(j = 1, \ldots, p)$ by $\phi_j = \phi_j(X_j)$, the ACE problem is

$$\frac{\mathrm{E}[(\theta - \sum \phi_j)^2]}{\mathrm{E}[\theta^2]} = \min_{\phi_1, \ldots, \phi_p, \theta},$$

where minimization is carried out over all centered and square integrable transformations θ and ϕ_j:

$$\mathrm{E}[\theta] = 0, \quad \mathrm{E}[\theta^2] < \infty, \qquad \mathrm{E}[\phi_j] = 0, \quad \mathrm{E}[\phi_j^2] < \infty .$$

The ACE problem has essentially the structure of a canonical correlation problem. This is easily seen by writing canonical correlation analysis of variables U_j and V_k in an asymmetric way:

$$\frac{\mathrm{E}[(\sum_k \beta_k V_k - \sum_j \alpha_j U_j)^2]}{\mathrm{E}[(\sum_k \beta_k V_k)^2]} = \min_{\alpha_1, \alpha_2, \ldots, \beta_1, \beta_2, \ldots}$$

where minimization is carried out over all coefficients α_j and β_k. Thus, the two problems differ only in the linear spaces employed: $\{\theta| \ldots\}$ and $\{\sum \phi_j | \ldots\}$ for ACE, and $\{\sum \beta_k V_k \mid \beta_k \in \mathbb{R}\}$ and $\{\sum \alpha_j U_j \mid \alpha_j \in \mathbb{R}\}$ for canonical correlation analysis.

2 The ACE Algorithm

To solve the ACE problem, the following is an obvious alternating algorithm: Minimize the criterion first w.r.t. θ, then w.r.t. $(\phi_1, \ldots, \phi_p)$, then loop back and iterate. Both steps are doable: For given $(\phi_1, \ldots, \phi_p)$, the minimizing θ is found as follows:

$$\theta \leftarrow \mathrm{E}[\sum \phi_j | Y] \; ; \quad \theta \leftarrow \theta / \, \mathrm{E}[\theta^2]^{1/2} \; .$$

Minimizing with regard to $(\phi_1, \ldots, \phi_p)$ for given θ is a little trickier since it amounts to fitting an additive regression model. One often uses the so-called "backfitting" algorithm, essentially a version of Gauss–Seidel: Minimize with regard to one ϕ_j at a time, and loop over $j = 1, 2, \ldots, p$ repeatedly till convergence (see Buja, Hastie and Tibshirani (1989) for an analysis of the algorithm). In practice, a single loop is preferred over looping till convergence before one returns to updating θ.

For the present purposes, we will assume that for given θ a minimizing additive fit $\sum \phi_j$ can be obtained. This is equivalent to assuming that projections P_X and P_Y onto the spaces of functions θ and $\sum \phi_j$, respectively, exist and can be realized at least conceptually if not computationally:

$$\begin{aligned} P_Y(Z) &= \mathrm{argmin}_\theta \; \mathrm{E}[(Z - \theta)^2] \, , \\ P_X(Z) &= \mathrm{argmin}_{\sum \phi_j} \; \mathrm{E}[(Z - \sum \phi_j)^2] \, . \end{aligned}$$

Existence of these projections can be guaranteed under sufficient conditions laid out in Breiman and Friedman (1985, Appendix). The projection P_Y is essentially the conditional expectation operator, while P_X is a general projection for whose existence certain technical conditions have to be met.

With these assumptions and idealizations, the ACE algorithm is just the power algorithm applied to the iterated projection $P_X P_Y$:

$$\theta \leftarrow P_Y P_X \theta \; , \quad \theta \leftarrow \theta / \, \mathrm{E}[\theta^2]^{1/2} \; ,$$

where the second part is just a standardization to prevent convergence to zero.

3 Estimation with Penalized Least Squares

In practice, the transformations θ and ϕ_j are estimated from data with the help of smoothers, i.e., nonparametric function estimators. Among smoothers, two types are particularly attractive on mathematical grounds: Regression splines and smoothing splines. They are defined as minimizers of an unpenalized and a penalized least squares criterion, respectively. This is in contrast to most other types of smoothers that do not permit a definition in terms of an optimization problem.

Regression splines, being just OLS regressions, yield a particularly simple version of ACE estimation. The problem reduces to a canonical correlation analysis of two sets of variables, the first consisting of a spline basis of θ-transformations, and the second of the union of spline bases of ϕ_j-transformations. Hence, if plain least squares smoothers are used as building blocks, the ACE method is well understood, and the ACE algorithm is unnecessary since conventional numerical linear algebra can be used for finding solutions.

This is not so when the building blocks are penalized least squares smoothers such as smoothing splines. The projections P_X and P_Y are then replaced with mappings that are no longer projections.

We introduce some notation for penalized least squares: Generic quadratic penalties will be denoted by $J(\theta)$ and $J_j(\phi_j)$. The most common examples are the penalties that define cubic smoothing splines: $J(\theta) = \lambda \int (\theta''(y))^2 dy$ and $J_j(\phi_j) = \lambda_j \int (\phi_j''(x_j))^2 dx_j$.

In random variable language, penalized regression of a variable Z on Y is hence defined generically by

$$S_Y(Z) = \operatorname{argmin}_\theta \ \mathrm{E}[(Z-\theta)^2] + J(\theta) .$$

Similarly, the additive penalized regression of a variable Z on the variables $X_1, \ldots, X_p$ is defined by

$$S_X(Z) = \operatorname{argmin}_{\sum \phi_j} \ \mathrm{E}[(Z - \sum \phi_j)^2] + \sum J_j(\phi_j) .$$

Actual estimators can be obtained by taking expectations with regard to the empirical distribution of a sample of data.

In order to assure that the penalized LS estimators S_Y and S_X exist, one has to make some technical assumptions which we generously ignore. We have used random variable notation in order to save a cumbersome index for cases (subjects, units). This does not imply that penalized regression is sensible for populations, except in the context of bias calculations. As the sample size increases, the amounts (λ and λ_j) of penalization should go to zero at a proper rate in order to assure consistency of the curve estimates for a convenient class of smooth functions. Since consistency is none of our concerns, we take the freedom to simplify notation in whichever way. In particular, since choice of the amount of smoothing isn't any of our concerns either, we will think of the penalization amounts as absorbed in the penalties.

4 The ACE Algorithm with Penalized LS

The ACE algorithm with regard to S_X and S_Y is

$$\theta \leftarrow S_Y S_X \theta ,$$

interleaved with standardizations to prevent the iterations from converging to zero. Although it is intuitive that this is a proper extension of ACE, it is less intuitive to write down a criterion that is minimized by this algorithm.

Hastie and Tibshirani (1990, p 181) have some partial answers to this question. They write down a criterion which is equivalent to the following penalized fraction of unexplained variance:

$$\frac{\mathrm{E}[(\theta - \sum \phi_j)^2] \ + \ J(\theta) \ + \ \sum J_j(\phi_j)}{\mathrm{E}[\theta^2]} \ .$$

A curious thing happens: Hastie and Tibshirani perform calculations which show that there exists only an indirect connection between the operator $S_Y S_X$ and this criterion. Essentially, they show that there is a mismatch between S_Y and the penalty $J(\theta)$ in the criterion. In order to match the result of the power algorithm for $S_Y S_X$ and the minimizing solution of the above criterion, it is necessary to multiply $J(\theta)$ with a factor that depends on the largest eigenvalue of $S_Y S_X$ (l.c., p 181). The implication is that the above is not the criterion for the penalized ACE algorithm.

There is a simple fix to the problem. The correct criterion is the following:

$$\frac{\mathrm{E}[(\theta - \sum \phi_j)^2] \ + \ J(\theta) \ + \ \sum J_j(\phi_j)}{\mathrm{E}[\theta^2] \ + \ J(\theta)} \ .$$

The stationary solutions of this criterion with regard to θ are exactly the eigensolutions of the operator $S_Y S_X$. The penalty in the denominator is essential to make this happen.

We could simply redo the calculations of Hastie and Tibshirani to show that this is true. The stationary equations turn out to be equivalent to

$$\sum \phi_j \ = \ S_X \theta \ , \qquad \gamma\theta \ = \ S_Y \sum \phi_j \ ,$$

which yields the eigenequation $\gamma\theta = S_Y S_X \theta$ with eigenvalue γ. This provides the short answer but leaves open the question whether there is a deeper necessity that leads to this criterion. Indeed, there is more to it than a guess. The way we arrived at the solution is by simple geometric considerations which are common knowledge and which work for both ordinary and penalized ACE if properly applied. We give a short description of this geometry next.

5 Geometrizing Penalized ACE

The essence of penalized ACE geometry can be captured in terms of the case $p = 1$. Because this affords considerable notational relief, we restrict consideration to this simple case.

The entities we are operating on are hence the functions θ and ϕ. It makes sense to define a product space of pairs of transformations:

$$H \ = \ \left\{ \begin{pmatrix} \phi \\ \theta \end{pmatrix} \middle| \ldots \right\}$$

The symbols $|\ldots\}$ are our way of cheerfully ignoring all technicalities relating to Sobolev-type specifications. For obvious reasons, we also introduce the component subspaces:

$$H_X = \left\{ \begin{pmatrix} \phi \\ 0 \end{pmatrix} \middle| \ldots \right\}, \quad H_Y = \left\{ \begin{pmatrix} 0 \\ \theta \end{pmatrix} \middle| \ldots \right\}$$

(Notation: We write $(\phi, \theta)^\top = \begin{pmatrix} \phi \\ \theta \end{pmatrix}$ in-line.)

Lifting the ACE algorithm from function space to product space is essential when penalized building blocks are used. At the level of function spaces, penalization destroys the projection properties of the alternating ACE steps. However, at the level of the product space H, these steps reclaim their status as projections. The task is therefore to cast the ACE steps as operations on H in a suitable manner.

The alternating ACE steps can be described as mapping a pair $(0, \theta)^\top$ to $(S_X\theta, 0)^\top$, and then mapping this pair to $(0, S_Y S_X \theta)^\top$. While this defines mappings that go back and forth between the subspaces H_X and H_Y, we need mappings that are defined on all of H and that have projection properties. This is achieved as follows: Define

$$T_X \begin{pmatrix} \phi \\ \theta \end{pmatrix} = \begin{pmatrix} \phi + S_X\theta \\ 0 \end{pmatrix},$$

$$T_Y \begin{pmatrix} \phi \\ \theta \end{pmatrix} = \begin{pmatrix} 0 \\ \theta + S_Y\phi \end{pmatrix}.$$

These maps describe ACE properly:

$$T_Y T_X \begin{pmatrix} 0 \\ \theta \end{pmatrix} = \begin{pmatrix} 0 \\ S_Y S_X \theta \end{pmatrix}, \quad T_X T_Y \begin{pmatrix} \phi \\ 0 \end{pmatrix} = \begin{pmatrix} S_X S_Y \phi \\ 0 \end{pmatrix},$$

The ranges of T_X and T_Y are H_X and H_Y, respectively. Clearly, T_X and T_Y are idempotent:

$$T_X T_X = T_X, \quad T_Y T_Y = T_Y.$$

It is therefore not too surprising that T_X and T_Y turn out to be orthogonal projections, that is, idempotent and symmetric linear maps, with regard to an appropriate inner product, or more precisely, with regard to a bilinear form that has most properties of an inner product. The suitable bilinear form can be derived from the following familiar looking quadratic form defined on H:

$$\left\| \begin{pmatrix} \theta \\ \phi \end{pmatrix} \right\|^2 = \mathrm{E}[(\theta + \phi)^2] + J(\theta) + J_1(\phi).$$

We write this form as a squared norm, although we allow the possibility that $\|(\ldots)\| = 0$ and $(\ldots) \neq \mathbf{0}$, which translates to the existence of perfect fits in the ACE regression. However, degeneracy has already been ruled out on the subspaces H_X and H_Y through the assumption that the penalized minimizations with regard to θ and ϕ, respectively, have unique solutions. In terms of this quadratic, the penalized fitting maps S_X and S_Y are defined by

$$\begin{aligned} S_X(\theta) &= \text{argmin}_\phi \left\| \begin{pmatrix} -\phi \\ \theta \end{pmatrix} \right\|^2 , \\ S_Y(\phi) &= \text{argmin}_\theta \left\| \begin{pmatrix} \phi \\ -\theta \end{pmatrix} \right\|^2 . \end{aligned}$$

In order to show symmetry of T_X and T_Y, we need the bilinear form $\langle .., .. \rangle$ associated with the quadratic form $\|\ldots\|^2$. The connection between the two is through the polarization formula: Letting $\mathbf{z} = (\theta, \phi)^\top$ and $\tilde{\mathbf{z}} = (\tilde{\theta}, \tilde{\phi})^\top$, we have

$$\langle \mathbf{z}, \tilde{\mathbf{z}} \rangle = \frac{\|\mathbf{z} + \tilde{\mathbf{z}}\|^2 - \|\mathbf{z} - \tilde{\mathbf{z}}\|^2}{4} .$$

Symmetry of T_X and T_Y with regard to $\langle .., .. \rangle$ means:

$$\langle T_X \mathbf{z}, \tilde{\mathbf{z}} \rangle = \langle \mathbf{z}, T_X \tilde{\mathbf{z}} \rangle , \qquad \langle T_Y \mathbf{z}, \tilde{\mathbf{z}} \rangle = \langle \mathbf{z}, T_Y \tilde{\mathbf{z}} \rangle .$$

To show this, we need to take a small detour and actually show that $I - T_X$ and $I - T_Y$ are symmetric, which is equivalent to the above. The reason is that, for example, $I - T_X$ relates more directly to the minimization problem that defines S_X:

$$(I - T_X)\begin{pmatrix} \phi \\ \theta \end{pmatrix} = \begin{pmatrix} -S_X\theta \\ \theta \end{pmatrix} ,$$

and hence

$$\left\| (I - T_X)\begin{pmatrix} \tilde{\phi} \\ \theta \end{pmatrix} \right\|^2 = \min_\phi \left\| \begin{pmatrix} -\phi \\ \theta \end{pmatrix} \right\|^2 , \quad \forall\, \tilde{\phi},\ \theta.$$

The minimization property of S_X can be expressed as

$$\left\| \begin{pmatrix} -S_X\theta \\ \theta \end{pmatrix} \right\|^2 \leq \left\| \begin{pmatrix} -S_X\theta + \tilde{\phi} \\ \theta \end{pmatrix} \right\|^2 , \quad \forall\, \tilde{\phi},\ \theta,$$

from which follows by standard arguments:

$$\left\langle \begin{pmatrix} -S_X\theta \\ \theta \end{pmatrix}, \begin{pmatrix} \tilde{\phi} \\ 0 \end{pmatrix} \right\rangle = 0 .$$

The implication is that the value of an expression of the form

$$\left\langle \begin{pmatrix} -S_X\theta \\ \theta \end{pmatrix}, \begin{pmatrix} \tilde{\phi} \\ \tilde{\theta} \end{pmatrix} \right\rangle$$

is independent of $\tilde{\phi}$. This can be used to show symmetry. We have

$$\left\langle (I - T_X)\begin{pmatrix} \phi \\ \theta \end{pmatrix}, \begin{pmatrix} \tilde{\phi} \\ \tilde{\theta} \end{pmatrix} \right\rangle = \left\langle \begin{pmatrix} -S_X\theta \\ \theta \end{pmatrix}, \begin{pmatrix} \tilde{\phi} \\ \tilde{\theta} \end{pmatrix} \right\rangle$$
$$= \left\langle \begin{pmatrix} -S_X\theta \\ \theta \end{pmatrix}, \begin{pmatrix} -S_X\tilde{\theta} \\ \tilde{\theta} \end{pmatrix} \right\rangle .$$

Since the last expression is symmetric in θ and $\tilde{\theta}$ (and independent of ϕ and $\tilde{\phi}$), it follows that $I - T_X$ is symmetric with regard to $\langle .., .. \rangle$.

6 A Penalized ACE Criterion

The previous section can be summarized as follows: We described penalized ACE in terms of two maps T_X and T_Y; we showed that, in the penalized least squares (semi-)metric on H, these maps are the orthogonal projections onto the spaces H_X and H_Y, respectively. We can now proceed by writing down a criterion whose stationary solutions are found by projecting alternatingly with T_X and T_Y. One such criterion (there are many equivalent ones) is

$$\frac{\|\mathbf{y} - \mathbf{x}\|^2}{\|\mathbf{y}\|^2} = \min_{\mathbf{x} \in H_X,\, \mathbf{y} \in H_Y},$$

which expresses the idea of searching for the smallest angle between a unit length element of H_Y and its projection onto H_X. This criterion translates into the following:

$$\frac{\mathrm{E}[(\theta - \phi)^2] + J(\theta) + J_1(\phi)}{\mathrm{E}[\theta^2] + J(\theta)} = \min_{\phi,\, \theta} .$$

Obviously, this can be adapted so T_X fits a full additive model $\sum \phi_j$ as opposed to a single function ϕ. Apart from this, we have arrived at the final form of the problem: We now understand that penalized ACE requires a modified R^2-criterion where both numerator *and* denominator are penalized.

There is a final point to be made: In the practical application of a power algorithm, the exact mode of renormalization is quite irrelevant; for example, the iterates of a penalized ACE algorithm can easily be renormalized with unpenalized variance or any other measure of spread. The point of renormalization of iterates is to prevent loss of machine precision due to underflow, which leaves much arbitrariness in the choice of renormalization. However, this arbitrariness does not imply that any old measure of spread can serve as a denominator in a generalized R^2-criterion. When attempting to find a criterion whose stationary solutions are exactly the limits of a given power algorithm, only very particular choices of normalizing denominators are possible. In the present note, the derivation of the appropriate form of normalization was in terms of a geometric framework to which basic principles of alternating projections could be applied.

7 Discussion

The thoughts developed in this note are formulated in terms of an extension of ACE. However, the implications are much broader.

First, the framework of ACE can be extended to generalized canonical analysis with more than two blocks of transformations. Along these lines, the additive principal components of Donnell et al. (1994) can be extended with penalized curve estimators and a corresponding penalized variance criterion. Similarly, many scaling methods used in social sciences (such as Gifi (1990)) can be extended with penalty methods. The payoff could be considerable, in particular when the number of estimated transformations is large in relation to the sample size.

Second, a recent line of research called "functional data analysis" has emerged, concerned with the estimation of functionals of signals, as opposed to transformations of variables. A pertinent reference is Leurgans et al. (1993) who propose penalized canonical correlation analysis as a means for analyzing data consisting of sets of pairs of signals. Their method is based on a penalized correlation criterion that can be shown to be equivalent in structure to the penalized R^2-criterion used here. Similarly, other types of penalized functional multivariate analysis can be conceived. One more example is penalized linear discriminant analysis (Hastie et al. (1995)), but this is not the end of it.

8 Appendix: Closed Sums of Subspaces

8.1 Motivation

In this appendix we look into a technicality that is at the heart of all additive modeling, including ACE (Breiman and Friedman (1985)), GAMs (generalize additive models, Hastie and Tibshirani (1990)) and PPR (projection pursuit regression, Friedman and Stuetzle (1981)): If an additive model of the form $\sum_j \phi(X_j)$ is to be fitted to a population (that is, to a multivariate distribution as opposed to data), an assumption has to be made that this is theoretically possible. The problem is that no best fit $\sum \phi(X_j)$ may exist. This odd situation can arise when the space of such additive functions is topologically not closed. If we form the spaces $K = \{\phi_1(X_1)| \ldots\}$, $L = \{\phi_2(X_2)| \ldots\}$, $M = \{\phi_3(X_3)| \ldots\}$, ..., of centered and square integrable variables ($\mathrm{E}\phi_j(X_j) = 0$, $\mathrm{E}[\phi_j(X_j)^2] < \infty$), the problem boils down to establishing conditions under which the sum $L + K + M + \ldots$ of subspaces is closed. We can assume that the individual spaces are indeed closed. Halmos (1957, section I.15) gives a geometric construction of two closed subspaces whose sum is not closed. Below, we give an example of a pair of random variables X_1 and X_2 whose sum of L_2-spaces is not closed.

There is a particular reason why I am pleased to be given the opportunity to place this material in this Festschrift: Peter Huber teased me repeatedly for my cavalier treatment of this technical aspect of additive modeling. The best example for what he had in mind may have been our handwaving in Buja, Hastie and Tibshirani (1989, section 3.3.1): "..., which will be closed under some technical assumptions." Subsequently Peter Huber proved some closedness results of great generality (Huber (1991); one proof is attributed to H. Rieder). However, in the application of his theory to the obvious example of the multivariate normal distribution, Peter Huber encounters technical hardship and use of most likely inessential peculiarities of the normal. He states: "The proof uses several properties of the multivariate normal distribution in a serious fashion."

In this appendix, I would like to point out that there exist other, simple albeit less general, techniques for establishing closedness. Their advantage is straight applicability to standard examples of multivariate distributions. The conditions are used in Breiman and Friedman (1985) although the closedness result is not explicit in their paper. Furthermore, Breiman and Friedman impose the unnecessary additional condition that the subspaces intersect only at zero.

8.2 A sufficient closedness condition for sums of subspaces

We start with a theorem for subspaces in a general Hilbert space and then specialize to subspaces generated by transformations of random variables.

Consider a finite number of closed subspaces K, L, M, ... of some larger Hilbert space with inner product $\langle .., .. \rangle$ and associated norm $\|..\|$. Denote orthogonal projections onto subspaces K, L, ... by P_K, P_L, ... A restriction of a projection P_K to a subspace L will be denoted $P_K|L$; thus $P_K|L : L \to K$.

Theorem 8.1 *If all restricted projections $P_K|L$, $P_K|M$, $P_L|M$, ... are compact maps, then the sum $K + L + M \ldots$ is closed.*

We note that $P_L|K$ is the adjoint map of $P_K|L$ because for $x \in K$ and $y \in L$ we have $\langle P_L x, y \rangle = \langle x, y \rangle = \langle x, P_K y \rangle$. Hence if $P_K|L$ is compact, so is $P_L|K$.

Intuitively, the compactness condition prevents two subspaces from approaching each other in more than finitely many dimensions. A non-closed sum of two subspaces arises exactly when the two subspaces approach each other arbitrarily closely but do not contain the limit in their intersection. Such behavior is not possible when the closest directions are limited to finitely many dimensions. The compactness condition is only sufficient but not necessary, but it generalizes to sums of finitely many subspaces, which is a major convenience.

A sufficient condition for compactness is the Hilbert–Schmidt property: Given an orthonormal basis y_j of L, $P_K|L$ is said to have this property if

$$\sum_j \|P_K y_j\|^2 < \infty .$$

This holds, for example, if K and L are the spaces of square integrable functions of two variates X and Y with joint bivariate density $f_{X,Y}$, marginal densities

f_X and f_Y, satisfying

$$\int\int \frac{f_{X,Y}(x,y)^2}{f_X(x)f_Y(y)}dxdy \;<\; \infty.$$

This condition, for a bivariate normal with correlation $-1 < \rho < +1$ and standardized marginals, is verified easily:

$$\frac{f_{X,Y}(x,y)^2}{f_X(x)f_Y(y)} \;\propto\; \exp\Big(-\frac{(1-\rho)^2(x+y)^2+(1+\rho)^2(x-y)^2}{4(1-\rho^2)}\Big),$$

has a finite integral.

8.3 Geometry of pairs of subspaces

The geometry of a pair of subspaces is best understood in terms of their principal directions and angles: Two unit vectors $x \in K$ and $y \in L$ are said to be principal directions if they are stationary w.r.t. $\langle x, y\rangle$. This is equivalent to:

$$P_K(y) = cx \;, \quad P_L(x) = cy \;.$$

We can assume w.l.o.g. $c \geq 0$. Obviously $\langle x, y\rangle = c \leq 1$. Therefore, $c = \cos(\alpha)$, where α is called a principal angle. If x_i, y_i $(i = 1, 2)$ are two pairs of principal directions and if $c_1 \neq c_2$, then $\langle x_i, y_j\rangle = 0$ for $i \neq j$.

The largest c (if it exists) is the cosine of the smallest angle α that is taken on between K and L. If $K \cap L \neq 0$, any $x = y \in K \cap L$ is a pair of principal directions forming a zero angle.

If $P_K|L$ is compact, a singular value expansion exists of the form:

$$P_K(y) \;=\; \sum c_i\langle y, y_i\rangle x_i$$

for $y \in L$, where each pair x_i, y_i forms principal directions, and the singular value c_i is the principal cosine. Both $(x_i)_i$ and $(y_i)_i$ form orthonormal systems, and at least one of them is complete in its respective subspace. The values c_i can be assumed sorted in descending order and their only accumulation point is zero. Thus, for any $\epsilon > 0$ there exist only finitely many (orthogonal) principal directions with angle smaller than ϵ. In particular, if the intersection of K and L is infinite dimensional, $P_K|L$ cannot be compact.

The Hilbert–Schmidt property is equivalent to $\sum c_i^2 < \infty$. Because for a bivariate normal with standardized margins and correlation ρ we have $c_i = \rho^i$, we see again that the projections are Hilbert–Schmidt and the sum of subspaces is closed. Other examples of such expansions can be found in Buja (1990) and the references therein.

Here is a condition under which the sum of two subspaces is not closed: The principal cosines c_i accumulate at $+1$, but $+1$ is not a principal cosine. If this is the case, one can select a series x_i and y_i of principal directions such that $c_i \rightarrow 1$, $c_i < 1$ and $\sum(1-c_i^2) < \infty$. One can then form $z = \sum(y_i - c_i x_i)$ which is summable and in the closure of $K+L$. Following Halmos (1957, section I.15) one can show that z is not in $K+L$: assuming $z = x+y$ ($x \in K$, $y \in L$) we have

$$\begin{aligned} 0 &< 1 - c_i^2 = \langle z, y_i - c_i x_i \rangle = \langle x + y, y_i - c_i x_i \rangle = \langle y, y_i - c_i x_i \rangle \\ &= \langle \sum \langle y, y_j \rangle y_j, y_i - c_i x_i \rangle = \langle \langle y, y_i \rangle y_i, y_i - c_i x_i \rangle = \langle y, y_i \rangle (1 - c_i^2) , \end{aligned}$$

hence $\langle y, y_i \rangle = 1$ which is impossible because $\sum \langle y, y_i \rangle^2 < \infty$.

An example where this applies is in Buja (1990, p 1061): Consider the degenerate uniform distribution on a circle in the plane and its elliptic distortions, parametrized by the correlation ρ. The spaces K and L are again the subspaces of centered square integrable transformations $\phi(X)$ and $\theta(Y)$. The principal cosines, as a function of ρ, are given by $c_i = |\cos(i \cos^{-1}(\rho))|$. Thus, whenever $\cos^{-1}(\rho)$ is an irrational multiple of π, the values c_i are dense in the interval $[0,1]$. In particular, the value $+1$ (and hence angle zero) is approached arbitrarily closely, but never reached.

8.4 Proof of Theorem 8.1

The following lemmas must be standard. A form of Lemma 1 is due to Kober (1939). H. Rieder also gave a proof, which he handed to P.J. Huber (1991).

Lemma 8.2 *If $K \cap L = 0$, then $K+L$ is closed iff there exists some C (≥ 1) such that for all $x \in K$ and $y \in L$,*

$$\|x\|^2 + \|y\|^2 \leq C \cdot \|x+y\|^2 \tag{8.1}$$

PROOF Endow $K \times L$ with the norm $(\|x\|^2 + \|y\|^2)^{1/2}$. The summation mapping $S(x,y) = x+y$, $K \times L \to K+L$ is bounded, 1-1 and onto. Condition (8.1) is equivalent to boundedness of the inverse of S. The Lemma follows from the open mapping theorem and the fact that $K+L$ is closed iff it is complete. ////

Lemma 8.3 *A sufficient condition for (8.1) is the existence of some $\epsilon > 0$ such that for all nonzero $x \in K$, $y \in L$,*

$$\cos(x,y) \leq 1 - \epsilon. \tag{8.2}$$

PROOF From the assumption of the lemma we get

$$\langle x, y \rangle \leq (1-\epsilon) \cdot \|x\| \|y\| \leq (1-\epsilon) \cdot (\|x\|^2 + \|y\|^2)/2,$$

which can be reshaped to

$$\epsilon \cdot (\|x\|^2 + \|y\|^2) \leq \|x - y\|^2.$$

This is equivalent to (8.1) with $C = 1/\epsilon$ and a change of sign in y. ////

Lemma 8.4 *If $K \cap L = 0$ and $P_K|L$ is compact, then (8.2) is satisfied.*

PROOF Follows from remarks made earlier. The largest singular value of $P_K|L$ is the largest cosine between the two subspaces. ////

Lemma 8.5 *If $P_K|L$ is compact, then $K+L$ is closed.*

PROOF If the intersection $K \cap L$ is nonzero, form $\bar{L} = L \cap (L \cap K)^{\perp}$, where $(\ldots)^{\perp}$ is the orthogonal complement of a subspace. We have $K + L = K + \bar{L}$ and $K \cap \bar{L} = 0$, and $P_K|\bar{L}$ is also compact. Thus Lemma 8.4 applies to K and $\bar{L}$. ////

PROOF OF THEOREM 8.1 By induction over the number of subspaces. The beginning of the induction is given by Lemma 8.5. Assume $L + M + \ldots$ is closed; we show that the same is true for $K + L + M + \ldots$. We prove that $P_K|(L+M+\ldots)$ is compact. Let u_m be a bounded sequence in $L+M+\ldots$. We need to show that there exists a convergent subsequence of $P_K u_m$. Because $L+M+\ldots$ is closed by assumption and because the summation mapping from $L \times M \times \ldots$ to $L+M+\ldots$ is bounded and onto, it is open by the open mapping theorem. Hence there exist bounded sequences $y_m \in L$, $z_m \in M, \ldots$ such that $u_m = y_m + z_m + \ldots$. By assumption, $P_K|L$ is compact, so there exists a convergent subsequence of $P_K y_m$, for which we use again the same notation. Repeating the selection process for $P_K|M$ and z_m, and all remaining subspaces, we obtain a convergent subsequence $P_K u_m$. ////

References

[1] Breiman, L. and Friedman, J.H. (1985): Estimating optimal transformation for multiple regression and correlation. *J. Amer. Statist. Assoc.* **80** 580–598.

[2] Buja, A. (1990): Remarks on functional canonical variates, alternating least squares methods and ACE. *Ann. Statist.* **18** 1032–1069.

[3] Buja, A., Hastie, T. and Tibshirani, R. (1989): Linear smoothers and additive models. *Ann. Statist.* **17** 453–555.

[4] Donnell, D.J., Buja, A. and Stuetzle, W. (1994): Analysis of additive dependencies and concurvities using smallest additive principal components. *Ann. Statist.* **22** 1635–1668.

[5] Friedman, J.H. and Stuetzle, W. (1981): Projection pursuit regression. *J. Amer. Statist. Assoc.* **76** 817–823.

[6] Gifi, A. (1990): *Nonlinear Multivariate Analysis.* Wiley, Chichester.

[7] Halmos, P.R. (1957): *Introduction to Hilbert Space.* Chelsea Publishing, New York.

[8] Hastie, T.J., Buja, A. and Tibshirani, R.J. (1995): Penalized discriminant analysis. *Ann. Statist.* **23** 73–102.

[9] Hastie, T.J. and Tibshirani, R.J. (1990): *Generalized Additive Models.* Chapman & Hall, London.

[10] Huber, P.J. (1991): *Closedness of spaces of ridge functions.* Technical report PJH-91-1, Dept. of Mathematics, M.I.T.

[11] Kober, H. (1939): A theorem on Banach spaces. *Compositio Math.* **7** 135–140.

[12] Leurgans, S., Moyeed, R. and Silverman, B. (1993): Canonical correlation analysis when the data are curves. *J. Roy. Statist. Soc. B* **55** 725–740.

Nonparametric Estimation of Global Functionals of Conditional Quantiles

Probal Chaudhuri *
Indian Statistical Institute, Calcutta

Kjell Doksum †
Univerisity of California, Berkeley

Alexander Samarov ‡
University of Massachusetts and M.I.T.

Abstract

For fixed $\alpha \in (0,1)$, the quantile regression function gives the αth quantile $\theta_\alpha(\mathbf{x})$ in the conditional distribution of a response variable Y given the value $\mathbf{X} = \mathbf{x}$ of a vector of covariates. It can be used to measure the effect of covariates not only in the center of a population, but also in the upper and lower tails. When there are many covariates, the curse of dimensionality makes accurate estimation of the quantile regression function difficult. A functional that escapes this curse, at least asymptotically, and summarizes key features of the quantile specific relationship between $\mathbf{X}$ and Y is the vector β_α of weighted expected values of the vector of partial derivatives of the quantile function $\theta_\alpha(\mathbf{x})$. In a nonparametric setting, β_α can be regarded as the vector of quantile specific nonparametric regression coefficients while in semiparametric transformation and single index models, β_α gives the direction of the parameter vector in the parametric part of the model. We show that, under suitable regularity conditions, the estimate of β_α obtained by using the locally polynomial quantile estimate of Chaudhuri (1991a), is $\sqrt{n}$ consistent and asymptotically normal with asymptotic variance equal to the variance of the influence function of the functional β_α. We discuss how the estimates of β_α can be used for model diagnostics and in the construction of a link function estimates in general single index models.

Key words and phrases: Average derivative estimate, transformation model, projection pursuit model, index model, heteroscedasticity, reduction of dimensionality, quantile specific regression coefficients.

AMS 1991 subject classifications: Primary 62J02; secondary 62G99.

* Partially supported by a grant from Indian Statistical Institute

† Partially supported by NSF Grant DMS-93-07403

‡ Partially supported by NSF Grant DMS-93–06245 and by a grant from IFSRC at MIT

1 Introduction

The quantile regression function is defined as the αth quantile $\theta_\alpha(\mathbf{x})$ in the conditional distribution $F_{Y|\mathbf{X}}(y|\mathbf{x})$ of a response variable Y given the value $\mathbf{X} = \mathbf{x}$ of a d-vector of covariates: $\theta_\alpha(\mathbf{x}) = \inf\{y : F_{Y|\mathbf{X}}(y|\mathbf{x}) \geq \alpha\}$ for $0 < \alpha < 1$. It has the advantage, over the commonly used mean regression, that by considering different α, it can be used to measure the effect of covariates not only in the center of a population, but also in the upper and lower tails. For instance, the effect of a covariate can be very different for high and low income groups. Thus, in the latest presidential election the Democrats produced data showing that between 1980 and 1992, there was an increase in the number of people in the high salary category as well as the number of people in the low salary category. This phenomena could be demonstrated by computing the $\alpha = .90$ quantile regression function $\theta_{.90}(x)$ of salary Y as a function of the covariate $x =$ time and comparing it with the $\alpha = .10$ quantile regression function $\theta_{.10}(x)$. An increasing $\theta_{.90}(x)$ and a decreasing $\theta_{.10}(x)$ would correspond to the Democrats hypothesis that "the rich got richer and the poor got poorer" during the Republican administration.

Similarly, in survival analysis it is of interest to study the effect of a covariate on high risk individuals as well the effect on median and low risk individuals. Thus we would be interested in the quantiles $\theta_{.1}(\mathbf{x})$, $\theta_{.5}(\mathbf{x})$ and $\theta_{.9}(\mathbf{x})$ of the survival time Y given a vector $\mathbf{x}$ of covariates.

The statistical literature includes heteroscedastic models that capture the phenomena in the above salary and survival analysis examples; for example, see Bickel (1978) and Carroll and Ruppert (1988). However, this literature focuses on the estimation of the parameters in the mean regression function $\mu(\mathbf{x}) = \mathrm{E}(Y|\mathbf{x})$. Similarly, nonparametric regression techniques typically focus on the estimation of the mean conditional response $E(Y|\mathbf{x})$ and its derivatives which do not reveal dependence on the covariates in the lower and upper tails of the response distribution. Among the papers which address these last issues are Koenker and Bassett (1978), Efron (1991), and Koenker, Portnoy, and Ng (1994). In this paper we will consider measures of quantile dependent regression effects. First we discuss their use in a nonparametric setting and next consider their use in semiparametric models such as the transformation model and the single index models.

1.1 Nonparametric quantile regression coefficients

In linear statistical inference the partial derivatives $\partial\mu(\mathbf{x})/\partial x_i$, which are assumed to be constant and are called regression coefficients, are of primary interest since they measure how much the mean response is changed as the ith covariate is perturbed while the other covariates are held fixed. Similarly, we will consider the quantile dependent regression coefficient curves

$$\theta'_{\alpha i}(\mathbf{x}) = \partial\theta_\alpha(\mathbf{x})/\partial x_i, \quad i = 1, \ldots, d$$

which measure how much the αth response quantile is changed as the ith covariate is perturbed while the other covariates are held fixed.

We consider the nonparametric setting where estimation of the gradient vector $\nabla\theta_\alpha(\mathbf{x}) = (\theta'_{\alpha 1}(\mathbf{x}), \ldots, \theta'_{\alpha d}(\mathbf{x}))$ is subject to the "curse of dimensionality". That is, accurate pointwise estimation is difficult with the sample sizes usually available in practice because of the sparsity of data in subsets of $\mathbb{R}^d$ even for moderately large values of d. Thus, to obtain more accurately estimable functionals we will focus on the average gradient vector

$$\boldsymbol{\beta}_\alpha = (\beta_{\alpha 1}, \ldots, \beta_{\alpha d}) = \mathrm{E}(\nabla\theta_\alpha(\mathbf{X})).$$

The vector $\boldsymbol{\beta}_\alpha$, which gives a concise summary of quantile specific regression effects, will be called the vector of *(nonparametric) quantile regression coefficients.* Note that $\beta_{\alpha i}$ gives the average change in the quantile of the response as the ith covariate is perturbed while the other covariates are held fixed. Note also that in the linear model $Y = \sum_{j=1}^d \gamma_j X_j + \epsilon$, the vector $\boldsymbol{\beta}_\alpha$ coincides with the vector $\boldsymbol{\gamma}$ of regression coefficients.

The following simple example illustrates the use of the quantile specific regression coefficients.

Example 1.1 From Bailar (1991), we get Table 1 which gives the first, middle and third quartiles of statistics professor salaries for the academic year 1991–92. Departments of Biostatistics and Colleges of Education were excluded. The explanatory variable x is the number of years in the rank of full professor. From Table 1 we see the different trends over time in the three quartiles. Table 2 illustrates the quantile regression coefficient curves for $\alpha = .25,\ .5,\ .75$, and gives the estimated nonparametric quantile regression coefficients

$$(\hat{\beta}_{.25}, \hat{\beta}_{.5}, \hat{\beta}_{.75}) = (0.31, 0.67, 1.01),$$

computed as a weighted average of $\hat{\theta}'_\alpha(x)$ using the weights $\hat{p}(x)$. Again, these coefficients reveal a big difference in the effect of the covariate between the three quantiles. ////

x	2	5	8	11	14	17	20	23	25^+
$\hat{\theta}_{.25}(x)$	50.1	51.5	56.7	54.5	55.5	56.0	60.5	60.6	54.8
$\hat{\theta}_{.50}(x)$	54.0	62.2	63.8	61.5	62.8	69.0	70.9	66.9	62.2
$\hat{\theta}_{.75}(x)$	61.9	71.4	71.8	72.4	75.7	77.7	76.9	80.6	83.4
n_x	79	69	48	65	63	52	30	27	36

Table 1: Quartiles of salaries (in thousands of dollars) of Statistics Professors 1991–1992. x is the number of years as Full Professor. n_x is the sample size. $\sum n_x = 469$.

x	2-5	5-8	8-11	11-14	14-17	17-20	20-23	23-25$^+$	$\hat{\beta}_\alpha$
$\hat{\theta}'_{.25}(x)$	0.47	1.73	-0.73	0.33	0.17	1.5	0.03	-1.93	0.31
$\hat{\theta}'_{.50}(x)$	2.73	0.53	-0.77	0.43	2.07	0.63	-1.33	-1.57	0.67
$\hat{\theta}'_{.75}(x)$	3.17	0.13	0.20	1.1	0.67	-0.27	1.23	0.93	1.01
$\hat{p}(x)$	.18	.14	.14	.16	.14	.10	.07	.08	

Table 2: Quartile specific rates of change in salaries of Statistics Professors as seniority increases. $\hat{p}(x)$ is the proportion of people in the indicated category.

1.2 Transformation models

Many models in statistics, in particular in survival analysis, can be written in the form of a transformation model

$$h(Y) = \sum_{j=1}^{d} X_j \gamma_j + \epsilon, \tag{1.1}$$

where Y is survival time, $\mathbf{X} = (X_1, \cdots, X_d)$ is a vector of covariates, $\boldsymbol{\gamma} = (\gamma_1, \cdots, \gamma_d)$ is a vector of regression coefficients, ϵ is a residual independent of $\mathbf{X}$, and h is an increasing function specific to the model being considered.

For instance, the Cox proportional hazard model is of this form with $h(y) = \log\{-\log[1 - F_0(y)]\}$ and the distribution F_ϵ of ϵ is equal to the extreme value distribution $1 - \exp\{-\exp\{t\}\}$. Here F_0 is an unknown continuous distribution function referred to as the baseline distribution: it is the distribution of Y when the γ's are all zero. Dabrowska and Doksum (1987) considered the estimation of $\theta_\alpha(\mathbf{x})$ in this model. Similarly, the proportional odds rate model is of the form (1.1) with $h(y) = \log\{F_0(y)/[1 - F_0(y)]\}$ and F_ϵ the logistic distribution $1/[1 + \exp\{-t\}]$. See Doksum and Gasko (1990) for the details and history of these two and similar models. A third important survival analysis model of the form (1.1) is the accelerated failure time model where $h(y) = \log y$ and F_ϵ is unknown.

In the above three models, the first two have unknown h and known F_ϵ, while the third has known h and unknown F_ϵ. Other models of the form (1.1) have parametric h and F_ϵ. For instance, Box and Cox (1964) and Bickel and Doksum (1981) have h equal to a power transformation and let F_ϵ depend on a scale parameter. Box and Cox consider normal F_ϵ while Bickel and Doksum consider robustness questions for more general F_ϵ.

We consider model (1.1) with both h and F_ϵ unknown, and assume that h is continuous and strictly monotone and F_ϵ is continuous. Since h is unknown, $\boldsymbol{\gamma}$ is only identifiable up to a multiplicative constant; in other words, only the direction of $\boldsymbol{\gamma}$ is identifiable.

We drop the assumption that $\mathbf{X}$ and ϵ are independent and add instead a weaker assumption that the conditional quantile e_α of ϵ given $\mathbf{X} = \mathbf{x}$ does not depend on $\mathbf{x}$. Then, using the notation $g = h^{-1}$,

$$\theta_\alpha(\mathbf{x}) = g(\mathbf{x}^\top \gamma + e_\alpha) \text{ and } \boldsymbol{\beta}_\alpha = c_\alpha \boldsymbol{\gamma}, \quad \text{where } c_\alpha = \mathrm{E}\left[g'(\mathbf{X}^\top \gamma + e_\alpha)\right].$$

It follows that $\boldsymbol{\beta}_\alpha$ has the same direction as $\boldsymbol{\gamma}$, and we may without loss of generality estimate $\boldsymbol{\beta}_\alpha$. (Note that $(\beta_{\alpha i}/\beta_{\alpha j}) = (\gamma_i/\gamma_j)$ so that $\beta_{\alpha i}$ and $\beta_{\alpha j}$ give the relative importance of the covariates X_i and X_j.) The implication of this is that, for instance, the coefficients in the Cox model can be given an interpretation similar to the usual intuitive idea of what regression coefficients are: the Cox regression coefficients give the average change in a quantile (e.g., median) survival time as the ith covariate is perturbed while the others are held fixed. The quantile regression vector $\boldsymbol{\beta}_\alpha$ is a unifying concept that represents the coefficient vectors in the standard linear model, the Cox model, the proportional odds rate model, the accelerated failure time model, etc.

Remark 1.2 Let $\boldsymbol{\delta}_\alpha = \boldsymbol{\beta}_\alpha/|\boldsymbol{\beta}_\alpha|$ where $|\cdot|$ is the Euclidean norm. In model (1.1), $\boldsymbol{\delta}_\alpha = \boldsymbol{\delta}$ does not depend on α, and $\boldsymbol{\delta}$ represents the direction of $\boldsymbol{\gamma}$ so that estimates of $\boldsymbol{\delta}_\alpha$ obtained at grid points $\alpha_1, \cdots, \alpha_k$ can be combined into an estimate of $\boldsymbol{\delta}$ by computing their weighted average. Conversely, if $\boldsymbol{\delta}_{\alpha_1} \neq \boldsymbol{\delta}_{\alpha_2}$ for two different values of α, then the model (1.1) with $\mathbf{X}$ independent of ϵ does not hold, which suggests that the conditional quantile approach can also be used for model diagnostics; see Section 3. ////

Remark 1.3 We obtain an estimating equation for $g = h^{-1}$ by introducing $Z = \sum_{j=1}^{d} X_j \beta_{\alpha j}$ and noting that, if we let $\xi_\alpha(Z)$ denote the αth quantile in the conditional distribution of Y given Z, then g can be expressed as

$$g(Z) = \xi_\alpha\left(c_\alpha(Z - \epsilon_\alpha)\right),$$

and we can estimate the "shape" of g and h using an estimate of the αth quantile function $\xi_\alpha(Z)$ (note that g is identifiable up to a location and scale of its argument). ////

1.3 Single index models

An important semiparametric regression class of models is projection pursuit regression, which has been used by a number of authors (e.g., Friedman and Tukey (1974), Huber (1985)) while analyzing high dimensional data in an attempt to cope with the curse of dimensionality. The one term projection pursuit model, which gives the first step in projection pursuit regression, has the form

$$Y = g(\boldsymbol{\gamma}^\top \mathbf{X}) + \epsilon, \tag{1.2}$$

where $\boldsymbol{\gamma}$ is a d-dimensional parameter vector (the projection vector), ϵ denotes random error, and g is a smooth real valued function of a real variable. Stigler (1986, pp. 283–290) pointed out that Francis Galton used a projection pursuit type analysis while computing "mid-parents' heights" in course of his analysis of the data on the heights of a group of parents and their adult children in the late 19th century. Note that when (1.2) holds, we must have $\theta_\alpha(\mathbf{x}) = g(\boldsymbol{\gamma}^\top \mathbf{x}) + e_\alpha(\mathbf{x})$, where $e_\alpha(\mathbf{x})$ is the αth quantile in the conditional distribution of ϵ given $\mathbf{X} = \mathbf{x}$. Therefore, if $e_\alpha(\mathbf{x})$ is a constant free from $\mathbf{x}$ for some $0 < \alpha < 1$, the gradient vector $\nabla\theta_\alpha(\mathbf{x})$ will be equal to a scalar multiple of $\boldsymbol{\gamma}$ for all $\mathbf{x}$. Consequently, an estimate of $\boldsymbol{\beta}_\alpha$ gives an estimate of the projection direction $|\boldsymbol{\gamma}|^{-1}\boldsymbol{\gamma}$. Note that when the smooth function g is completely unspecified, only the direction of $\boldsymbol{\gamma}$ (and not its magnitude) is identifiable as in the transformation model (1.1).

In recent econometric literature, there is a considerable interest in the so called single index model (see, e.g., Han (1987), Powell et al. (1989), Newey and Ruud (1991), Sherman (1993)) defined by

$$Y = \phi(\boldsymbol{\gamma}^\top \mathbf{X}, \epsilon), \tag{1.3}$$

where ϵ is a random error independent of $\mathbf{X}$, and ϕ, which is a real valued function of two real variables, is typically assumed to be monotonic in both of its arguments. Li and Duan (1991) considered a very similar model in their regression analysis under link violations. They did not assume any monotonicity condition on the unknown link function ϕ. However, their sliced inverse regression approach for estimating the direction of $\boldsymbol{\gamma}$ requires some strong distributional assumptions (e.g., elliptic symmetry) on the regressor (in addition to independence between $\mathbf{X}$ and ϵ that may be hard to justify in practice (see also Section 3).

Hårdle and Stoker (1989) and Samarov (1993) investigated procedures for estimating the direction of $\boldsymbol{\gamma}$ in (1.2) and (1.3) respectively, using estimates of the gradient of the conditional mean of Y given $\mathbf{X} = \mathbf{x}$. Their approach requires neither the elliptic symmetry of the regressors nor the monotonicity of ϕ. However, the use of the conditional mean of the response makes the procedure non-robust, and it does not allow for the estimation of the function ϕ in (1.3).

It is important to note that all these earlier approaches require independence between the errors ϵ and the regressor $\mathbf{X}$, thus imposing a strong homoscedasticity condition which might be unrealistic in many practical situations. The approach of this paper allows one to weaken this assumption and only requires that, for some $0 < \alpha < 1$, the αth conditional quantile $e_\alpha(\mathbf{x})$ is a constant free from $\mathbf{x}$, which is a kind of a centering assumption for the distribution of the error ϵ.

Note that model (1.1) is a special case of model (1.3), and model (1.2) is not a special case of model (1.3) unless g is assumed to be monotonic. We will also drop the assumption of monotonicity of ϕ with respect to its first argument and assume that ϕ is strictly increasing in its second argument. Note that this will cover

(i) the regression model with product error $Y = e\psi(\boldsymbol{\gamma}^\top X)$, where ϕ is smooth and positive,

(ii) the heteroscedastic projection pursuit model $Y = g(\gamma^{\top}X) + e\psi(\gamma^{\top}X)$, where g is smooth and ϕ is smooth and positive, and

(iii) the heteroscedastic projection pursuit model with transformation

$$h(Y) = \psi_1(\gamma^{\top}\mathbf{X}) + e\psi_2(\gamma^{\top}\mathbf{X}),$$

where ϕ_1 is smooth, h is smooth and monotonic, and ϕ_2 is smooth and positive.

Now in model (1.3) with ϕ monotonic only in its second argument,

$$\theta_\alpha(\mathbf{x}) = \phi\{\gamma^{\top}\mathbf{x}, e_\alpha(\mathbf{x})\},$$

and if there exists $0 < \alpha < 1$ such that $e_\alpha(\mathbf{x})$ is a constant free from $\mathbf{x}$, $\nabla\theta_\alpha(\mathbf{x})$ will again be a scalar multiple of γ for all $\mathbf{x}$. Hence, an estimate for $\boldsymbol{\beta}_\alpha$ can be used to estimate the direction of γ in this case too.

The rest of the paper is organized as follows. In the next section we consider nonparametric estimation of the average gradient functional $\boldsymbol{\beta}_\alpha$. A discussion of diagnostic applications and of estimation of the link function in the model (1.3) are given in Section 3. The proofs can be found in Chaudhuri, Doksum, and Samarov (1994).

2 Estimation and Main Results

Let $(\mathbf{X}_1, Y_1), \cdots, (\mathbf{X}_n, Y_n)$ be n independent random vectors distributed as $(\mathbf{X}, Y)$, $\mathbf{X} \in \mathbb{R}^d$, $Y \in \mathbb{R}^1$. For fixed $0 < \alpha < 1$, let $\theta(\mathbf{x})$ be the conditional αth quantile of Y given $\mathbf{X} = \mathbf{x}$, and let $f(\mathbf{x})$ denote the density of $\mathbf{X}$. We want to estimate

$$\boldsymbol{\beta} = \int \{\nabla\theta(\mathbf{x})\}\, w(\mathbf{x})\, f(\mathbf{x})\, d\mathbf{x}, \tag{2.1}$$

where the dependence of $\theta(\mathbf{x})$ and $\boldsymbol{\beta}$ on α is suppressed when it does not cause an ambiguity, and $w(\mathbf{x})$ is a smooth weight function with a compact support within the interior of the support of $f(\mathbf{x})$. This weight function is introduced to obtain functionals and estimates that are not overly influenced by outlying $\mathbf{x}$ values (high leverage points). It allows our functional $\boldsymbol{\beta}$ to focus on quantile dependent regression effects without being unduly influenced by the tail behaviour of $f(\mathbf{x})$. It also takes care of boundary effects that occur in nonparametric smoothing. The weight function does not alter the fact that $\boldsymbol{\beta}$ has the same direction as γ in the single index model with an unknown monotonic link.

We will consider here the direct plug-in estimator

$$\hat{\boldsymbol{\beta}} = n^{-1} \sum \left\{\nabla\hat{\theta}(\mathbf{X}_i)\right\}\, w(\mathbf{X}_i), \tag{2.2}$$

where $\nabla\hat{\theta}(\mathbf{X}_i)$ is a nonparametric estimator of the gradient of the conditional quantile $\theta(\mathbf{x})$ at $\mathbf{x} = \mathbf{X}_i$. In Chaudhuri et al. (1994) we also consider another estimator of $\boldsymbol{\beta}$ based on integration by parts in (2.1).

While various nonparametric estimators of conditional quantiles could be used in (2.2), including kernel, nearest neighbor, and spline estimators; e.g., see Truong (1989), Bhattacharya and Gangopadhyay (1990), Koenker et al. (1994), Dabrowska (1992), we will consider here the locally polynomial estimators (see Chaudhuri (1991a,b)). The reason is that in order to develop asymptotic results for $\hat{\beta}_1$ and $\hat{\beta}_2$, we need to consider local polynomials in d variables with arbitrary degrees, and Chaudhuri's results provide Bahadur-type expansions of estimators of $\theta(\mathbf{x})$ as well as of estimators of $\nabla\theta(\mathbf{x})$ which can be readily adapted for our purposes here.

Consider a positive real sequence $\delta_n \to 0$, which will be chosen more explicitly later. Let $C_n(\mathbf{X}_i)$ be a cube in $\mathbb{R}^d$ centered at $\mathbf{X}_i$ with side legth $2\delta_n$, and let $S_n(\mathbf{X}_i)$ be the index set defined by

$$S_n(\mathbf{X}_i) = \{j : 1 \le j \le n, j \ne i, \mathbf{X}_j \in C_n(\mathbf{X}_i)\} \quad \text{and} \quad N_n(\mathbf{X}_i) = \#(S_n(\mathbf{X}_i)).$$

For $\mathbf{u} = (u_1, \ldots, u_d)$, a d-dimensional vector of nonnegative integers, set $[\mathbf{u}] = u_1 + \ldots + u_d$. Let A be the set of all d-dimensional vectors $\mathbf{u}$ with nonnegative integer components such that $[\mathbf{u}] \le k$ for some integer $k \ge 0$. Let $s(A) = \#(A)$ and let $\mathbf{c} = (c_\mathbf{u})_{\mathbf{u}\in A}$ be a vector of dimension $s(A)$. Also, given $\mathbf{X}_1, \mathbf{X}_2 \in \mathbb{R}^d$, define $P_n(\mathbf{c}, \mathbf{X}_1, \mathbf{X}_2)$ to be the following polynomial: $\sum_{\mathbf{u}\in A} c_\mathbf{u}[(\mathbf{X}_1 - \mathbf{X}_2)/\delta_n]^\mathbf{u}$—where, if $\mathbf{z} \in \mathbb{R}^d$ and $\mathbf{u} \in A$, we set $\mathbf{z}^\mathbf{u} = \prod_{i=1}^d z_i^{u_i}$ with the convention that $0^0 = 1$. Let $\hat{\mathbf{c}}_n(\mathbf{X}_i)$ be a minimizer of

$$\sum_{j\in S_n(\mathbf{X}_i)} \rho_\alpha(Y_j - P_n(\mathbf{c}, \mathbf{X}_i, \mathbf{X}_j)), \tag{2.3}$$

where $\rho_\alpha(s) = |s| + (2\alpha - 1)s$. Since $0 < \alpha < 1$, $\rho_\alpha(s)$ tends to ∞ as $|s| \to \infty$, and so the above minimization problem always has a solution, see Chaudhuri (1991a, b) for more on its uniqueness and other properties. We now set $\hat{\theta}(\mathbf{X}_i) = \hat{c}_{n,0}(\mathbf{X}_i)$ and $\nabla\hat{\theta}(\mathbf{X}_i) = \hat{\mathbf{c}}_{n,1}(\mathbf{X}_i)/\delta_n$.

Note that (2.3) defines a leave-one-out estimator; i.e., $\hat{\mathbf{c}}_n(\mathbf{X}_i)$ does not involve Y_i. This simplifies the use of the conditioning argument at various places in the proof.

In what follows, the asymptotic relations such as $\mathbf{a} = O(1), o(1), O_p(1)$, or $o_p(1)$, applied to a vector $\mathbf{a}$, will be understood componentwise.

Let V be an open convex set in $\mathbb{R}^d$. We will say that a function $m : \mathbb{R}^d \to \mathbb{R}^1$ has the order of smoothness p on V with $p = l + \gamma$, where $l \ge 0$ is an integer and $0 < \gamma \le 1$, and will write $m \in H_p(V)$, if

(i) partial derivatives $D^\mathbf{u} m(\mathbf{x}) := \partial^{[\mathbf{u}]} m(\mathbf{x})/\partial x_1^{u_1} \ldots \partial x_d^{u_d}$ exist and are continuous for all $\mathbf{x} \in V$ and $[\mathbf{u}] \le l$

(ii) there exists a constant $C > 0$ such that

$$|D^\mathbf{u} m(\mathbf{x}_1) - D^\mathbf{u} m(\mathbf{x}_2)| \le C|\mathbf{x}_1 - \mathbf{x}_2|^\gamma \text{ for all } \mathbf{x}_1, \mathbf{x}_2 \in V \text{ and } [\mathbf{u}] = l.$$

Condition 1 : The marginal density $f(\mathbf{x})$ of $\mathbf{X}$ is positive on V and f is an element of $H_{1+\gamma}(V)$.

Condition 2 : The weight function w is supported on a compact set with nonempty interior, $\text{supp}(w) \subset V$ and $w \in H_{1+\gamma}(V)$.

Condition 3 : The conditional density $f_{\epsilon|\mathbf{X}}(e|\mathbf{x})$ of $\epsilon = Y - \theta(\mathbf{X})$ given $\mathbf{X} = \mathbf{x}$, considered as a function of $\mathbf{x}$, belongs to $H_{1+\gamma}(V)$ for all e in a neighborhood of zero (zero being the αth quantile of the conditional distribution). Further, the conditional density is positive for $e = 0$ for all values of $\mathbf{x} \in V$, and its first partial derivative w.r.t. e exists continuously for values of e in a neighborhood of zero for all $\mathbf{x} \in V$.

Condition 4 : The conditional αth quantile function $\theta(\mathbf{x})$ of Y given $\mathbf{X} = \mathbf{x}$ has the order of smoothness p; i.e., $\theta(\mathbf{x}) \in H_p(V)$, with $p > 3 + 3d/2$.

Condition 4 implies that for every $\mathbf{x} \in V$, $k = [p]$, and all n sufficiently large, $\theta(\mathbf{x} + \mathbf{t}\delta_n)$ can be approximated by the k-order Taylor polynomial

$$\theta_n^*(\mathbf{x} + \mathbf{t}\delta_n, \mathbf{x}) = \sum_{\mathbf{u} \in A} c_{n,\mathbf{u}}(\mathbf{x}) \mathbf{t}^{\mathbf{u}}, \tag{2.4}$$

with the coefficients $\mathbf{c}_{n,\mathbf{u}}(\mathbf{x}) = (\mathbf{u}!)^{-1} D^{\mathbf{u}} \theta(\mathbf{x}) \delta_n^{[\mathbf{u}]}$, where $\mathbf{u}! = u_1! \ldots u_d!$, and the remainder $r(\mathbf{t}\delta_n, \mathbf{x}) = \theta(\mathbf{x} + \mathbf{t}\delta) - \theta_n^*(\mathbf{x} + \mathbf{t}\delta_n, \mathbf{x})$ satisfies the inequality

$$|r(\mathbf{t}\delta_n, \mathbf{x})| \le C(|\mathbf{t}|\delta_n)^p, \tag{2.5}$$

uniformly over $|\mathbf{t}| \le 1$ and $\mathbf{x} \in V$.

Theorem 2.1 *Assume that conditions 1, 2, 3 and 4 hold and let the order of the polynomial in (2.3) be $k = [p]$, where p is as in condition 4. Assume also that the "bandwidth" δ_n in the definition (2.3) of the conditional quantile estimator is such that*

$$\delta_n \asymp n^{-\kappa} \quad \text{with} \quad \frac{1}{2(p-1)} < \kappa < \frac{1}{4+3d}. \tag{2.6}$$

Then, as $n \to \infty$,

$$\begin{aligned} \hat{\boldsymbol{\beta}} &= o_p(n^{-1/2}) \\ &+ \frac{1}{n} \sum_{i=1}^{n} w(\mathbf{X}_i) \nabla \theta(\mathbf{X}_i) - (\alpha - \mathrm{I}\{\epsilon_i \le 0\}) \frac{\nabla w(\mathbf{X}_i) + w(\mathbf{X}_i) \ell(\mathbf{X}_i)}{f_{Y|\mathbf{X}}(\theta(\mathbf{X}_i)|\mathbf{X}_i)}, \end{aligned} \tag{2.7}$$

where $\epsilon_i = Y_i - \theta(\mathbf{X}_i)$, $\ell(\mathbf{X}) = -\nabla f(\mathbf{X})/f(\mathbf{X})$, and $\mathrm{I}\{\cdot\}$ denotes the indicator function.

PROOF The proof of the theorem is given in Chaudhuri et al. (1994). ////

3 Discussion

Efficiency considerations. The theorem in Section 2 means that the estimator $\hat{\boldsymbol{\beta}}$ is, using the terminology of Bickel et al. (1994), asymptotically linear with the influence function

$$\mathrm{IF}_\alpha(\mathbf{X}, Y) = w(\mathbf{X})\nabla\theta(\mathbf{X}) - (\alpha - \mathrm{I}\{\epsilon \le 0\})\frac{\nabla w(\mathbf{X}) + w(\mathbf{X})\ell(\mathbf{X})}{f_{Y|\mathbf{X}}(\theta(\mathbf{X})|\mathbf{X})} - \boldsymbol{\beta}, \quad (3.1)$$

and hence is asymptotically normal with mean zero and covariance matrix $\mathrm{Var}(\mathrm{IF}_\alpha(\mathbf{X}, Y))$. A straightforward computation shows that $\mathrm{IF}_\alpha(\mathbf{x}, y)$ is, in fact, the efficient influence function, that is, it coincides with the influence function of the functional $\boldsymbol{\beta}$, so that Proposition 3.3.1 of Bickel et al. (1994) implies that, under additional regularity conditions (which remain to be worked out) guaranteeing pathwise differentiability of the functional $\boldsymbol{\beta}$, the estimator $\hat{\boldsymbol{\beta}}$ is asymptotically efficient in the class of regular estimators.

Note that the asymptotic efficiency of the nonparametric estimator $\hat{\boldsymbol{\beta}}$ of the functional $\boldsymbol{\beta}$ does not imply its efficiency as an estimate of the coefficients γ in the semiparametric models (1.1)-(1.3), cf. Klaassen (1992), Horowitz (1993), and Klein and Spady (1993). Even though the estimator $\hat{\boldsymbol{\beta}}$ will not typically be fully efficient in models (1.1)-(1.3), the fact that they are $\sqrt{n}$ consistent means that it can serve as an initial estimator for various "one-step" and other "improved" estimators in those models, see Klaasen (1992), Bickel et al. (1994).

To illustrate the efficiency properties of our estimator, we consider the following simple example.

Example 3.1 Consider the transformation model (1.1) with $h(Y) = a + bX + \epsilon$, where X and ϵ are independent, h is increasing and differentiable, and X is normal $\mathcal{N}(\mu_1, \sigma_1^2)$ (cf. Doksum (1987)). In this case $\theta'_\alpha(x) = b/h'(\theta_\alpha(x))$, $\ell(x) = (x - \mu_1)/\sigma_1^2$ and $f_{Y|x}(\theta_\alpha(x)|x) = f_\epsilon(e_\alpha)h'(\theta_\alpha(x))$, where e_α is the αth quantile of ϵ. We have from (2.7) that the asymptotic variance of $\hat{\beta}_1$ (and $\hat{\beta}_2$) is

$$\frac{\alpha(1-\alpha)}{\sigma_1^2 n f_\epsilon^2(e_\alpha)} \mathrm{E}\Big(\frac{w(X)(X-\mu_1)/\sigma_1 + w'(X)}{h'(\theta_\alpha(X))}\Big)^2 + \frac{b^2}{n}\mathrm{Var}\Big(\frac{w(X)}{h'(\theta_\alpha(X))}\Big).$$

If we take $w(x)$ equal to one except in the extreme tails of the density of X, then, to a very close approximation, this asymptotic variance is equal to

$$[\alpha(1-\alpha)/n\sigma_1^2 f_\epsilon^2(e_\alpha)][\mathrm{E}\{(X-\mu_1)/\sigma_1 h'(\theta_\alpha(X))\}^2] + \frac{b^2}{n}\mathrm{Var}\Big(\frac{1}{h'(\theta_\alpha(X))}\Big).$$

In the case when $h(y) = y$, we have $\beta_\alpha = b$ and this expression reduces to

$$\frac{\alpha(1-\alpha)}{n\sigma_1^2 f_\epsilon^2(e_\alpha)},$$

which we recognize as the asymptotic variance of the quantile slope estimator in the linear model, see Koenker and Basset (1978). This means that our estimator, which is constructed without knowing h, is nearly as efficient in this case as the Koenker–Basset estimator which uses the linearity of $h(y)$. We also note that for this model the asymptotic variance of the Hărdle–Stoker estimator $\hat{\beta}_{HS}$ of $E\left(w(X)\frac{\partial}{\partial X}\,\mathrm{E}(Y|X)\right)$ $(= b$ here) is equal to $\sigma_\epsilon^2/n\sigma_1^2$. Therefore, the asymptotic efficiency of our estimator of b relative to the Hărdle–Stoker estimator is in this case

$$\frac{\sigma_\epsilon^2 f_\epsilon^2(e_\alpha)}{\alpha(1-\alpha)},$$

which is equal to the relative asymptotic efficiency of the sample α-quantile estimator vs. the sample mean, which may be greater or less than one depending on α and the distribution of ϵ. ////

The choice of bandwidth. Note that the choice (2.6) of the bandwidth δ_n "undersmooths" when compared to the optimal curve estimation bandwidth $\delta_n \asymp n^{-(2p+d)^{-1}}$, cf. Chaudhuri (1991a,b). The "undersmoothing" is needed to make the bias of the estimator of the order $o(n^{-1/2})$; the variance attains the order $1/n$ because of the averaging over different $\mathbf{X}_i$'s. As long as the bandwidth δ_n satisfies condition (2.6), the choice of bandwidth only has a second order effect on the mean squared error (MSE) of $\hat{\beta}$. In the case of average derivative estimation of γ in model (1.2), Hărdle and Tsybakov (1993) have used the second order term in the MSE to obtain an expression for the asymptotically optimal bandwidth. Note that in their approach, undersmoothing is also needed to obtain the desired asymptotic results. Recently, Hărdle, Hall and Ichimura (1993) have investigated simultaneous estimation of the optimal bandwidth and the vector γ in model (1.2). In the nonparametric case, a natural approach which requires few assumptions, is to use cross-validation based on the criteria function ρ_α or to choose the bandwidth which maximizes the empirical, leave-one-out, version of the quantile explanatory power

$$1 - \frac{\mathrm{E}[\rho_\alpha(Y - \theta(\mathbf{X}))]}{\mathrm{E}[\rho_\alpha(Y - \mu_Y)]}$$

of the covariates (c.f., Doksum and Samarov (1994). We will address this key question of bandwidth selection in detail in our future work.

Estimating the "link" functions in semiparametric models. Assume now that in the semiparametric models (1.1)-(1.3), for a given $0 < \alpha < 1$, the conditional α-quantile of ϵ given $\mathbf{X} = \mathbf{x}$ is constant in $\mathbf{x}$; that is, $e_\alpha(\mathbf{x}) = e_\alpha$. Set $Z = \gamma^\top \mathbf{X}$, and denote by $\xi_\alpha(z)$ the conditional αth quantile of Y given $Z = z$. Then we have $\xi_\alpha(z) = h^{-1}(z + e_\alpha)$ in model (1.1), $\xi_\alpha(z) = g(z) + e_\alpha$ in model (1.2), and $\xi_\alpha(z) = \phi(z, e_\alpha)$ in model (1.3). So, after getting an estimate of the direction of γ, one can project the observed $\mathbf{X}$'s on that estimated direction and then use those real valued projections to construct non-parametric estimates of h, g and ϕ in model (1.1), (1.2) and (1.3) respectively (keeping in mind the identifiability constraints in each of these models). This can be viewed as dimensionality reduction before constructing nonparametric estimates of the functional parameters in the models (1.1), (1.2) and (1.3). Under suitable regularity conditions,it is easy to construct an estimate $\hat{\xi}_\alpha(z)$ of $\xi_\alpha(z)$ that will converge at the rate $O_p(n^{-2/5})$, which is the usual rate for nonparametric pointwise estimation of a function of a single real variable. Properties of some nonparametric estimates of the conditional quantile function $\xi_\alpha(z)$ constructed following the above strategy will be investigated in detail in a separate paper. Nonparametric estimates of an unknown monotone transformation in regression models similar to (1.1) can be found in Doksum (1987), Cuzick (1988), Horowitz (1993) and Duan and Ye (1994).

Model diagnostics. The nonparametric estimates of the average derivatives of conditional quantiles, or quantile regression coefficients, $\boldsymbol{\beta}_\alpha$ lead to some useful model diagnostic techniques, cf. related works on heteroscedasticity by Hendricks and Koenker (1992) and Koenker, et al. (1992). Note first that if condition 4 in Section 2 holds for several conditional quantiles $\boldsymbol{\theta}_{\alpha_1}(\mathbf{x}), \ldots, \boldsymbol{\theta}_{\alpha_k}(\mathbf{x})$, where $0 < \alpha_1 < \alpha_2 < \ldots < \alpha_k < 1$, Theorem 2.1 implies that our estimates of $\boldsymbol{\beta}_{\alpha_1}, \ldots, \boldsymbol{\beta}_{\alpha_k}$ are jointly asymptotically normal with mean zero and covariance matrix equal to the covariance matrix of $\mathrm{IF}_{\alpha_1}(\mathbf{X}, Y), \ldots, \mathrm{IF}_{\alpha_k}(\mathbf{X}, Y)$, see (3.1). Using the asymptotic normal distribution of estimators of $\boldsymbol{\beta}_{\alpha_1}, \ldots, \boldsymbol{\beta}_{\alpha_k}$, we can construct asymptotic tests of their equality when $d = dim(X) = 1$, and thereby test homoscedasticity.

In the models (1.1)–(1.3) in the presence of strong homoscedasticity; that is, when ϵ and $\mathbf{X}$ are independent, $\nabla \boldsymbol{\theta}_\alpha(\mathbf{x})$ will be proportional to the parameter vector γ for all α and $\mathbf{x}$, and hence the estimated directions of $\nabla \boldsymbol{\theta}_\alpha(\mathbf{x})$'s for different values of α and $\mathbf{x}$ should be closely aligned, and so should be the estimates of quantile regression coefficients $\boldsymbol{\beta}_\alpha$ for different α's. Using again the joint asymptotic normality of the estimates of $\boldsymbol{\beta}_{\alpha_j}$ for $j = 1, \ldots, k$, we can construct asymptotic tests of homoscedasticity for the models (1.1)–(1.3) by testing the hypothesis of identical directions of $\boldsymbol{\beta}_{\alpha_j}$'s.

Further diagnostic information can be obtained by using nonparametric estimates of the $d \times d$-matrix functional

$$\Gamma_\alpha = \mathrm{E}\, w(\mathbf{X}) \nabla \boldsymbol{\theta}_\alpha(\mathbf{X}) \nabla \boldsymbol{\theta}_\alpha(\mathbf{X})^\top, \tag{3.2}$$

which can be estimated in a way essentially similar to $\boldsymbol{\beta}_\alpha$ (asymptotic properties of the estimates of (3.2) will be considered in a separate paper). In particular, the validity of the single index models (1.3) can be tested by testing that the rank of Γ_α is one. More generally, Γ_α can be used to identify the linear subspace spanned by the vectors γ_j, $j = 1, \ldots, k$ in the general dimensionality reduction (or multiple index) model $Y = G(\mathbf{x}^\top \gamma_1, \ldots, \mathbf{x}^\top \gamma_k, \epsilon)$ of Li (1991): just note that, provided the function G is monotone in ϵ and the αth conditional quantile of ϵ given $\mathbf{X}$ is free from $\mathbf{X}$, this subspace coincides with the subspace of those eigenvectors of Γ_α which have nonzero eigenvalues, cf. Samarov (1993).

Summary and further work. The quantile regression function $\theta_\alpha(\mathbf{x})$, for $0 < \alpha < 1$, is useful for studying quantile dependent regression effects. We have introduced the expected weighted gradient vector $\boldsymbol{\beta}_\alpha = \mathrm{E}[w(\mathbf{X})\nabla\theta_\alpha(\mathbf{X})]$ as a vector of quantile specific regression coefficients in an attempt to produce a functional which captures the essentials of quantile specific regression effects and whose estimators, at least asymptotically, do not suffer from the curse of dimensionality. In linear models, $\boldsymbol{\beta}_\alpha$ reduces to the regression coefficient vector, and in the semiparametric index model $Y = \psi(\mathbf{X}^\top\boldsymbol{\gamma}, \epsilon)$ with the link function ψ monotone in the second argument and αth quantile of ϵ free from $\mathbf{x}$, $\boldsymbol{\beta}_\alpha$ coincides with the direction of the regression coefficient vector $\boldsymbol{\gamma}$. We use the estimators of $\theta_\alpha(\mathbf{x})$ and its derivatives studied by Chaudhuri (1991a,b) to construct an estimator $\hat{\boldsymbol{\beta}}_\alpha$ of $\boldsymbol{\beta}_\alpha$ and we show that this estimator is asymptotically normal at the $\sqrt{n}$ rate with an asymptotic variance that equals the asymptotic variance of the nonparametric efficient influence function.

A number of important issues remains to be addressed:

(i) The finite sample size performance of the estimators has to be investigated using Monte Carlo methods. This would include an investigation of bandwidth selection rules used in $\hat{\beta}$.

(ii) Statistical properties of the estimates of the link function in models (1.1), (1.2), and (1.3) remain to be more fully investigated. In particular, the estimates of $\xi_\alpha(z)$ mentioned earlier in this section which converge at the rate $O_p(n^{-2/5})$ are not necessarily monotone. We need to establish asymptotic results for the isotonic versions of our estimators.

(iii) While Example 3.1 suggests that the loss in efficiency of our nonparametric estimators, when applied to some parametric models, may be not very large, it is of interest to find out how close the asymptotic variance of $\hat{\boldsymbol{\beta}}_\alpha$ is to the asymptotic efficiency bounds in the semiparametric models (1.1), (1.2), and (1.3).

(iv) In our examples of transformation model, we included some important models used in survival analysis. We are currently working on extending results of this paper to censored data, see Dabrowska (1992).

References

[1] Bailar, B.A. (1991): Salary survey of U.S. colleges and universities offering degrees in statistics. *Amstat News* **182** 3.

[2] Bhattacharya, P. and Gangopadhyay, A. (1990): Kernel and nearest-neighbor estimation of a conditional quantile. *Ann. Statist.* **18** 1400–1415.

[3] Bickel, P.J. (1978): Using residuals robustly I: Tests for heteroscedasticity, nonlinearity. *Ann. Statist.* **6** 266–91.

[4] Bickel, P.J. and Doksum, K.A. (1981): An analysis of transformation revisited. *J. Amer. Statist. Assoc.* **76** 296–311.

[5] Bickel, P.J., Klaassen, C., Ritov, Y. and Wellner, J. (1994): *Efficient and Adaptive Estimation for Semiparametric Models*. John Hopkins Univ. Press, Baltimore.

[6] Box, G.E.P. and Cox, D.R. (1964): An analysis of transformations. *J. Roy. Statist. Soc. B* **26** 211–252.

[7] Carroll, R.J. and Ruppert, D. (1988): *Transformation and Weighting in Regression*. Chapman & Hall, New York.

[8] Chaudhuri, P. (1991a): Nonparametric estimates of regression quantiles and their local Bahadur representation. *Ann. Statist.* **19** 760–777.

[9] Chaudhuri, P. (1991b): Global nonparametric estimation of conditional quantile functions and their derivatives. *J. Multivariate Analysis* **39** 246–269.

[10] Chaudhuri, P., Doksum, K. and Samarov, A. (1994): Nonparametric estimation of global funtionals based on quantile regression. Submitted to *Ann. Statist.*.

[11] Cuzick, J. (1988): Rank regression. *Ann. Statist.* **16** 1369–1389.

[12] Dabrowska, D. (1992): Nonparametric quantile regression with censored data. *Sankhya A* **54**.

[13] Dabrowska, D. and Doksum, K. (1987): Estimates and confidence intervals for median and mean life in the proportional hazard model. *Biometrika* **74** 799–807.

[14] Doksum, K. (1987): An extension of partial likelihood methods for proportional hazard models to general transformation models. *Ann. Statist.* **15** 325–345.

[15] Doksum, K. and Gasko, M. (1990): On a correspondence between models in binary regression analysis and in survival analysis. *International Statistical Review* **58** 243–252.

[16] Doksum, K. and Samarov, A. (1994): Nonparametric estimation of global functionals and a measure of the explanatory power of covariates in regression. *Ann. Statist.*, to appear.

[17] Duan, N. and Ye, J. (1994): *Nonparametric $\sqrt{n}$ Consistent Estimation for the General Transformation Models*. Tech. Report #138, Grad. School of Bus., Univ. of Chicago.

[18] Efron, B. (1991): Regression percentiles using asymmetric square error loss. *Statistica Sinica* **1** 93–125.

[19] Friedman, J. and Tukey, J. (1974): A projection pursuit algorithm for exploratory data anlysis. *IEEE Transactions on Computers* **C-23** 881–889.

[20] Han, A. (1987): A non-parametric analysis of transformations. *J. of Econometrics* **35** 191–209.

[21] Härdle, W., Hall, P. and Ichimura, H.: Optimal smoothing in single-index models. *Ann. Statist.* **21** 157–178.

[22] Härdle, W. and Stoker, T. (1989): Investigating smooth multiple regression by the method of average derivatives. *J. Amer. Statist. Assoc.* **84** 986–995.

[23] Härdle, W. and Tsybakov, A.B. (1993): How sensitive are average derivatives? *J. of Econometrics* **58** 31–48.

[24] Hendricks, W. and Koenker, R. (1992): Hierarchical spline model for conditional quantiles and the demand for electricity. *J. Amer. Statist. Assoc.* **87** 58–68.

[25] Horowitz, J.L. (1993): *Semiparametric Estimation of a Regression Model with an Unknown Transformation of the Dependent Variable.* Tech. Report., Dept. of Statistics, Univ. of Iowa.

[26] Huber, P. (1985): Projection pursuit. *Ann. Statist.* **13** 435–525.

[27] Klaassen, C. (1992): *Efficient Estimation in the Clayton-Cuzick Model for Survival Data.* Tech. Report., Univ. of Amsterdam.

[28] Klein, R. and Spady, R. (1993): An efficient semiparametric estimator for binary response models. *Econometrica* **61** 387–421.

[29] Koenker, R. and Basset, G. (1978): Regression quantiles. *Econometrica* **46** 33–50.

[30] Koenker, R., Ng, P. and Portnoy, S. (1994): Quantile smoothing splines. *Biometrika* **81** 673–680.

[31] Li, K.-C. (1991): Sliced inverse regression for dimension reduction. *J. Amer. Statist. Assoc.* **86** 316–342.

[32] Li, K.-C. and Duan, N. (1991): Slicing regression: A link-free regression method. *Ann. Statist.* **19** 505–530.

[33] Newey, W.K. and Ruud, P.A. (1991): *Density Weighted Linear Least Squares.* Technical Report, M.I.T.

[34] Powell, J., Stock, J. and Stoker, T. (1989): Semiparametric estimation of index coefficients. *Econometrica* **57** 1403–1430.

[35] Prakasa Rao, B.L.S. (1983): *Nonparametric Function Estimation.* Academic Press.

[36] Samarov, A. (1993): Exploring regression structure using nonparametric functional estimation. *J. Amer. Statist. Assoc.* **88** 836–849.

[37] Sherman, R. (1993): The limiting distribution of the maximum rank correlation estimator. *Econometrica* **61** 123–137.

[38] Stigler, S.M. (1986): The history of statistics: *The Measurement of Uncertainty before* 1900. Belknap Press, Cambridge, MA.

[39] Truong, Y. (1989): Asymptotic properties of kernel estimates based on local medians. *Ann. Statist.* **17** 606–617.

High Breakdown Point Estimators in Logistic Regression

Andreas Christmann
University of Hamburg

Abstract

Estimators with high finite sample breakdown points are of special interest in robust statistics. However, in contrast to estimation in linear regression models the breakdown point approach have not yet received much attention in logistic regression models. Although various robust estimators have been proposed in logistic regression models, their breakdown points are often not yet known. Here it is shown for logistic regression models with binary data that there is no estimator with a high finite sample breakdown point, provided the estimator has to fulfill a weak condition. However, in logistic regression models with large strata modifications of Rousseeuw's least median of squares estimator and least trimmed squares estimator have finite sample breakdown points of approximately 1/2. Both estimators are strongly consistent under a large supermodel of the logistic regression model. Existing programs can be used to compute such estimates.

Key words and phrases: High breakdown point, least median of squares, least trimmed squares, least median of weighted squares, least trimmed weighted squares, logistic regression, outliers, overdispersion, robust regression.
AMS 1991 subject classifications: 62F35; 62F10.

1 Introduction

Classical estimators for the unknown parameter vector in logistic regression models are the maximum likelihood estimator (Cox and Snell (1989)) and the weighted least squares estimator (Berkson (1944)). However, it is well-known that outliers can have a strong impact on these non-robust estimators. Therefore, many robust alternatives have been proposed. Pregibon (1982) considered resistant estimators. Stefanski, Carroll and Ruppert (1986) and Künsch, Stefanski and Carroll (1989) investigated M-estimators, and Vidmar, McKean and Hettmansperger (1992) proposed rank-based methods. Morgenthaler (1992) treats least-absolute-deviations fits. In the author's doctoral thesis an estimator is investigated which minimizes a sum of squared Hellinger distances between certain binomial distributions.

There exist different concepts of robustness and many optimality criteria for estimators, for example Huber (1981, chapter 1) and Hampel, Ronchetti, Rousseeuw and Stahel (1986, chapter 1.3). However, optimality criteria shall not be considered here. It can be argued that good robust estimators should fulfill some desirable properties. For example, Davies (1994, p 362) investigates linear regression models and mentions five desirable properties which a regression functional should have. One of these properties is a high breakdown point (perhaps calculated in terms of metrics) at each regression distribution.

However, in contrast to estimation in linear regression models the breakdown point approach have not yet received much attention in logistic regression models. From a practical point of view, one desirable property of robust estimators is a high finite sample breakdown point as defined by Donoho and Huber (1983). This concept has the advantage of simplicity and it is often found suitable by non-statisticians. As far as we know, the breakdown points of the robust estimators mentioned above for the unknown parameter vector in logistic regression models are not yet known. Hence, there are the following questions:

- Do there exist estimators with high finite sample breakdown points?
- Are such estimators strongly consistent under a reasonable large supermodel of the logistic model?
- How can one compute such estimators (at least approximately)?

The aim of this paper is to offer some answers. In Section 2 it is shown for logistic regression with binary data that there is no estimator with a high finite sample breakdown point in the sense of Donoho and Huber (1983), if the estimator has to fulfill a weak condition. In Section 3 a supermodel H of the logistic regression model with large strata is defined. Many outlier generating models are submodels of H. In Section 4 the least median of weighted squares estimator $\hat{\theta}_{LMWS}$ and the least trimmed weighted squares estimator $\hat{\theta}_{LTWS}$ are defined. Both estimators have a high finite sample breakdown point and are modifications of the well-known estimators LMS and LTS proposed by Rousseeuw (1984) in linear regression models. Section 5 shows that both estimators are strongly consistent under H. An example from cancer research is treated in Section 6, and Section 7 gives a discussion.

2 Result for Binary Response Variables

Let $x_i \in \mathbb{R}^p$, $i = 1, ..., n$. Under the logistic regression model there are n independent random variables Y_i which are binomial distributed with index m_i and probability $\pi_i(\theta)$, $i = 1, ..., n$. Here, m_i is a known positive integer and $\pi_i(\theta) = 1/\{1 + \exp(-x_i^\top \theta)\}$. The vector of parameters $\theta \in \mathbb{R}^p$ is unknown. In this paper we shall assume that all $x_i \neq 0$, and that the design matrix $X = (x_1, ..., x_n)^\top$ has full column rank $p \leq n$. Define $z_i = (x_i^\top, y_i)^\top$, $i = 1, ..., n$.

Let T_n be an estimator for θ. In logistic regression models the finite sample breakdown point of T_n in the sense of Donoho and Huber (1983) can be defined by $\varepsilon_n^*(T_n; z_1, ..., z_n) = k_n/n$, where

$$k_n = \min\left\{ k \;\middle|\; \max_{i_1,\ldots,i_k} \sup_{a_1,\ldots,a_k} ||T_n(w_1, ..., w_n) - T_n(z_1, ..., z_n)|| = \infty \right\}. \tag{2.1}$$

Here the sample $(w_1, ..., w_n)$ is obtained by replacing k data points $z_{i_1}, ..., z_{i_k}$ by arbitrary values $a_i \in \mathbb{R}^p \times \{0, ..., m_i\}, i = 1, ..., k$. In (2.1), $||t||$ denotes the Euclidean norm of a vector t.

In the following part of this section, let $m_1 = ... = m_n = 1$. The large strata case shall be treated in Section 4.

Assume that there exists a hyperplane through zero in $\mathbb{R}^p$ such that all points x_i lie on the same side of it. For example, this is satisfied if one of the components of x_i has constant sign, which happens in the case of regression with an intercept term. Consider an estimator T_n for θ which has the following property:

$$\text{if } y_1 = ... = y_n \text{ then } [\pi_i(T_n(z_1, ..., z_n)) = y_i \text{ for all } i = 1, ..., n]. \tag{2.2}$$

As all constants m_i are equal to one, it holds that $y_1 = \ldots = y_n = 0$ or $y_1 = \ldots = y_n = 1$; that is, there are only failures or only successes. Under the logistic regression model the probability of this event is positive and equal to $\prod\{1 - \pi_i(\theta)\}$ or $\prod\{\pi_i(\theta)\}$, respectively, where $\theta \in \mathbb{R}^p$. It follows from (2.2) that $||T_n(z_1, ..., z_n)|| = \infty$. This is true, for example for many M-estimators including the maximum likelihood estimator. Albert and Anderson (1984) and Santner and Duffy (1986) give conditions which guarantee that the maximum likelihood estimate has finite Euclidean norm.

If an estimator T_n for θ fulfills (2.2) and if $||T_n(z_1, ..., z_n)|| < \infty$, then

$$\varepsilon_n^*(T_n; z_1, ..., z_n) \leq \frac{\min\{\sum y_i, n - \sum y_i\}}{n}. \tag{2.3}$$

Of course, the above inequality is often extremely rough. But it follows from (2.3), that the finite sample breakdown point ε_n^* of any estimator T_n with property (2.2) will often depend on the sample, and ε_n^* can be as small as possible.

3 A Supermodel

Let $m_1, ..., m_n$ be positive integers. Define the sets $M_i = \{0, ..., m_i\}$, and denote the set of all subsets of M_i by $\mathcal{A}_i$, $i = 1, ..., n$. Let $h = [\![n/2]\!] + [\![(p+1)/2]\!]$, where $[\![r]\!]$ stands for the largest integer less than or equal to r. Denote by $\Lambda(z) = 1/\{1+\exp(-z)\}$, for $z \in \mathbb{R}$, the cumulative distribution function of the logistic distribution. Let $X = (x_1, ..., x_n)^\top$ be a fixed $n \times p$ design matrix with rank $p \leq n$, where $x_i \neq 0$, $1 \leq i \leq n$.

Model H: *Let y_{i,m_i} be observations of (not necessarily independent) random variables Y_{i,m_i} on the measurable spaces $(M_i, \mathcal{A}_i)$, $1 \le i \le n$. Let*

$$\begin{aligned} P_{i,m_i} &= 1/(2m_i)\,, & \text{if} \quad & Y_{i,m_i} = 0\,, \\ &= Y_{i,m_i}/m_i\,, & \text{if} \quad & 1 \le Y_{i,m_i} \le m_i - 1\,, \\ &= 1 - 1/(2m_i)\,, & \text{if} \quad & Y_{i,m_i} = m_i\,, \quad 1 \le i \le n. \end{aligned} \tag{3.1}$$

It is assumed that there exist $\pi_i \in (0,1)$, $1 \le i \le n$, such that

$$(P_{1,m_1}, \ldots, P_{n,m_n}) \rightarrow (\pi_1, \ldots, \pi_n) \quad \text{almost surely, if} \quad \min_i(m_i) \rightarrow \infty\,, \tag{3.2}$$

and that there exists exactly one vector $\theta_0 \in \mathbb{R}^p$ such that

$$|\{i; \pi_i = \Lambda(x_i^\top \theta_0)\}| \ge h > |\{i; \pi_i = \Lambda(x_i^\top \theta)\}| \quad \text{for all} \quad \theta \ne \theta_0\,. \tag{3.3}$$

Note that (3.2) guarantees that the relative frequencies P_{i,m_i} converge almost surely to some constants π_i, $1 \le i \le n$, which can be interpreted as 'true' success probabilities. The assumption (3.3) assures that at least $h > n/2$ of the quantities π_i can be modelled according to a logistic regression model, and that its parameter vector θ_0 is identifiable. However, the *distribution* of the corresponding h random variables Y_{i,m_i} need not to be binomial as long as (3.2) is valid.

In the presence of multiple outliers it is often unrealistic to assume that all outliers are realisations of independent random variables, see also Davies and Gather (1993, p. 782). Therefore, model H allows that not all random variables Y_{i,m_i} are independent.

For logistic regression models the strong law of large numbers guarantees the validity of (3.2), and (3.3) is true by definition of $\Lambda(x_i^\top \theta_0)$ and $\text{rank}\,(X) = p$.

Model H contains some models proposed in the literature. McCullagh and Nelder (1989, pp 124ff) and Liang and McCullagh (1993), among others, consider models with over- and underdispersion. Overdispersion can result in the occurrence of outliers with respect to the logistic regression model. Model H allows the following type of over- and underdispersion. Let k_i be fixed positive integers such that all values of m_i are multiples of k_i, and let $\sigma_i^2 \in (0, \infty)$, $1 \le i \le n$, be fixed constants. Assume that all random variables Y_{i,m_i} are the sum of m_i/k_i independent and identically distributed random variables on the set $\{0, \ldots, k_i\}$, which have expectation $k_i\pi_i$ and variance $k_i\pi_i(1-\pi_i)\sigma_i^2$. Then, $\mathrm{E}(Y_{i,m_i}) = m_i\pi_i$ and $\mathrm{Var}(Y_{i,m_i}) = m_i\pi_i(1-\pi_i)\sigma_i^2$. The strong law of large numbers guarantees the validity of (3.2).

An important class of submodels of H, which themselves are supermodels of logistic regression models, has the following structure: at least h of the n random variables Y_{i,m_i} are independent and are distributed according to a logistic regression model. The remaining random variables, which need not to be independent as long as (3.2) is satisfied, have a binomial distribution with parameters m_i and $\pi_i \ne \Lambda(x_i^\top \theta_0)$. For example, it is possible to model contaminants with respect to the logistic model by specifying $\pi_i = G(x_i^\top \theta)$, where $G : \mathbb{R} \to (0,1)$, $G \neq \Lambda$, is an appropriate, strictly monotone function. Copas (1988) considers $G = \gamma + (1-2\gamma)\Lambda$, where $\gamma \in (0, 1/2)$. Location shift and scale modification can be modelled by $\pi_i = \Lambda(x_i^\top\theta + d)$, $d \neq 0$, and $\pi_i = \Lambda(\sigma\, x_i^\top \theta)$, $\sigma \notin \{0, 1\}$, respectively.

4 Definition of the Estimators

In this section we consider model H and large strata; that is, all values of m_i are reasonably large and $m_+ = \sum m_i \to \infty$, $m_i/m_+ \to c_i \in (0,1)$, but n and p are fixed.

Define $V_{i,m_i} = \{m_i P_{i,m_i}(1-P_{i,m_i})\}^{1/2}$, $i = 1, ..., n$. Under the logistic model with large strata the empirical logit transform $\Lambda^{-1}(P_{i,m_i})$ is approximately normally distributed with mean $x_i^\top\theta$ and variance $[m_i\pi_i(\theta)\{1-\pi_i(\theta)\}]^{-1}$. This variance can be consistently estimated by V_{i,m_i}^{-2}, see Cox&Snell (1989, pp. 31 ff). Therefore, define $\tilde{Y}_{i,m_i} = V_{i,m_i}\Lambda^{-1}(P_{i,m_i})$ and $\tilde{X}_{i,m_i} = V_{i,m_i}x_i$. Note that P_{i,m_i} and $\tilde{Z}_{i,m_i} = (\tilde{X}_{i,m_i}^\top, \tilde{Y}_{i,m_i})^\top$ are well defined. Davis (1985) investigates many modifications of P_{i,m_i}. The weighted least squares estimate minimizes $\sum(\tilde{y}_{i,m_i} - \tilde{x}_{i,m_i}^\top\theta)^2$ with respect to θ. For the large strata model the weighted least squares estimator is consistent, asymptotically normal, and efficient; see Davis (1985).

Christmann (1994a) defines the *least median of weighted squares estimator* $\hat{\theta}_{LMWS}$ for θ in the logistic regression model by Rousseeuw's least median of squares estimator for the transformed random vectors $\tilde{Z}_{i,m_i}$, $1 \leq i \leq n$. Hence

$$\min_\theta \left(r^2(\theta)\right)_{h:n} = (r^2(\hat{\theta}_{LMWS}))_{h:n}\,. \tag{4.1}$$

Here, $r^2(\theta)_{1:n} \leq r^2(\theta)_{2:n} \leq \ldots \leq r^2(\theta)_{n:n}$ are the ordered squared 'residuals' $r_i(\theta) = \tilde{y}_{i,m_i} - \tilde{x}_{i,m_i}^\top\theta$. Analogously, define the *least trimmed weighted squares estimator* $\hat{\theta}_{LTWS}$ for θ as any solution of

$$\min_\theta \sum_{i=1}^{h} \left(r^2(\theta)\right)_{i:n} = \sum_{i=1}^{h}\left(r^2(\hat{\theta}_{LTWS})\right)_{i:n}\,. \tag{4.2}$$

We shall say that $\tilde{z}_{1,m_1}, ..., \tilde{z}_{n,m_n}$ are in general position if any p of them determine θ uniquely; see Rousseeuw (1984, p. 872).

By construction of $\hat{\theta}_{LMWS}$ and $\hat{\theta}_{LTWS}$ the key results derived by Rousseeuw (1984) and Rousseeuw and Leroy (1987, pp. 112 ff) about LMS and LTS estimates remain valid for $\hat{\theta}_{LMWS}$ and $\hat{\theta}_{LTWS}$.

Proposition 4.1 *(a) There always exists a solution of (4.1) and (4.2).*

(b) If $p > 1$ and the transformed observations $\tilde{z}_{1,m_1}, ..., \tilde{z}_{n,m_n}$ are in general position, then the breakdown point of $\hat{\theta}_{LMWS}$ and $\hat{\theta}_{LTWS}$ in the sense of Donoho and Huber (1983) is $(\llbracket (n-p)/2 \rrbracket + 1)/n$.

(c) *If* $p > 1$ *and there exists some* θ *such that strictly more than* $(n+p-1)/2$ *of the transformed observations* $\tilde{z}_{1,m_1}, \ldots, \tilde{z}_{n,m_n}$ *satisfy* $\tilde{y}_{i,m_i} = \tilde{x}_{i,m_i}^{\top}\theta$ *exactly and are in general position, then* $\hat{\theta}_{LMWS} = \theta$ *and* $\hat{\theta}_{LTWS} = \theta$ *whatever the other observations are ("exact fit property") .*

5 Asymptotical Results

Under weak conditions $\hat{\theta}_{LMWS}$ and $\hat{\theta}_{LTWS}$ converge almost surely to θ_0, if the supermodel H is valid.

Theorem 5.1 *Assume that* $m_i/m_+ \to c_i \in (0,1)$, *if* $\min_i(m_i) \to \infty$, *for* $1 \le i \le n$, *and that all* $h \times p$ *matrices* X^* *which contain* h *rows of* X *have rank* p. *Then* $\hat{\theta}_{LMWS}$ *and* $\hat{\theta}_{LTWS}$ *converge almost surely to* θ_0 *under model H, if* $\min_i(m_i) \to \infty$.

A proof is given in Christmann (1993).

In general the asymptotic distributions of the above high breakdown point estimators are not yet known. However, consider the following special situation. Assume that Y_{i,m_i}, $1 \le i \le n$, satisfy model H, that *exactly* h of them follow a logistic regression model, but assume $P_{i,m_i} \to \pi_i$ almost surely for the remaining random variables, where $\pi_i \ne \Lambda(x_i^{\top}\theta_0)$. For this particular case, the next theorem shows that $\hat{\theta}_{LTWS}$ based on the whole data has the same asymptotic normal distribution as the weighted least squares estimator based on those random variables which come from the logistic model.

Corollary 5.2 *Assume that the assumptions of Theorem 5.1 are satisfied. Let* Y_{i,m_i}, $1 \le i \le n$, *be random variables according to model H, such that the set* $J = \{i; \pi_i = \Lambda(x_i^{\top}\theta_0)\}$ *has exactly* h *elements and that* Y_{i,m_i}, $i \in J$, *are independent and have binomial distributions with* m_i *and* $\pi_i = \Lambda(x_i^{\top}\theta_0)$. *Then* $m_J^{1/2}(\hat{\theta}_{LTWS} - \theta_0)$ *converges in distribution to* $N(0, W_J^{-1})$, *if* $\min_i(m_i) \to \infty$, *where* $m_J = \sum_{i\in J} m_i$, *and* $W_J = \sum_{i \in J} c_i\Lambda(x_i^{\top}\theta_0)\{1-\Lambda(x_i^{\top}\theta_0)\}\, x_i x_i^{\top}$.

A proof is given in Christmann (1993).

6 Example

Consider the data set given in Table 1 from cancer research, c.f. Thompson and Funderlic (1981). Cox (1987) analyses this data set by threshold dose-response models and by a simple logistic regression model with background; that is, $\pi_i = \lambda + (1-\lambda)\Lambda(x_i^{\top}\theta)\}$, where $\theta \in \mathbb{R}^p$ and $0 < \lambda < 1$ are unknown. There are good biological reasons to use such models. Bedrick and Hill (1990) test whether there are outliers with respect to a simple logistic regression model without background in this data set.

i	$x_i^\top$	y_i	m_i	i	$x_i^\top$	y_i	m_i
1	(1, 0·301)	19	19	6	(1, −1·208)	4	20
2	(1, 0·000)	18	20	7	(1, −1·509)	0	16
3	(1, −0·301)	19	19	8	(1, −1·807)	0	19
4	(1, −0·602)	14	21	9	(1, −2·108)	0	40
5	(1, −0·903)	15	19	10	(1, −2·710)	2	81

Table 1: Cancer data set; i, case number

Assume that a simple logistic regression model without background has been postulated before observing these data. The parameters of this model are estimated by three methods: maximum likelihood (ML), weighted least squares (WLS), and least median of weighted squares (LMWS), see Table 2. Additionally, these estimates are computed for the first nine observations only because y_{10} seems to be an outlier with respect to this model.

	whole data set			without case 10		
	ML	WLS	LMWS	ML	WLS	LMWS
Intercept	3·23	2·47	2·58	3·98	2·94	2·76
	(0·48)	(0·39)		(0·63)	(0·47)	
Slope	3·42	2·56	3·43	4·34	3·28	3·43
	(0·43)	(0·31)		(0·63)	(0·49)	

Estimated standard errors are given in parenthesis.

Table 2: Estimation results for cancer data set

Both, the ML estimate and the WLS estimate considerably depend on case 10, whereas the LMWS estimate is rather stable for this data set. The "residuals" $r_{10}(\hat{\theta}_{LMWS}) = 4{\cdot}2$ and $r_{10}(\hat{\theta}_{WLS}) = 1{\cdot}1$ computed from the whole data set differ strikingly. The WLS estimate and the LMWS estimate are similar for the reduced data set.

Two possible reasons for the occurrence of outliers in logistic regression models are typing errors, and that the number of failures has been erroneously reported instead of the number of successes; c.f. Copas (1988). Consider now a modification of the above data set for which

$$x_9 = (1, -21{\cdot}08)^\top, \;\; y_9 = 40, \;\; x_{10} = (1, -27{\cdot}10)^\top, \;\; y_{10} = 79.$$

These observations are extreme outliers in y- and in x-direction with respect to the simple logistic regression model. The ML, WLS, and LMWS estimates for the modified data set are given in Table 3. These estimates are also computed for a reduced data set which does not contain the last two cases.

	whole data set			without cases 9 and 10		
	ML	WLS	LMWS	ML	WLS	LMWS
Intercept	0·29	0·42	2·58	3·88	2·89	2·58
	(0·17)	(0·27)		(0·64)	(0·49)	
Slope	−0·13	−0·12	3·43	4·22	3·20	3·43
	(0·02)	(0·03)		(0·66)	(0·55)	

Estimated standard errors are given in parenthesis.

Table 3: Estimation results for the modified data set

Table 3 shows the strong impact of the last two cases on the ML and WLS estimates. Even the sign of the estimated slope has changed for these estimates by the occurrence of two outliers. The "residuals" $r_i(\hat{\theta}_{WLS})$ are 0·97 for $i = 9$ and −0·08 for $i = 10$. In contrast to that, the corresponding "residuals" based on $\hat{\theta}_{LMWS}$ are 52·0 and 131·2.

7 Discussion

The aim of this paper is to give some answers whether there exist estimators with high finite sample breakdown point in the sense of Donoho and Huber (1983) in logistic regression models. If the estimator has to fulfill a weak and reasonable condition it is shown that the answer is negative if all response variables Y_i are binary. In contrast to this, two high breakdown point estimators denoted by $\hat{\theta}_{LMWS}$ and $\hat{\theta}_{LTWS}$ exist and are strongly consistent under a general supermodel of the logistic regression model with large strata. These estimators are defined as the well-known least median of squares estimator and the least trimmed squares estimator (Rousseeuw (1984), Rousseeuw and Leroy (1987)) after the data are transformed by the empirical logit transformation. Both estimators have the advantage that they can be computed using existing software. In a first step one computes the transformed response vector and the transformed design matrix. Then one can use currently available software to compute LMS- or LTS-estimates; for example, the progam PROGRESS developed by Rousseeuw and Leroy (1987), LMSREG and LTSREG in S-Plus, or the subroutines HYLMSE and HYLTSE given by Marazzi (1993).

Davis (1985) shows that the weighted least squares estimator (Berkson (1944)) may be inconsistent in logistic regression models, if m_i is bounded, but $n \to \infty$. We conjecture that this may happen for $\hat{\theta}_{LMWS}$ and $\hat{\theta}_{LTWS}$, too, because both estimators are robustifications of the weighted least squares estimator. Note that the maximum likelihood estimator for θ is consistent and asymptotically normal in logistic regression models if $n \to \infty$ and if weak assumptions are satisfied; c.f. Fahrmeir&Kaufmann (1985). Therefore, it should be investigated whether similar robustifications of the ML estimator may be appropriate in this case.

The results of this paper can be generalized. It is possible to consider other values of h. However, this will often reduce the finite sample breakdown point. Making obvious modifications the results of this paper can also be derived for categorical regression models, which use another continuous, strictly increasing distribution function $F\colon \mathbb{R} \to (0,1)$ instead of Λ. For example, this can be done for the probit model, which uses the cumulative distribution function of the standard normal distribution instead of Λ. Christmann (1994b) gives some results about high breakdown point estimation in other regression models; for example, Poisson regression with large groups.

The asymptotic distributions of $\hat{\theta}_{LMWS}$ and of $\hat{\theta}_{LTWS}$ under the logistic regression model or under reasonable supermodels are not yet known. Note that the transformed random variables $\tilde{Y}_{i,m_i}$, $1 \leq i \leq n$, can have discrete and asymmetric distributions and need not be independent under H. Hence general results proved by Davies (1990) about the asymptotics of S-estimators in linear regression models are not directly applicable for $\hat{\theta}_{LMWS}$ and $\hat{\theta}_{LTWS}$. However, Theorem 5.1 only shows consistency of these estimators for large values of m_i where n and p are fixed, while Davies (1990) permits the general case.

Of course, there should be more efficient estimators than those considered here. An investigation of general S-estimators (Rousseeuw and Yohai (1984), Davies (1990)), MM-estimators (Yohai (1987)) and generalized S-estimators (Croux, Rousseeuw and Hössjer (1993)) for data transformed via the empirical logit transform is beyond the scope of the paper.

References

[1] Albert, A. and Anderson, J.A. (1984): On the existence of maximum likelihood estimates in logistic regression models. *Biometrika* **71** 1–10.

[2] Bedrick, E.J. and Hill, J.R. (1990): Outlier tests for logistic regression: A conditional approach. *Biometrika* **77** 815–27.

[3] Berkson, J. (1944): Application of the logistic function to bio-assay. *J. Amer. Statist. Assoc.* **39** 357–365.

[4] Christmann, A. (1993): *Strong consistency of the least median of weighted squares estimator for large strata.* Research paper 93/12. Univ. of Dortmund, Dept. of Statistics, submitted.

[5] Christmann, A. (1994a): Least median of weighted squares in logistic regression with large strata. *Biometrika* **81** 413–417.

[6] Christmann, A. (1994b): *High Breakdown Point Estimation for Certain Regression Models.* Research paper 94/1. Univ. of Dortmund, Dept. of Statistics. Submitted.

[7] Copas, J.B. (1988): Binary regression models for contaminated data. With discussion. *J. Roy. Statist. Soc. B* **50** 225–265.

[8] Cox, C. (1987): Threshold dose-response models in toxicology. *Biometrics* **43** 511–523.

[9] Cox, D.R. and Snell, E.J. (1989): *The Analysis of Binary Data*. 2nd ed. Chapman & Hall, London.

[10] Croux, C., Rousseeuw, P.J. and Hössjer, O. (1993): *Generalized S-estimators*. Report no. 93-03, Department of Mathematics and Computer Science, Univ. of Antwerp.

[11] Davies, P.L. (1990): The asymptotics of S-estimators in the linear regression model. *Ann. Statist.* **18** 1651–1675.

[12] Davies, L. (1994): Desirable properties, breakdown and efficiency in the linear model. *Statist. Probab. Lett.* **19** 361–370.

[13] Davies, L. and Gather, U. (1993): The identification of multiple outliers. (With discussion.) *J. Amer. Statist. Assoc.* **88** 782–801.

[14] Davis, L.J. (1985): Consistency and asymptotic normality of the minimum logit chi-squared estimator when the number of design points is large. *Ann. Statist.* **13** 947–957.

[15] Donoho, D.L. and Huber, P.J. (1983): The notion of breakdown point. In *A Festschrift for Erich L. Lehmann* (P.J. Bickel, K.A. Docksum, and J.L. Hodges, Jr., eds.), 157–184. Wadsworth, Belmont, CA.

[16] Fahrmeir, L. and Kaufmann, H. (1985): Consistency and asymptotic normality of the maximum likelihood estimator in generalized linear models. *Ann. Statist.* **13** 342–368.

[17] Hampel, F.R., Rousseeuw, P.J., Ronchetti, E.M. and Stahel, W.A. (1986): *Robust Statistics—The Approach Based on Influence Functions*. Wiley, New York.

[18] Huber, P.J. (1981): *Robust Statistics*. Wiley, New York.

[19] Künsch, H.R., Stefanski, L.A. and Carroll, R.J. (1989): Conditionally unbiased bounded-influence estimation in general regression models, with applications to generalized linear models. *J. Amer. Statist. Assoc.* **84** 460–466.

[20] Liang, K.-Y. and McCullagh, P. (1993): Case studies in binary dispersion. *Biometrics* **49** 623–630.

[21] Marazzi, A. (1993): *Algorithms, Routines, and S Functions for Robust Statistics*. Wadsworth, Belmont, CA.

[22] McCullagh, P. and Nelder, J.A. (1989): *Generalized Linear Models*. 2nd ed. Chapman & Hall, London.

[23] Morgenthaler, S. (1992): Least-absolute-deviations fits for generalized linear models. *Biometrika* **79** 747–754.

[24] Pregibon, D. (1982): Resistant fits for some commonly used logistic models with medical applications. *Biometrics* **38** 485-498.

[25] Rousseeuw, P.J. (1984): Least median of squares regression. *J. Amer. Statist. Assoc.* **79** 871–880.

[26] Rousseeuw, P.J. and Leroy, A.M. (1987): *Robust Regression and Outlier Detection.* Wiley, New York.

[27] Rousseeuw, P.J. and Yohai, V. (1984): Robust regression by means of S-estimators. In *Robust and Nonlinear Time Series Analysis*, (J. Franke, W. Härdle, and R. D. Martin, eds.), Lecture Notes in Statistics No. 26, 256–272. Springer-Verlag, New York.

[28] Santner, T.J. and Duffy, D.E. (1986): A note on A. Albert and J.A. Anderson's conditions for the existence of maximum likelihood estimates in logistic regression models. *Biometrika* **73** 755–758.

[29] Stefanski, L.A., Carroll, R.J. and Ruppert, D. (1986): Optimally bounded score functions for generalized linear models with applications to logistic regression. *Biometrika* **73** 413–424.

[30] Thompson, W.A. and Funderlic, R.E. (1981): A simple threshold model for the classical bioassay problem. In *Measurement of Risks* (G.G. Berg and H.D. Maillie, eds.), 511–533. Plenum, New York.

[31] Vidmar, T.J., McKean, J.W. and Hettmansperger, T.P. (1992): Robust procedures for drug combination problems with quantal responses. *Appl. Statist.* **41** 299–315.

[32] Yohai, V.J. (1987): High breakdown-point and high efficiency robust estimates for regression. *Ann. Statist.* **15** 642–656.

The Guinea Pig of Multiple Regression

Yadolah Dodge
University of Neuchâtel

Abstract

The role played by a 21×4 data set in the development (progress and regress) of multiple regression is discussed. While this 'famous' data set has been (and is being) used as a guinea pig [1] for almost every method of estimation introduced in the regression market, it appears that no one has questioned the origin and correctness in the last 30 years. An attempt is made to clarify some points in this regard. In particular, it is argued here that this data set is in fact a subset of a longer data set of which the rest is missing.

Key words and phrases: Data, estimation, meta-analysis, multiple regression, outliers, robustness.
AMS 1991 subject classifications: Primary 62J05; secondary 62F35.

1 Introduction

Statistics is the science of making sense out of data. For various types of data different statistical techniques have been developed for extracting relevant information, for determining what the data 'mean'. Real data are produced by the thousands on a daily basis all over the planet earth. Some are collected routinely and some after carefully designed experiments in the context of prespecified hypotheses. Of all these data some are classified as confidential and remain secret for ever, while others, less fortunate, end up in public domain. Of those published small data sets, some remain unanalyzed, some get to be analyzed just once, and some are analyzed over and over again: the latter play the role of the guinea pig in the many fields of applied statistical science.

Fisher's (1936) Iris data is an example of such. The data give measurements of four flower parts on 50 specimens of each of three species of iris. They were collected (but not published in full) by E. Anderson, and most famously analyzed by Sir Ronald Fisher. His interest was to develop a methodology for discriminating between the species on the basis of the four measurements. Later the data became the guinea pig of all kinds of multivariate methods such as component, correspondence and factorial analysis, to mention a few.

[1] A guinea pig is a small animal used in the laboratory experiments for testing different types of drugs. Often, after an experiment, the animal on which the drug is tested does not survive.

There is another data set which played (and continues to play) a similar role in the last 30 years in the area of multiple regression. It was introduced as an example to illustrate certain computational procedures of multiple regression using the least squares method by Brownlee (1960, Section 17.12, pp. 491–492). Since then the data set have been used at least 90 distinct papers and books on multiple linear regression.

The purpose of this paper is to trace the history of this 21×4 data set and to document the instances in which it has been used as an example of statistical methodology. We put forward and try to substantiate the hypothesis that this is a fragment of a larger data set. In conclusion, we wonder if it has affected the development of multiple regression since the manner in which the fragment has been selected is entirely unknown.

In Section 2 we introduce the data and its artificial evolution; in Section 3 we put forward the first three analyses made on the data set. Section 4 discusses the first robust analysis on the data. A list of equations and outliers found by different scientists involved with the analysis of this data set forms Section 5 and in Section 6 we make an attempt to shed some light as to the origin of the data set.

2 The Data and its Evolution

Brownlee (1960) introduced a 21x4 data set as an example to illustrate certain computational procedures for multiple regression in *Statistical Theory and Methodology in Science and Engineering*, John Wiley and Sons, New York. The data as he listed them, are reproduced in Table 1. Brownlee (1960, p. 491) describes the data as follows:

It was obtained from 21 days of operation of a plant for the oxidation of ammonia to nitric acid. The first column represents the rate of operation of the plant. The nitric oxides produced are absorbed in a countercurrent absorption tower. The third variable is the concentration of acid circulating, minus 50, times 10: that is, 89 corresponds to 58.9 per cent acid. The second column is the temperature of cooling water circulated through coils in the absorption tower. The dependent variable y is 10 times the percentage of the ingoing ammonia to the plant that escapes from the absorption column unabsorbed; that is, an (inverse) measure of the over-all efficiency of the plant.

In the second edition of the same book, Brownlee (1965, p. 454) make some minor changes in the data. He replaces *It was obtained*, by *There were obtained*, and also *first column* by x_1 and *second column* by x_2. These changes can be explained by the introduction of mathematical symbols.

In both cases, the author does not provide any further scientific description as to when, how, where, under what conditions and by whom the data were collected.

Draper and Smith (1966, p. 204) describe the data as follows:

A set of data given on p. 454 of K.A. Brownlee's Statistical Theory and

Observation Number	Air Flow x_1	Cooling Water Inlet Temperature x_2	Acid Concentration x_3	Stack Loss y
1	80	27	89	42
2	80	27	88	37
3	75	25	90	37
4	62	24	87	28
5	62	22	87	18
6	62	23	87	18
7	62	24	93	19
8	62	24	93	20
9	58	23	87	15
10	58	18	80	14
11	58	18	89	14
12	58	17	88	13
13	58	18	82	11
14	58	19	93	12
15	50	18	89	8
16	50	18	86	7
17	50	19	72	8
18	50	19	79	8
19	50	20	80	9
20	56	20	82	15
21	70	20	91	15

Table 1: Brownlee's (1960, p. 402) data

Methodology in Science and Engineering (second edition), 1965, Wiley, New York, is reproduced below. The figures are from 21 days of operation of a plant oxidizing NH_3 to HNO_3 and the variables are:

X_1 *= rate of operation,*

X_2 *= temperature of the cooling water in the coils of the absorbing tower for the nitric oxides,*

X_3 *= concentration of HNO_3 in the absorbing liquid (coded by minus 50, times 10),*

Y *= the percent of the ingoing NH_3 that is lost by escaping in the unabsorbed nitric oxides ($\times 10$). This is an inverse measure of the yield of HNO_3 for the plant.*

They add *Investigation of plant operations indicates that the following sets of runs can be considered as replicates: (1,2), (4,5,6), (7,8),(11,12), and (18,19). While the runs in each set are not exact replicates, the points are sufficiently close to each other in the X-space for them to be used as such.*

Daniel and Wood (1971, Chapter 5) provide the data with some additional information:

In looking for a multifactor example using real data, *we were struck by the fact that no clean-cut, undesigned data were on hand. By 'clean cut' we mean 'meeting all the necessary statistical requirements without any complicating circumstances'. When we tried to analyze textbook examples, we found unforeseen complications in each. One of the more interesting problems is presented here, not only as an example of fitting an equation with three independent variables, but also to point out some of the precautions that are necessary.*

The example is from Brownlee (Section 13.12). It is also presented in Draper and Smith(Chapter 6). The data represent 21 **successive** *days of operation of a plant oxidizing ammonia to nitric acid. Factor x_1 is the flow of air to the plant. Factor x_2 is the temperature of the cooling water entering the countercurrent nitric oxide absorption tower. Factor x_3 is the concentration of nitric acid in the absorbing liquid(coded for ease of hand calculation by subtracting 50 and then multiplying by 10). The response, y, is 10 times the percentage of the ingoing ammonia that is lost as unabsorbed nitric oxides; it is an indirect measure of the yield of nitric acid.*

From 1960 to 1965 the description of data has been somewhat polished by Brownlee. In 1966 Draper and Smith give the impression that they really investigated the plant. The 1971 work of Daniel and Wood leaves the reader with little doubt that the data set is in fact 'real' and that it represents 21 'successive' days of operation.

Here are some examples in the statistical literature in which the words 'real' data and 'successive' days have been taken seriously.

Denby and Mallows (1977) write: *Our first example relates to a set of 21 observations given by Brownlee.* They add: *The 21 observations relate to measurements taken on* **successive** *days.*

Atkinson (1981) writes: *For an example with* **real** *data we look at a much analyzed example originally given by Brownlee (1965, p. 454)...*

Rey (1983) writes: *There are essentially two points recommending this data set; it is a set of* **real** *data and this set has been examined without too many prejudices as to what it contains.*

And finally, Preece (1986) writes that: *I know that I am unhappy with the few popular examples that receive recurrent attention in the literature. I am here thinking of, for example, the stackloss data of Brownlee (1965, p. 454), analyses of which have seen produced that ignore the* **time- sequence** *(time again!) despite indications of a time-lag in the variates; indeed analysis of these data is 'utterly useless' (Barnard, 1982) because of the lack of an experimental design for the investigation that produced them.*

The words 'real data' and 'successive' become fairly established for this 21×4 data set even though Brownlee does not state this implicitly or explicitly in describing the data set on two occasions.

3 Fitting the Least Squares Equations

Brownlee (1960) uses the classical *least squares* method to model his 21×4 data set. Brownlee's goal is to demonstrate with this data how to perform the necessary computation (sums, sums of squares, sums of products, and so on up to 9 decimal points), and how to test the hypotheses related to the parameters. Brownlee first finds the regression equation of y on x_1, x_2 and x_3 as

$$\hat{y} = -39.919 + 0.7156x_1 + 1.2953x_2 - 0.1521x_3 \tag{3.1}$$

He then makes separate t tests on individual parameters and finds the contribution from the x_3 variable non-significant. He hence drops it from the model and gives as final equation

$$\hat{y} = -50.359 + 0.671159x_1 + 1.295338x_2 \tag{3.2}$$

Draper and Smith (1966, pp. 204–215) give the data as an exercise to be used for selecting the 'best' regression equation. They provide the reader with computer output for all 7 possible runs. Their solution to the exercise (theoretically the same as (3.2)) can be found on page 337 with

$$\hat{y} = -50.358836 + 0.6711545x_1 + 1.295351x_2 \tag{3.3}$$

as the prediction equation. They then add the remark that this fitted equation is least satisfactory at observation 21.

Daniel and Wood (1971, Chapter 5) use the data not only as an example of fitting an equation with three independent variables, but also to point out some of the precautions that are necessary. Using graphical techniques they observe that the normal plot of the residuals indicates one very low point (21) and that the residuals versus $\hat{y}$ plot is irregular in an unfamiliar way. The residuals of the points (1,2,3,4,21) are widely separated from the others, and the remaining sixteen residuals seem to show some sort of subsidiary trends in sets.

During a complete chapter, the authors try to find a model which avoids these irregularities. They note that such irregularities may especially appear in four possible cases:

1. *A small minority of points, perhaps only one or two, may be entirely erroneous.*

2. *The model does not have the right functional form. The large residuals may be reflecting 'function bias'.*

3. *The random errors may be quite irregular, not at all 'normal'.*

4. *Some points may reflect changing conditions; after large changes in an influential factor the system may not yet have come to equilibrium under the new conditions.*

They note: *Point 21 was the last piece of data to be taken and reflects* **a jump** *in x_1 (from 56 to 70) that might signal lack of equilibrium, but removal of 21 has not sufficed to make everything satisfactory.*

After 8 passes (runs), they decided to eliminate point 21 and drop x_3 from the model. This action provides them with highest F and R^2 statistics. When x_3 has been dropped, they observe that there are three sets of near replicates in the data set. These are observations (5,6,7,8), (10,11,12,13,14) and (15,16,17,18,19). With this extra information they found a significant lack of fit which they suppose to come from either choosing a wrong equation or the presence of some **bad** data points.

They then take a longer look at the data in the order of appearance of the **successive** points. They add: *It is distressing to see y drop from 42 to 37 with no change in x_1 or x_2 (points 1 and 2). It is also upsetting to see y hold at 37 (point 3) while x_1 drops from 80 to 75. Some light begins to appear on scanning the next five points, since x_1 is steady and y drops (in two days) to 18 and then holds fairly steady for four days.* **There appears to be a lag of about one day in the response to a change in x_1.**

At this stage, they finally conclude that: *when x_1 was above 60 the plant required about a day to come to equilibrium. We decide, therefore, to drop points 1, 3, 4 and 21 permanently, since they correspond to transient states that cannot possibly be represented by the same equation as the 'lined-out' points.*

After nearly 43 runs they finally chose the following equation which best represents the remaining 17 points

$$\hat{y} = -15.6 - 0.006x_1 + 0.53x_2 + 0.0068x_1^2 \tag{3.4}$$

They still ask if observation 2 (which is now far away from the others in the x_i's space) should be retained. But computation without this datum only changes the equation slightly and they decide to keep it. It is interesting to note that the 4 bad points were not detected by using formal statistical tests, but by inspection of the raw data.

Observations (1,3,4,21) are declared *outliers* by the authors. The data, now called the **stack loss data**, takes a new identity: **a real data set with four outliers**.

Still not happy with what they have found, they add: *Nor do we think it worthwhile to press further. The data have been handled rather more rudely than is permissible without consultation with those who produced them. But the plant is long since shut down, and the process surely obsolete. Our methodological points have been made.*

The authors returned to the problem for the last time on page 138 and recommend yet another model for the data by clumping some data points. They point out that *this is the fault of the data* and only x_1 effects y with coefficient 0.97. After removal of the four transitional points we have but **six** really different conditions, with points 5–8, 10–14, and 15–19 judged to be local nests with their error. The six remaining points are well fitted by:

$$\hat{y} = -41.2 + 0.97x_1 \tag{3.5}$$

They add: *We have overdone this study, hoping that the readers will understand our purposes and that they will regard charitably our final reduction of the problem almost to absurdity.* With this declaration one might well have hoped that the authors have finally slaughtered the guinea pig.

In spite of their pioneering analysis, which became a milestone for the next twenty years in the area of data analysis, Daniel and Wood were unable to find an appropriate model for the data. However, thanks to them, Brownlee's data became known to many statisticians who never read the original book by Brownlee in which the 'phoney-real' data (Stigler, 1977) were presented as a simple example to illustrate the use of multiple regression.

4 Robust Era and the Resurrection of the Guinea Pig

Daniel and Wood acknowledged the fact that the data have been handled rather rudely and without consultation with the originators of the data. However, and somewhat ironically, the data set find a new life with the innovation of robust methods into multiple linear regression. Following the basic methodology for Monte Carlo studies of robust estimators which were established at Princeton (1970–1971), Andrews (1974) presented his *sine*-estimate for the problem of multiple linear regression and with his method of estimation he gave the following equation for the stack loss data

$$y = -37.2 + 0.82x_1 + 0.52x_2 - 0.07x_3. \tag{4.6}$$

He remarks that: *Most researchers do not have the insight and perseverance of these authors (Daniel and Wood)*,and concluded that *the robust fitting procedure (sine-estimate) has* **immediately** *and* **routinely** *led to the identification of 4 questionable points. The fit is independent of these points.*

This feature of a robust procedure; that is, the automatic detection of outliers, impressed many statisticians. For example:

Hocking (1976): *Andrews proposed robust methods for multivariate regression and provided an illustration using an example from Daniel and Wood. It is of interest to note that both references reach, essentially, the same conclusions.*

Hogg (1979b): *It is very impressive to note that the robust method on all 21 points provides essentially the same estimates as does least squares using the 17 observations, with the 4 bad points set aside.*

Hogg (1979a): *Moreover, the robust methods found these points automatically without having the insight of Daniel and Wood.*

Mallows (1979): *The use of robust methods leads to considerable gains in efficiency, because many fewer computer runs are needed in the development of adequate models. Compare, for example, the one-pass robust regression fit of the*

Brownlee Stack Loss Data that was obtained by Andrews, which identified the same four outliers that Daniel and Wood decided to omit after eight runs of a least squares program.

Stirling (1984): *Andrews showed that robust estimation of the parameters showed up four outliers that had been found after detailed analysis by Daniel and Wood.*

Rivest (1989): *L'analyse du stack loss data constitue le premier exemple concret où une simple analyse robuste amène des résultats semblables à une analyse poussée à l'aide de la méthode des moindres carrés.*

Nietzsche (1883): *Also sprach Zarathustra.*

In 1964 Peter Huber presented the robust estimation method for the estimation of a location parameter. In 1968 Huber's method was extended to the regression situation. However, with Andrew's 1974 paper, robust data analysis reached a turning point and became automatic and routine procedure in multiple regression. The stack loss data became its new born plastic guinea pig on which the battery of newly invented methodology were to be tried. The four questionable observations, called by Atkinson (1985) the *canonical set of observations (1,3,4,21)*, were to be rediscovered in at least 22 distinct articles between 1974 to 1994. Moreover, the least squares method found itself under constant fire because it could not detect the four questionable outliers in a routine and automatic way.

In what follows we try to put forward the result of two decades of work on the stack loss data, work which created a feeling of uneasiness in the users due to the multiplicity of methods in the area of multiple linear regression.

5 Multiplication of the Analyses

When a statistician proposes a new method, or extends or modifies an old one, the proposer is obliged to the credibility of the method on a known data set. For some statisticians the best method is the one which detected the 'four known outliers' (*It is a paradigm of robust estimation that observations (1,3,4,21) are outliers*, Atkinson, 1982a). Though some statisticians never abandoned the idea of either one or no outliers, for others, if a method detected more than four outliers in the stack loss data it is even more respectable. The intriguing consequence of this research is the open question of why, after all these analyses, the union of all outliers found by different scientists leaves only 5 inliers in a set of data consisting of exactly 21 observations.

Each method introduced in the regression market brought with it a brand new equation. Since the 'true' equation for the stack loss data was not known the 'best' equation was the most recently published. After all how could any one reject the latest one!

To give an idea of the diversity of the equations found for the same data set, we display the equations taken from this impressive literature with three explanatory variables included in the model. They are listed by decreasing order of the constant term. Since all these equations describe the same phenomena, it is natural to ask, which one, if any, represents the true state of problem!

1. $\hat{y} = -43.9 + 0.883x_1 + 0.836x_2 - 0.105x_3$
   (Aitkin and Wilson, 1980) Mixture Model

2. $\hat{y} = -43.70 + 0.889x_1 + 0.817x_2 - 0.1071x_3$
   (Daniel and Wood, 1971) Least Squares without Datum 21

3. $\hat{y} = -42.83 + 0.93x_1 + 0.63x_2 - 0.10x_3$
   (Ruppert and Carroll, 1980) L-Estimate

4. $\hat{y} = -42.8 + 0.81x_1 + 0.89x_2 - 0.09x_3$
   (Moberg, Ramberg and Randles, 1980) Hill-Holland version of Sine–Estimate

5. $\hat{y} = -42.6 + 0.955x_1 + 0.560x_2 - 0.108x_3$
   (Aitkin and Wilson, 1980) Mixture Model

6. $\hat{y} = -42.45 + 0.9566x_1 + 0.5556x_2 - 0.1088x_3$
   (Chambers and Heathcote, 1981) Least Squares without Data 4 and 21

7. $\hat{y} = -42.13 + 0.817x_1 + 1.016x_2 - 0.126x_3$
   (Staudte and Sheather, 1990) Robust Weighted Least Squares Based on DFITS

8. $\hat{y} = -42.096 + 0.983x_1 + 0.660x_2 - 0.153x_3$
   (Antoch and Jureckova, 1985) L-Estimate

9. $\hat{y} = -42.003 + 0.931x_1 + 0.635x_2 - 0.114x_3$
   (Antoch and Jureckova, 1985) L-Estimate

10. $\hat{y} = -41.990 + 0.93352x_1 + 0.61946x_2 - 0.11278x_3$
    (Rey, 1977) Sine-Estimate

11. $\hat{y} = -41.78 + 0.9529x_1 + 0.5459x_2 - 0.1119x_3$
    (Chambers and Heathcote, 1981) Tukey's Biweight without Data 4 and 21

12. $\hat{y} = -41.749 + 0.7995x_1 + 1.0639x_2 - 0.1310x_3$
    (Wiens, 1992) Bounded Influence Estimation

13. $\hat{y} = -41.551 + 0.93911x_1 + 0.58026x_2 - 0.11295x_3$
    (Rey, 1977) Sine-Estimate

14. $\hat{y} = -41.19 + 0.811x_1 + 1.010x_2 - 0.133x_3$
    (Osborne, 1985) Huber's M-Estimate

15. $\hat{y} = -41.17 + 0.8133x_1 + 1.000x_2 - 0.1324x_3$
    (Birkes and Dodge, 1993) Huber's M-Estimate

16. $\hat{y} = -41.07 + 0.7962x_1 + 1.0562x_2 - 0.1355x_3$
(Wiens, 1992) Huber's M-Estimate

17. $\hat{y} = -41.0 + 0.83x_1 + 0.91x_2 - 0.13x_3$
(Ruppert and Carroll, 1980) Huber's M-Estimate

18. $\hat{y} = -40.93 + 0.84x_1 + 0.80x_2 - 0.105x_3$
(Staudte and Sheather, 1990) Robust Weighted Least Squares Based on DFITS

19. $\hat{y} = -40.90 + 0.852x_1 + 0.865x_2 - 0.128x_3$
(Welsh, 1987) L-Estimate

20. $\hat{y} = -40.82 + 0.84x_1 + 0.54x_2 - 0.04x_3$
(Lange, Little and Taylor, 1989) MLE when Student Errors

21. $\hat{y} = -40.79 + 0.851x_1 + 0.869x_2 - 0.129x_3$
(Welsh, 1987) L-Estimate

22. $\hat{y} = -40.71 + 0.81x_1 + 0.97x_2 - 0.13x_3$
(Lange, Little and Taylor, 1989) MLE when Student Errors

23. $\hat{y} = -40.66 + 0.750x_1 + 1.201x_2 - 0.143x_3$
(Osborne, 1985) Huber's M-Estimate

24. $\hat{y} = -40.62 + 0.6861x_1 + 1.312x_2 - 0.1273x_3$
(Birkes and Dodge, 1993) Ridge

25. $\hat{y} = -40.56 + 0.7668x_1 + 1.129x_2 - 0.1392x_3$
(Chambers and Heathcote, 1981) Tukey's Biweight

26. $\hat{y} = -40.37 + 0.72x_1 + 0.96x_2 - 0.07x_3$
(Ruppert and Carroll, 1980) L-Estimate

27. $\hat{y} = -40.19 + 0.825x_1 + 0.827x_2 - 0.112x_3$
(Osborne, 1985) Huber's M-Estimate

28. $\hat{y} = -40.16 + 0.8155x_1 + 0.888x_2 - 0.1202x_3$
(Birkes and Dodge, 1993) Non-Parametric

29. $\hat{y} = -40.07 + 0.86x_1 + 0.75x_2 - 0.12x_3$
(Lange, Little and Taylor, 1989) MLE when Student Errors

30. $\hat{y} = -39.95 + 0.80x_1 + 0.90x_2 - 0.11x_3$
(Hettmansperger and McKean, 1977) R-Estimate

31. $\hat{y} = -39.919 + 0.7156x_1 + 1.2953x_2 - 0.1521x_3$
(Brownlee, 1960) Least Squares

32. $\hat{y} = -39.70 + 0.83x_1 + 0.58x_2 - 0.06x_3$
(Hettmansperger and McKean, 1977) Least Absolute Deviations

33. $\hat{y} = -39.33 + 0.8288x_1 + 0.7590x_2 - 0.1087x_3$
(Wiens, 1992) Huber's M-Estimate

34. $\hat{y} = -39.25 + 0.75x_1 - 0.50x_2 + 0.0x_3$
(Tichavsky, 1991) Least Median of Squares

35. $\hat{y} = -39.21 + 0.8297x_1 + 0.7512x_2 - 0.1088x_3$
(Chambers and Heathcote, 1981) Huber's M-Estimate

36. $\hat{y} = -39.13 + 0.85x_1 + 0.66x_2 - 0.10x_3$
(Lange, Little and Taylor, 1989) MLE when Student Errors

37. $\hat{y} = -38.82 + 0.8326x_1 + 0.7174x_2 - 0.1075x_3$
(Wiens, 1992) Bounded Influence Estimation

38. $\hat{y} = -38.805 + 0.82643x_1 + 0.64760x_2 - 0.08577x_3$
(Rey, 1977) Least Power (p=1.2)

39. $\hat{y} = -38.69 + 0.9014x_1 + 0.5039x_2 - 0.1047x_3$
(Chambers and Heathcote, 1981) Huber's M-Estimate without Data 4 and 21

40. $\hat{y} = -38.680 + 0.83436x_1 + 0.71358x_2 - 0.10941x_3$
(Lenth, 1978) Huber's M-Estimate

41. $\hat{y} = -38.62 + 0.85x_1 + 0.49x_2 - 0.07x_3$
(Lange, Little and Taylor, 1989) MLE when Student Errors

42. $\hat{y} = -38.589 + 0.82594x_1 + 0.69431x_2 - 0.09961x_3$
(Rey, 1983) Fair's M-Estimate

43. $\hat{y} = -38.51 + 0.85x_1 + 0.67x_2 - 0.11x_3$
(Hettmansperger and McKean, 1977) R-Estimate

44. $\hat{y} = -38.50 + 0.85x_1 + 0.49x_2 - 0.07x_3$
(Lange, Little and Taylor, 1989) MLE when Student Errors

45. $\hat{y} = -38.3 + 0.785x_1 + 0.591x_2 - 0.005x_3$
(Aitkin and Wilson, 1980) Mixture Model

46. $\hat{y} = -38.158 + 0.83800x_1 + 0.66290x_2 - 0.10631x_3$
(Rey, 1977) Huber's M-Estimate

47. $\hat{y} = -38.1 + 0.85x_1 + 0.56x_2 - 0.09x_3$
(Moberg, Ramberg and Randles, 1980) Adaptive Regresssion Procedure

48. $\hat{y} = -38.0 + 0.794x_1 + 0.588x_2 - 0.006x_3$
(Aitkin and Wilson, 1980) Mixture Model

49. $\hat{y} = -38.0 + 0.85x_1 + 0.72x_2 - 0.13x_3$
(Hettmansperger, 1984) R-Estimate

50. $\hat{y} = -37.9 + 0.78x_1 + 0.57x_2 - 0.06x_3$
(Moberg, Ramberg and Randles, 1980) Hill–Holland Version of Sine–Estimate without Data 1, 3, 4 and 21

51. $\hat{y} = -37.72 + 0.8529x_1 + 0.4610x_2 - 0.07372x_3$
(Stirling, 1984) M-Estimate

52. $\hat{y} = -37.6 + 0.80x_1 + 0.58x_2 - 0.07x_3$
(Andrews, 1974) Least Squares without Data 1, 3, 4 and 21

53. $\hat{y} = -37.35 + 0.8301x_1 + 0.4918x_2 - 0.0711x_3$
(Chambers and Heathcote, 1981) Huber's M-Estimate without Data 1, 3, 4 and 21

54. $\hat{y} = -37.334 + 0.81018x_1 + 0.54199x_2 - 0.07037x_3$
(Rey, 1977) Sine-Estimate

55. $\hat{y} = -37.2 + 0.82x_1 + 0.52x_2 - 0.07x_3$
(Andrews, 1974) SINE-Estimate

56. $\hat{y} = -36.488 + 0.84882x_1 + 0.43577x_2 - 0.079419x_3$
(Lenth, 1978) Sine-Estimate

57. $\hat{y} = -36.28 + 0.8383x_1 + 0.4420x_2 - 0.0751x_3$
(Chambers and Heathcote, 1981) Extension of Least Squares without Data 4 and 21

58. $\hat{y} = -35.9 + 0.82x_1 + 0.43x_2 - 0.07x_3$
(Moberg, Ramberg and Randles, 1980) Least Median of Squares without Data 1, 3, 4, 21

59. $\hat{y} = -35.84 + 0.7470x_1 + 0.5675x_2 - 0.0493x_3$
(Chambers and Heathcote, 1981) Extension of Least Squares

60. $\hat{y} = -34.5 + 0.714x_1 + 0.357x_2 + 0.000x_3$
(Rousseeuw and Leroy, 1987) Least Median of Squares

Although most of these methods revealed the same four outliers (1, 3, 4, 21), a closer look of these analyses shows that there are in fact at least 26 distinct sets of detected outliers. We provide below the chronological list of these discoveries (without repetition).

1. **21** (Draper and Smith, 1966) Least Squares

2. **1, 3, 4, 21** (Daniel and Wood, 1971) Least Squares + Plots

3. **2, 3, 4, 13, 14, 20, 21** (Denby and Mallows, 1977) Huber's M-Estimate

4. **1, 2, 3, 4, 21** (Andrews and Pregibon, 1978) Andrews and Pregibon's Diagnostic

5. **1, 3, 4, 13, 21** (Lenth, 1978) Sine-Estimate + Huber's M-Estimate
6. **1, 2, 4, 21** (Cook, 1979) Cook's Distance
7. **4, 9, 21** (Ruppert and Carroll, 1980) L-Estimate
8. **1, 3, 9, 21** (Ruppert and Carroll, 1980) L-Estimate
9. **4, 21** (Chambers and Heathcote, 1981) Extension of Least Squares + Transformation
10. **1, 2, 3** (Gray and Ling, 1984) Clustering on the Modified Hat Matrix
11. **1, 2, 3, 4** (Gray and Ling, 1984) Clustering on the Modified Hat Matrix
12. **4, 9, 17, 21** (Antoch and Jureckova, 1985) L-Estimate
13. **2, 4, 9, 17, 21** (Antoch and Jureckova, 1985) L-Estimate
14. **1, 2, 4, 17, 21** (Carroll and Ruppert, 1985) Pseudo-MLE Diagnostics
15. **2, 3, 4, 16, 21** (Carroll and Ruppert, 1985) Bounded Influence Transformation
16. **2, 3, 4, 21** (Carroll and Ruppert, 1985) Bounded Influence Transformation
17. **1, 3, 4, 17, 21** (Narula and Wellington, 1985) Least Absolute Deviations
18. **3, 4, 9, 21** (Welsh, 1987) L-Estimate
19. **3, 4, 21** (Welsh, 1987) L-Estimate
20. **No outliers** (Parker, 1988) Least Absolute Deviations + Transformation
21. **1, 2, 21** (Cléroux, Helbling and Ranger, 1989) Diagnostics Based on Multiple Correlation
22. **1, 3** (Gray, 1989) Internal Norm
23. **2, 21** (Gray, 1989) Internal Norm
24. **1, 2, 3, 21** (Hadi, 1992) Minimum Volume Ellipsoid
25. **1, 2, 3, 4, 8, 13, 14, 20, 21** (Hawkins, 1994) Least Trimmed Squares
26. **1, 2, 3, 10, 15, 16, 18, 19, 21** (Hawkins, 1994) Minimum Covariance Determinant

If we take the union of these sets of outliers we found as final outliers the observations **1, 2, 3, 4, 8, 9, 10, 13, 14, 15, 16, 17, 18, 19, 20, 21**, and as inliers the observations **5, 6, 7, 11, 12**.

6 The Origin of the Data

Some light on the origin of these data appeared with a remark of Barnard (1982, p. 27) on Atkinson's (1982a) paper read before the Royal Statistical Society at the meeting organized by the Research Section on Wednesday, October 14th, 1981:

There are not many people, I guess, by now who know the origin of the stack loss data that were used by Brownlee. They were, in fact, produced during the war when Brownlee, as a chemist, was sent to the Statistical Section of the Ministry of Supply to learn about statistics. It was necessary, among other things, to discuss multiple regression. You will find in the pamphlet issued by the Ministry of Supply (which later grew into Brownlee's book) that the data in question were cited as an instance of how utterly useless it is to try to analyze past records that were not generated by experimental design. It is quite remarkable that, so far as I know, in all discussions of this set of data, this fact is not mentioned, in spite of the fact that no discussion of possible outliers can be regarded as adequate that does not bring in the qualitative context as well as purely quantitative considerations.

Atkinson replied to Barnard as follows:

In Brownlee (1948) the comments on the difficulties of interpreting observational data, which Professor Barnard mentions, follow the analysis of data on sticks of cordite. A set of stack loss data is analyzed on pp. 72–83. Records of 139 days are mentioned, although the full data are not given. The three explanatory variables are different from those in Brownlee (1965), with air temperature replacing **gas flow**.

There is thus another data set with 139 observations which is also called *stack loss data*. One may ask what is the relationship between these two versions of the stack loss data, both introduced by the same author, once in a pamphlet issued by Ministry of Supply which later grew into Brownlee's first book, and once in his 1960 book. Our hypothesis is that the 1960 stack loss data set is in fact a subset of 1946 data set. To solve this conundrum we first present the data with its description which appeared in Brownlee's (1946) book entitled *Industrial Experimentation* (Atkinson, in his comment, refers to the third edition).

Brownlee (1946) in the Chapter called *Multiple Correlation* introduced the example as follows (p. 55):

In the example we shall consider the stack loss from a Pressure Oxidation Plant for making nitric acid (about 55%–60%) by the oxidation of ammonia with air. The resulting nitric gases are absorbed by being passed upwards through a bubble cap absorption column, nitric acid being circulated down the column. The absorption is exothermic; that is, gives rise to the liberation of heat, and the temperature tends to rise. The absorption of the nitric gases is less efficient at higher temperatures, so to keep the temperature down water is circulated through cooling coils in the column. The column also tends to cool itself by radiation, etc., to its surroundings. A small quantity of the nitric gases fails to be absorbed and escapes by the stack: this we will regard as the dependent variable. Its minimization is, of course, of prime economic importance and we will first *determine its dependence upon the two independent variables air temperature, x_a, and cooling water temperature, x_w. In a subsequent section we will consider a third independent variable, strength of the nitric acid being circulated in the absorption column.*

He continues to add (p. 56):

It makes the computation lighter, particularly if we are not using a calculating machine, to use arbitrary zeros and to multiply by 10 or 100, etc., if necessary, to remove the decimal points. Thus in the present instance the variables used are actually:

$y = 10$ (*stack loss in* $gm/M^3 - 3.0$).

$x_w =$ *water temperature in* $^0\,C - 20$.

$x_a =$ *air temperature in* $^0\,F - 50$.

For each set of observations on the three variables, y, x_w, and x_a, we form the three pairs of products, yx_w, x_wx_a, and x_ay. Thus we have a table of 139 sets as below:

Date	y	x_w	x_a	x_wx_a	x_ay	x_wy
30.12.43	–20	–6	–11	66	220	120
29	–9	1	–3	–3	27	–9
28	–18	1	–2	–2	36	–18
27	–6	–1	–1	1	9	6
25	–3	2	–3	–6	6	–6
22	–5	–5	–10	50	50	25
22	–20	2	–20	–40	400	–40
21	+19	5	+14	70	266	+95
18	–4	–1	–14	14	56	4
.	.	.	.	.	.	.
.	.	.	.	.	.	.
.	.	.	.	.	.	.
	1479	406	1	3533	17340	12797

Table 2: Brownlee's (1946, p. 56) data

Only this 9 row segment of a larger table is published by Brownlee.

In what follows an attempt is made to bring convincing arguments for proving the similarity of these two data sets and hence to show that the 21 observations were taken in an arbitrary (or perhaps carefully selected) manner from the 139 observations. A possible reason is confidentiality of publishing data, especially during the war. Let us examine different aspects of the problem:

- **The plant**

  In presenting the two data sets Brownlee describes a plant which produces nitric acid by the oxidation of the ammonia with air. The two descriptions of the fabrication process seem to be identical. In 1946 Brownlee provides more detailed information on the process of making nitric acid from the oxidation of ammonia than in 1960, but the changes in the description of the plant from 1960 to 1965 are much more revolutionary than the one from 1946 to 1960. Probably the two sets of stack loss data were taken from the same plant.

- **Successive days**

Although, the 1946 data were not published in full, sums, sums of squares and products are available. Notice that the data are well ordered in time, but that the days are not successive! There are no data for some days as for the 26th or for the 24th of December, and the 22nd of December is twice represented (and Christmas day once!). Moreover the data are not given in increasing ordering but by decreasing (the first data in the list comes from the end of the year 1943 and the 139th data may come from the beginning of the month of August of the same year, assuming about one measure is taken daily)! If the stack loss data of 1960 were collected and published on a similar manner, the arguments based on the hypothesis of *successive days* used by Daniel and Wood (1971) for detecting outliers is not correct.

- **The variables**

Now let us have a look at the name of the variables. In 1946 we have the following variables:

x_a: air temperature; x_w: water temperature; x_s: strength of acid; y_{46}: stack loss.

In 1960 we have the following variables:

x_1: air flow; x_2: cooling water inlet temperature; x_3: acid concentration; y_{60}: stack loss.

The similarity between these names is strong. The variables y_{60} and y_{46} have exactly the same name. The variables x_3 and x_2 of 1960 correspond clearly respectively to the variables x_s and x_w of 1946. Perhaps one can see a minor difference between x_1 and x_a.

While this may reflect the evolution of scientific language between 1946 and 1960, as can already be seen for the variable x_s of 1946 which is called *strength of acid* and its corresponding variable x_3 of 1960 called *acid concentration*. We are therefore tempted to think that these two sets of variable are the same.

- **Measurements**

To see that the two variables x_1 and x_2 of Table 1 and x_a and x_w of Table 2 represent the same variables in both occasion and their differences is the result of timing (period) in which the data is recorded, let us use Brownlee's description for the first two explanatory variables. Brownlee (1946) notes that x_w = water temperature in $^0C - 20^0C$ and that x_a = air temperature in $^0F - 50^0F$. Now let: $x_1 = x_a + 50$ and $x_2 = x_w + 20$.

It would just mean that in 1960 Brownlee did not subtract 20, respectively 50, from the measurements of temperature of air, respectively of water. In doing so we obtain for x_1 and x_2 the following values for the 9 available observations taken in the winter of 1943.

Obs.	y	x_2	x_1
1	−20	14	39
2	−9	21	47
3	−18	21	48
4	−6	19	49
5	−3	22	47
6	−5	15	40
7	−20	22	30
8	19	25	64
9	−4	19	36

Table 3: Transformed Brownlee's values of 1946

Take now, for example the observation number 8 of Table 3 (1946) for which the observed value of $y = 19$ (the only positive value in this table). The corresponding values for y are $x_1 = 64$, and $x_2 = 25$. In Table 1 (1960), for the stack loss of $y = 19$ we find for $x_1 = 62$ and $x_2 = 24$.

Another argument which supports this assumption can be found in the computation of some basic statistics. Brownlee (1946) didn't give the 139 data in full, but at the end of p. 56, he gave the sums of squares of the observations for the variables x_a, x_w and y_{46}. It allowed us to calculate the results given in Table 4.

1946 (139 observations)	1960 (21 observations)
$mean(x_a) = 0.0072$	$mean(x_1) = 60.43$
$mean(x_w) = 2.92$	$mean(x_2) = 21.095$
$stdev(x_a) = 8.73$	$stdev(x_1) = 9.17$
$stdev(x_w) = 4.06$	$stdev(x_2) = 3.161$
$corr(x_a, x_w) = 0.7222$	$corr(x_1, x_2) = 0.786$

Table 4: Basic statistics for 1946 and 1960 data sets

If the 21 observations were (random) part of the 139, we could expect that two corresponding variables (without taking account the transformation) would present similar means and similar standard deviations. Since

$mean(X+b) = mean(X)+b$ and that $stdev(X+b) = stdev(X)$ we see that our hypothesis of stating that the 1960 data set is a subset of 1946 one cannot be rejected. The standard deviation of x_a is very close to the standard deviation of x_1 and the standard deviation of x_w is very close of the standard deviation of x_2. The mean of x_w is about 20 less than the mean of x_2 and the mean of x_a is about 50 less than the mean of x_1 (in fact 60, due perhaps to the fact that the 21 observations come mostly from the summer of 1943). Moreover the correlation between x_a and x_w is not far away from the correlation between x_1 and x_2. Since correlation is invariant under linear transformation, we are comforted in our idea.

- **Repeated Transformations and Confidentiality**

  The fact that Brownlee transformed the variables between 1946 and 1960 can be explained for the ease of computation but also for the confidentiality, especially relevant as this data set was produced and collected during the war. In fact Brownlee seems to transform the variables quite often. For example in the same book *Industrial Experimentation* he talks again about a plant which produces nitric acid from ammonia in a Chapter called *Miscellaneous Aspects of Analysis of Variance.* In this occasion, the response variable is clearly expressed in a completely different way (p. 127 of the fourth version, 1949):

  The data in Table 12.9 relates to the conversion efficiency measured daily of a pressure oxidation plant for the manufacture of nitric acid from ammonia. After 7 days the plant is stopped temporarily and the catalyst subjected to a re-vivifying treatment. A constant has been subtracted from all the results and the decimal point removed.

Run No.	1	2	3	4	5	6	7	Totals
Day 1	60	64	14	30	67	72	35	342
2	62	63	46	41	42	38	35	327
3	58	63	46	40	43	46	33	329
4	52	36	39	41	38	47	46	299
5	31	34	42	37	37	38	47	266
6	23	32	43	17	33	60	47	255
7	26	27	57	12	26	41	38	227
Totals	312	319	287	218	286	342	281	2045

Table 5: Table 12.9 of Brownlee (1949, p. 127)

- **Jumps and Outliers**

  The presence of certain *jumps* detected by Daniel and Wood (1971) without any analysis and only by looking carefully at the data set that led them to find the four *outliers* could well be explained by the way that Brownlee has selected the 21 observations out of the full data set. For example he could have chosen some data produced in winter (in December), some in autumn (in October) and some mostly in summer (in August), and the changing conditions may be the cause of these jumps. Moreover, the multiplicity of the outliers found in such a small data set could be another consequence of the many missing pieces in the data set. It was really *the fault of the data* as they stated loudly.

- **Regression Lines with 1946 Data Set**

  At the end of their analysis with observations 2,9,20 and the mean values of 5–8, 10–14 and 15–19 Daniel and Wood find the equation (3.5) with only x_1 as independent variable to be $\hat{y} = -41.2 + 0.97x_1$. If we take from 1946 data set observations 2,4,5,6,8 and 9 the resulting linear equation with the transformed data would be $\hat{y} = -41.2 + 0.85x_1$ which is strangely close to their findings. Moreover with two independent variables of 1946 data set and the observations 1,2,5,6,8 and 9 we find $\hat{y} = -55.3 + 0.65x_1 + 1.13x_2$ which is similar to the findings of Brownlee himself of 1960 with 21 observations after dropping the x_3 from the model; that is, $\hat{y} = -50.36 + 0.67x_1 + 1.30x_2$.

Putting into consideration the foregoing different aspects of the problem in question, our conclusion is that the 21 observations published in 1960 are in fact a part of the 1946 data set with 139 observations. In other words if a bird walks like a crow, looks like a crow and sings like a crow, one may conclude that it is a crow.

In such a situation we may describe the stack loss data 'phoney-real' data, because though it has some connection with the real world the liasion is murky and could fool many scientists. The stack loss data has well served its original author as a illustrative example, but it probably would have been wiser not to use it as a basis of comparison for all regression methods.

7 Final Comments

Our search for the original Memorandum which was the basis of Brownlee's (1946) first book continues. We have written to many people whom we thought may have some information regarding this data set. So far no success beyond the fact that it became clear to us that all data sets taken during the war on explosive matters were kept confidential up to 1993. Instead of drawing any minor or major conclusion from this paper, we decided to furnish the reader with some of our personal communications with those who could tell us something about this data set.

Barnard (personal communication): *This refers to his secondment to SR 17 during the war from his earlier employment as chemist in the Royal Ordnance factory at Bishopton, near Glasgow. I think he collected the stack loss data from operations in that factory; and I have what is now a vague recollection that the stack in question would have been a tall chimney.*

Daniel (personal communication): *Brownlee published the data in 1945 or 1946, not in 1960. And they were probably taken during the war and dug out by him later. He once told me, probably in the fifties, that the plant had been long since demolished.*

Irving (1994): *We can always imagine a better detail than the one we can remember. The correct detail is rarely, exactly, what happened: the most truthful detail is what* **could** *have happened, or what* **should** *have.*

Draper (personal communication): *If we had known at the time that so much effort was going to go into re-analyzing these data, we no doubt would have been very very specific and written a lot more!*

Daniel (personal communication): *The last chapter in the 'Fitting' book gave some indications of what we thought a real job of fitting an equation should look like. We failed. Seventeen data in the final equation: Ridiculous. We did after all get in touch with one of the originators of the data. He liked it and rightly remarked: 'Twenty years too late.'*

Larntz (1989): *I doubt that this will be the last time we see the stack loss data of Brownlee.*

Shakespeare: *Much ado about nothing.*

Acknowledgements

I would like to thank the following people for providing us with useful documents: Mr. C.A. Rowsell of Ministry of Defence (U.K.), Mr. W.A. Noblett of Cambridge University Library, Mrs. Kay Roper of University of Cambridge, Dept. of Pure Mathematics, Mrs. S.A. Riseborough of HMSO, Professor Marc Hallin of Université Libre de Bruxelles, and W.J.J. Rey. My special gratitude goes to Cuthbert Daniel, George Barnard, David Cox and Norman Draper for their kind letters and personal discussions; to Hans Nyquist, Joe Whittaker, Steve Fienberg, Stephan Senn, Stephan Stigler for reading the paper and providing helpful comments. Finally, I am specially indebted to Valentin Rousson, my assistant, who collected and summarized all the papers with great care.

References

[1] Aitkin, M. and Wilson, G.T. (1980): Mixture models, outliers, and the EM algorithm. *Technometrics* **22** 325–331.

[2] Andrews, D.F. (1974): A robust method for multiple linear regression. *Technometrics* **16** 523–531.

[3] Andrews, D.F. and Pregibon, D. (1978): Finding the outliers that matter. *J. Roy. Statist. Soc. B* **40** 85–93.

[4] Antille, G. and El May, H. (1992): The use of slices in the LMS and the method of density slices: Foundation and comparison. In *Computational Statistics* (Y. Dodge and J. Whittaker, eds.), 441–445. Physica- Verlag.

[5] Antoch, J. and Jureckova, J. (1985): Trimmed least squares estimator resistant to leverage points. *CSQ* **4** 329–339.

[6] Atkinson, A.C. (1980): Examples showing the use of two graphical displays for the detection of influential and outlying observations in regression. In *Compstat 1980* (M.M. Barritt and D. Wishart, eds.), 276–282. Physica-Verlag, Vienna.

[7] Atkinson, A.C. (1981): Two graphical displays for outlying and influential observations in regression. *Biometrika* **68** 13–20.

[8] Atkinson, A.C. (1982a): Regression diagnostics, transformations and constructed variables. *J. Roy. Statist. Soc. B* **44** 1–36.

[9] Atkinson, A.C. (1982b): Robust and diagnostic regression analyses. *Commun. in Statist., Theory and Methods* **11** 2559–2571.

[10] Atkinson, A.C. (1985): *Plots, Transformations, and Regression*, 129–136, 261–262 and 266–268. Clarendon Press, Oxford.

[11] Atkinson, A.C. (1986): Diagnostic tests for transformations. *Technometrics* **28** 29–37.

[12] Atkinson, A.C. (1986): Masking unmasked. *Biometrika* **73** 533–541.

[13] Atkinson, A.C. and Weisberg, S. (1991): Simulated annealing for the detection of multiple outliers using least squares and least median of squares fitting. In *Directions in Robust Statistics and Diagnostics - Part I.* (W. Stahel and S. Weisberg eds.), 7–20. Springer-Verlag, New York.

[14] Bassett, J.R. and Koenker, R. (1982): An empirical quantile function for linear models. *J. Amer. Statist. Assoc.* **77** 407–415.

[15] Birkes, D. and Dodge, Y. (1993): *Alternative Methods of Regression*, 177–179. John Wiley, New York.

[16] Brownlee, K.A. (1960): *Statistical Theory and Methodology in Science and Engineering*, 491–500. John Wiley, New York.

[17] Carmody, T.J. (1988): Diagnostics for multivariate smoothing splines. *J. Statist. Plann. Inference* **19** 171–186.

[18] Carroll, R.J. and Ruppert, D. (1985): Transformations in regression: A robust analysis. *Technometrics* **27** 1–12.

[19] Chaloner, K. and Brant, R. (1988): A bayesian approach to outlier detection and residual analysis. *Biometrika* **75** 651-659.

[20] Chalton, D.O. and Troskie, C.G. (1992): Q plots, a graphical aid for regression analysis. *Commun. in Statist., Theory and Methods* **21** 625–636.

[21] Chambers, J.M., Cleveland, W.S., Kleiner, B. and Tukey, P.A. (1983): *Graphical Methods for Data Analysis*, 188, 309 and 376. Wadsworth & Brooks/Cole Publishing Company, California.

[22] Chambers, R.L. and Heathcote, C.R. (1981): On the estimation of slope and the identification of outliers in linear regression. *Biometrika* **68** 21–33.

[23] Chatterjee, S. and Hadi, A.S. (1988): Impact of simultaneous omission of a variable and an observation on a linear regression equation. *Comput. Statist. & Data Analysis* **6** 129-144.

[24] Chatterjee, S. and Hadi, A.S. (1988): *Sensitivity Analysis in Linear Regression*, 226–233. John Wiley, New York.

[25] Cheng, K.S. and Hettmansperger, T.P. (1983): Weighted least squares rank estimates. *Commun. in Statist., Theory and Methods* **12** 1069–1086.

[26] Clark, D.I. and Osborne, M.R. (1986): Finite algorithms for Huber's M-estimator. *SIAM, J. Sci. Statist. Comput.* **7** 72–85.

[27] Cléroux, R., Helbling, J.-M. and Ranger, N. (1986): Some methods of detecting multivariate outliers. *CSQ* **3** 177–195.

[28] Cléroux, R., Helbling, J.-M. and Ranger, N. (1989): Influential subsets diagnostics based on multiple correlation. *CSQ* **2** 99–117.

[29] Cook, R.D. (1979): Influential observations in linear regression. *J. Amer. Statist. Assoc.* **74** 169–174.

[30] Crisp, A. and Burridge, J. (1993): A note on the uniqueness of M-estimators in robust regression. *Canadian J. Statist.* **21** 205–208.

[31] Daniel, C. and Wood, F.S. (1980): *Fitting Equations to Data*, (60–82 and 138–139). John Wiley, New York, 2nd edition.

[32] Dempster, A.P. and Gasko-Green, M. (1981): New tools for residuals analysis. *Ann. Statist.* **9** 945–959.

[33] Denby, L. and Mallows, C.L. (1977): Two diagnostic displays for robust regression analysis. *Technometrics* **19** 1–13.

[34] Dodge, Y. (1984): Robust estimation of regression coefficients by minimizing a convex combination of least squares and least absolute deviations. *CSQ* **1** 139–153.

[35] Dodge, Y., Antoch, J. and Jureckova, J. (1991): Computational Aspects of adaptative combination of least squares and least absolute deviations estimators. *Comput. Statist. & Data Analysis* **12** 87–99.

[36] Dollinger, M.B. and Staudte, R.G. (1990): The construction of equileverage designs for multiple linear regression. *Austral. J. Statist.* **32** 99–118.

[37] Draper, N.R. and Smith, H. (1966): *Applied Regression Analysis* 204–215 and 337. John Wiley, New York.

[38] Eubank, R.L. (1988): *Spline Smoothing and Nonparametric Regression*, 338–347. Marcel Dekker, New York.

[39] Flack, V.F. and Flores, R.A. (1989): Using simulated envelopes in the evaluation of normal probability plots of regression analysis. *Technometrics* **31** 219–225.

[40] Gray, J.B. (1989): The internal norm approach to influence diagnostics. *Commun. in Statist., Theory and Methods* **18** 943–958.

[41] Gray, J.B. (1993): Approximating the internal norm influence measure in linear regression. *Commun. in Statist., Simul. and Comput.* **22** 117–135.

[42] Gray, J.B. and Ling, R.F. (1984): K-clustering as a detection tool for influential subsets in regression. *Technometrics* **26** 305–318.

[43] Hadi, A.S. (1992): Identifying multiple outliers in multivariate data. *J. Roy. Statist. Soc. B*, **54** 761–771.

[44] Hampel, F.R., Ronchetti, E.M., Rousseeuw, P.J. and Stahel, W.A. (1986): *Robust Statistics—The Approach Based on Influence Functions*, 330–332. John Wiley, New York.

[45] Hand, D.J., Daly, F., Lunn, A.D., McConway, K.J. and Ostrowski, E. (1994): *A Handbook of Small Data Sets*, 156–157. Chapmann & Hall, London.

[46] Hawkins, D.M. (1980): *Identification of Outliers*, 97–99. Chapman & Hall, New York.

[47] Hawkins, D.M. (1993): The feasible set algorithm for least median of squares regression. *Comput. Statist. & Data Analysis* **16** 81–101.

[48] Hawkins, D.M. (1994): The feasible solution algorithm for least trimmed squares regression. *Comput. Statist. & Data Analysis* **17** 185–196.

[49] Hawkins, D.M. (1994): The feasible solution algorithm for the minimum covariance determinant estimator in multivariate data. *Comput. Statist. & Data Analysis* **17** 197–210.

[50] Hawkins, D.M., Bradu, D. and Kass, V. (1984): Location of several outliers in multiple-regression data using elemental sets. *Technometrics* **26** 197–208.

[51] Hettmansperger, T.P. (1984): *Statistical Inference Based on Ranks*, 266–270. John Wiley, New York.

[52] Hettmansperger, T.P. and McKean, J.W. (1977): A robust alternative based on ranks to least squares in analyzing linear models. *Technometrics* **19** 275–284.

[53] Hogg, R.V. (1979a): An introduction to robust estimation. In *Robustness in Statistics*. (R.L. Launer and G.N. Wilkinson eds.), 1–17. Academic Press, New York.

[54] Hogg, R.V. (1979b): Statistical robustness: one view of its use in applications today. *The American Statistician* **33** 108–115.

[55] Iman, R.C. and Conover, W.J. (1979): The use of the rank transform in regression. *Technometrics* **21** 499–509.

[56] Lange, K.L., Little, R.J.A. and Taylor, J.M.G. (1989): Robust statistical modeling using the T distribution. *J. Amer. Statist. Assoc.* **84** 881–896.

[57] Lenth, R.V. (1978): *A Computational Procedure for Robust Multiple Regression*. Technical Report **53**, Univ. Iowa.

[58] Li, G. (1985): Robust regression. In *Exploring Data Tables, Trends and Shapes*. (D. Hoaglin, F. Mosteller, and J. Tukey eds.), 281–343. John Wiley, New York.

[59] Lye, J.N. (1991): Implications of uncorrelatedness and independence in linear regression analysis. *Commun. in Statist., Theory and Methods* **20** 2407-2430.

[60] Marasinghe, M.G. (1985): A multistage procedure for detecting several outliers in linear regression. *Technometrics* **27** 395–399.

[61] Marazzi, A. (1994): Developpement dun logiciel de modélisation robuste. In *XXVIe Journées de Statistique*.

[62] Moberg, T.F., Ramberg, J.S. and Randles, R.H. (1980): An adaptative multiple regression procedure based on M-estimators. *Technometrics* **22** 213–224.

[63] Narula, S.N. and Wellington, J.F. (1985): Interior analysis for the minimum sum of absolute errors regression. *Technometrics* **27** 181–188.

[64] Narula, S.C. and Wellington, J.F. (1989): On the robustness of the simple linear minimum sum of absolute errors regression. In *Robust Regression - Analysis and Applications*. (K.D. Lawrence, and J.L. Arthur eds.), 129–141. Marcel Dekker, New York.

[65] Neykov, N.M. and Neytchev, P.N. (1991): Least median of squares, least trimmed squares and S estimation by means of BMDP3R and BMDPAR. *CSQ* **4** 281–293.

[66] Osborne, M.R. (1985): *Finite Algorithms in Optimization and Data Analysis*, 267–270. John Wiley, New York.

[67] Osborne, M.R. (1992): An effective method for computing regression quantiles. *IMA, J. of Numerical Analysis* **12** 151–166.

[68] Osborne, M.R. and Watson, G.A. (1985): An analysis of the total approximation problem in separate norms, and an algorithm for the total L1 problem. *SIAM, J. Sci. Statist. Comput.* **6** 410–424.

[69] Parker, I. (1988): Transformations and influential observations in minimum sum of absolute errors regression. *Technometrics* **30** 215–220.

[70] Paul, S.R. (1983): Sequential detection of unusual points in regression. *The Statistician* **32** 417–424.

[71] Paul, S.R. and Fung, K.Y. (1991): A generalized extreme studentized residual multiple-outlier-detection procedure in linear regression. *Technometrics* **33** 339–348.

[72] Portnoy, S. (1987): Using regression fractiles to identify outliers. In *Statistical Data Analysis Based on the L1-Norm and Related Methods.* (Y. Dodge ed.), 345–356. Elsevier Science Publishers B.V., North Holland.

[73] Rey, W.J.J. (1977): *M-Estimators in Robust Regression, a Case Study*. M.B.L.E. Research Laboratory, Brussels.

[74] Rey, W.J.J. (1978): *Robust Statistical Methods.* Lecture Notes in Mathematics #690, 68–72. Springer-Verlag, Berlin.

[75] Rey, W.J.J. (1983): *Introduction to Robust and Quasi-Robust Statistical Methods*, 152–162. Springer-Verlag, Berlin.

[76] Rivest, L.-P. (1988): A new scale step for hubers M-estimators in multiple regression. *SIAM, J. Sci. Statist. Comput.* **9** 164–169.

[77] Rivest, L.-P. (1989): De lunicité des estimateurs robustes en régression lorsque le paramétre déchelle et le paramétre de la régression sont estimés simultanément. *Canadian J. Statist.* **17** 141–153.

[78] Rousseeuw, P.J. and Leroy A.M. (1987): *Robust Regression & Outlier Detection*, 76–78. John Wiley, New York.

[79] Rousseeuw, P.J. and Van Zomeren, B.C. (1990): Unmasking multivariate outliers and leverage points. *J. Amer. Statist. Assoc.* **85** 633–651.

[80] Ruppert, D. and Carroll, R.J. (1980): Trimmed least squares estimation in the linear model. *J. Amer. Statist. Assoc.* **75** 828–838.

[81] Salahuddin and Hawkes, A.G. (1991): Cross-validation in stepwise regression. *Commun. in Statist., Theory and Methods* **20** 1163–1182.

[82] S-Plus for Windows (1993): *Reference Manual, Vol.1*, 16. Statistical Sciences, Inc., Seattle, Washington.

[83] Staudte, R.G. and Sheather, S.J. (1990): *Robust Estimation and Testing*, 215–218 and 222–223. John Wiley, New York.

[84] Stirling, W.D. (1984): Iteratively reweighted least squares for models with a linear part. *Applied Statistics* **33** 7–17.

[85] Tichavsky, P. (1991): Algorithms for and geometrical characterization of solutions in the LMS and the LTS linear regression. *CSQ* **2** 139–151.

[86] Tiku, M.L., Tan, W.Y. and Balakrishnan, N. (1986): *Robust Inference*, 177–178. Marcel Dekker, New York.

[87] Tukey, P.A. (1983): Graphical methods. In *Statistical Data Analysis*. Proceedings of Symposia in Applied Mathematics **28** 8–48. American Mathematical Society.

[88] Welsh, A.H. (1987): The trimmed mean in the linear model. *Ann. Statist.* **15** 20–36.

[89] Wetherill, G.B. (1986): *Regression Analysis with Applications*, 149–153. Chapman & Hall, London.

[90] Wiens, D.P. (1992): A note of the computation of robust, bounded influence estimates and test statistics in regression. *Comput. Statist. & Data Analysis* **13** 211–220.

[91] Wolke, R. and Schwetlick, H. (1988): Iteratively reweighted least squares: algorithms, convergence analysis and numerical comparisons. *SIAM, J. Sci. Statist. Comput.* **9** 907–921.

Additional References

[92] Barnard, G.A. (1982): Contribution to discussion of the paper Regression Diagnostics, Transformations and Constructed Variables, by A.C. Atkinson. *J. Roy. Statist. Soc. B* **44** 1–36.

[93] Brownlee, K.A. (1946): *Industrial Experimentation*. HMSO, London.

[94] Brownlee, K.A. (1948): *Industrial Experimentation*. HMSO, London, 3rd edition.

[95] Brownlee, K.A. (1949): *Industrial Experimentation*. HMSO, London, 4th edition.

[96] Brownlee, K.A. (1965): *Statistical Theory and Methodology in Science and Engineering*. John Wiley, New York, 2nd edition.

[97] Fisher, R.A. (1936): The Use of Multiple Measurements in Taxonomic Problems. *Annals of Eugenics* **7** 179–184.

[98] Hocking, R.R. (1976): The Analysis and Selection of Variables in Linear Regression. *Biometrics* **32** 1–49.

[99] Huber (1964): Robust Estimation of Location Parameter. *Ann. Math. Statist.* **35** 73–101.

[100] Larntz, K., (1989): Contribution to the book reviews. *J. Amer. Statist. Assoc.* **84** 624.

[101] Mallows, C.L. (1979): Robust Methods - Some Examples of their Use. *The American Statistician* **33** 179–184.

[102] Preece, D.A. (1986): Illustrative Examples: Illustratives of what? *The Statistician* **35** 33–44.

[103] Stigler, S.M. (1977): Do Robust Estimators Work with REAL DATA?. *Ann. Statist.* **5** 1055–1098.

When is a p-Value a Good Measure of Evidence?

Michael B. Dollinger*
Pacific Lutheran University, Tacoma

Elena Kulinskaya
La Trobe University, Melbourne

Robert G. Staudte†
La Trobe University, Melbourne

Abstract

Blyth and Staudte (1993, 1995a) introduced the notion of a level-α weight of evidence for an alternative to a hypothesis; it is an estimator of the ideal power function which gives for each experimental outcome a measure of how safe it is to reject the hypothesis in favour of the alternative. It is shown here that for a simple alternative to a one-sided hypothesis regarding a normal mean, the minimum risk level-α measure of evidence approaches the Neyman-Pearson level-α test critical function as the simple alternative moves to infinity. This result is not true for shift parameters of all distributions, as the Huber least-favourable example illustrates.

This result also suggests that for normal data when the p-value is compared with other measures of evidence of its own level, it will be found wanting, except for distant alternatives; plots of the corresponding risk functions confirm this.

Acceptability profiles (Blyth and Staudte (1993,1995a)) are derived from a family of measures of evidence in the same way that confidence intervals are derived from a family of tests. Acceptability profiles derived from p-values and other measures of evidence are compared.

Key words and phrases: Hypothesis testing; estimation; weight of evidence, acceptability profile; likelihood ratio estimator.
AMS 1991 subject classifications: Primary 62A20; secondary 62F03, 62F04, 62F11.

* Michael B. Dollinger would like to thank the Department of Mathematics, Agder College, Kristiansand, Norway, where he was visiting researcher during his work on this paper.

† This research was sponsored in part by a grant from the Australian Research Council.

1 Introduction to Level-α Measures of Evidence

Scientists often carry out a formal test of a hypothesis in order to obtain a measure of evidence (the p-value) against it. While there are numerous critiques of the p-value, it continues to be used because there have not been any obviously superior measures nor critieria for judging its merit as a measure of evidence. Recently Blyth and Staudte (1995a) introduced a method for comparing measures of evidence for an alternative to a hypothesis within the Neyman-Pearson framework (1928). In this setting it is possible to demonstrate that the p-value is only a good measure of evidence against alternatives which are 'distant' from the hypothesis, at least for many standard statistical problems, and moreover there are demonstrably better measures of evidence.

Assume that the data X have a density $p_\theta(x)$ indexed by $\theta \in \Theta$, and that the parameter space Θ is the disjoint union of a hypothesis Θ_0 and an alternative Θ_1, either of which may be simple or composite. A test critical function $T = T(X)$ takes on values 0 or 1 with the interpretation that it equals 1 when it is asserted that $\theta \in \Theta_1$, and 0 when it is asserted that $\theta \in \Theta_0$. (It is assumed that any randomization devices are included in the random vector X.)

In the Neyman-Pearson framework the Type I error of asserting $\theta \in \Theta_1$, when it is not so, is considered to be so important that a bound α is placed on the Type I error probability: $\sup_{\Theta_0} \mathrm{E}_\theta T(X) \leq \alpha$. Subject to this bound, it is desired to find a T which minimizes in some way the Type II error probabilities $\mathrm{E}_\theta[1 - T(X)]$, for $\theta \in \Theta_1$.

The power function of a test T is defined for each θ to be $\Pi(\theta) = \mathrm{E}_\theta T(X)$. It gives the probability of rejecting Θ_0 for each value of θ. Ideally it would equal 0 for $\theta \in \Theta_0$ and 1 for $\theta \in \Theta_1$. While the statistician is interested in the power function of a test T, the scientist is interested in the probability, given $X = x$, that the hypothesis would be rejected in a repetition of the experiment. That is, given what has been already observed, what is the evidence that the alternative would be chosen in a repetition of the experiment? Thus the scientist is interested in the estimated power $\Pi(\tilde{\theta}(X))$, where $\tilde{\theta}$ is an estimator of θ, and the estimated power is a weight of evidence for the alternative.

These considerations motivate the definition of a measure (or weight) of evidence for the alternative as an estimator $T(X)$ of the ideal power function, where T could be a test critical function, but in general will take values anywhere in the closed unit interval. With squared error loss, it is required that the Type I risk be bounded $\sup_{\Theta_0} \mathrm{E}_\theta[T(X) - 0]^2 \leq \alpha$, in which case the measure of evidence is called a level-α estimator (or level-α measure of evidence for the alternative). Note in passing that a level-α test critical function is also a level-α estimator. Note also that estimated power functions $T_{EP}(X) = \Pi(\tilde{\theta}(X))$ could also, with appropriate choice of critical regions which are not necessarily of level-α for testing, be level-α estimators in this sense.

To summarize, this framework allows one to generalize test critical functions and estimated power functions to a class of measures of evidence for the alternative, all sharing the same desirable property of having a bound on the Type I error risk. In keeping with the Neyman-Pearson philosophy, subject to this bound, one tries to find an estimator which minimizes in some way the Type II risks $\mathrm{E}_\theta[1 - T(X)]^2$, for $\theta \in \Theta_1$. Another interpretation (Blyth and Staudte (1995a)) of these estimators is that they measure *how* safe it is to reject the hypothesis at the given level. Evidence is used to help make a decision, in contrast to the reverse order in which a p-value measure of evidence is obtained as a byproduct of evaluating a test decision rule.

A level-α estimator of the ideal power function gives a measure of how safely one can reject at the given level, and is thus a *guarded* measure of evidence for the alternative. Other *unguarded* (because they do not place a bound on the risk under the hypothesis) measures of evidence for hypotheses are obtained in a similar way using squared error loss by Schaafsma (1989), Schaafsma et al. (1989), Hwang et al. (1992) and Casella and Wells (1993).

While squared error loss does not yield risks which have probabilistic interpretations (unless the estimators take on only 0 and 1), it does have the compensating property that it yields reasonable answers in a variety of settings. For example, for simple hypothesis $p(x) = p_0(x) \equiv p_{\theta_0}(x)$ against simple alternative $p(x) = p_1(x) \equiv p_{\theta_1}(x)$ the minimum risk level-α estimator is given by $T(x) = 1/[1 + \kappa\lambda(x)]$, where $\lambda(x) = p_0(x)/p_1(x)$ and κ is chosen to obtain any desired level α. This result can be extended (Blyth and Staudte (1995a) to composite hypothesis against a simple alternative in the same way Lehmann (1986, pp. 104–105) extended the Neyman-Pearson Lemma to a composite hypothesis.

It was suggested in Blyth and Staudte (1995a), and will be demonstrated in Section 2.1.1 below, that for normal models the minimum risk level-α estimator of a one-sided composite hypothesis versus a simple alternative approaches the uniformly most powerful level-α test critical function as the alternative approaches infinity; this fact helps explain why the p-value is a good measure of evidence for distant alternatives, see Section 2.1.3. However, as shown in Sections 2.2 and 2.3, for some other models the limiting optimal weights of evidence do not converge to the optimal test critical function.

2 Test Critical Functions Can Arise as Limiting Measures of Evidence

It is shown in this section that the limit of minimum risk level-α measures of evidence for a simple alternative to a fixed hypothesis may (Section 2.1.1) or may not (Sections 2.2 and 2.3) approach the Neyman-Pearson level-α test critical function as the alternative moves to ∞, depending on the model. The normal model is discussed in Section 2.1, while Sections 2.2 and 2.3 discuss the extreme-value and Huber least favourable distributions, respectively.

Throughout Section 2, graphs are presented using the level $\alpha = 1/3$. The level $\alpha = 1/3$ is important here because it allows comparisons with the p-value estimator which has this level. The p-value measure of evidence is defined by $T_{PV}(x) = 1 - \alpha(x)$, where $\alpha(x)$ is the smallest level for which rejection would occur. For the hypothesis $\Theta_0 = (\infty, 0]$, the level is $\alpha(x) = \Pr_0\{X \geq x\}$. Now $T_{PV}(X) = \Phi(X)$ has the uniform distribution on the unit interval when $\theta = 0$. Therefore, $T_{PV}(X)$ has level $\alpha = \mathrm{E}_0[T_{PV}(X) - 0]^2 = 1/3$. However, neither the p-value measure of evidence nor the level $\alpha = 1/3$ are recommended in practice, as discussed in Section 3.

2.1 Example 1: Weights of evidence for hypotheses for a shift in the normal distribution

Assume X has the density given by $p_\theta(x) = \varphi(x-\theta) = \exp\{-(x-\theta)^2/2\}/\sqrt{2\pi}$. Consider the simple hypothesis $\Theta_0 = \{0\}$ and, for any $\theta_1 > 0$, the simple alternative $\Theta_1 = \{\theta_1\}$. The minimum risk level-α estimator is given by $T_{\alpha,\theta_1}(x) = 1/[1 + \kappa\lambda_{\theta_1}(x)]$, where $\lambda_{\theta_1}(x) = \exp\{\theta_1^2/2 - \theta_1 x\}$, and $\kappa = \kappa(\alpha, \theta_1)$ is chosen so as to achieve an exact level α. While a unique κ exists for any fixed α and θ_1, calculation of specific values of κ requires an iterative search using numerical integration routines; see Section 2.1.2 below.

2.1.1 Limiting measures of evidence

Some examples of minimum risk measures of evidence T_{α,θ_1} for $\alpha = 1/3$ and various choices of θ_1 are shown in Figure 1. Note that each is continuously increasing from 0 to 1 in x. Moreover, by Lemma 2 of Blyth and Staudte (1995a), each T_{α,θ_1} is minimum risk of the same level for the composite hypothesis $\Theta_0 = (-\infty, 0]$ against the respective alternative θ_1.

Examination of Figure 1 indicates that as $\theta_1 \to 0$, these estimators everywhere approach $\sqrt{\alpha}$. This limit is the 'no data' level-α estimator $T_{\alpha,0} \equiv \sqrt{\alpha}$, because as the alternative approaches the hypothesis the weight of evidence for a fixed x becomes unable to distinguish between them; that is, it behaves as if it did not have x available at all.

As $\theta_1 \to \infty$ these estimators appear to approach the level $\alpha = 1/3$ test critical function $I_{[z_{1-\alpha},\infty)}(x) = I_{[.4307,\infty)}(x)$ which is uniformly most powerful for $\Theta_0 = (-\infty, 0]$ against the composite alternative $\Theta_1 = (0, \infty)$. These observations suggest the following theorem.

Theorem 2.1 *Let T_{α,θ_1} be the minimum risk level-α estimator for the normal composite hypothesis $\Theta_0 = (-\infty, 0]$ against the simple alternative θ_1. Then*

(i) $T_{\alpha,\theta_1}(x) \to \sqrt{\alpha}$ as $\theta_1 \to 0$;
(ii) for $x \neq z_{1-\alpha} = \Phi^{-1}(1-\alpha)$, $T_{\alpha,\theta_1}(x) \to I_{[z_{1-\alpha},\infty)}(x)$ as $\theta_1 \to \infty$; and
(iii) $T_{\alpha,\theta_1}(z_{1-\alpha}) \to e/(e+1)$ as $\theta_1 \to \infty$.

PROOF

(i) The proof of (i) is straightforward.

(ii) The optimal level-α estimator $T_{\alpha,\theta_1}(x) = 1/[1 + \kappa\lambda_{\theta_1}(x)]$, where $\lambda_{\theta_1}(x) = \exp\{\theta_1^2/2 - \theta_1 x\}$, and $\kappa = \kappa(\alpha, \theta_1)$ is chosen so $\mathrm{E}_0\, T^2_{\alpha,\theta_1}(X) = \alpha$. By

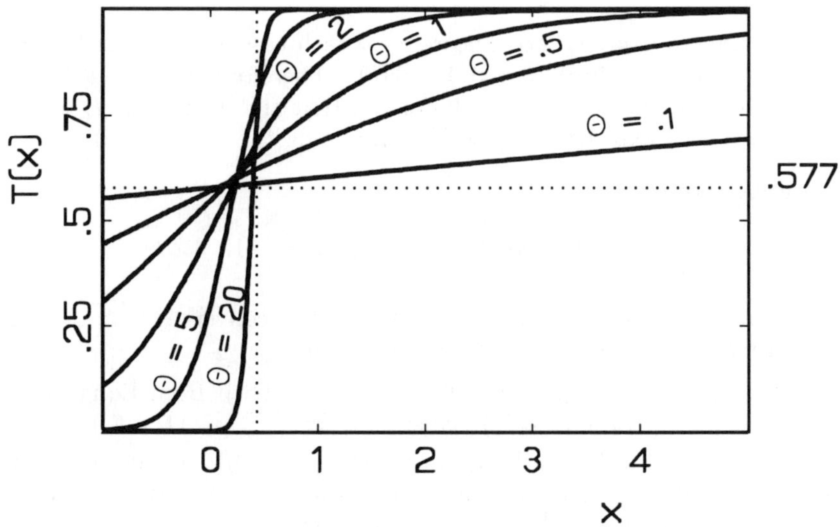

Figure 1: Graphs for the normal model of various level-1/3 measures of evidence for alternative θ_1 to hypothesis $\theta \leq 0$, as θ_1 takes on values .1, .5, 1, 2, 5 and 20, respectively. The vertical line indicates where the level-1/3 Neyman-Pearson critical function jumps from 0 to 1, namely at $z_{1-1/3} = .4307$.

means of the 1–1 transformation $c = \theta_1/2 + (\ln \kappa)/\theta_1$ the estimator can be rewritten $T_{\alpha,\theta_1}(x) = 1/[1+\exp\{\theta_1(c-x)\}]$, where the constant $c = c(\alpha, \theta_1)$ satisfies

$$\alpha = \int \varphi(x)dx \Big/ [1 + \exp\{\theta_1(c - x)\}]^2. \tag{2.1}$$

By making the change of variable $u = x - c$ this integral can be written as a sum

$$\alpha = \int_{u<0} \varphi(u+c)du \Big/ [1 + e^{-\theta_1 u}]^2 + \int_{u>0} \varphi(u+c)du \Big/ [1 + e^{-\theta_1 u}]^2.$$

The integral over $u < 0$ converges to 0 as $\theta_1 \to \infty$ because its integrand does, regardless of the behaviour of $c = c(\alpha, \theta_1)$. Therefore the integral over $u > 0$ converges to α.

Next consider the behaviour of $c(\alpha, \theta_1)$, as $\theta_1 \to \infty$, where it is tacitly assumed that $0 < \alpha < 1$. It can be verified that $c(\alpha, \theta_1)$ is bounded above, because otherwise the right hand side of (2.1) will be less than α for sufficiently large θ_1. Similarly, $c(\alpha, \theta_1)$ is bounded below for sufficiently large θ_1.

If $c = c(\alpha, \theta_1)$ does not have a limit as $\theta_1 \to \infty$ then by passing to two subsequences we may obtain different limit points $a < b$. For each of these subsequences the integrand of the integral over $u > 0$ converges to a bounded function $\varphi(u + a)$, $\varphi(u + b)$, respectively. Application of the Dominated Convergence Theorem for each case leads to the following contradiction: $1-\Phi(a)=\alpha=1-\Phi(b)$. Therefore $c=c(\alpha,\theta_1)\rightarrow c_\infty$ as $\theta_1\rightarrow\infty$ and $c_\infty=z_{1-\alpha}$. It follows that the estimator $T_{\alpha,\theta_1}(x)=\lim_{\theta_1\rightarrow\infty}1/[1+\exp\{\theta_1(z_{1-\alpha}-x)\}]$, completing the proof of (ii).

(iii) It will be shown that

$$\lim_{\theta_1\rightarrow\infty}\theta_1\{c(\alpha,\theta_1)-z_{1-\alpha}\}=-1, \tag{2.2}$$

which makes precise the rate at which $c(\alpha,\theta_1)$ converges to $z_{1-\alpha}$. The conclusion of (iii) follows directly from this limit.

Let $f(x,\theta_1)=[1+\exp\{\theta_1(c(\alpha,\theta_1)-x)\}]^{-2}$. Then from Equation 2.1, the constant $c(\alpha,\theta_1)=c(\theta_1)$ satisfies $\alpha=\int_{-\infty}^{\infty}\varphi(x)f(x,\theta_1)dx$. The notation $g(\theta_1)=\mathrm{o}(\theta_1^{-1})$ will mean that $\lim_{\theta_1\rightarrow\infty}\theta_1 g(\theta_1)=0$. A sequence of approximations to this last integral, with each approximation having error no more than $\mathrm{o}(\theta_1^{-1})$, will establish Equation (2.2).

Points a and b, depending on θ_1, can be chosen to satisfy the three conditions: $\int_{-\infty}^{a}f(x,\theta_1)dx=\mathrm{o}(\theta_1^{-1})$; $f(x,\theta_1)=1-\mathrm{o}(\theta_1^{-1})$ for $x>b$; and $b-a=\mathrm{O}(\theta_1^{-1})$. Then an integration of the first term in the second line below by partial fractions yields the identities:

$$\begin{aligned}
\alpha &= \int_{-\infty}^{a}\varphi(x)f(x,\theta_1)dx+\int_a^b\varphi(x)f(x,\theta_1)dx+\int_b^\infty\varphi(x)f(x,\theta_1)dx\\
&= \varphi(\xi)\int_{-\infty}^{b}f(x,\theta_1)dx+\int_b^\infty\varphi(x)dx+\mathrm{o}(\theta_1^{-1}) \qquad \text{where } \xi\epsilon(a,b)\\
&= \varphi(\xi)\{b-c(\theta_1)-\theta_1^{-1}\}+\int_b^\infty\varphi(x)dx+\mathrm{o}(\theta_1^{-1})\\
&= \int_{c(\theta_1)+\theta_1^{-1}}^{b}\varphi(\xi)dx+\int_b^\infty\varphi(x)dx+\mathrm{o}(\theta_1^{-1})\\
&= \int_{c(\theta_1)+\theta_1^{-1}}^{b}\varphi(x)dx+\int_b^\infty\varphi(x)dx+\mathrm{o}(\theta_1^{-1})\\
&= 1-\Phi(c(\theta_1)+\theta_1^{-1})+\mathrm{o}(\theta_1^{-1}).
\end{aligned}$$

Thus $\Phi(c(\theta_1)+\theta_1^{-1})=1-\alpha+\mathrm{o}(\theta_1^{-1})$, and inverting this equation yields $c(\theta_1)+\theta_1^{-1}=z_{1-\alpha}+\mathrm{o}(\theta_1^{-1})$. Then Equation (2.2) follows directly. ////

Corollary 2.2 *If $X_1,\ldots,X_n$ are independent, each with density $\varphi(x-\theta)$, then the optimal level-α measures of evidence for $\Theta=(-\infty,0]$ against the fixed alternative θ_1 converge, as $n\rightarrow\infty$, to the level-α Neyman-Pearson test critical function.*

Proof The sufficient statistic $\overline{X}$ has the normal distribution with mean θ and variance $1/n$, so by Lemma 1 of Blyth and Staudte (1995a) the minimum risk level-α estimator of $\theta = 0$ against $\theta = \theta_1$ has the form $T_{MR,\alpha}(\overline{x}) = 1/[1 + \kappa_n \lambda_n(\overline{x})]$, where $\lambda_n(\overline{x})$ is the ratio of likelihoods of the sample mean under this simple hypothesis and alternative, respectively. By Lemma 2 of Blyth and Staudte (1995a) this same estimator is the minimum risk level-α estimator of $\Theta = (-\infty, 0]$ against θ_1. By writing $\kappa_n = \exp\{-n\theta_1^2/2 + \theta_1 n^{1/2} c_n\}$, the defining equation for c_n and hence κ_n becomes

$$\alpha = \int \frac{n^{1/2}\varphi(n^{1/2}\overline{x})d\overline{x}}{[1 + \exp\{n^{1/2}\theta_1(c_n - n^{1/2}\overline{x})\}]^2}.$$

After a change of variable ($u = n^{1/2}\overline{x}$) this expression is identical to that for α in the proof of Theorem 2.1 (ii), with θ_1 replaced by the product $n^{1/2}\theta_1$; therefore as $n \to \infty$ for fixed θ_1, the limit of $T_{MR,\alpha}(\overline{x})$ is the Neyman-Pearson critical function and $z_{1-\alpha} = \lim_{n\to\infty} c_n$. ////

Remark 2.3 This Corollary suggests that for large samples from a distribution for which a central limit theorem holds, the optimal minimum risk estimator of $\Theta_0 = (-\infty, 0]$ against the fixed alternative θ_1 is in the form of a Neyman-Pearson critical function. ////

Remark 2.4 Perhaps more interesting is the fact that for alternatives $\theta_n = \theta_1/n^{1/2}$ approaching the hypothesis, the proof of this Corollary shows that the limiting constant $c_n \to c(\alpha, \theta_1)$. Hence for contiguous alternatives the limiting behaviour of measures of evidence based on the sample mean does not degenerate into test critical functions, but stays within the class of continuous measures of evidence. ////

2.1.2 Calculating the constants $c(\alpha, \theta_1)$ for the normal mean

The constants $c(\alpha, \theta_1)$ (or equivalently $\kappa(\alpha, \theta_1)$) play the same role for level-α measures of evidence as critical points do for tests. They are of special interest for hypotheses concerning a normal mean, and hence are studied in some detail here. Note that the weight of evidence at the point $x = c(\alpha, \theta_1)$ is $T(x) = 1/2$.

The optimal (minimum risk) level-α estimator $T_{\alpha,\theta_1}(x) = 1/[1 + \kappa\lambda_{\theta_1}(x)]$, where $\lambda_{\theta_1}(x) = \exp\{\theta_1^2/2 - \theta_1 x\}$, and $\kappa = \kappa(\alpha, \theta_1)$ is chosen so that $\mathrm{E}_0\, T^2_{\alpha,\theta_1}(X) = \alpha$. While κ can be found by iterative search using numerical integration routines, this requires some care because for large θ_1 the constant κ is of order $e^{-\theta_1^2/2}$. This follows from the result of Theorem 2.1 (ii) which shows $c = \theta_1/2 + (\ln \kappa)/\theta_1$ has a finite limit $z_{1-\alpha}$ as $\theta_1 \to \infty$. The proof of Theorem 2.1 (iii) gives specific information about the rate at which c approaches the limiting value $z_{1-\alpha}$ as $\theta_1 \to \infty$.

The behaviour near 0 is quite different. For $\theta_1 \to 0$, the constant κ approaches the limit $\alpha^{-1/2} - 1$, so $|c| \to \infty$ at the rate $1/\theta$ unless $\alpha = .25$, and then $c \to 0$. Figure 2 illustrates these three distinct types of behaviour of $c = c(\alpha, \theta_1)$ as a function of θ_1 which occur for $\alpha < .25, = .25$ and $> .25$.

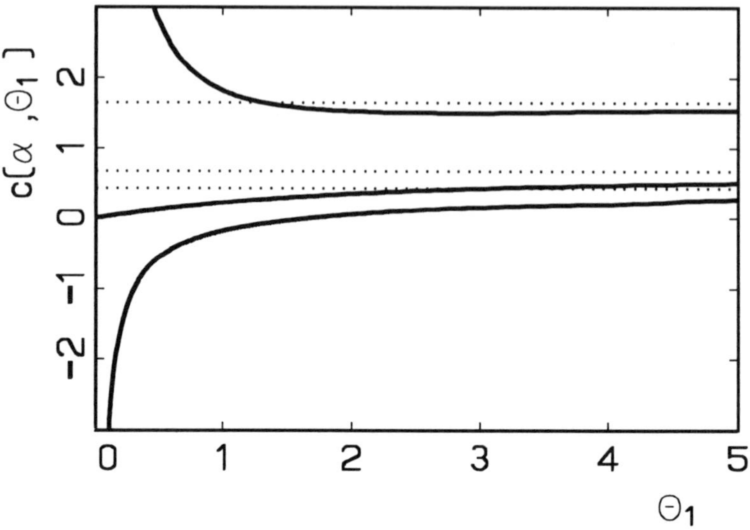

Figure 2: Graph of $c(\alpha, \theta_1)$ as a function of θ_1 for the normal mean and selected values $\alpha = .05$, $.25$ and $1/3$. The horizontal lines indicate the asymptotes which are critical values for the respective Neyman-Pearson tests.

θ \ α	.01	.05	.1	.25	.33(3)
.01	219.731	124.484	77.122	.0023	−31.1892
.05	43.987	24.927	15.449	.0125	−6.2314
.2	11.155	6.356	3.960	.0496	−1.5328
1	2.967	1.818	1.219	.2176	−.1884
3	2.241	1.493	1.093	.4236	.1539
10	2.252	1.563	1.196	.5823	.3359
100	2.317	1.635	1.272	.6646	.4208
500	2.324	1.643	1.280	.6725	.4287
∞	2.326	1.645	1.282	.6745	.4307

Table 1: Constants $c = c(\alpha, \theta_1)$ required to obtain level-α minimum risk estimators $T_{MR,\theta_1}(x) = 1/[1 + \exp\{\theta_1(c - x)\}]$ for the normal mean hypothesis $\theta \leq 0$ against alternative $\theta = \theta_1$.

Table 1 gives values of $c(\alpha, \theta_1)$ for $\alpha = .01, .05, .1, .25$ and $1/3$, and selected values of θ_1. For $\theta_1 < 5$ these constants were computed using standard quadrature and optimization routines on the software package GAUSS. These values were compared with the results obtained by finding solutions of a partial sums approximation to the integral $\alpha = \mathrm{E}_0[T_{\alpha,\theta_1}(X) - 0]^2$, as described in Theorem 2.6, and which may also be of use in determining c for other models.

First note that $T_{\alpha,\theta_1}(X) = 1/[1+\mathcal{R}]$, where $\mathcal{R} = \lambda_{\theta_1}(x)/\lambda_{\theta_1}(c)$ is the ratio of likelihood ratios for the simple hypothesis against simple alternative. Therefore the integral equation $\alpha = \mathrm{E}_0[1/(1+\mathcal{R})^2]$ must be solved for c. Lemma 2.5 gives a good approximation to $1/(1+y)^2$, for $|y| < 1$. If $|y| > 1$, the same expansion can be applied to $(1+y)^{-2}$ after rewriting it $y^{-2}/(1+y^{-1})^2$.

Lemma 2.5 *(i) For y with $|y| < 1$ and each positive integer m define the partial sum $S_m(y) = \left\{\sum_0^m (-1)^k y^k\right\}^2$. Then $S_m(y) = 1/(1+y)^2 + \mathrm{O}(y^{m+1})$, where the remainder term depends on m only through the exponent on y.*

(ii) The sums defined in (i) can be calculated by rewriting the square as a double sum and simplifying to $S_m(y) = \sum_0^{2m} (-y)^j \min\{j+1, 2m-j+1\}$.

Theorem 2.6 *Let $\mathcal{R} = \mathcal{R}(c, \theta_1, x) = \lambda_{\theta_1}(x)/\lambda_{\theta_1}(c)$ where $\lambda_{\theta_1}(x)$ is the likelihood ratio of null to alternative likelihood at θ_1, and $c = c(\alpha, \theta_1)$ is the critical constant for a level-α estimator. Then*

$$\alpha = \sum_0^{2m} (-1)^j \min\{j+1, 2m-j+1\} \left[\int_{x<c} \mathcal{R}^{-j-2} p_0(x)dx + \int_{x>c} \mathcal{R}^j p_0(x)dx\right] + R_m$$

and $R_m = \mathrm{O}\left(\int_{x<c} \mathcal{R}^{-m-3} p_0(x)dx\right) + \mathrm{O}\left(\int_{x>c} \mathcal{R}^{m+1} p_0(x)dx\right)$.

PROOF The ratio $\mathcal{R} > 1$ for $x < c$ and $\mathcal{R} < 1$ for $x > c$ so the proof follows directly from Lemma 2.5. ////

Corollary 2.7 *For the normal model we have $p_0(x) = \varphi(x)$ and $\mathcal{R} = e^{\theta_1(c-x)}$. For $x > 0$, define the ratio $r(x) = \{1 - \Phi(x)\}/\varphi(x)$. Using the relation $p_0(x)\mathcal{R}^j = \varphi(x + j\theta_1)\varphi(c)/\varphi(c + j\theta_1)$ one may write*

$$\alpha = \sum_0^{2m} (-1)^j \min\{j+1, 2m-j+1\} \left[r\{(j+2)\theta_1 - c\} + r\{c + j\theta_1\}\right] + R_m.$$

For any fixed m and θ_1 the remainder R_m can be set equal to 0 and a solution $c_m(\alpha, \theta_1)$ found which is an approximation to the desired $c(\alpha, \theta_1)$. For $x > 5$ the ratio $r(x)$ is replaced by a number of terms n in the series expansion for $r(x)$ given in Feller (1950, page 179). The calculations for Table 1 were carried out with $m = 100$ and $n = 10$.

As a further check on accuracy, the constants in Table 1 were verified using the software package *Mathematica*.

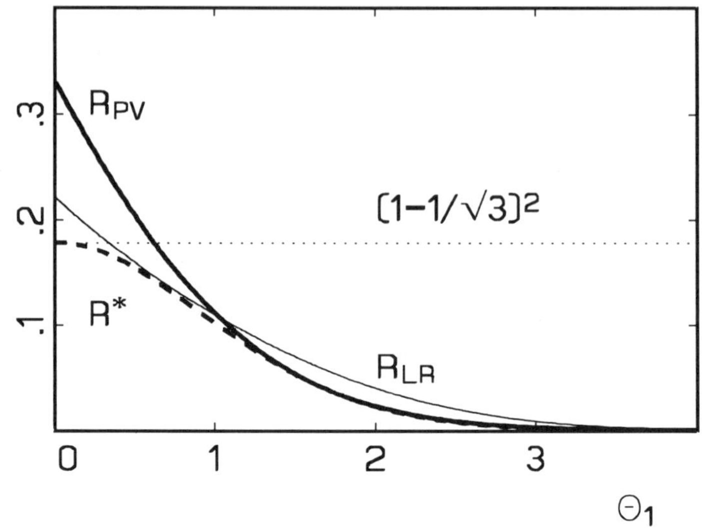

Figure 3: Graph of risk functions of level- $1/3$ estimators for the normal mean together with the envelope risk as a dashed line.

2.1.3 Comparison of risk functions for the normal mean

The envelope risk for a level- α measure of evidence for a hypothesis Θ_0 against an alternative Θ_1 is the mapping $\theta_1 \to \inf_T \mathrm{E}_{\theta_1}[T(X)-1]^2$, $\theta_1 \in \Theta_1$, where the infimum is over all level- α estimators of Θ_0 against Θ_1.

For composite hypothesis $\Theta_0 = (-\infty, 0]$ against composite alternative $\Theta_1 = (0,\infty)$, the envelope risk is readily calculated in principle, because the minimum risk level- α estimator for Θ_0 against any fixed simple alternative θ_1 is known to be of the form $T_{MR,\theta_1}(x) = 1/[1+\kappa\lambda_{\theta_1}(x)]$, where $\lambda_{\theta_1}(x) = \exp\{\theta_1^2/2 - \theta_1 x\}$, and $\kappa = \kappa(\alpha, \theta_1)$ is chosen so that $\mathrm{E}_0\, T^2_{\alpha,\theta_1}(X) = \alpha$. The critical constants κ are found by the methods in the previous section, and the envelope risk by computing for each θ_1 the risk $R^*(\theta_1) = \mathrm{E}_{\theta_1}[T_{MR,\theta_1}(X) - 1]^2$. The plot of $R^*(\theta_1)$ versus θ_1 is shown as a dashed curve in Figure 3 for the case $\alpha = 1/3$.

We will find and also plot the risk function of the p-value estimator and that of a likelihood ratio estimator. The p-value estimator $T_{PV}(X) = 1 - \alpha(X)$ has risk function $R_{PV}(\theta) = \mathrm{E}_\theta[T_{PV}(X) - 1]^2$, which is also plotted in Figure 3. Note that it approaches the envelope risk as $\theta \to \infty$, but is very large for small values of θ. However there are other estimators which are uniformly close to the envelope risk.

For example the likelihood ratio estimator (Blyth and Staudte (1993)) for the hypothesis $\Theta_0 = (-\infty, 0]$ against composite alternative $\Theta_1 = (0,\infty)$ is defined by $T_{LR}(x) = 1/\{1 + \kappa\lambda(x)\}$, where $\lambda(x) = \sup_{\Theta_0} p_\theta(x)/\sup_{\Theta_1} p_\theta(x)$ and κ is chosen to obtain level α. For $\alpha = 1/3$, the constant $\kappa = .7809$. The graph of the risk function for this estimator in Figure 3 shows that it has much smaller risk than the p-value estimator for small θ and not much larger risk for large θ.

As mentioned in the Introduction the scientist is interested in $\Pi(\tilde{\theta}(X))$, where $\tilde{\theta}$ is an estimator of θ, because it is an indication of the probability of rejecting the hypothesis in a repetition of the experiment. If we take $\Pi(\theta)$ to be the power function $\Pi(\theta) = \Phi(\theta - z_{1-\beta})$ of the most powerful test $I_{[z_{1-\beta},\infty)}(X)$ and $\tilde{\theta}(X) = X$, then the composition $\Pi(\tilde{\theta}(X))$ is $T_{EP}(X) = \Phi(X - z_{1-\beta})$. It has level α as an estimator subject to squared-error loss by appropriate choice of β. For a level α=1/3 estimator, $\beta = 1/2$. Thus for this value of α only, the estimated power and the p-value estimators are identical.

2.2 Example 2: Weights of evidence for hypotheses for a shift in the extreme-value distribution

Assume X has the extreme-value density which is given for each x by $p_\theta(x) = f(x-\theta)$, where $f(x) = \exp\{x - e^x\}$, and θ is an unknown shift parameter. Consider the simple hypothesis $\Theta_0 = \{0\}$ and, for any $\theta_1 > 0$, the simple alternative $\Theta_1 = \{\theta_1\}$. The minimum risk level-α estimator is of the form $T_{\alpha,\theta_1}(x) = 1/[1 + \kappa\lambda_{\theta_1}(x)]$, where $\lambda_{\theta_1}(x) = \exp\{\theta_1 - e^x(1 - e^{-\theta_1})\}$, and $\kappa = \kappa(\alpha, \theta_1)$ is chosen so as to achieve an exact level α.

There exists a unique $c = c(\alpha, \theta_1)$ for which $\kappa(\alpha, \theta_1) = 1/\lambda_c(\theta_1)$. Therefore the product $\kappa\lambda_x(\theta_1) = \lambda_x(\theta_1)/\lambda_c(\theta_1)$ is the likelihood ratio of θ_1 given x relative to its value given c. Assuming that $c = c(\alpha, \theta_1)$ has a limit c_∞ as $\theta_1 \to \infty$, the product $\kappa(\alpha, \theta_1)\lambda_x(\theta_1)$ approaches $\exp\{e^{c_\infty} - e^x\}$ for every x.

Applying the Dominated Convergence Theorem to the right hand side of $\alpha = \mathrm{E}_0[1 + \kappa(\alpha, \theta_1)\lambda_x(\theta_1)]^{-2}$ and then making a change of variable $y = \exp\{-e^x\}$ leads to $c_\infty = \ln\{\ln(\alpha^{-1} - 1)\}$. We earlier assumed the existence of this limit to find it; however if the limit did not exist, there would be two subsequences as $\theta_1 \to \infty$ for which distinct limits existed, and arguing as before in the Proof of Theorem 2.1 this would lead to a contradiction.

Unlike the result of Theorem 2.1, however, in this case the limiting weight of evidence is not a test critical function, but rather the continuous function $T_{\alpha,\infty}(x) = 1/[1 + (\alpha^{-1} - 1)\exp\{-e^x\}]$. The level α Neyman-Pearson test of the hypothesis $\Theta_0 = \{0\}$ against the simple alternative $\Theta_1 = \{\theta_1\}$ has critical function with a jump at $\ln\{\ln(\alpha^{-1})\}$. Figure 4 shows the level-α weights of evidence for various choices of θ_1 when $\alpha = 1/3$, their limiting value as $\theta_1 \to \infty$, and the level 1/3 Neyman-Pearson test critical function.

For small α the limiting (as $\theta \to \infty$) measure of evidence and the Neyman-Pearson test critical function would not be so dissimilar, but they would still be distinct.

2.3 Example 3: Weights of evidence for hypotheses for a shift in the Huber least favourable distribution

Assume X has the Huber least favourable density, which, for any x and shift parameter θ, reads $p_\theta(x) = f_b(x-\theta)$ where $f_b(x) = c_b \exp(-x^2/2)$ for $|x| \le b$ and

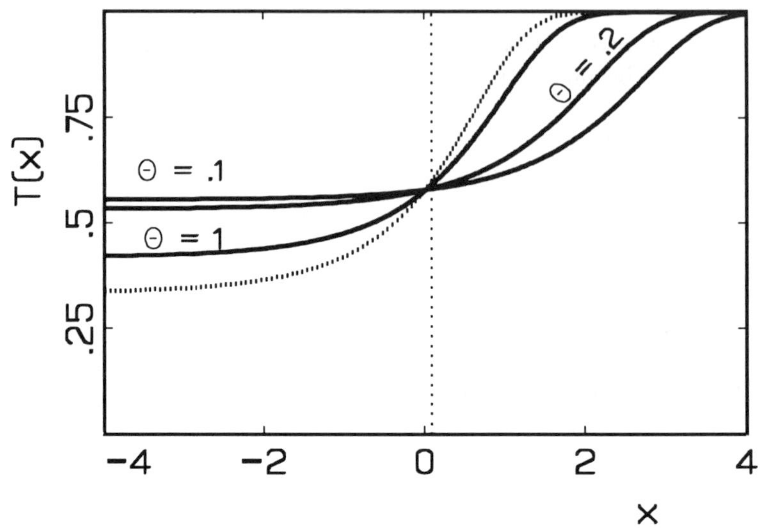

Figure 4: Graphs for the extreme-value distribution of various level-1/3 measures of evidence for alternative θ_1 to hypothesis $\theta \leq 0$, as θ_1 takes on values .1, .2, 1 and ∞, respectively. The latter limiting case is shown as a dotted curve. The vertical line indicates where the level-1/3 Neyman-Pearson critical function jumps from 0 to 1, namely at $\ln(\ln(3)) = .094$.

$f_b(x) = c_b \exp(b^2/2 - b|x|)$ for $|x| > b$. The corresponding maximum likelihood estimator is the M-estimator with ψ function $\psi_b(x) = \min\{\max(-b, x), b\}$; cf. Huber (1964).

Consider the simple hypothesis $\Theta_0 = \{0\}$, and the simple alternative $\Theta_1 = \{\theta_1\}$, where $\theta_1 > 0$. The minimum risk level-α estimator for these hypotheses is found, as in the previous examples, and plotted in Figure 5 for $\alpha = 1/3$, $b = 1$, and $\theta_1 = 1$, 2, 3 and ∞. Note that for any fixed θ_1 the weight of evidence for θ_1 is constant for small enough x, and also for large enough x, reflecting the insensitivity of the likelihood ratio to outliers.

Also note that the limiting estimator as $\theta_1 \to \infty$ is continuous, just as in the extreme-value example. This raises the question of whether it is the boundedness of the ψ function (and hence the influence function) which determines if the limiting measure of evidence is continuous or a Neyman-Pearson test.

2.4 Comparing acceptability profiles

Acceptability profiles were introduced (under the name confidence profiles) in Blyth and Staudte (1993, 1995a) as an alternative to confidence intervals. Acceptability profiles are derived from a family of level-α measures of evidence in the same way that confidence intervals are obtained from a family of tests. Assuming that for each θ_0 one has a level-α estimator T_{θ_0} of $\Theta_0 = \{\theta_0\}$ against the two-sided alternative $\Theta_1 = \{\theta : \theta \neq \theta_0\}$, then given $X = x$

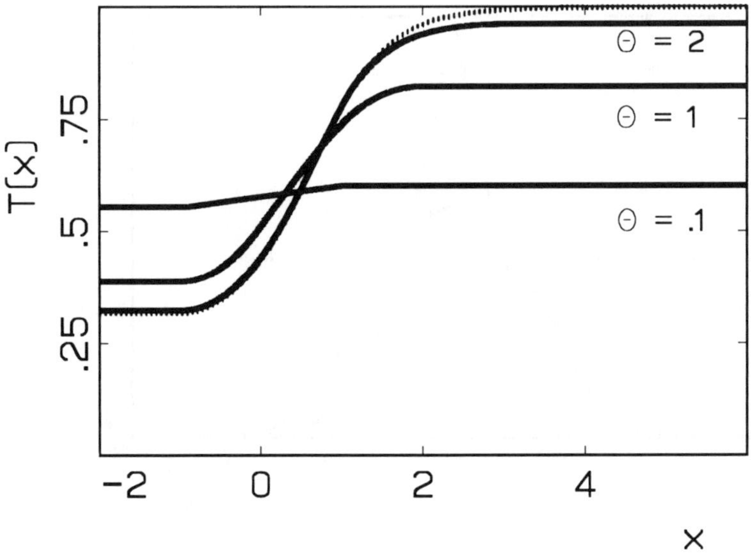

Figure 5: Graphs for the Huber least favourable distribution of various level-1/3 measures of evidence for alternative θ_1 to hypothesis $\theta \leq 0$, as θ_1 takes on values $.1$, 1, 2 and ∞, respectively. The latter limiting case is shown as a dotted curve.

the level-α acceptability of θ_0 is defined by the lack of evidence against it: $A_x(\theta_0) = 1 - T_{\theta_0}(x)$. The level-$\alpha$ acceptability profile is the graph of the map $\theta \to A_x(\theta)$. This is exactly what one does with critical functions for a family of tests to obtain the indicator function of a confidence interval.

Assume X has the normal distribution with unknown mean θ and variance 1. The p-value of X for a two-sided test of $\Theta_0 = \{\theta_0\}$ against $\Theta_1 = \{\theta : \theta \neq \theta_0\}$ is $\alpha(x) = 2\Phi(-|x - \theta_0|)$. The p-value measure of evidence is defined by $T_{PV}(X) = 1 - \alpha(X) = 2\Phi(|X - \theta_0|) - 1$, and it has level $\alpha = \mathrm{E}_{\theta_0}[T_{PV}(X) - 0]^2 = 1/3$. Therefore the level 1/3 acceptability profile corresponding to the p-value is the mapping $\theta_0 \to 2\Phi(-|x - \theta_0|)$ which is shown in Figure 6 as a dashed curve.

Also shown for comparison is the horizontal line at $1 - \sqrt{\alpha}$ determined by the level $\alpha = 1/3$ 'no data' measure of evidence $T \equiv \sqrt{1/3}$. Values of θ_0 for which the p-value profile exceeds this line are more acceptable at level 1/3 after observing $X = x$ than they were before the experiment.

It can be shown using the Cauchy-Schwarz Inequality that for a level-α profile, the average acceptability at the true value is greater than or equal to the no-data profile: that is, $\mathrm{E}_{\theta_0} A_X(\theta_0) \geq 1 - \sqrt{\alpha}$. Moreover, the average acceptability of a false value θ_0 is bounded above by the square root of the risk of the corresponding weight of evidence: $\mathrm{E}_\theta A_X(\theta_0) \leq \sqrt{R_{T_{\theta_0}}(\theta)}$. This is motivation for defining acceptability profiles in terms of weights of evidence with uniformly low risks.

The curved solid line in Figure 6 is the level 1/3 acceptability profile derived from the family of level-1/3 likelihood ratio estimators, one for each θ_0 and the

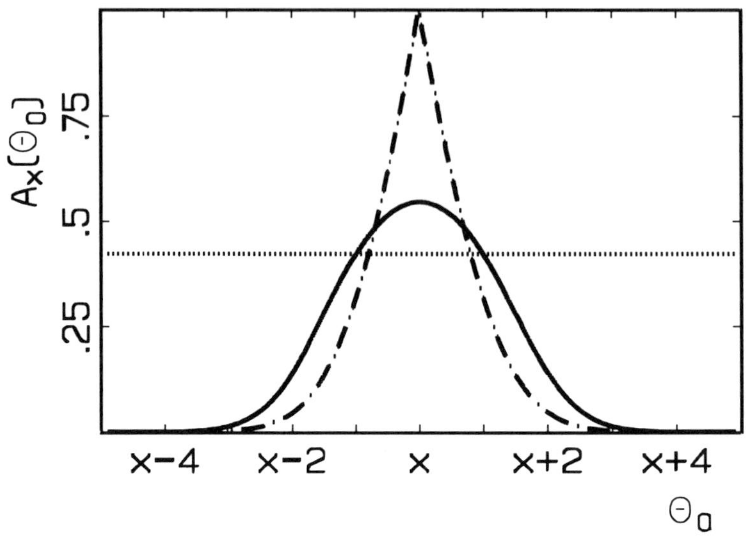

Figure 6: Level 1/3 acceptability profiles for a normal mean θ_0, given $X = x$. The horizontal line shows the level 1/3 'no data' profile, the dashed line the profile derived from the p-value, and the continuous curve is the profile derived from the two-sided likelihood ratio estimator.

2-sided alternative to it. It is derived in Blyth and Staudte (1995b), Section 4, and will be rewritten here as

$$A_x(\theta_0) = 1 - 1/\left[1 + \exp\{(c_\alpha^2 - z^2)/2\}\right]$$

where $z = (x - \theta_0)$, and c_α is chosen to obtain level α. This means solving for c_α in $\int \varphi(z)dz/[1 + \exp\{(c_\alpha^2 - z^2)/2]^2 = \alpha$. Examples are $c_{.01} = 2.65$, $c_{.05} = 2.05$ and $c_{1/3} = .60$. The reader can supply the indicator function of the two-sided level 1/3 confidence interval for θ, which can be seen as a crude approximation to the profiles in Figure 6.

As mentioned earlier, the average acceptability of a true θ_0 will exceed the horizontal line in Figure 6 for any level 1/3-profile. The p-value profile has very low average acceptability for θ far from the true value, at the expense of higher than necessary average acceptability of false values near the true value. The average acceptability of false values for the likelihood ratio profile is more uniformly near the minimum possible for every value.

These results are immediately applicable to finding a level-α profile for the mean θ of a normal(θ,1) distribution, in the case of n observations. The profile is exactly as for the case $n = 1$, with the same c_α and $z = x - \theta$ replaced by $n^{-1/2}(\overline{x} - \theta)$. The profile has a maximum at $\overline{x}$ and, as for confidence intervals, narrows at the rate $n^{-1/2}$.

3 Discussion

The level $\alpha = 1/3$ was used throughout this paper because it was desired to make comparisons with the p-value measure of evidence and the acceptability profile derived from it. However, this level is not recommended in practice: as stated in Theorem 2.1 (i), and can be verified for many other models, the level-α minimum risk estimator $T_{MR,\theta_1}(X)$ approaches the 'no-data' estimator $\sqrt{\alpha}$ as the alternative approaches the hypothesis. It therefore has Type II risk $\mathrm{E}_{\theta_1}[T_{MR,\theta_1}(X) - 1]^2 \to [1 - \sqrt{\alpha}]^2$. This is smaller than the maximum Type I risk α only if $\alpha \leq 1/4$. Thus to guarantee that the risk under the alternative near the hypothesis is at least as large as the risk under the hypothesis, a level $\alpha \leq 1/4$ should be adopted.

The fact that test critical functions can arise as limiting measures of evidence for the normal model, as illustrated in Section 2.1, is paralleled for binomial and extended hypergeometric models, when allowance is made for discreteness. However, it is false for other examples, as illustrated in Sections 2.2 and 2.3. We are seeking conditions which identify these two separate classes of models.

The limiting behaviour of measures of evidence for the examples considered here demonstrate that a general theory of inference based on weights of evidence for hypotheses is possible and desirable, partly because the p-value derived from a test of these hypotheses is a good measure only when the alternative is far from the hypothesis.

References

[1] Blyth, C.R. and Staudte, R.G. (1993): Hypothesis estimation and confidence profiles. In *Proceedings of the 49th Session of the International Statistical Institute* **1** 135–136.

[2] Blyth, C.R. and Staudte, R.G. (1995a): Estimating statistical hypotheses. *Statist. Probab. Lett.* **23** 45–52.

[3] Blyth, C.R. and Staudte, R.G. (1995b): Hypothesis estimation and acceptability profiles. *Parisankhyan Samikkha*, to appear.

[4] Casella, G. and Wells, M.T. (1993): *Comparing p-Values to Neyman-Pearson Tests.* Technical Report, Cornell Univ. Biometrics Unit Technical Report.

[5] Feller, W. (1950): *An Introduction to Probability Theory and Its Applications.* Wiley, New York.

[6] Hwang, J.T., Casella, G.,Robert, C., Wells, M.T. and Farrell, R.H. (1992): Estimation of accuracy in testing. *Ann. Statist.* **20** 490–509.

[7] Lehmann, E.L. (1986): *Testing Statistical Hypotheses.* Wiley, New York.

[8] Neyman, J. and Pearson, E.S. (1928): On the use and interpretation of certain test criteria for purposes of statistical inference. *Biometrika* **20** 175–240, 263–294.

[9] Schaafsma, W. (1989): Discussing the truth or falsity of a statistical hypothesis h and its negation a. In *International Workshop on Theory and Practice in Data Analysis*, number GDR REPORT R-MATH-01/89, 150–160. Berlin.

[10] Schaafsma, W., Tolboom, J. and van der Meulen, B. (1989): Discussing truth or falsity by computing a q-value. In *Statistical Data Analysis and Inference* (Y. Dodge ed.), 85–100. North-Holland, Amsterdam.

Geometric Properties of Principal Curves in the Plane*

Tom Duchamp
Department of Mathematics
University of Washington

Werner Stuetzle
Department of Statistics
University of Washington

Abstract

Principal curves were introduced to formalize the notion of "a curve passing through the middle of a dataset". Vaguely speaking, a curve is said to pass through the middle of a dataset if every point on the curve is the average of the observations projecting onto it. This idea can be made precise by defining principal curves for probability densities. Principal curves can be regarded as a generalization of linear principal components—if a principal curve happens to be a straight line, then it is a principal component. In this paper we study principal curves in the plane. We show that principal curves are solutions of a differential equation. By solving this differential equation, we find principal curves for uniform densities on rectangles and annuli. There are oscillating solutions besides the obvious straight and circular ones, indicating that principal curves in general will not be unique. If a density has several principal curves, they have to cross, a property somewhat analogous to the orthogonality of principal components. Finally, we study principal curves for spherical and elliptical distributions.

Key words and phrases: Principal curves, least-squares, curve fitting, nonlinear regression.
AMS 1991 subject classifications: Primary 62G07; secondary 62J02, 62H25, 62H30.

1 Introduction

The problem of fitting one or two-dimensional manifolds to point sets in two, respectively three, dimensions occurs in a variety of contexts, such as modeling of object boundaries in two or three-dimensional images (Banfield and Raftery (1992), Brinkley (1985), Martin et al. (1993), Schudy and Ballard (1978, 1979), Sheehan et al. (1992)), and reconstruction of objects from range data (Fang and Gossard (1992), Goshtasby (1992), Hoppe et al. (1992, 1993), Muraki (1991),

* This research was supported by DOE grant DE-FG06-85-ER25006 and NSF grants CCR-8957323 and DMS-9103002.

Solina and Bajcsy (1990), Vemuri et al. (1986)). It is typically not reasonable to assume that the unknown manifold is of a simple parametric form, like an ellipsoid. Fitting methods need to be flexible and able to accommodate a variety of shapes.

Manifold fitting is fundamentally different from regression. It is worthwhile to contrast the respective goals for the case of two-dimensional data. In regression, we are given points $(x_1, y_1), \ldots, (x_n, y_n)$. The goal is to find a *function* $f(x)$ summarizing the dependence of the response variable Y on the predictor variable X; the two variables are thus treated asymmetrically. Under the assumption that f is linear, a common choice is the least squares straight line. There has been a large amount of research on nonparametric regression methods that make only very general assumptions about the nature of f.

In manifold fitting, we are also given points $\mathbf{x}_1, \ldots \mathbf{x}_n \in \mathbb{R}^2$. The goal is to find a *one-dimensional manifold* Γ summarizing the association between the variables X_1 and X_2. The two variables are treated symmetrically. It is usually clear from the context whether the manifold should be topologically a circle or a closed interval. Under the assumption that Γ is a straight line, a common choice is the largest principal component. Nonparametric methods not relying on the linearity assumption have typically been crafted on an ad-hoc basis. This is not satisfactory, and a theoretical underpinning would be desirable.

To statistically analyze the behavior of fitting methods we need a stochastic model that is thought to give rise to the data. A simple model is to assume that the data points are i.i.d. observations of a two-dimensional random vector distributed according to some (unknown) density p. Then two questions arise:

(1) Which characteristic of the density are we trying to estimate, and why?

(2) How can we estimate this characteristic?

In the context of regression, answers to those questions are well understood. We usually estimate the conditional expectation $\mathrm{E}(Y \mid x)$, because it minimizes the expected squared prediction error $\mathrm{E}(Y - f(X))^2$ among all functions f. There are many approaches to estimating $\mathrm{E}(Y \mid x)$, often based on local averaging. The amount of averaging or, more generally, the complexity of the model is typically chosen to minimize an estimate of expected squared prediction error, like the cross-validated residual sum of squares.

In the context of manifold estimation the situation is not as clear cut—there are no generally accepted answers to questions (1) and (2) above. In this paper we discuss answers based on the concept of *principal curves.* Principal curves were introduced in Hastie (1984) and Hastie and Stuetzle (1989) to formalize the notion of "a curve passing through the middle of a dataset". Vaguely speaking, a curve Γ is said to pass through the middle of a dataset if every point $\mathbf{x}$ on the curve is the average of the observations projecting onto it.

To make this idea precise, Hastie and Stuetzle (1989) define principal curves for probability densities.

Definition of principal curves: Let $\mathbf{X}$ denote a two-dimensional random vector distributed according to a probability density p, and let $\Gamma \subset \mathbb{R}^2$ be a smoothly embedded closed interval (*arc*) or circle (*loop*). For each point $\mathbf{x} \in \mathbb{R}^2$, let $d(\mathbf{x}, \Gamma)$ denote the distance from $\mathbf{x}$ to Γ. Because Γ is compact, for each $\mathbf{x} \in \mathbb{R}^2$ the distance $d(\mathbf{x}, \Gamma)$ is realized by at least one point of Γ. Of course, there may be several such points; a point $\mathbf{x}$ with several closest points on the curve is called an *ambiguity point.* The *projection map*

$$\pi_\Gamma : \mathbb{R}^2 \to \Gamma,$$

is the map which assigns to each $\mathbf{x} \in \mathbb{R}^2$ a point $\pi_\Gamma(\mathbf{x}) \in \Gamma$ realizing the distance from $\mathbf{x}$ to Γ,

$$d(\mathbf{x}, \Gamma) = \|x - \pi_\Gamma(\mathbf{x})\|.$$

Notice that the map π_Γ is well-defined only on the complement of the set of ambiguity points of Γ. But the set of ambiguity points has Lebesgue measure zero (see Hastie and Stuetzle (1989)) and can be ignored in probability calculations. It is not difficult to show that π_Γ is continuous on the complement of the set of ambiguity points.

The vague notion that every point on the curve should be the average of the observations projecting onto it can now be formalized:

Definition 1.1 *(Hastie and Stuetzle (1989)) A curve* Γ *is called* self-consistent *or a* principal curve *of a density* p *if* $\mathrm{E}(\mathbf{X} \mid \pi_\Gamma(\mathbf{X}) = \mathbf{x}) = \mathbf{x}$ *for almost every* $\mathbf{x} \in \Gamma$.

The notion of projection also leads to a natural definition of the distance between a random vector $\mathbf{X}$ or its associated density, and a curve Γ:

$$d^2(\mathbf{X}, \Gamma) = \mathrm{E}(\|\mathbf{X} - \pi_\Gamma(\mathbf{X})\|^2).$$

Principal curves as generalizations of linear principal components: Besides formalizing the notion of "a curve passing through the middle of a dataset", principal curves share two properties with linear principal components, which make them appear as a natural generalization (Hastie and Stuetzle (1989)):

- If a principal curve happens to be a straight line, then it is a (linear) principal component.

- Principal curves are critical points of the distance function in the variational sense: Let Γ be a principal curve, and let Γ_t be a smooth family of curves with $\Gamma_0 = \Gamma$, then

$$\frac{d}{dt} d^2(\mathbf{X}, \Gamma_t)\Big|_{t=0} = 0$$

  Linear principal components share this property if Γ_t is restricted to be a smooth family of straight lines. The largest principal component minimizes the distance to $\mathbf{X}$, the smallest principal component maximizes the distance (among all lines passing through the mean), and the others are saddlepoints.

Summary of results: The goal of this paper is to further contribute to the theoretical understanding of principal curves. We now present an informal synopsis of our results:

1. Suppose that $\Gamma \subset \Omega$ is a principal arc for a density p with compact support Ω. Then Γ satisfies the following *transversality conditions*: (1) the endpoints of Γ lie on the boundary of Ω; (2) Γ intersects the boundary orthogonally; (3) the endpoints of Γ are (weakly) convex points of Ω.

2. If Γ_1 and Γ_2 both are principal curves for a density p, they cannot be linearly separable.

3. Suppose that Γ is a principal curve for a density p. Under appropriate conditions on Γ and p, the curvature of Γ can be expressed in terms of certain conditional moments of p. Principal curves that satisfy these conditions are called *regular*.

4. Regular principal curves are solutions of a system of ordinary differential equations. These equations can be used to calculate principal curves for uniform densities on rectangles and annuli. There are oscillating solutions besides the obvious straight and circular ones, indicating that principal curves in general will not be unique.

5. Any two regular principal curves of a density intersect.

6. The only regular principal arcs for a radially symmetric density are straight lines through the center.

7. While the uniform density on a circular disk has a circular principal curve, the uniform density on an ellipse does not have an elliptical principal curve.

Notation and conventions: The following notation is used throughout the paper: L_Γ denotes the length of Γ; Λ denotes either the closed interval $[0, L_\Gamma]$ when Γ is an arc, or the circle of circumference L_Γ when Γ is a loop; $\mathbf{x} = \mathbf{x}(s)$ denotes an arc length parameterization of Γ. The unit tangent and normal vector fields to Γ are written $\mathbf{T}(s)$ and $\mathbf{N}(s)$, respectively, and oriented so that the pair $(\mathbf{T}(s), \mathbf{N}(s))$ is consistent with the standard orientation of $\mathbb{R}^2$. The angle between the positive x-axis and $\mathbf{T}(s)$ is denoted by $\theta(s)$.

The map

$$\lambda : \begin{cases} \mathbb{R}^2 & \to \Lambda \\ \mathbf{x} & \mapsto s = \lambda(\mathbf{x}) \end{cases}$$

defined by the formula $\pi_\Gamma(\mathbf{x}) = \mathbf{x}(\lambda(\mathbf{x}))$ is called the *projection index* (see Hastie and Stuetzle (1989)).

Finally, throughout this paper $\Omega \subset \mathbb{R}^2$ denotes a compact, connected region with smooth boundary $\partial\Omega$. The density p is assumed to be supported on Ω, strictly positive on the interior of Ω, and smooth on all of Ω.

2 Transversality Conditions

The goal of this section is to study the endpoints of principal arcs.

Lemma 2.1 . *If* $\Gamma \subset \Omega$ *is a principal curve of* p, *then the line connecting any point* $\mathbf{x} \in \Omega$ *to its projection* $\pi_\Gamma(\mathbf{x})$ *intersects* Γ *orthogonally:*

$$\langle(\mathbf{x} - \pi_\Gamma(\mathbf{x})), \mathbf{T}(\pi_\Gamma(\mathbf{x}))\rangle = 0 \qquad \forall \mathbf{x} \in \Omega.$$

PROOF Choose a point $\mathbf{x} \in \Omega$. If $\pi_\Gamma(\mathbf{x})$ is an interior point of Γ then the identity follows from the fact that $\pi_\Gamma(\mathbf{x})$ is a point of Γ realizing the distance between $\mathbf{x}$ and Γ.

Suppose that $\mathbf{x}_0 = \pi_\Gamma(\mathbf{x})$ is an endpoint of Γ. Without loss of generality, we may assume that the orientation of Γ has been chosen so that $\mathbf{T}_0 \equiv \mathbf{T}(\mathbf{x}_0)$ is outward-pointing. Note that the inequality $\langle(\mathbf{y} - \mathbf{x}_0), \mathbf{T}_0\rangle \geq 0$ is satisfied for all $\mathbf{y} \in \Omega$ such that $\pi_\Gamma(\mathbf{y}) = \mathbf{x}_0$. For otherwise, the distance from $\mathbf{y}$ to Γ would be strictly less than $\|\mathbf{y} - \mathbf{x}_0\|$. Consequently, the subset $\pi_\Gamma^{-1}(\mathbf{x}_0) \cap \Omega$ is contained in the half-plane $H = \{\mathbf{x} \in \mathbb{R}^2 \mid \langle(\mathbf{x} - \mathbf{x}_0), \mathbf{T}_0\rangle \geq 0\}$.

Suppose that the proposition is false. Then $\langle(\mathbf{y} - \mathbf{x}_0), \mathbf{T}_0\rangle > 0$ for some point $\mathbf{y} \in \Omega$ with $\pi(\mathbf{y}) = \mathbf{x}_0$. Since p is continuous on Ω and strictly positive on the interior of Ω, there is an open set $Q \subset H$, such that (i) $\pi_\Gamma(Q) = \mathbf{x}_0$, (ii) $p > 0$ on Q and (iii) $\langle(\mathbf{y} - \mathbf{x}_0), \mathbf{T}_0\rangle > 0$ for all $\mathbf{y} \in Q$. But these conditions together imply that the point $\mathrm{E}(\mathbf{X} \mid \pi_\Gamma(\mathbf{X}))$ is contained in the interior of H, violating the self-consistency condition $\mathrm{E}(\mathbf{X} \mid \pi_\Gamma(\mathbf{X}) = \mathbf{x}_0) = \mathbf{x}_0$ ($\mathbf{x}_0$ is on the boundary of H). ////

The following proposition is an immediate consequence of Lemma 2.1:

Proposition 2.2 . *If* $\Gamma \subset \Omega$ *is a principal arc then* Γ *satisfies the following transversality conditions:*

1. *the endpoints of* Γ *lie on the boundary* $\partial\Omega$;
2. Γ *intersects* $\partial\Omega$ *orthogonally;*
3. *the endpoints of* Γ *are (weakly) convex points of* $\partial\Omega$.

3 Normal coordinates

Suppose that Γ is a principal curve for p and that there are no ambiguity points in the support of p. To interpret self-consistency as a curvature condition we need the notion of *normal coordinates.*

Definition 3.1 . *The* normal coordinate map *of* Γ *is the map* $\nu_\Gamma : \Lambda \times \mathbb{R} \to \mathbb{R}^2$ *defined by the formula*

$$\nu_\Gamma(s, v) = \mathbf{x}(s) + v\mathbf{N}(s)$$

and the normal coordinate transformation *is the map* $\mu_\Gamma : \Omega \longrightarrow \Lambda \times \mathbb{R}$ *defined by the formula*

$$\mu_\Gamma(\mathbf{x}) = (\lambda(\mathbf{x}), \langle \mathbf{x} - \mathbf{x}(\lambda(\mathbf{x})), \mathbf{N}(\lambda(\mathbf{x}))\rangle) .$$

The components (s, v) *of* $\mu_\Gamma(\mathbf{x})$ *are called the* normal coordinates *of* $\mathbf{x}$.

By virtue of our assumption that Ω does not contain ambiguity points of Γ, the normal map is a left inverse of the normal coordinate transformation μ_Γ:

$$\nu_\Gamma \circ \mu_\Gamma = id_\Omega.$$

We can now state a formal definition of regularity:

Definition 3.2 . *A smooth curve* $\Gamma \subset \Omega$ *is called* regular *if the following conditions are satisfied:*

1. Ω *contains no ambiguity points of* Γ.
2. *The map* $\mu_\Gamma : \Omega \longrightarrow \Lambda \times \mathbb{R}$ *is a diffeomorphism onto its image.*
3. *The image* $\mu_\Gamma(\Omega)$ *is of the form*

$$\mu_\Gamma(\Omega) = \{(s, v) \in \Lambda \times \mathbb{R} \,|\, v_-(s) \leq v \leq v_+(s)\},$$

where v_- *and* v_+ *are smooth and* $v_-(s) < 0 < v_+(s)$ *on the interior of* Λ.

Remark 3.3 Regularity condition (3) implies that for all $\mathbf{x}$ on the boundary of Ω the line segment joining $\mathbf{x}$ to $\pi_\Gamma(\mathbf{x})$ is not tangential to the boundary at $\mathbf{x}$. ////

Figure 1 shows a regular arc and a regular loop. Regularity implies that for any $s \in \Lambda$ the set $\mathcal{V}(s) = \{v \,|\, (s, v) \in \mu_\Gamma(\Omega)\}$ of points in Ω projecting onto $\mathbf{x}(s)$ is an interval:

$$\mathcal{V}(s) = \{\mathbf{x}(s) + v\mathbf{N}(s) \,|\, v_-(s) < v < v_+(s)\}.$$

For later reference, we now calculate the Jacobian determinant of the normal map. Recall that the *curvature function* $\kappa = \kappa(s)$ is given by the formula

$$\kappa = \frac{d\theta}{ds} .$$

Let $\imath$ and $\jmath$ denote the standard unit vectors in $\mathbb{R}^2$. Differentiation with respect to s of the identities

$$\mathbf{T} = \cos(\theta)\,\imath + \sin(\theta)\,\jmath \text{ and } \mathbf{N} = -\sin(\theta)\,\imath + \cos(\theta)\,\jmath$$

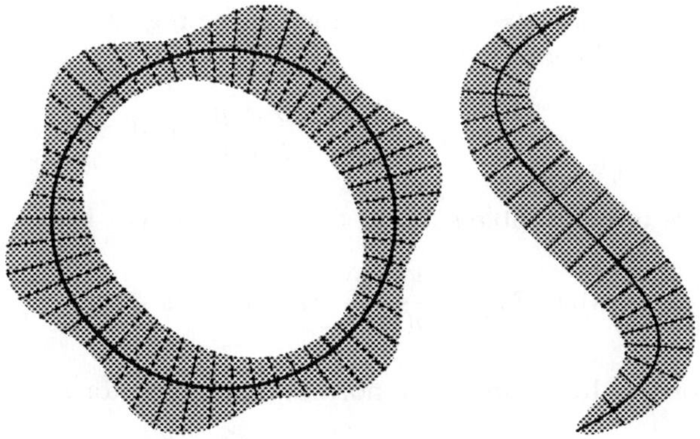

Figure 1: A regular loop and a regular arc.

yields the *Frenet formulas*

$$\frac{d\mathbf{T}}{ds} = \kappa\mathbf{N} \quad \text{and} \quad \frac{d\mathbf{N}}{ds} = -\kappa\mathbf{T}\,.$$

Then $\dfrac{\partial\nu_\Gamma(s,v)}{\partial s} = \mathbf{x}'(s) + v\mathbf{N}'(s) = (1 - v\kappa(s))\,\mathbf{T}(s)$ and $\dfrac{\partial\nu_\Gamma(s,v)}{\partial v} = \mathbf{N}(s)$. The Jacobian determinant $\dfrac{\partial(x,y)}{\partial(s,v)}$ of the normal coordinate map is now easily computed:

$$\begin{aligned}\frac{\partial(x,y)}{\partial(s,v)} &= \left|\frac{\partial\nu_\Gamma(s,v)}{\partial s} \times \frac{\partial\nu_\Gamma(s,v)}{\partial v}\right| \\ &= |(1 - v\kappa(s))\mathbf{T}(s) \times \mathbf{N}(s)| = 1 - v\kappa(s)\,.\end{aligned} \tag{3.1}$$

4 Self-Consistency as a Curvature Condition

Suppose that Γ is a regular principal curve for a density p. We will now derive a relation between the curvature of Γ and certain conditional moments of p. We do this by rewriting the self-consistency condition in normal coordinates.

Recall that Γ is called self-consistent if $\mathrm{E}(\mathbf{X} \mid \pi_\Gamma(\mathbf{X}) = \mathbf{x}) = \mathbf{x}$ for almost all $\mathbf{x} \in \Gamma$. By definition of conditional expectation, this is equivalent to the condition that

$$\int_{\pi_\Gamma^{-1}(A)} \mathbf{x}\, p(\mathbf{x})\ \mathrm{d}\mathbf{x} = \int_{\pi_\Gamma^{-1}(A)} \pi_\Gamma(\mathbf{x})\, p(\mathbf{x})\ \mathrm{d}\mathbf{x}$$

or

$$\int_{\pi_\Gamma^{-1}(A)} (\mathbf{x} - \pi_\Gamma(\mathbf{x}))\, p(\mathbf{x})\ \mathrm{d}\mathbf{x} = 0 \tag{4.2}$$

for all measurable $A \subset \Gamma$. Because Γ is assumed to be regular, Equation (4.2) can be rewritten in normal coordinates:

$$\int_{\{(s,v)\in\mu_\Gamma(\Omega)\mid s\in A\}} v\, p(\mathbf{x}(s)+v\mathbf{N}(s))\, \frac{\partial(x,y)}{\partial(s,v)}\, dv\, ds = 0\,,$$

where A now denotes a measurable subset of Λ. This implies that

$$\int_{\mathcal{V}(s)} v\, p(\mathbf{x}(s)+v\mathbf{N}(s))\, \frac{\partial(x,y)}{\partial(s,v)}\, dv = 0 \qquad s-\text{a.e.}$$

Substituting (3.1) for the Jacobian of the normal map gives, for all $s \in \Lambda$,

$$\int_{\mathcal{V}(s)} v\, p(\mathbf{x}(s)+v\mathbf{N}(s))\, dv - \kappa(s)\int_{\mathcal{V}(s)} v^2\, p(\mathbf{x}(s)+v\mathbf{N}(s))\, dv = 0\,. \tag{4.3}$$

Let

$$\mu_\perp(s) = \frac{\int_{\mathcal{V}(s)} v\, p(\mathbf{x}(s)+v\mathbf{N}(s))\, dv}{\int_{\mathcal{V}(s)} p(\mathbf{x}(s)+v\mathbf{N}(s))\, dv}$$

denote the mean of the *transverse density* on $\mathcal{V}(s)$ induced by p, and let $\sigma_\perp^2(s)$ denote its variance. Equation (4.3) can then be rewritten as

$$\kappa(s) = \frac{\mu_\perp(s)}{\mu_\perp^2(s)+\sigma_\perp^2(s)}\,. \tag{4.4}$$

Equation (4.4) relates the curvature of Γ to first and second moments of the density induced on the normal line to the curve at s. Consider the case where $p(\mathbf{x})$ is uniform. If $\mathbf{x}(s)$ happens to coincide with the transverse mean, that is, the center of the normal line segment, then the curvature $\kappa(s)$ has to vanish. Otherwise, the center of curvature is in the direction of the transverse mean. This makes intuitive sense: Consider an infinitesimal segment of the curve centered at $\mathbf{x}(s)$. As shown in Figure 2, the set of points projecting onto this segment is wedge shaped. If the center of the normal line segment falls below $\mathbf{x}(s)$, then the part of the wedge below the curve is longer than the part above the curve. In order for the mean of the segment to fall on the curve, the segment thus has to open up as we go upwards, meaning the curvature of Γ has to be negative.

5 Self-Consistency as a Differential Equation

Equation (4.3) can be thought of as a differential equation satisfied by principal curves. Unfortunately, it cannot be used to find principal curves because the integration boundaries in the moment integrals themselves depend on the curve. In order to use a differential equation for finding principal curves, we proceed differently: We define a differential equation in such a way that its solution curves, *when they are regular*, are actually principal curves. This differential equation expresses curvature in terms of *transverse moments* of p.

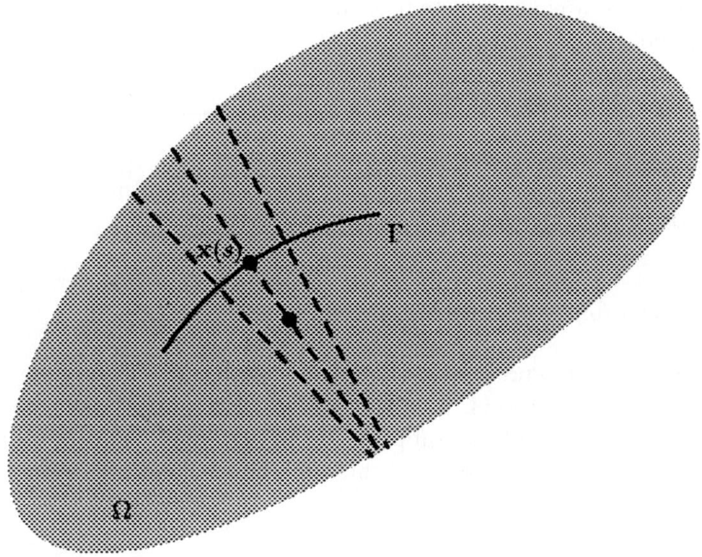

Figure 2: The center of curvature of a principal curve Γ points in the direction of the transverse mean.

Definition 5.1 . *The* k-th transverse moment *of the density* p *at* $(\mathbf{x}, \theta)$ *is the function* $\mu_k : \Omega \times S^1 \to \mathbb{R}$ *defined by the formula*

$$\mu_k(\mathbf{x}, \theta)) = \int_{v_-(\mathbf{x},\theta)}^{v_+(\mathbf{x},\theta)} v^k p(\mathbf{x} + v\mathbf{N}(\theta))\, dv\,.$$

The integration boundaries $v_+ = v_+(\mathbf{x}, \theta)$ *and* $v_- = v_-(\mathbf{x}, \theta)$ *are determined by the condition that* $\mathbf{x} + v_+\mathbf{N}$ *and* $\mathbf{x} + v_-\mathbf{N}$ *be the boundaries of the line segment around* $\mathbf{x}$ *obtained by intersecting the line* $\{\mathbf{x} + v\mathbf{N} \,:\, v \in \mathbb{R}\}$ *with the support* Ω *of* p. *This line segment is called the* transverse line segment *at* $(\mathbf{x}, \theta)$ *and is denoted by* $\ell(\mathbf{x}, \theta)$. *Similarly* $\mathcal{V}(\mathbf{x}, \theta) = \{v \,:\, v_-(\mathbf{x}, \theta) \le v \le v_+(\mathbf{x}, \theta)\}$ *is called the* transverse interval *of* Ω *at* $(\mathbf{x}, \theta)$.

The mean and variance of the *transverse density*

$$p^{\perp}_{(\mathbf{x},\theta)}(v) = \frac{p(\mathbf{x} + v\,\mathbf{N}(\theta))}{\mu_0(\mathbf{x}, \theta)}, \quad v_-(\mathbf{x}, \theta) \le v \le v_+(\mathbf{x}, \theta)$$

can be written in terms of transverse moments:

$$\mu_\perp(\mathbf{x}, \theta) = \frac{\mu_1(\mathbf{x}, \theta)}{\mu_0(\mathbf{x}, \theta)}$$

$$\sigma^2_\perp(\mathbf{x}, \theta) = \frac{\mu_2(\mathbf{x}, \theta)}{\mu_0(\mathbf{x}, \theta)} - \mu^2_\perp(\mathbf{x}, \theta)\,.$$

Consider now the following system of first order differential equations, which we call the *principal curve equations*:

$$\frac{d\mathbf{x}}{ds} = \cos(\theta)\,\imath + \sin(\theta)\,\jmath\,; \quad \frac{d\theta}{ds} = \frac{\mu_1(\mathbf{x},\theta)}{\mu_2(\mathbf{x},\theta)} = \frac{\mu_\perp(\mathbf{x},\theta)}{\mu_\perp^2(\mathbf{x},\theta) + \sigma_\perp^2(\mathbf{x},\theta)}\,. \tag{5.5}$$

Clearly, any solution of this system satisfies the self-consistency condition (4.4). This simple observation yields the following theorem.

Theorem 5.2 . *A regular curve* Γ *is a principal curve of the density* p *if and only if it is a solution of the system of equations (5.5).*

Remark 5.3 It is worth noting that the system (5.5) may be singular along the boundary of Ω. Suppose that Γ is a principal arc. By virtue of the transversality conditions, the boundary of the support at the endpoints of Γ is convex. If it is strictly convex, all transverse moments μ_k vanish, and the self-consistency condition degenerates at the boundary. ////

6 Principal Curves for Uniform Densities

Our first example is the uniform density on the rectangular strip

$$\Omega_{a,b} = \{(x,y)\ :\ 0 \le x \le b\,, \quad -a/2 \le y \le a/2\}$$

of length b and width a.

For symmetry reasons, the horizontal line $y = 0$ and the vertical line $x = b/2$ are principal curves. If the region is square, then the same is true for the two diagonals. We will shortly see, however, that for long strips many other regular principal curves exist.

Without loss of generality, any regular principal curve for the strip can be assumed to be of the form $y = f(x)$. Curves of a more general type are excluded because they necessarily have ambiguity points in $\Omega_{a,b}$, and curves of the form $x = f(y)$ are dealt with by interchanging the x and y axes. Curves which enter and/or leave the strip are also excluded by the regularity requirement.

A solution to the system of differential equations (5.5) is uniquely determined by the boundary values $f(0)$ and $f'(0)$. To satisfy the transversality conditions, any principle curve has to intersect the left boundary of the rectangle orthogonally, meaning $f'(0) = 0$. The requirement that the support of the density must not contain any centers of curvature places a restriction on $f(0) = y_0$:

$$-\frac{a}{6} < y_0 < \frac{a}{6}$$

Ignore for the moment the fact that the rectangle has a right boundary and consider the infinite strip. We can obtain closed form expressions for the transverse moments and find explicit solutions of the system (5.5) in terms of elliptic functions (see Duchamp and Stuetzle (1993) for details). Solutions are periodic, and the period is monotonically increasing in $|y_0|$. A given starting value y_0 leads to a principal curve for the rectangle if the length b is an integer multiple of the half-period. Numerical calculations show that in order for a non-linear principal curve to exist, the aspect ratio b/a must be in an interval $I_n \approx (0.9069\ n, 0.978\ n)$, for some n. The number of such intervals containing a given value b/a increases approximately linearly with b/a. Consequently, rectangles with large aspect ratio have a large number of principal curves.

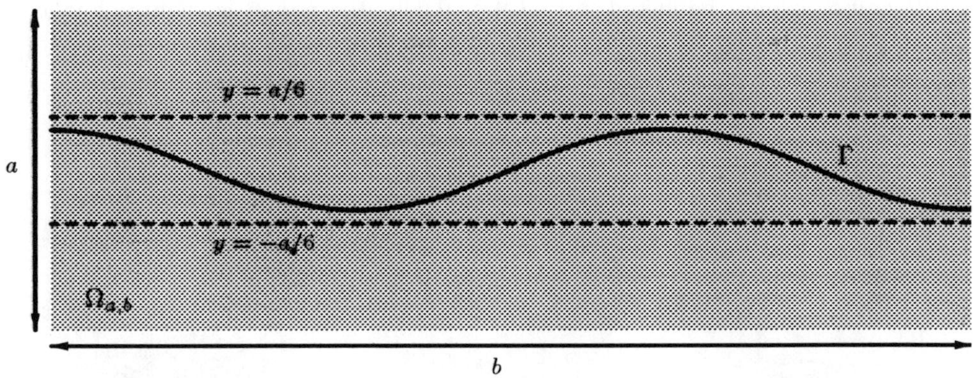

Figure 3: An oscillating principal curve for the uniform distribution on a rectangle

Consider next the uniform distribution on the annulus

$$\Omega_{R_1,R_2} = \{(r,\phi)\ :\ R_1 \leq r \leq R_2\},$$

where (r,ϕ) are polar coordinates. For symmetry reasons there has to be a circular principal curve. Using Equation (4.4) relating curvature to conditional moments of the density, its radius can be shown to be

$$r_{circ} = \frac{2\,(R_1^2 + R_1 R_2 + R_2^2)}{3\,(R_1 + R_2)}. \tag{6.6}$$

It is easy to see that *all* regular curves for the annulus are of the form $r = f(\phi)$. After a possible rotation, we may assume that f attains a local minimum at $\phi = 0$. The regularity condition requires that no center of curvature of Γ lie in the annulus. A simple calculation at a local minimum and a local maximum of f shows that $f(\phi)$ must lie between $R_{min} = (2R_1 + R_2/3$ and $R_{max} = (R_1 + 2R_2)/3$.

Because we did not succeed in finding an analytic solution of the principal curve equations (5.5) for the case of the annulus, we used a variable-step 4th-order Runge-Kutta method (Runge-Kutta-Fehlberg).

Our experimental results indicate that solutions are periodic, with period $T(R_1, R_2, r_0)$ depending on the initial value r_0. In order for a solution to be a principal loop, it has to be closed, meaning that $T(R_1, R_2, r_0) = 2\pi/n$, $n = 1, 2, 3, \ldots$. This will be the case for a discrete set of values r_0. Consider, for example, the annulus $\Omega_{0.45,1}$. In addition to the circular principal loop $r = r_{circ} \approx 0.760$, there is one other principal loop, given by the initial conditions $r(0) \approx 0.646$, $r'(0) = 0$. This principle loop has period $T = \pi/4$ and oscillates between the values $r = 0.646$ and $r = 0.874$. Initial conditions of the form $r'(0) = 0$ and $r(0) < 0.646$ give curves whose periods are slightly greater than $\pi/2$ and so cannot be closed; for $r(0) > 0.646$, the period is slightly smaller than $\pi/2$ and again the curve cannot close (see Figure 4).

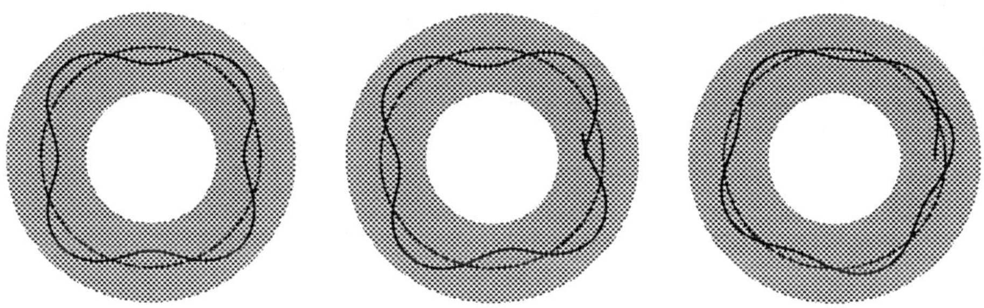

Figure 4: Three solutions of the principal curve equations on the annulus $\Omega_{0.45,1}$. From left to right the initial conditions are: $r(0) = 0.646$, $r(0) = 0.634$ and $r(0) = 0.690$. Only the first is a closed curve, and thus a principal loop

Computing principal curves for a number of annuli with different aspect ratios suggests that the number of principal loops increases with increasing aspect ratio, while the period $T(R_1, R_2, r_0)$ decreases with increasing aspect ratio (see Figure 5).

7 Principal Curves Cross

In analogy to linear principal components, densities in general will have multiple principal curves. These curves satisfy conditions that are vaguely analogous to the orthogonality of linear principal components.

Proposition 7.1 *Suppose that Γ_1 and Γ_2 are principal curves for a density p. Then they are not linearly separable.*

PROOF Let $\mathbf{X}$ be a random vector with density p, and let Γ be a principal curve. Then

$$\mathrm{E}(\mathbf{X}) = \mathrm{E}_\Gamma\ \mathrm{E}(\mathbf{X} \mid \pi_\Gamma(\mathbf{X}) = \gamma) = \mathrm{E}(\pi_\Gamma(\mathbf{X}))$$

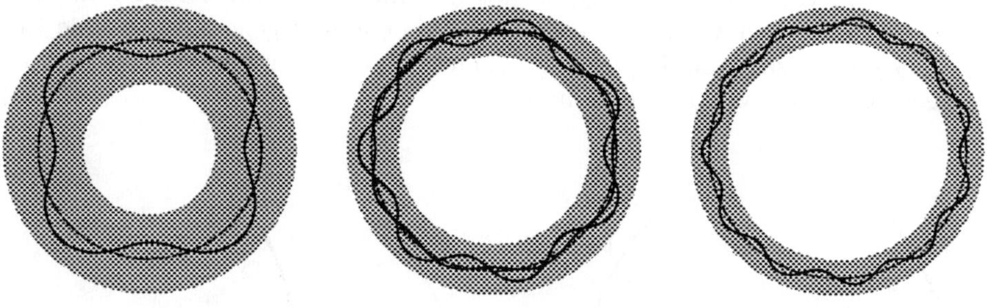

Figure 5: As the aspect ratio increases, the period of principal curves decreases. From left to right are the three annuli $\Omega_{0.45,1}$, $\Omega_{0.65,1}$ and $\Omega_{0.75,1}$ with some principal curves on each. Two principal curves are shown for the middle annulus. The right annulus also supports several principal curves (not shown).

As $\mathrm{E}(\pi_\Gamma(\mathbf{X}))$ is in the convex hull of Γ, the convex hulls of any two principal curves have to intersect. ////

We next show that, under mild convexity conditions on Ω, any two regular principal curves of a density must cross.

Theorem 7.2 . *Let Γ_1 and Γ_2 be two regular principal curves of the density p. If Γ_1 has endpoints, suppose that $\partial\Omega$ is strongly convex at those endpoints. Then Γ_1 and Γ_2 intersect.*

PROOF The proof is by contradiction and proceeds in two steps. Suppose that Γ_1 and Γ_2 do not intersect. Then there must be points $\mathbf{x}_1 \in \Gamma_1$ and $\mathbf{x}_2 \in \Gamma_2$ for which $\mathrm{dist}(\Gamma_1, \Gamma_2) = \mathrm{dist}(\mathbf{x}_1, \mathbf{x}_2) > 0$. According to Lemma 7.3 below these points must lie in the interior of the respective curves. This implies that the line L joining $\mathbf{x}_1$ and $\mathbf{x}_2$ intersects both curves orthogonally, and that the transverse distributions coincide. We will use the self-consistency condition, which is satisfied by both curves, to locate their centers of curvature on L.

Let $\mathcal{V} \subset L$ denote the connected component of the intersection of L with Ω containing $\mathbf{x}_1$ and $\mathbf{x}_2$. Let $\mathbf{x}_0 \in \ell$ be the mean of the transverse density induced on $\mathcal{V}$, and let $\sigma_\perp^2$ be its variance. Parameterize $\mathcal{V} = \{\mathbf{x}_0 + u\mathbf{N}_1 \mid a < u < b\}$, and set $\mathbf{x}_i = \mathbf{x}_0 + u_i\mathbf{N}_1$.

Writing the self-consistency condition (4.4) in the form $\kappa_i = -u_i/(u_i^2 + \sigma_\perp^2)$ shows that the u-coordinate of the center of curvature of Γ_i is given by the formula

$$c_i = u_i - \frac{u_i^2 + \sigma_\perp^2}{u_i} = -\frac{\sigma_\perp^2}{u_i}. \tag{7.7}$$

First consider the case where u_1 and u_2 have opposite signs, as shown in Figure 6(a). In this case, $d(\mathbf{x}_1, \mathbf{x}_2)$ is not a minimum. Next consider the case where u_1 and u_2 have the same sign. Without loss of generality assume that $u_1 < u_2 \leq 0$, as shown in Figure 6(b). Equation (7.7) then implies that $c_1 < c_2$, so again $d(\mathbf{x}_1, \mathbf{x}_2)$ is not a minimum[1]. ////

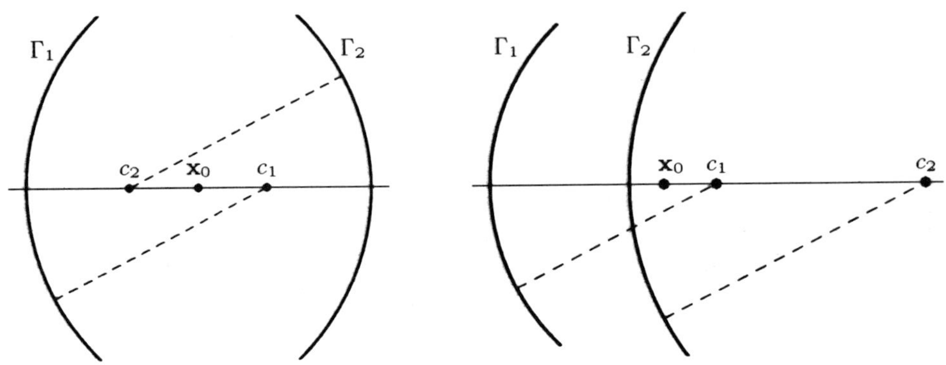

Figure 6: (a) Curves Γ_1, Γ_2 intersect the normal L on opposite sides of the mean; (b) Curves intersect the normal on the same side of the mean.

Lemma 7.3 *Let Γ_1, Γ_2 be as in Theorem 7.2, and assume that $\Gamma_1 \cap \Gamma_2 = \emptyset$. Then the distance between Γ_1 and Γ_2 is realized at some interior points $\mathbf{x}_1 \in \Gamma_1$ and $\mathbf{x}_2 \in \Gamma_2$.*

PROOF Note first that the projection of $\mathbf{x}_2$ onto Γ_1 is $\mathbf{x}_1$. This shows that $\mathbf{x}_1$ cannot be an endpoint of Γ_1, for, by virtue of the strong convexity assumption, the only points in Ω projecting onto the endpoints of Γ_1 are the endpoints themselves.

We now show that $\mathbf{x}_2$ cannot be an endpoint of Γ_2. Let $\mathbf{x}_e$ be an endpoint of Γ_2. By the transversality condition (2) of Proposition 2.2, Γ_2 intersects the boundary of Ω orthogonally at $\mathbf{x}_e$. By virtue of Remark 3.3, the vector from $\mathbf{x}_e$ to $\pi_{\Gamma_1}(\mathbf{x}_e)$ intersects the boundary transversely, and therefore its inner product with the (inward pointing) tangent vector of Γ_2 at $\mathbf{x}_e$ is positive. Consequently, x_e cannot be the point on Γ_2 closest to $\mathbf{x}_1$. ////

8 Principal Curves on the Disk and the Ellipse

As shown in Section 6, any spherically symmetric density has a circle as a principal loop. We now consider principal arcs for spherically symmetric densities:

[1] We wish to thank Andreas Buja for this observation.

Proposition 8.1 *Let p be a spherically symmetric density with compact support Ω —a disk centered at the origin. Then the only regular principal arcs are straight lines through the origin.*

PROOF Suppose Γ is a regular principal arc. Without loss of generality we may assume that Γ enters Ω at the point $(-r, 0)$, where r is the radius of Ω. The transversality conditions require that the tangent of Γ at $(-r, 0)$ be horizontal. The straight line $y = 0$ obviously is a principal arc that satisfies these conditions. We will use the fact that Γ is a solution to the principal curve equations (5.5) and thus is uniquely determined by a point and a tangent vector to conclude that Γ and the straight line $y = 0$ coincide. However, using a point and tangent vector on the boundary is too simplistic, because the principal curve equations are singular on the boundary. We thus have to argue as follows:

Suppose Γ is not a straight line. Then there has to be a point $\mathbf{x}_1$ with curvature $\kappa(\mathbf{x}_1) \neq 0$. Assume without loss of generality that $\kappa(\mathbf{x}_1) > 0$. This implies that that there must also be a point $\mathbf{x}_2$ with $\kappa(\mathbf{x}_2) < 0$ (otherwise, Γ would not cross the principal curve $y = 0$, contradicting Theorem 7.2). Therefore, there is a point $\mathbf{x}_0 \in \Gamma$ with $\kappa(\mathbf{x}_0) = 0$. As a consequence of the self-consistency condition (4.4) and spherical symmetry, Γ intersects the normal line $\{\mathbf{x}_0 + v\mathbf{N}(\mathbf{x}_0)\}$ in the center of its intersection with Ω. Thus the tangent line $\{\mathbf{x}_0 + v\mathbf{T}(\mathbf{x}_0)\}$ passes through the origin. Because Γ is a solution of the principal curve equations (5.5), it is uniquelly determined by $\mathbf{x}_0$ and $\mathbf{T}(\mathbf{x}_0)$. This means that it has to be a straight line through the origin, and we have reached a contradiction.////

Our understanding of principal curves for elliptical, non-spherical densities is not as complete. Assume that p is a uniform density on elliptical region Ω. Obviously, the major and minor axes of the ellipse are principal arcs. We do not know if there are any others. Given that spherically symmetric densities have circles as principal loops, one might suspect that elliptical densities have ellipses as principal loops. However, this is not true in general:

Proposition 8.2 *Let p be the uniform density on an ellipse that is not a circle. Then there is no elliptical principal loop.*

PROOF Assume that the ellipse $\Gamma = \{(x, y) \mid x^2 + y^2/b^2 = 1\}$ is a principal curve for the uniform distribution on a convex region Ω with smooth boundary $\partial\Omega$, which is symmetric with respect to both the x- and y-axes. Using the self-consistency condition, we determine a parameterization for the curve $\partial\Omega$ and show that it is not an ellipse.

Parametrize Γ as $\mathbf{f}(\theta) = (\cos(\theta), b\sin(\theta))$. Then $\partial\Omega$ can be written as $\mathbf{z}(\theta) = \mathbf{f}(\theta) + v_-(\theta)\mathbf{N}(\theta)$, where $v_-(\theta)$ is the negative distance along the normal between $\mathbf{f}(\theta)$ and $\partial\Omega$ (see Figure 7). Let $\mathbf{z}_+(\theta) = \mathbf{f}(\theta) + v_+(\theta)\mathbf{N}(\theta)$ be the point of intersection of the normal line with the x-axis.

It is not difficult to show that the set of ambiguity points of Γ is the interval of the x-axis joining the foci of the ellipse Γ. It follows that the inverse image $\mathcal{V}(\theta)$ of $\mathbf{f}(\theta)$ under the projection map $\pi_\Gamma : \Omega \to \Gamma$ is the interval joining $\mathbf{z}(\theta)$

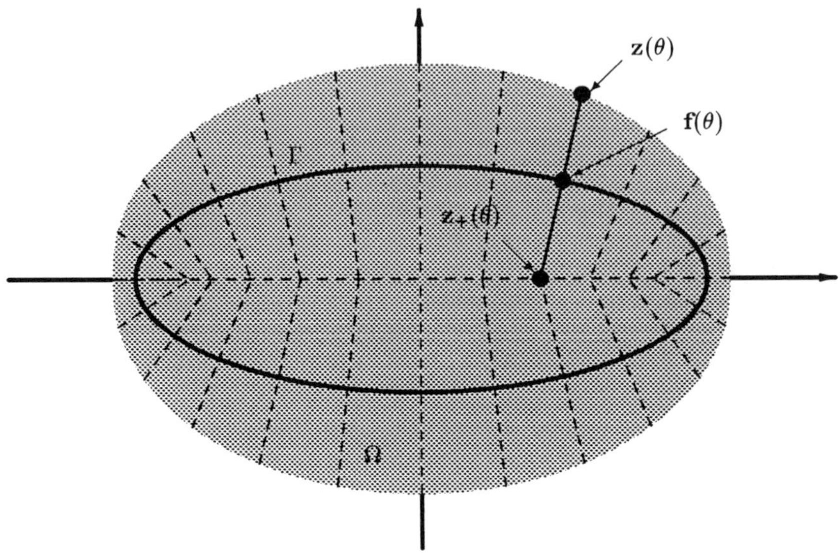

Figure 7:

and $\mathbf{z}_+(\theta)$. We now use Equation (4.3) and the fact that p is uniform to express the self-consistency condition as

$$(v_+(\theta)^2 - v_-(\theta)^2)/2 + (v_+(\theta)^3 - v_-(\theta)^3)/3\kappa(\theta) = 0. \tag{8.8}$$

Because Γ is assumed to be an ellipse, both $\kappa(\theta)$ and $v_+(\theta)$ are known, and Equation (8.8) can then be solved for $v_-(\theta)$.

It remains to show that $\mathbf{z}(\theta) = (x(\theta), y(\theta))$ is not an ellipse, unless $b = 1$. If the boundary were an ellipse, then its semi-major and semi-minor axes A, B would be determined by the formulas $(A, 0) = \mathbf{z}_-(0)$ and $(0, B) = \mathbf{z}_-(\pi/2)$; moreover, the error $e(\theta) = (x(\theta)/A)^2 + (y(\theta)/B)^2 - 1$ would be identically zero.

To show that $e(\theta)$ is not identically zero, it suffices to compute the second derivative of $e(\theta)$ at $\theta = 0$. One obtains the formula

$$e''(0) = 6b^2 \left(\frac{12b^2}{\left(6b^2 - 3 + \sqrt{3(1+2b^2)(3-2b^2)}\right)^2} - \frac{1}{2+b^2} \right)$$

It is not difficult, but messy, to give an analytic proof that $e''(0)$ is positive for $0 < b < 1$. The proof proceeds by simplifying the term in parentheses and isolating the numerator. Setting it equal to zero leads to the problem of finding the zeros of a fourth degree polynomial in b^2, which can be checked to have no zeros between 0 and 1. ////

9 Conclusion

Principal curves were introduced to formalize the notion of "a curve passing through the middle of a dataset". Vaguely speaking, a curve is said to pass through the middle of a dataset if every point on the curve is the average of the observations projecting onto it. This idea can be made precise by defining principal curves for probability densities. Principal curves can be regarded as a generalization of linear principal components—if a principal curve happens to be a straight line, then it is a principal component. In this paper we study principal curves in the plane. We show that principal curves are solutions of a differential equation. By solving this differential equation, we find principal curves for uniform densities on rectangles and annuli. There are oscillating solutions besides the obvious straight and circular ones, indicating that principal curves in general will not be unique. If a density has several principal curves, they have to cross, a property somewhat analogous to the orthogonality of principal components. Finally, we investigate principal curves for spherical and elliptical distributions.

In a companion paper (Duchamp and Stuetzle (1995)) we analyze the extremal properties of principal curves. Like principal components, principal curves are critical points of the expected squared distance to the data. However, the largest principal component minimizes the distance, whereas all principal curves are saddlepoints. This explains why cross-validation does not appear to be a viable method for choosing the complexity of principal curve estimates.

References

[1] Banfield, J.D. and Raftery, A.E. (1992): Ice floe identification in satellite images using mathematical morphology and clustering about principal curves. *J. Amer. Statist. Assoc.* **87** 7–16.

[2] Brinkley, J.F. (1985): Knowledge-driven ultrasonic three-dimensional organ modeling. *IEEE Trans. Pat. Anal. Mach. Intell.* **7**(4) 431–441.

[3] Duchamp, T. and Stuetzle, W. (1993): *The geometry of principal curves in the plane.* TR 250, Department of Statistics, University of Washington.

[4] Duchamp, T. and Stuetzle, W. (1995): Extremal properties of principal curves in the plane. Submitted to *Ann. Statist.*.

[5] Fang, L. and Gossard, D.C. (1992): Reconstruction of smooth parametric surfaces from unorganized data points. In *Curves and Surfaces in Computer Vision and Graphics 3* (J. Warren, ed.), **1830** 226–236, SPIE.

[6] Goshtasby, A. (1992): Surface reconstruction from scattered measurements. In *Curves and Surfaces in Computer Vision and Graphics 3* (J. Warren, ed.), **1830** 247–256, SPIE.

[7] Hastie, T.J. (1984): *Principal Curves and Surfaces*. Ph.D. Thesis, Stanford University.

[8] Hastie, T.J. and Stuetzle, W. (1989): Principal curves. *J. Amer. Statist. Assoc.* **84** 502–516.

[9] Hoppe, H., DeRose, T., Duchamp, T., McDonald, J. and Stuetzle, W. (1992): Surface reconstruction from unorganized points. *Computer Graphics* **26**(2) 71–78.

[10] Hoppe, H., DeRose, T., Duchamp, T., McDonald, J. and Stuetzle, W. (1993): Mesh optimization. *Computer Graphics* **27** 19–26.

[11] Martin, R.W., Bashein, G., Nessly, M.L. and Sheehan, F. (1993): Methodology for three-dimensional reconstruction of the left ventricle from transesophageal echocardiograms. *Ultrasound in Med. and Biol.* **19**(1) 27–38.

[12] Muraki, S. (1991): Volumetric shape description of range data using "blobby model". *Computer Graphics* **25**(4) 227–235.

[13] Schudy, R.B. and Ballard, D.H. (1978): *Model detection of cardiac chambers in ultrasound images*. Technical Report 12, Computer Science Department, Univ. of Rochester.

[14] Schudy, R.B. and Ballard, D.H. (1979): Towards an anatomical model of heart motion as seen in 4-d cardiac ultrasound data. In *Proceedings of the 6th Conference on Computer Applications in Radiology and Computer-Aided Analysis of Radiological Images*.

[15] Sheehan, F.H., Feneley, M.P., DeBruijn, N.P., Rankin, J.S., Davis, J.W., Bolson, E.L., Glass, P.S. and Clements, F.M. (1992): Quantitative analysis of regional wall thickening by transesophageal echocardiography. *Journal of Thoracic and Cardiovascular Surgery* **103**(2) 347–354.

[16] Solina, F. and Bajcsy, R. (1990): Recovery of parametric models from range images: The case for superquadrics with global deformations. *IEEE Trans. Pat. Anal. Mach. Intell.* **12**(2) 131–147.

[17] Vemuri, B.C., Mitiche, A. and Aggarwal, J.K. (1986): Curvature-based representation of objects from range data. *Image and Vision Computing* **4**(2) 107–114.

On Robust Estimation of Variograms in Geostatistics

Rudolf Dutter
University of Technology, Vienna

Abstract

The distribution of regionalized variables in the field of spatial statistics is mainly characterized by the variogram (or the covariance-function), which essentially is the variance of the differences of the variables in space. The usual estimate (empirical variance) of this function is highly non-robust.

Several alternative (*robust*) proposals for the estimation can be found in the literature. The definition often does not follow an obvious intuition (e.g., the estimator of Cressie and Hawkins). The present paper considers different estimators of the variogram (including the application of very new scale estimators). The investigation is mainly done by simulation in the one-dimensional case: data on regular and irregular grids. The results are rather surprising. The behavior of the estimators sometimes is unexpected. The main reasons seem to be the high dependence between the data values and the small sample size if the spatial distribution is irregular, both, however, correspond to practical situations often met.

Key words and phrases: Geostatistics, estimation of the variogram, simulation, robustness.
AMS 1991 subject classifications: 86A32, 62G35, 62H11, 65C05.

1 Introduction

Geostatistical tools have been developed for the analysis of phenomenons in space which show some spatial structure. A function $z(\boldsymbol{x})$ which gives the value of a parameter of such a phenomenon at each point (position) $\boldsymbol{x}$ in some region $D \subset \mathbb{R}^p$ in the p-dimensional space (p usually 1, 2 or 3) is called a *regionalized variable*. The distributional law of the corresponding random function $Z(\boldsymbol{x})$ is assumed to fulfill some stationarity conditions in order to be able to solve estimation problems. For our considerations in this paper we suppose second order stationarity; that is, the first two moments do not depend on the location $\boldsymbol{x}$ in space:

$$\mathrm{E}\{Z(\boldsymbol{x})\} = m(\boldsymbol{x}) = m,$$

$$\sigma^2(\boldsymbol{x}) = \mathrm{Var}\{Z(\boldsymbol{x})\} = \mathrm{E}\{(Z(\boldsymbol{x}) - m(\boldsymbol{x}))^2\} = \sigma^2,$$

$$C(\boldsymbol{x}, \boldsymbol{x}+\boldsymbol{h}) = \mathrm{E}\{(Z(\boldsymbol{x})-m(\boldsymbol{x}))(Z(\boldsymbol{x}+\boldsymbol{h})-m(\boldsymbol{x}+\boldsymbol{h}))\} = C(\boldsymbol{h}).$$

$C(\boldsymbol{x}, \boldsymbol{x}+\boldsymbol{h})$ stands for the covariance of $Z(\boldsymbol{x})$ and $Z(\boldsymbol{x}+\boldsymbol{h})$ which should depend only on the distance (or *lag*) $\boldsymbol{h}$ and not on $\boldsymbol{x}$. Note that $C(\boldsymbol{O}) = \sigma^2$.

In geostatistics, usually the *variogram function*

$$2\gamma(\boldsymbol{x}, \boldsymbol{x}+\boldsymbol{h}) = \mathrm{Var}\{Z(\boldsymbol{x})-Z(\boldsymbol{x}+\boldsymbol{h})\},$$

the variance of *increments* at distance $\boldsymbol{h}$, is used for measuring the variability in space. Under second order stationarity, this variogram will also not depend on $\boldsymbol{x}$:

$$\gamma(\boldsymbol{x}, \boldsymbol{x}+\boldsymbol{h}) = C(\boldsymbol{O})-C(\boldsymbol{h}) = \gamma(\boldsymbol{h}).$$

This *semi-variogram* $\gamma(\boldsymbol{h})$ is often synonymously named *variogram* when the usage is clear.

The variogram is the most important tool for the structural analysis of spatial phenomenons. It is the basis for further analysis, for example interpolation or extrapolation of values including the precision of an estimation.

In this paper we concentrate on the estimation of $\gamma(\boldsymbol{h})$. In the assumed case of second order stationarity it is defined by

$$\gamma(\boldsymbol{h}) = \frac{1}{2}\mathrm{Var}\{Z(\boldsymbol{x})-Z(\boldsymbol{x}+\boldsymbol{h})\} = \frac{1}{2}\mathrm{E}\{(Z(\boldsymbol{x})-Z(\boldsymbol{x}+\boldsymbol{h}))^2\}$$

which immediately suggests to estimate it, in case of approximately normally distributed Z's, by

$$\hat{\gamma}(\boldsymbol{h}) = \frac{1}{2N_h}\sum_{i=1}^{N_h}(z(\boldsymbol{x}_i)-z(\boldsymbol{x}_i+\boldsymbol{h}))^2,$$

where $\boldsymbol{x}_i$ and $\boldsymbol{x}_i+\boldsymbol{h}$, $i=1,\ldots,N_h$, denote the sample locations where the data have been taken, and N_h is the number of pairs of samples found at distance (and direction) $\boldsymbol{h}$.

Of course, this estimator of *variance* is highly non-robust. Several alternative proposals can be found in the literature. We try to investigate them by simulation where the main emphasis is on a small, practically relevant sampling scheme.

In Section 2 we define the considered estimators. Section 3 states the sampling situation and the simulation strategy. Section 4 presents results and Section 5 states some conclusions.

2 Estimators of the Variogram

From the properties of the variogram under specific assumptions we can deduce different ways of estimating the variogram. Thus several attempts to derive robust estimates for the variogram function have been published. Basically there is one point in common: they all calculate a more or less robust estimate of the quantity $E\{(Z(\boldsymbol{x})-Z(\boldsymbol{x}+\boldsymbol{h}))^2\}$ using the observations of $(Z(\boldsymbol{x})-Z(\boldsymbol{x}+\boldsymbol{h}))^2$ at discrete distances or distance classes. In order to summarize we may divide the approaches into three categories.

1. Estimate the mean or center of the population of squared differences. Define $Y(\boldsymbol{x}) = (Z(\boldsymbol{x})-Z(\boldsymbol{x}+\boldsymbol{h}))^2$, then $2\gamma(\boldsymbol{h}) = \mathrm{E}(Y)$. Thus the problem of estimating the variogram results in estimating the expected value of the random variables $Y(\boldsymbol{x})$. Under the assumption of the normality of the $Z(\boldsymbol{x})$, the $Y(\boldsymbol{x})$ follow χ^2-distributions of the form $2\gamma(\boldsymbol{h})\chi_1^2$.

2. Estimate the variance or scale parameter of the population of differences. The estimation of the variogram may then be done by estimating the variance of the random variables $Z(\boldsymbol{x})-Z(\boldsymbol{x}+\boldsymbol{h})$. An advantage of this approach is that under the given assumption of second-order stationarity, the $Z(\boldsymbol{x})-Z(\boldsymbol{x}+\boldsymbol{h})$ follow more often a symmetric distribution.

3. Use the relationship between the variogram and the autocorrelation. Given a second-order stationary function the variogram $\gamma(\boldsymbol{h})$ can be written as
$$\gamma(\boldsymbol{h}) = (1-\varrho(\boldsymbol{h}))\,\mathrm{Var}\{Z(\boldsymbol{x})\},$$
where $\varrho(\boldsymbol{h})$ denotes the autocorrelation at lag $\boldsymbol{h}$. An estimate of the variogram can then be determined by estimating the autocorrelation function $\varrho(\boldsymbol{h})$.

2.1 Usual estimator

It has been mentioned before that it is defined by

$$\hat{\gamma}(\boldsymbol{h}) = \frac{1}{2N_h}\sum_{i=1}^{N_h}(z(\boldsymbol{x}_i)-z(\boldsymbol{x}_i+\boldsymbol{h}))^2, \tag{2.1}$$

which is optimal at the multivariate normal distribution but is highly non-robust.

2.2 The Cressie–Hawkins estimator

Cressie and Hawkins (1980) deduce a robust estimator of the variogram by following the first approach. Their attempt is based on the assumption that the $Z(\boldsymbol{x})$ are gaussian, hence the random variables $(Z(\boldsymbol{x})-Z(\boldsymbol{x}+\boldsymbol{h}))^2$ have distributions of the form $2\gamma(\boldsymbol{h})\chi_1^2$. To overcome the problem that this distribution is highly skewed they suggest a fourth root power transform of the squared differences,
$$Y(\boldsymbol{x},\boldsymbol{h}) = \mid Z(\boldsymbol{x})-Z(\boldsymbol{x}+\boldsymbol{h})\mid^{\frac{1}{2}},$$
and estimate the center of symmetry, $\bar{Y}$, for each distance $\boldsymbol{h}$.

The conclusion of their examination of ten estimators on one actual (lognormal) and six simulated data sets was that maximum likelihood estimators

(M-estimates) perform best in respect to robustness and resistance. The simulation study was done in one dimension simulating traverses using the model $Z(t) = 0.6Z(t-1) + U(t)$, the $U(t)$ being independently identically distributed random variables from the following distributions: normal, Laplace, and four differently symmetrically contaminated normal.

However, as the arithmetic mean of the $Y(\boldsymbol{x})$ also showed well satisfying performance they advise to use an estimator of the form

$$2\gamma(\boldsymbol{h}) = (\frac{1}{N_h} \sum_{i=1}^{N_h} \mid z(\boldsymbol{x}_i) - z(\boldsymbol{x}_i + \boldsymbol{h}) \mid^{\frac{1}{2}})^4 / (0.457 + \frac{0.494}{N_h} + \frac{0.045}{N_h^2}).$$

For large N_h the correction factor can be approximated by $1/0.457$. The use of the arithmetic mean instead of iterative robust M-estimates has also been recommended because of the easier calculation procedure.

2.3 Dowd's proposals

Dowd (1984) pointed at some proposals for a more resistant estimation of the variogram. His attempts are based on resistant estimates of the scale of the symmetric distribution of the differences $Z(\boldsymbol{x}) - Z(\boldsymbol{x} + \boldsymbol{h})$.

The first one which he reports to be only 88% as efficient as the traditional experimental variogram for a normal distribution is given by

$$\hat{\gamma}_{D1}(\boldsymbol{h}) = \frac{1}{2} \left(\frac{1}{N_h} \sum_{i=1}^{N_h} \mid z(\boldsymbol{x}_i) - z(\boldsymbol{x}_i + \boldsymbol{h}) \mid \right)^2 .$$

The second estimator is again a scale estimator derived from the median deviation. To ensure consistency at the normal distribution a correction factor is used and the variogram values can be calculated by

$$2\hat{\gamma}_{D2}(\boldsymbol{h}) = 2.198\,(\text{median} \mid z(\boldsymbol{x}_i) - z(\boldsymbol{x}_i + \boldsymbol{h}) \mid)^2 .$$

The third proposal is a modification of the second one. Define $y_i(\boldsymbol{h}) = z(\boldsymbol{x}_i) - z(\boldsymbol{x}_i + \boldsymbol{h})$ and let $\tilde{y}(\boldsymbol{h})$ be the median of the $y_i(\boldsymbol{h})$. Then a very resistant estimator of the variogram is defined by

$$2\hat{\gamma}_{D3}(\boldsymbol{h}) = 2.198\,(\text{median} \mid y_i(\boldsymbol{h}) - \tilde{y}(\boldsymbol{h}) \mid)^2 .$$

This estimator is usually known as *MAD* (*median absolute deviation*), and we consider only the $1st$ and the $3rd$ proposals in the following.

A disadvantage of this approach is the low asymptotic efficiency of about 40% for a normal distribution. However, in a test based on two actual data sets, Dowd reports the best performance of this median absolute deviation estimator among several estimators including that from Cressie–Hawkins.

2.4 Huberized variograms

Based on robust and resistant estimation of the variance of the differences $Z(\boldsymbol{x}) - Z(\boldsymbol{x} + \boldsymbol{h})$, Armstrong and Delfiner deduce a robust estimator of the variogram (see again Dowd (1984), for a review). Thus the estimate of the variogram can be obtained iteratively from

$$\frac{1}{N_h} \sum_{i=1}^{N_h} \psi^2 \left(\frac{y_i}{\sqrt{2\hat{\gamma}_A(\boldsymbol{h})}} \right) = \beta,$$

where $y_i = z(\boldsymbol{x}_i) - z(\boldsymbol{x}_i + \boldsymbol{h})$. The function $\psi(t)$ is defined by

$$\psi(t) = \begin{cases} -c & t < -c \\ t & -c \leq t \leq c \\ c & t > c \end{cases}$$

for some constant $c > 0$, and $\beta = \mathrm{E}\left\{\psi^2 \left(y_i / \sqrt{2\hat{\gamma}_A(\boldsymbol{h})}\right)\right\}$ is determined on the assumption that

$$\frac{y_i}{\sqrt{2\hat{\gamma}_A(\boldsymbol{h})}}$$

follows a $\mathcal{N}(0,1)$-distribution.

2.5 Shortest half

This estimator is the one-dimensional version of the Minimum-Volume-Estimator (MVE, see Rousseeuw (1985) and Rousseeuw and Leroy (1987)). The short verbal definition is: Find the shortest interval that contains more than half of the observations. The midpoint of the interval defines the location estimate and the length of the interval is proportional to the scale estimate (consistency at the normal distribution!). We have used the reweighting version as proposed by Rousseeuw: Reject outliers found via the Shortest Half-estimator from the data and compute the usual variance from the remaining values.

2.6 S_n- and Q_n-scale estimators

Croux and Rousseeuw (1992, 1993) describe algorithms for two *Highly Robust Estimators of Scale* which were proposed in the Technical Report Rousseeuw and Croux (1991).

2.6.1 The estimator S_n

This estimator of scale of the observations $(y_1, \ldots, y_n)$ is defined by

$$S_n = 1.1926 \times \operatorname*{lomed}_{i=1,\ldots,n} \; \operatorname*{himed}_{j=1,\ldots,n} |y_i - y_j|$$

where *lomed* means the $[(n+1)/2]$-*th* order statistic and *himed* the $([n/2]+1)$-*th* one. It has 50% breakdown point but only 58% (gaussian) efficiency.

2.6.2 The estimator Q_n

This estimator of scale is defined by

$$Q_n = d_n \times 2.2219 \times \{|y_i - y_j|; i < j\}_{(k)}$$

where $k = \binom{h}{2}$ with $h = [\frac{n}{2}] + 1$ and d_n is a small sample correction factor. It also has 50% breakdown point, is computationally more difficult, but attains 82% efficiency.

3 The Simulation

A motivation for this simulation study was the practical situation in the 2- or 3-dimensional space where the spatial structure is of interest. Quite often, the investigation is started on a regular grid. However, to also understand the microvariability, it is additionally sampled on a smaller grid in a subregion. A typical situation in two dimensions is sketched in Figure 1.

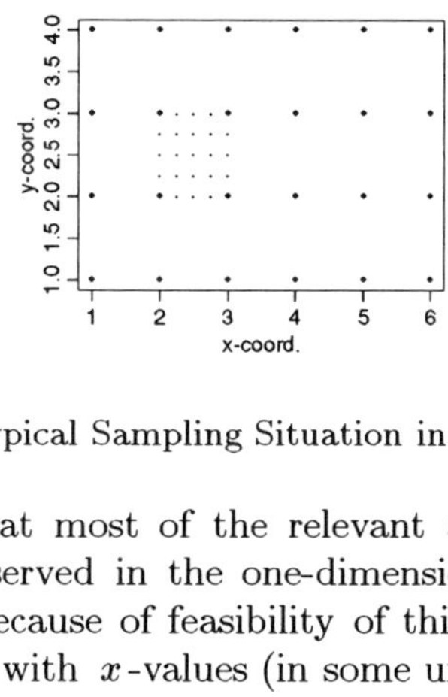

Figure 1: Typical Sampling Situation in 2 Dimensions

It is to be expected that most of the relevant aspects of the behavior of the estimators can be observed in the one-dimensional case (at least in case of isotropy). Therefore, because of feasibility of this study it has been chosen the one-dimensional space with x-values (in some units) of

$$0, 1, 2, \ldots, 39, 40, 40.3, 40.6, \ldots, 60.7, 61, 62, \ldots, 100$$

which means 150 sampling points. The density of points in the middle part is higher. Let us refer to this case as *Sampling Situation 1.*

For comparison with the regular grid case, we have also used *Sampling Situation 2* with x-values at

$$1, 2, \ldots, 100$$

which means 100 sampling points in the one-dimensional space.

3.1 Assumptions about the probability distribution

In the simplest situation, the data is supposed to be multinormally distributed with mean zero, variance $\sigma^2 = C$ and the covariance structure of two variables at spatial distance h given by

$$C(h) = C - \gamma(h)$$

where $\gamma(h)$ stands for the assumed (semi-)variogram. In particular, we have chosen the spherical model

$$\gamma(h) = \begin{cases} C(\frac{3}{2}\frac{h}{a} - \frac{1}{2}(\frac{h}{a})^3) & \text{for} \quad h \leq a \\ C & \text{for} \quad h > a. \end{cases}$$

The constant C also denotes the sill and a the range. Throughout the study, C is chosen to be equal to 1 and a equal 20. (See Figure 2.)

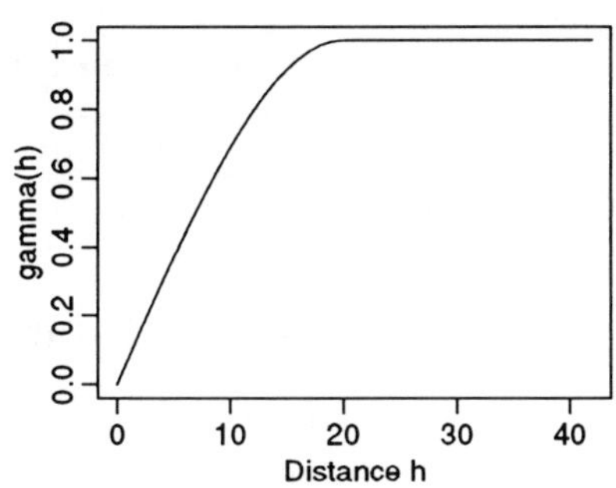

Figure 2: Spherical Variogram Model (Sill = 1, Range = 20)

Skewed distributions and data sets approximately normal in the middle part with some outliers are usual situations in geostatistical applications. It is the scope of this simulation study to generate such situations which can be compared with real applications.

The list of the distributions from which the simulated values are drawn and the corresponding labels to be referenced later on are given as follows.

N(0,1) – Normal distribution $\mathcal{N}(0,1)$.

LogN – Lognormal distribution.

LogNs – Lognormal distribution standardized, which means that the simulated values have been standardized to have zero mean and unit variance.

T3 – Student's t-distribution with 3 degrees of freedom. (Note that the variance is $p/(p-2) = 3$.)

NC551b – Normal with 5% contaminated values with heavy outliers (eventually easy to detect): $95\%\mathcal{N}(0,1) + 5\%\mathcal{N}(5,1)$. The independently generated values have probability .95 to be from $\mathcal{N}(0,1)$ and probability .05 to be from $\mathcal{N}(5,1)$; that is, normal distribution mean 5 and standard deviation 1. The contamination is done before (thereafter the letter b) applying the variogram (the *regionalization*). It may be argued that the outliers are then hard to detect.

NC551a – The $\mathcal{N}(0,1)$-distribution values are 'regionalized' and then, 5% of them are contaminated by replacing them by values $\mathcal{N}(5,1)$-distributed (i.e., adding a shift of 5). These outliers are easier to be detected.

NC533b – As NC551b, however, the contaminating distribution has shift 3 and standard deviation 3. These produced dispersed outliers may be less obvious.

NC533a – Similar to NC551a but contamination with $\mathcal{N}(3,3)$.

NC593b – As NC533b but heavy and dispersed outliers.

NC593a – As NC533a but heavy and dispersed outliers.

The different choices of the contaminating distributions for the contaminated normal models reflect different degrees of outliers.

3.2 Generation of the "regionalized data"

Denote now the vector of sample locations by $\boldsymbol{x} = (x_1, \ldots, x_n)^\top$ and the corresponding values of the regionalized variable by $\boldsymbol{z} = (z_1, \ldots, z_n)^\top$. Suppose that the probability distribution of the random function $\boldsymbol{Z}$ is

$$Z_i \sim \mathcal{N}(0, C), \qquad i = 1, \ldots, n,$$

and

$$\operatorname{Cov}(\boldsymbol{Z}) = \boldsymbol{C} = (C_{ij})$$

where

$$C_{ij} = C - \gamma(|\ x_i - x_j\ |).$$

The non-negative definite matrix $\boldsymbol{C}$ may be decomposed by

$$\boldsymbol{C} = RR^\top$$

and if the $(n \times 1)$-vector $\boldsymbol{y}$ is distributed $\mathcal{N}(\boldsymbol{0}, \boldsymbol{I})$, then

$$\boldsymbol{z} = R\boldsymbol{y}$$

satisfies the initial distributional assumption.

In this manner of *regionalization*, all the data for the simulation study (number of independent replications $n_{rep} = 100$) have been generated. We show a few examples of *Sampling Situation 1* (irregular grid) in Figure 3, namely independently distributed data without regionalization and dependent data (regionalized via the variogram γ) for the normal distribution, the t_3 and the extreme distribution NC593b.

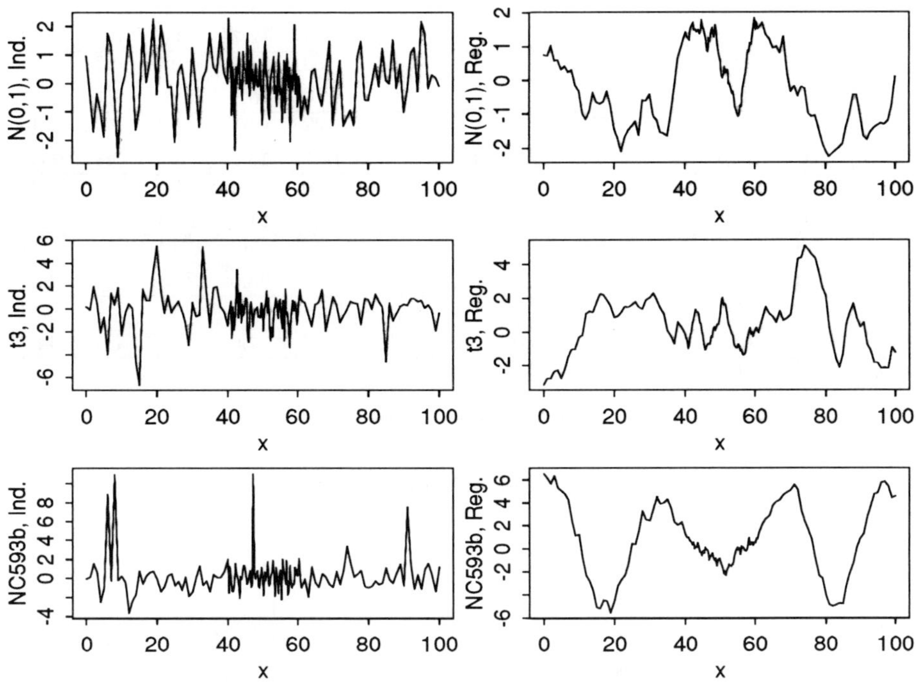

Figure 3: Some Simulated Data Sets

3.3 Estimation of the variogram

The *empirical* variograms $\widehat{\gamma}(h)$ for the different proposals of estimators have to be computed for different values of h, usually up to almost half the range of the x-values. For $\widehat{\gamma}(h)$ we need pairs of data values, say $(z(x_i), z(x_i + h))$ for $i = 1, \ldots, N_h$. In the regular *Sampling Situation 2* it is obvious that $\widehat{\gamma}(h)$ can be calculated only for certain discrete values, say,

$$h = 1, 2, \ldots, 50$$

where the corresponding numbers of available pairs are

$$N_h = 99, 98, \ldots, 51.$$

In the irregular case *Sampling Situation 1* it is possible to consider also values of h with a smaller incrementation. We have chosen the values of h with constant increment of .3. The values together with their possible numbers of pairs of data values are given in Table 1. Note the irregular numbers of pairs which are also graphed in Figure 4.

h	.3	.6	.9	1.2	1.5	1.8	2.1	2.4	2.7	3.0	3.3	3.6
N_h	70	69	68	67	66	65	64	63	62	136	62	61
h	3.9	4.2	4.5	4.8	5.1	5.4	5.7	6.0	6.3	6.6	...	8.7
N_h	60	59	58	57	56	55	54	122	54	53	...	46
h	9.0	9.3	...	11.7	12.0	12.3	...	14.7	15.0	15.3	...	17.7
N_h	108	46	...	38	94	38	...	30	80	30	...	22
h	18.0	18.3	...	20.7	21.0	21.3	...	23.7	24.0	24.3	...	26.7
N_h	66	22	...	14	52	14	...	14	49	14	...	14
h	27.0	27.3	...	29.7	30.0	30.3	...	32.7	33.0	33.3	...	35.7
N_h	46	14	...	14	43	14	...	14	40	14	...	14
h	36.0	36.3	...	38.7	39.0	39.3	...	41.7	42.0	42.3	...	44.7
N_h	37	14	...	14	34	14	...	14	34	12	...	12
h	45											
N_h	35											

Table 1: N_h and h in the Irregular Case

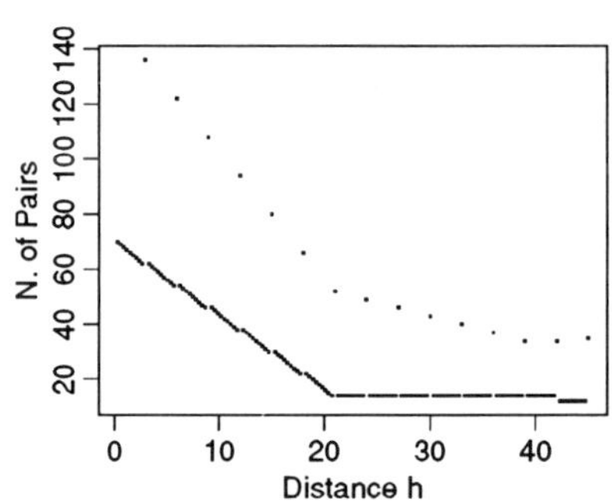

Figure 4: Number of Available Pairs N_h over h

4 Results

For each *Sampling Situation* (*1* and *2*) and the 10 considered distributions we have generated $n_{rep} = 100$ independent data sets and computed the variogram using the mentioned estimators for each data set. (A trial with 10.000 replicates has not produced any other conclusion than those stated here.)

4.1 The normal distribution case

Let us start with the regular case *Sampling Situation 2*. If we compute the value of the Usual Estimator (equation 2.1) for each $h = 1, \ldots, 49$ and plot the values, we obtain the left part of Figure 5, which is sometimes called *variogram cloud*. One might expect to see some central variogram estimate which, however, does not show up immediately. If one observes the scaling of this plot, it is quite surprising. Of course, the distribution of the values of the empirical $\hat{\gamma}$ is skewed

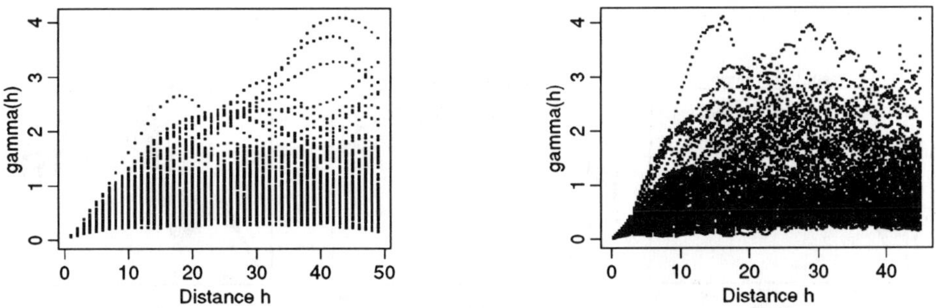

Figure 5: Normal Distribution, Usual Estimator

(χ^2-distribution!) but the scatterplot looks like showing many light and a few heavy outliers. The second part of Figure 5 gives the variogram cloud for the irregular case.

Let us now average all the computed $n_{rep} = 100$ variogram estimates. The *Sampling Situation 2* with selected 6 estimators is shown in Figure 6. The averaged values as well as the small confidence band with $\pm$ standard error are plotted. It was to be expected that all estimators find the correct variogram

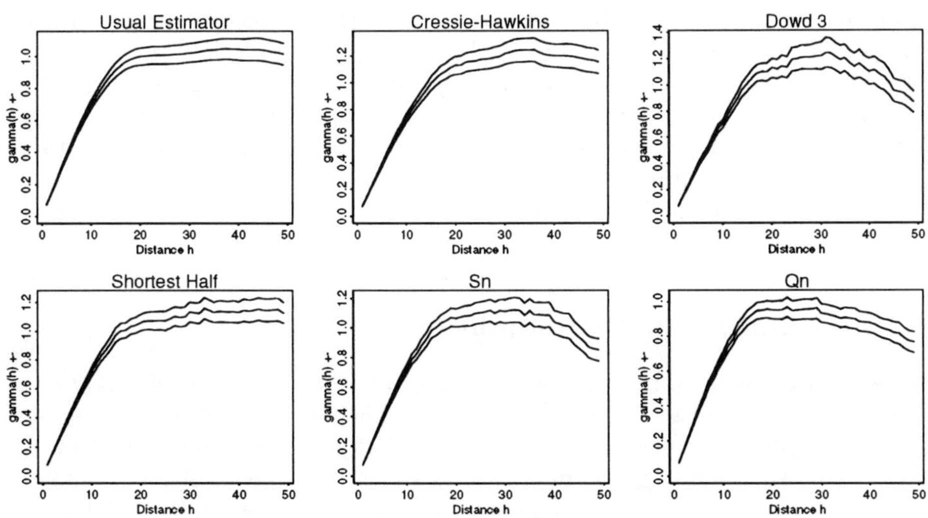

Figure 6: Normal Distribution: Regular Grid

within a certain confidence band, which is essentially the case. The graphs of Dowd$_3$, S_n and Q_n behave somewhat strangely for large distances, especially Q_n underestimates the variogram. Here we like to mention that we have checked carefully the algorithms and the programs for the estimators, also small sample corrections, especially S_n and Q_n have been thoroughly tested in many simulations.

Sampling Situation 1 is considered in Figure 7 where a similar behavior of the

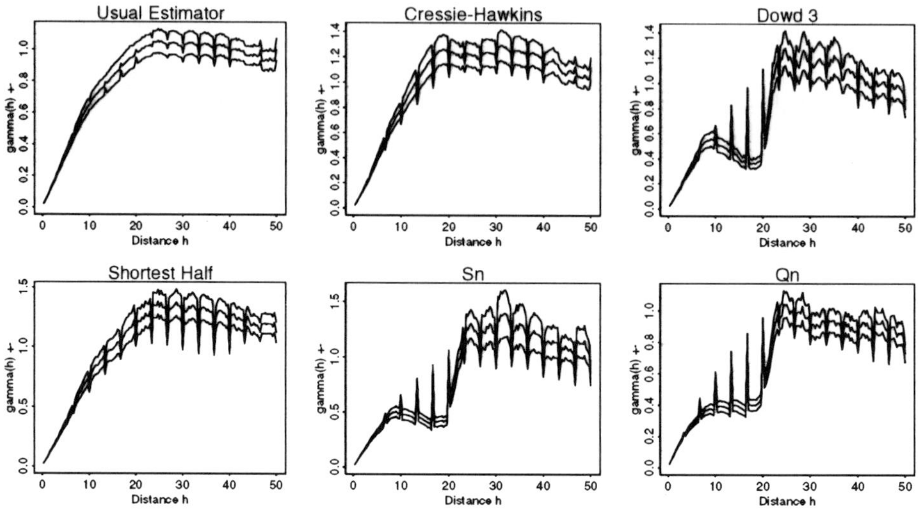

Figure 7: Normal Distribution: Irregular Grid

estimates should show up. However, we notice that the picture is very erratic. Every 10th value is at a reasonable place (comparable to those in Figure 6). The other values seem to be unreasonably different. Referring to Figure 4, it is clear that the sample sizes for these lags are very different and they have a high influence on some estimators. Nevertheless, one should realize that these irregularly spaced sample situations are usually met in practical applications. Here we know what should be the result and we have considered the ideal normal distribution as a starting point.

4.2 Regularly spaced samples

Let us now consider in more detail *Sampling Situation 2*, the regular grid case. Already here, we can see large differences in the behavior of the estimators. The report is again on the estimators Usual Estimator, Cressie–Hawkins, Dowd$_3$, Shortest Half, S_n and Q_n. Figures 8 to 12 show the behavior of the estimators in case of t_3-distributed data, lognormal-standardized, NC533b, NC593a and NC593b distributions.

The t_3-distribution (Figure 8) interestingly yields quite similar graphs for the considered estimators. Of course, the expected variance is not one but three. The lognormal-standardized distribution seems to provoke an underestimation of the variance by S_n and Q_n (Figure 9).

The contaminated normal distributions show more differences in the behavior of the estimators. Looking at the situation with the distribution NC533b Figure 10 reveals that all estimators compute higher values for the variogram

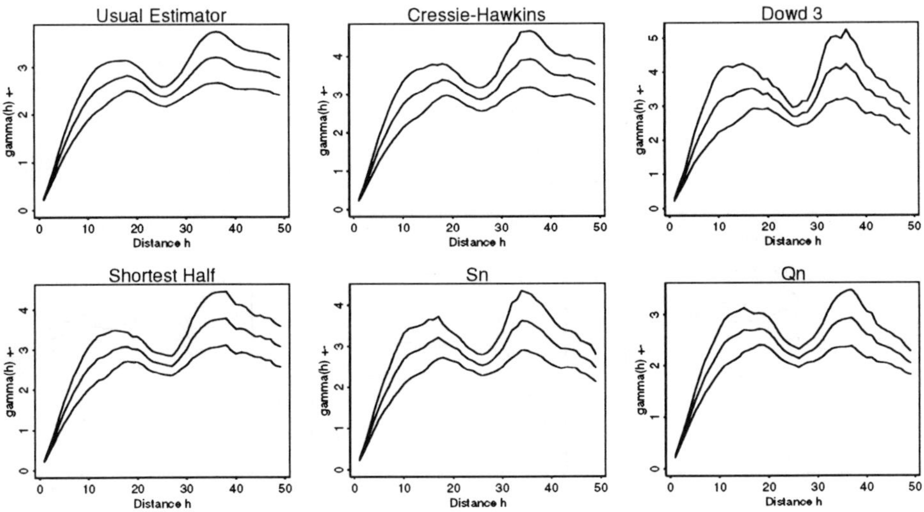

Figure 8: t_3 Distribution: Regular Grid

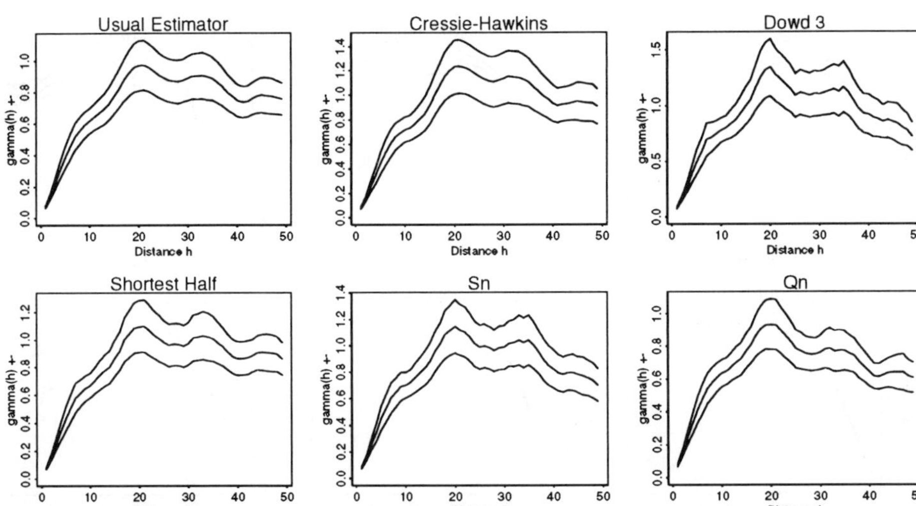

Figure 9: Distribution Lognormal-Stand.: Regular Grid

(the introduced outliers are hardly to be recognized). The best graph is given by Q_n.

The pictures based on the distribution NC593a and NC593b with heavy and dispersed outliers show obviously different behaviors (Figures 11 and 12).

In case of NC593a where outliers should be recognized quite easily, the Usual Estimator behaves extremely badly. The estimation of the variogram is 5 times too high, and it also shows a very high *nugget effect* which has not been intended

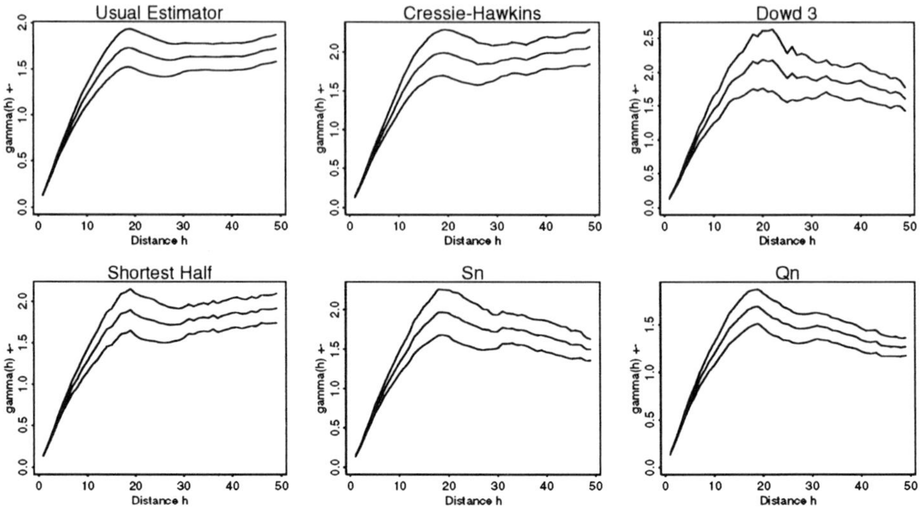

Figure 10: Distribution NC533b: Regular Grid

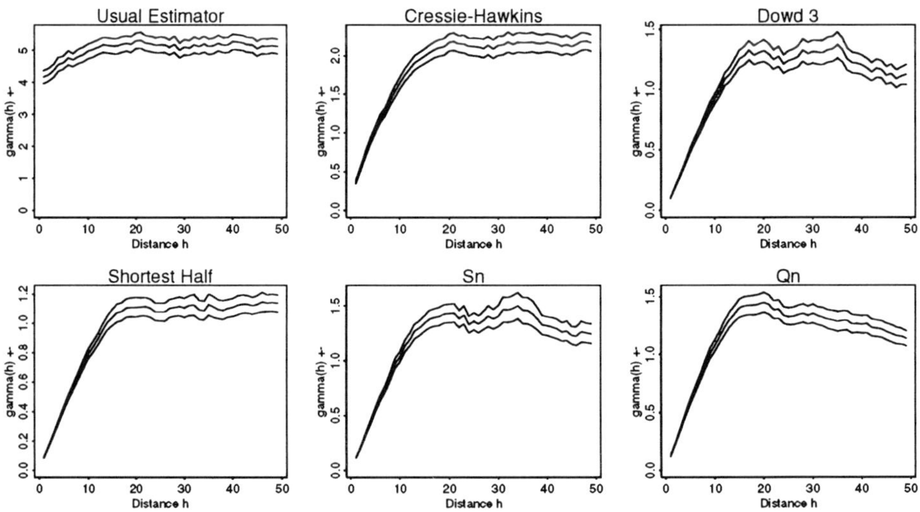

Figure 11: Distribution NC593a: Regular Grid

in the data. The Cressie–Hawkins Estimator finds a variance of about 2, only the other estimators manage quite well, best seems to be the Shortest Half.

The situation NC593b is very hard to handle for all involved estimators. They compute values for the variance from 4 to 6. Remarkably bad results are delivered from Dowd_3 but the (so-called) robust estimator by Cressie–Hawkins is not much better. The best behavior is shown by Q_n.

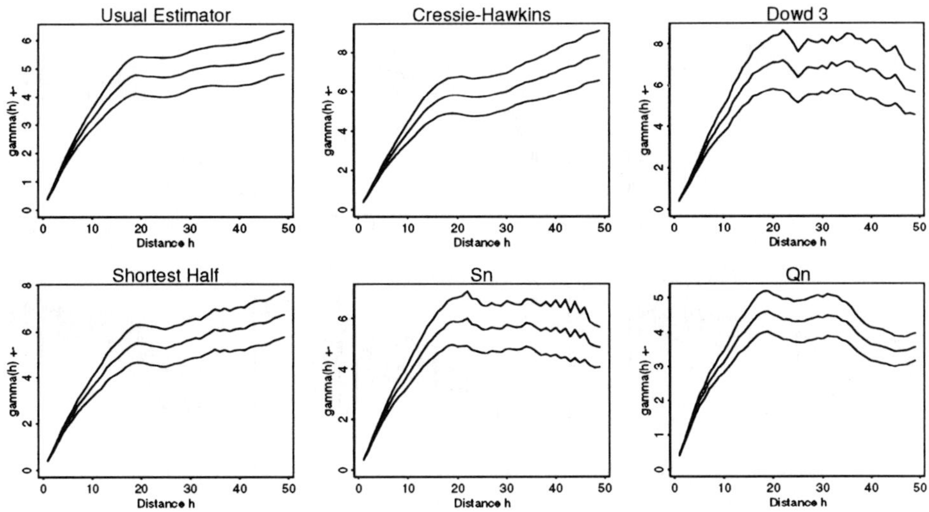

Figure 12: Distribution NC593b: Regular Grid

4.3 The irregular grid case

If we consider *Sampling Situation 1* and plot the values of the usual estimator, it already looks quite erratic in the normal distribution case (see Figure 7). Any other examined estimator reacts much worse, so the Cressie–Hawkins Estimator in Figure 7 and extremely badly Dowd$_3$ and also S_n and Q_n.

Now we report shortly the results on the distributions considered in the previous subsections and show the averaged variograms $\pm$ standard error in Figures 13 to 17.

The t_3-distribution situation mainly provokes the estimation of the real variance to be 3. However, the erratic behavior of Dowd$_3$, S_n and Q_n is remarkable. Q_n underestimates again. Similar statements can be given about the lognormal-standardized situation (Figure 14).

The contaminated normal distributions show a similar situation as in the regular distance case (Figures 15, 16 and 17), except that the values at *lags* not equal to multiples of 3 are often quite different. This can specially be observed for Dowd$_3$, S_n and Q_n, not so much for Shortest Half.

If one plots each 10th value of the estimated variograms in this irregular case, one obtains quite nice and smooth pictures. More simulation results can be found in Wurzer (1990), Harmacek (1992) and in a Technical Report.

5 Conclusions

Let us now try to summarize first on the regular grid case *Sampling Situation 2*. Table 2 displays that, in case of normal distribution, of course, all estimators show a good behavior, except that one might find Q_n is underestimating the

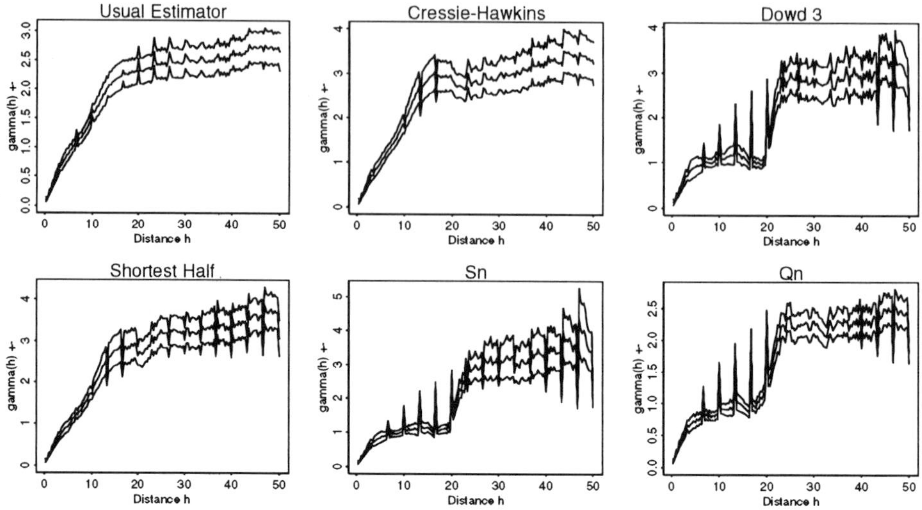

Figure 13: t_3 Distribution: Irregular Grid

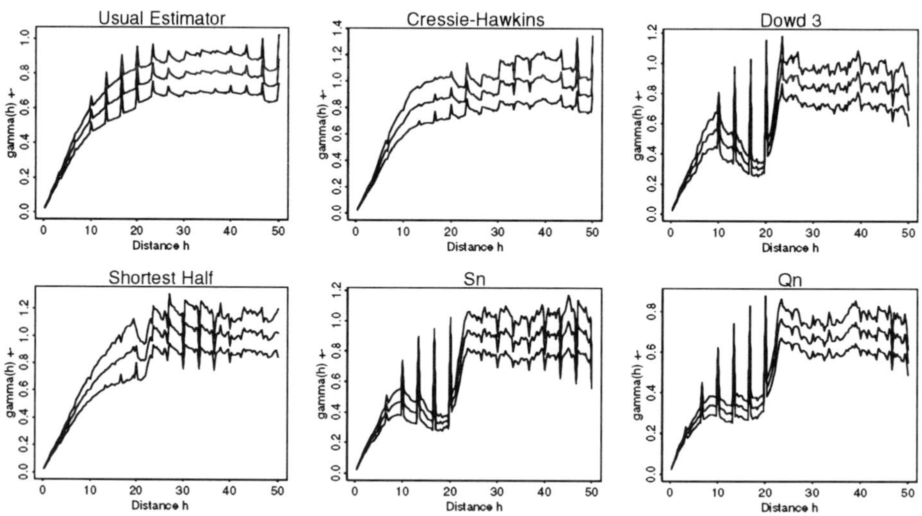

Figure 14: Distribution Lognormal-Stand.: Irregular Grid

variance (C$\downarrow$). The lognormal-standardized situation might be underestimated by S_n and Q_n because of the standardization (without actually having a normal distribution). C_0 in the column of Usual Estimators means the *nugget effect*. The question marks in the table indicate inability of interpretation.

From *Sampling Situation 1* we conclude:

- A good sampling plan is important.
- Pairs of samples should, whenever possible, used only from places *far*

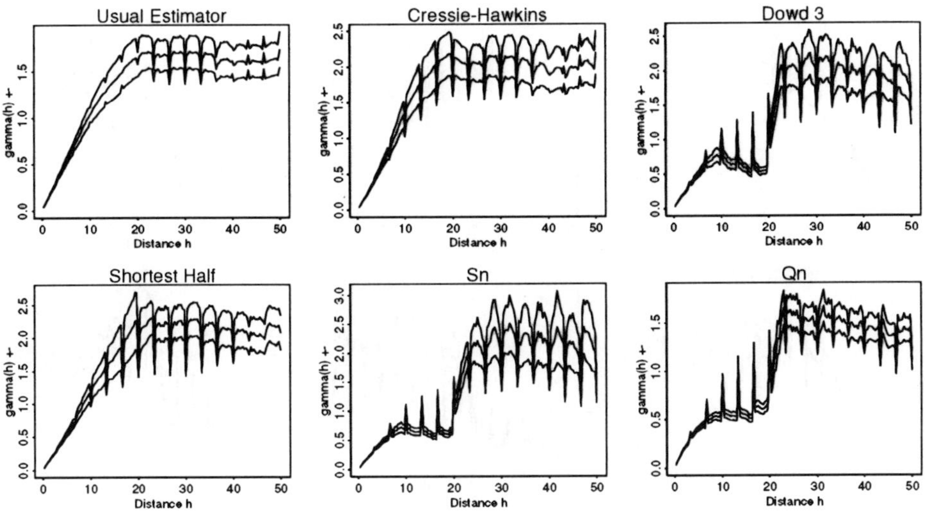

Figure 15: Distribution NC533b: Irregular Grid

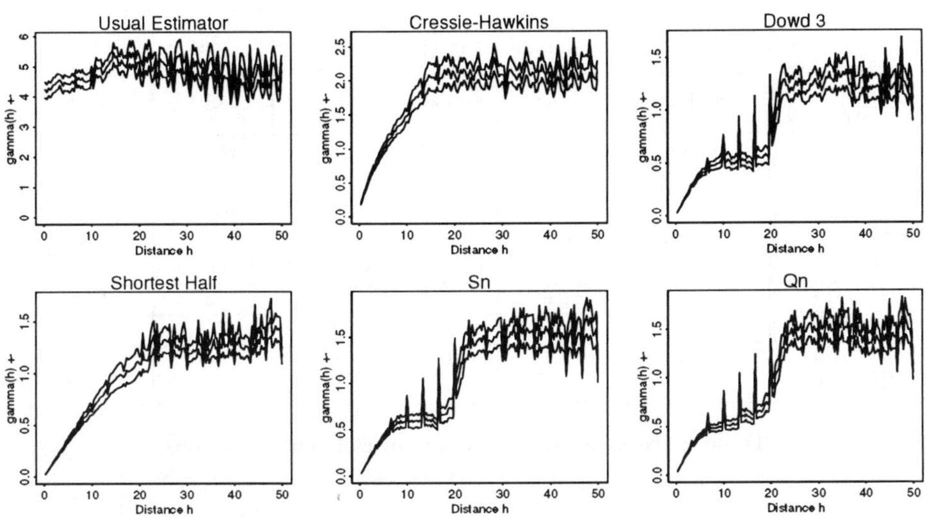

Figure 16: Distribution NC593a: Irregular Grid

enough of each other (*independence* relative to the dependence within a pair!).

- Small sample correction in estimators is essential.

From the second conclusion, in cases of sampling on a large grid and additionally in a small area on a small grid we could suggest: Compute the variogram for small h from the small grid and for larger h from the large grid *only* (and not using the small grid any more).

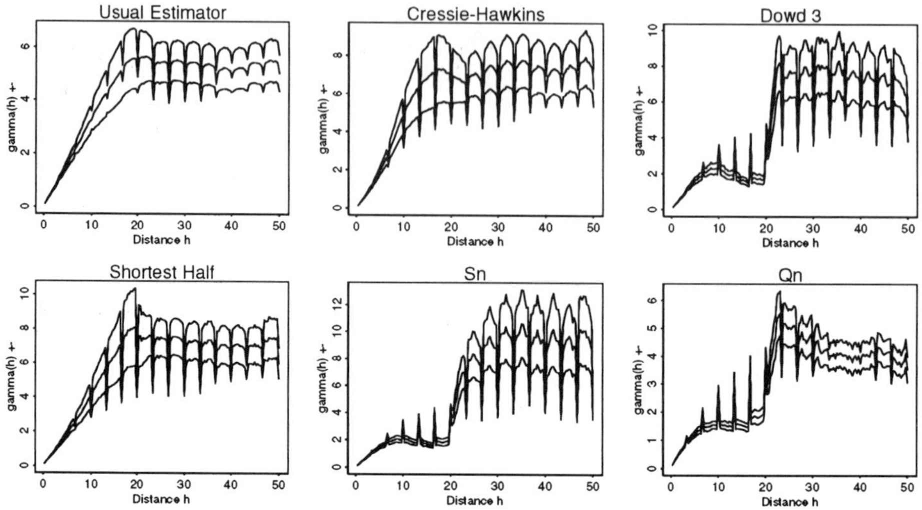

Figure 17: Distribution NC593b: Irregular Grid

	Usual Estimator	Cressie Hawkins	D_3 (MAD)	Short. Half	S_n	Q_n
Normal Distr.	perfect	o.k.	o.k.	o.k.	o.k.	$C \downarrow$
t_3	wavy	wavy	wavy	wavy	wavy	wavy
Lognorm. Stand.	wavy	wavy	wavy	wavy	$C \downarrow$ (stand.!)	$C \downarrow$ (stand.!)
NC533a	C = 1.7	C = 1.4	o.k.	o.k.	o.k.	o.k. ?
NC533b	C = 1.5 ?	C = 2.0	C = 2.0	C = 1.7	C = 1.6	C = 1.5 ?
NC551a	C = 2.1	1.6	1.2	1.1	1.3	1.3 ?
NC551b	2.0	2.1 ?	2.1	2.1	2.1	1.8
NC593a	5.2 ! $C_0 = 4$!	2.1	1.3	1.1	1.4	1.3
NC593b	5 !	6 !	6	5.7	5.5	4.2 ?

Table 2: Behavior of Estimators in Different Situations

References

[1] Cressie, N. and Hawkins, D.M. (1980): Robust estimation of the variogram. *Math. Geol.* **12**(2) 115–125.

[2] Croux, C. and Rousseeuw, P.J. (1992): Time-efficient algorithms for two highly robust estimators of scale. In *Computational Statistics* (Y. Dodge and J. Whittaker, eds.), Vol. 1, 411–428. Physica-Verlag, Heidelberg.

[3] Dowd, P.A. (1984): The variogram and kriging: Robust and resistent estimators. In *Geostatistics for Natural Resource Characterization* (G. Verly et al., eds.), 91–106. D. Reidel, Dordrecht.

[4] Dutter, R. (1985): *Geostatistik. Eine Einführung mit Anwendungen.* B.G. Teubner, Stuttgart.

[5] Harmacek, P. (1992): *Geostatistics: Robust Estimation for the Semivariogram—A Simulation Study*. Master's Thesis, Inst. f. Statistik u. Wahrscheinlichkeitstheorie, Technical University, Vienna. In German.

[6] Journel, A.G. and Huijbregts, Ch.J. (1978): *Mining Geostatistics*. Academic Press, New York.

[7] Matheron, G. (1971): *The Theory of Regionalized Variables and Its Applications*. Les Cahiers du Centre de Morphologie Mathématique de Paris, Fontainebleau, Frankreich.

[8] Rousseeuw, P.J. (1983): Multivariate estimation with high breakdown point. In *Mathematical Statistics and Applications* Fourth Pannonian Symposium on Mathematical Statistics and Probability, Bad Tatzmannsdorf, Austria, September 4–10, 1983 (I. Vincze, W. Grossmann, G. Pflug and W. Wertz, eds.), Vol. B, 283–297, Budapest, 1985. Akadémiai Kiadó.

[9] Rousseeuw, P.J. and Croux,C. (1991): *Alternatives to the median absolute deviation.* Technical Report 91–43, Universitaire Instelling Antwerpen, Antwerpen, Belgium.

[10] Rousseeuw, P.J. and Croux, C. (1993). Alternatives to the median absolute deviation. *J. Amer. Statist. Assoc.* **88**(424) 1273–1283.

[11] Rousseeuw, P.J. and Leroy, A.M. (1987): *Robust Regression and Outlier Detection.* John Wiley, New York.

[12] Wurzer, F. (1990): *Geostatistics: Exploratory, Resistant, and Robust Approaches.* Ph.D. Thesis, Technical University, Vienna.

Advances in Nonparametric Function Estimation

Theo Gasser
University of Zürich

Abstract

This paper has three aims: First, to demonstrate the need for nonparametric curve fitting techniques and to provide an introduction to such statistical techniques. Second, whatever method we want to apply, selection of a smoothing parameter or bandwidth is required. It would be desirable to determine the appropriate bandwidth from the data themselves, tuned to the problem at hand. Considerable progress has been made in recent years, most notably by the introduction of "plug-in" selectors, and the essence of this development is summarized. Third, a variety of function estimators has been discussed in the literature. Quite recently, local polynomial fitting stirred a lot of interest, seemed to become the ultimate answer. The potential pitfalls of these estimators and some remedies are discussed.

Key words and phrases: Curve estimation; nonparametric regression; bandwidth choice, kernel estimators; local polynomials.
AMS 1991 subject classifications: Primary 62G07; secondary 62G20.

1 Introduction

Regression analysis is a classical tool of statistics, and classically such an analysis is based on some prespecified parametric model. For reasons of simplicity, it would be desirable to work with a linear model, perhaps after some transformation of variable(s). The term "nonlinear models" already implies some of the difficulties with them: they have no positive definition, but convey only the information that a linear model is not appropriate. For one predictor x and an outcome variable y, the nonlinear parametric regression problem is defined as follows:

$$y_i = F(x_i, \underline{\theta}) + \varepsilon_i, \qquad i = 1, \ldots, n$$

where F is some fixed model, parametrized by the vector $\underline{\theta}$, and ε_i are residuals with $\mathrm{E}(\varepsilon_i) = 0$ and variance σ^2, possibly heteroscedastic. While estimating $\underline{\theta}$ may not always be easy due to the nonlinear nature of the problem, the main problem is the selection of F. While statisticians hope for knowledge-based suggestions from the field of application, our colleagues from other fields often consider model selection to be entirely the responsibility of the statistician. This results usually either in using models from the literature or in a trial-and-error search. In biomedicine, where curves from many subjects—and often for different groups like patients and controls—arise, this strategy may be far from successful.

On the other hand, the area becomes more and more important due to technical progress (electrodes, sensors, etc. leading in a semi-automatic way to curve information) and to using longitudinal rather than cross-sectional designs. Then, a nonparametric approach might be an attractive alternative:

$$y_i = r(x_i) + \varepsilon_i, \qquad i = 1, \ldots, n$$

Here, the function to be estimated is assumed to be a smooth function (e.g., $r \in C^{p+1}$). In addition to r itself, its derivatives are often of interest (usually r', r''), and also some characteristic features such as location and size of extrema, inflection points etc. (often used as "descriptive parameters" in subsequent analysis). The analysis of human growth curves proved to be a good example (Figure 1) for comparing a parametric model considered the best in the literature and a nonparametric kernel fit.

The girl shown is extremely regularly growing so that the lack of fit can be judged by eye. A principal discrepancy between model and data is that biologically a so-called mid-growth spurt occurs at about age 7 which is not part of the data whereas the pubertal spurt at about age 12 is characterized adequately by the model. The kernel fit brings out both features well. All common nonparametric fitting methods are equivalent to a weighted moving average of the data, and a weight function (or: kernel) and a smoothing parameter (or: bandwidth) has to be selected when applying them.

This paper is intended to give a brief overview on recent developments in function estimation for the non-specialist. It introduces first nonparametric function fitting by the kernel method , which has properties typical of many other methods (Section 2). The selection of the bandwidth from the data is treated in Section 3 both in theoretical and algorithmic terms. Section 3 discusses pros and cons of various fitting methods, and in particular properties of local plynomial fitting.

2 Kernel Regression Estimation

Given data (x_i, y_i), a kernel regression estimator $\hat{r}_\nu(x_0)$ for $r^{(\nu)}(x_0)$, $0 \leq x_0 \leq 1$ $(\nu = 0, 1, 2, \ldots)$ is defined as follows (Gasser and Müller (1979, 1984)):

$$\hat{r}_\nu(x_0) = \frac{1}{h^{\nu+1}} \sum_{i=1}^{n} y_i(x_i) \int_{s_{i-1}}^{s_i} K_\nu\left(\frac{x_0 - u}{h}\right) du$$

where:
$s_i =$ $(x_i + x_{i+1})/2$
$h =$ bandwidth $0 \leq h \leq 1/2$
$K_\nu =$ kernel, satisfying certain moment conditions

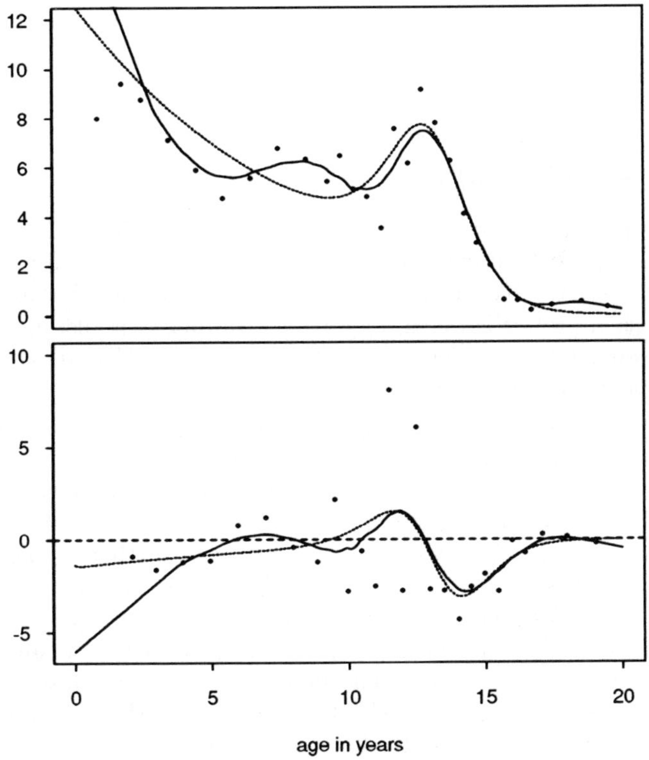

Figure 1: Height velocity (above) and acceleration (below) for a regularly growing girl (very low residual variance). Triangles: raw standardized increments, dashed lines: derivations of parametric Preece–Baines fit to r itself, solid lines: kernel fits to r using derivative kernels.

The design $x_1, \ldots, x_n$ may be fixed or random, and in both cases may be characterized by a design density f_x . Henceforth, we will consider estimation for $x_0 \in [0,1]$. The kernel K_ν has to satisfy the following moment conditions:

$$\int_{-1}^{1} K_\nu(x) x^j dx = \begin{cases} 0, & j = 0, \ldots, \nu-1, \nu+1, \ldots, k-1 \\ (-1)^\nu \nu!, & j = \nu \\ \beta \neq 0, & j = k \end{cases}$$

and is called a kernel of order (ν, k) $(k = \nu+2, \nu+4, \ldots)$. Moment conditions arise in a similar way for wavelets (Donoho et al. 1994). Based on an asymptotic analysis with $h \to 0$, $n\,h \to \infty$, the leading terms of bias and variance are as follows:

$$\text{Bias}(\widehat{r}_\nu(x_0); h) \quad = \quad \frac{(-1)^k \nu!\, r^{(k)}(x_0)\, h^{k-\nu} M_k}{k!} \;(1 + \text{o}(1)),$$

$$\mathrm{Var}(\widehat{r}_\nu(x_0); h) = \frac{\nu!^2 \sigma^2 V}{n\, h^{2\nu+1} f_x(x_0)} (1 + o(1))$$

where:

$$M_k = \int_{-1}^{1} K_\nu(x) x^k dx$$
$$V = \int_{-1}^{1} K_\nu(x)^2 dx$$

The derivation of these results shows that the asymptotic variance approximates the finite sample behaviour well, and that the agreement for the bias is satisfactory as long as the bandwidth is not too large (confirmed by simulations). The validity of asymptotics is much poorer for local polynomials, including the Nadaraya–Watson estimator (Seifert and Gasser 1994). Various convergence properties can be obtained under mild conditions.

Application of this estimator needs specification of a kernel K_ν and a global bandwidth h or a local bandwidth $h(x_0)$. Bandwidth selection has attracted a lot of attention and will be treated in Section 3. Kernel functions minimizing either the asymptotic variance or the asymptotic mean integreated square error (MISE) have been derived in Gasser et al. (1985) and further studied in Granovsky and Müller (1989). They have an explicit representation as polynomials and the integral in the estimator does thus not pose any practical problem. The above analysis is valid for $x_0 \in [h, 1-h]$, whereas lower order terms dominate at the boundary $[0, h)$ or $(1-h, 1]$. This can be avoided by the use of boundary kernels with asymmetric support which allow estimating $r^{(\nu)}$ on $[0,1]$ (Gasser et al. 1985). Such kernels should also be used when estimating a density with compact or semi-compact support. The application of the method was facilitated by the introduction of a fast algorithm: whereas a conventional implementation needs roughly n^2 operations, a fast updating algorithm needs only $O(n)$ operations. This is of particular importance for automatically collected mass data , or when repeatedly evaluating a fit, such as in bootstrapping.

3 Bandwidth Selection

The choice of bandwidth h is decisive for the appearance and quality of the fit. A visual trial-and-error search will often lead to an adequate fit, but it is time-consuming and subjective, a matter of concern to many scientists.

If successful, the estimation of the optimal bandwidth from the data would solve many problems at the same time. The formulae for the asymptotically optimal global bandwidth h_{asy}

$$h_{\text{asy}} = \left(\frac{k!^2\,(2\nu+1) V\, \sigma^2 \displaystyle\int 1/f_x(x) w(x)\, dx}{2\, n\, (k-\nu) M_k^2 \displaystyle\int \left(r^{(k)}(x)\right)^2 w(x)\, dx} \right)^{1/(2k+1)}$$

or local bandwidth $h_{\text{asy}}(x_0)$

$$h_{\text{asy}}(x_0) = \left(\frac{k!^2\,(2\nu+1)V\,\sigma^2 f_x(x_0)^{-1}}{2\,n\,(k-\nu)M_k^2\left(r^{(k)}(x_0)\right)^2} \right)^{1/(2k+1)}$$

might illustrate the problem (w is some weight function). A good large sample choice is essentially determined by the residual variance σ^2 and a derivative of the regression function, the curvature r'' in case of estimating the regression function r itself with a second order kernel K. The use of asymptotic formula was not considered practical for a long time, and research concentrated on methods based on cross-validation, unbiased risk estimation and related selectors (see Härdle 1990, and the literature cited there). All these residual-based methods for bandwidth selection have the same asymptotic behaviour (Härdle et al. 1988). Despite some nice optimality results for cross-validation—evidently too weak in retrospective—practical results (applied or simulated) were rather disappointing due to a substantial variability in the bandwidth chosen. Independently, a number of researchers started to study alternatives (in the context of density estimation see e.g. Hall et al. 1991; Sheather and Jones 1991). From simulations we knew that cross-validation leads to too variable results, stemming from the fact that the cross-validation function to be minimized is somewhat noisy, has often more than one local minimum and often no clearly discernible minimum due to a decline to the left.

The situation seemed to improve when a $\sqrt{n}$-consistent estimator of σ^2 was found, which is easy to compute (Gasser et al. 1986). The residual variance is part of the asymptotically optimal bandwidth h_{asy}, and this led us finally to base bandwidth selection on h_{asy}, by suggesting an iterative allgorithm for estimating $\int (r''(x))^2\,dx$ (Gasser et al. 1991)

(i) Start with a $h_0 = 1/n$, close to interpolation.

(ii) Iterate $i = 1, \ldots, i^*$

$$h_i = \left(\frac{V\hat{\sigma}^2 \displaystyle\int 1/f_x(x)w(x)\,dx}{n\,M_k^2 \displaystyle\int \left(\hat{r}_2''(x; n^{1/10}h_{i-1})\right)^2 w(x)\,dx} \right)^{1/(2k+1)}$$

(iii) Stop for $i^* = 11$ and put $\hat{h}_{\text{opt}} = h_{11}$.

The number of iterations and the choice of the inflation factor $n^{1/10}$ for the pilot bandwidth $g(h) = n^{1/10}h$ are based on theoretical asymptotic arguments, which illustrates well even the practical usefulness of an asymptotic analysis. The following relative rates of convergence can be derived for $\hat{h}_{\text{opt}}$ with respect to the MISE-optimal bandwidth h_{opt}:

$$\hat{h}_{\text{opt}} = h_{\text{opt}}(1 + O(n^{-q/5}) + O_P(n^{-1/2}))$$

whereas variance achieves the parametric rate of convergence bias is below with $q = 1$ or 2, depending on the order of kernel used for estimating r''. Hall and Marron (1991) showed that the optimal overall rate is $O_P(n^{-1/2})$, whereas cross-validation achieves only a rate of $O_P(n^{-1/10})$.

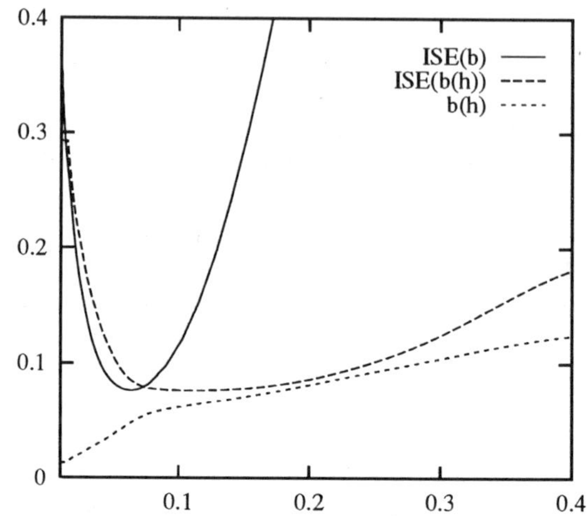

Figure 2: Example of integrated squared error ISE(h) (solid line), dependence of ISE on pilot bandwidth g (long dashes) and dependence of bandwidth h on pilot bandwidth g (short dashes).

Figure 2 illustrates why plug-in estimators work so well in practice: whereas the dependence of mean square error on the bandwidth h is crucial, its dependence on the pilot bandwidth g remains attractive over a broad range.

This procedure can easily be extended to a non-equally spaced design, since estimation of $\int 1/f_x(x)\,dx$ is relatively straight-forward, and also to using higher order kernels. While selecting the bandwidth for estimating derivatives r' or r'' does not present principal obstacles, the actual procedure needs some extra care. When applying any bandwidth selector in a situation with correlated errors ε_i, a severe—usually negative—bias in the bandwidth selected arises (Hart 1991). However, by estimating the following functional s (replacing σ^2) from the data

$$s = \sum_{k=-\infty}^{\infty} R_k\,,$$

(R_k = covariance with lag k) the plug-in approach can be successfully extended to a situation signal plus stationary residuals (Herrmann et al. 1992). This is remarkable, since no parametric model is assumed either for the regression function or for the time series.

Figure 3 shows that the plug-in appproach also works well for fitting two-dimensional data. Here, and in other more complicated models than univariate function fitting, the plug-in selection offers its greatest virtues compared to a

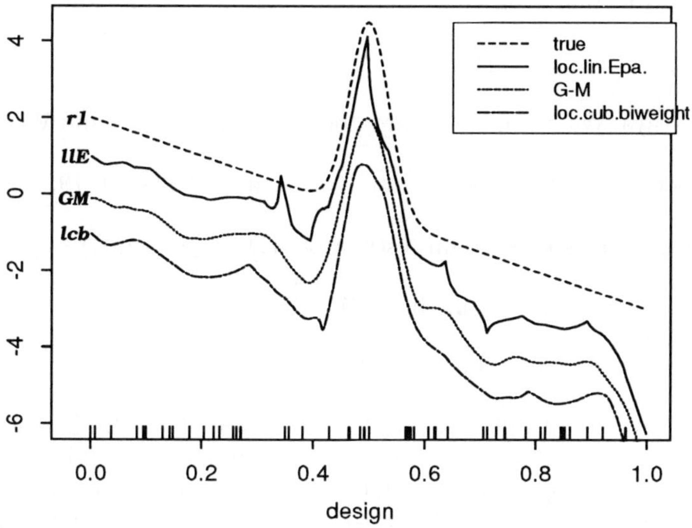

Figure 3: Example of plug-in kernel fit to two-dimensional simulated data with two peaks. Left above: true function, right above: noisy data, left below: kernel fit using the estimated plug-in bandwidth pair, right below: true optimal kernel fit.

visual trial-and-error search for an appropriate smoothing parameter which often becomes practically impossible.

Intuitively, a local bandwidth selection seems to be even a more attractive procedure, reducing the bandwidth where structure appears, and extending it in flat parts, thereby optimizing locally the variance-bias-compromise. On the other hand, the dimensionality of the problem increases dramatically, and a local procedure could become non-conservative by locking into residual structure. Both arguments are right: As shown by Müller and Stadtmüller (1987) the potential gains of a local procedure can in fact be realized asymptotically. On the other hand, a naive implementation of a local plug-in selector, based on the asymptotically optimal local bandwidth is not even consistent (Brockmann et al. 1993). A rather sophisticated local adaptation, guarding against overfitting, performs well. This was also true for the non-smooth examples constructed for wavelet shrinkage by Donoho et al. (1994), where global linear methods did badly. Therefore, the good performance of wavelet shrinkage might be more due to local adaptation than to its nonlinear properties. Nevertheless, a price has to be paid in terms of the relative rate of convergence: the optimal rate drops from $O_P(n^{-1/2})$ for global selection (Hall and Marron 1991) to $O_P(n^{-2/9})$ for local selection (Wang and Gasser 1994).

4 Choice of Estimator

A variety of function estimators has been suggested in the literature, and there has been a vivid discussion about the pros and cons of different methods. Well-known methods are the following:

1. Nadaraya–Watson kernel estimator (Nadaraya (1964) and Watson (1964))

2. Gasser–Müller kernel estimator (Gasser and Müller (1979, 1984))

3. k-nearest-neighbor (k-nn) estimators (Devroye (1978))

4. Smoothing splines (Reinsch (1967))

5. Local polynomials (Macauley (1931))

Estimators 3 and 4 differ essentially in the way, how the bandwidth adapts to the local design density $f_x(x_0)$: a k-nn estimator has a bandwidth which is approximately proportional to $f_x(x_0)^{-1}$ and thus eliminates approximately the dependence. Smoothing splines behave asymptotically as a kernel estimator with a smoothing parameter proportional to $f_x(x_0)^{-1/4}$.

It can be shown that no such rule is uniformly optimal for all combinations of design and regression functions. We, therefore, studied minimax properties by introducing the following design-dependent bandwidth scheme (Jennen–Steinmetz and Gasser (1988)):

$$h(x_0) = h_0\, f_x(x_0)^{-\alpha} \quad (0 \leq \alpha \leq 1)$$

with $\alpha = 0, 1/4, 1$ as particularly interesting schemes. The analysis showed that a fixed, design-independent bandwidth; i.e., $\alpha = 0$, is asymptotically minimax optimal.

The other estimators mentioned above, as well as further estimators not given, differ mainly in the way, how weights are assigned to individual design points. The Nadaraya–Watson estimator

$$\hat{r}(x_0) = \sum_{i=1}^{n} \left\{ K\left(\frac{x_i - x_0}{h}\right) \Big/ \sum_{j=1}^{n} K\left(\frac{x_i - x_0}{h}\right) \right\} y_i$$

just evaluates kernel function and does not take into account the spacings between different design points. Local polynomials are obtained as solution to the following least squares problem

$$\sum_{i=1}^{n} \left(Y_i - \sum_{j=0}^{p} \beta_j\, (x_i - x_0)^j \right)^2 K\left(\frac{x_i - x_0}{h}\right) = \min!$$

The Nadaraya–Watson estimator can also be interpreted as a local constant fit ($p = 0$). Fan and Gijbels (1992) have shown that for estimating r itself odd-order polynomials should be used and not even-order polynomials. The local linear fit has an explicit representation as a weighted average:

$$\hat{r}(x_0) = \sum_{i=1}^{n} \left\{ \frac{s_2 - s_1(x_i - x_0)}{s_0 s_2 - s_1^2} K\left(\frac{x_i - x_0}{h}\right) \right\} y_i$$

where s_j are j-th moments with respect to K (Fan (1993)). The weights of the local linear fit account for an unequal spacing between design points, similar to the Gasser–Müller kernel estimator, but different from the Nadaraya–Watson estimator. The leading terms of bias are in comparison as follows ($k = 2, \nu = 0$):

$$\text{Bias}(\hat{r}(x_0); h) = \begin{cases} c_1 h^2 r''(x_0)(1 + o(1)) \\ c_1 h^2 (r''(x_0) + 2r'(x_0) f_x'(x_0)/f_x(x_0))(1 + o(1)) \end{cases}$$

The local linear fit and the Gasser–Müller estimator are the same (given above) while the Nadaraya–Watson estimator (below) has a disturbing additional term. The leading term for the variance is

$$\text{Var}(\hat{r}(x_0); h) = \frac{c_2 \sigma^2}{n\, h\, f_x(x_0)} \;(1 + o(1))$$

which holds for the local linear fit, for the Nadaraya–Watson estimator, but for the Gasser–Müller estimator only in case of fixed design. For random design, its variance is increased by a factor 1.5 (Chu and Marron (1991)). This comparison makes local polynomials an attractive choice, and in fact, Fan (1993) showed minimax optimality for local linear fits among all linear estimators. Other attractive features are an automatic adjustment to the boundary, without using boundary kernels, and a fast updating algorithms needing $O(n)$ operations only (Seifert, et al. (1994)). This is the good message, and the bad message is that in general the conditional finite sample variance is unbounded and the unconditional finite sample variance is infinite (Seifert and Gasser (1994)). This is true for random design, where local polynomials should be particularly adequate, and the reason can be seen in the definition given above: the moments of the denominator can become close to zero with a substantial probability in randomly occurring sparse regions.

The effect can be easily seen in a simulated fit (Figure 4) where the optimal Epanechnikov weights and the realization-wise optimal band have been used. The variance problems become visually evident in a somewhat rugged visual appearance of the local linear fit.

For these reasons Fan (1993) had to use a modified linear estimator in his analysis. The modification consists in adding a term $1/n^2$ to prevent vanishing of the denominator. While this avoids an infinite variance, it is hardly a practically appealing solution. We have proposed and studied two methods which should combine a good finite sample and a good asymptotic behaviour (Seifert and Gasser (1994)):

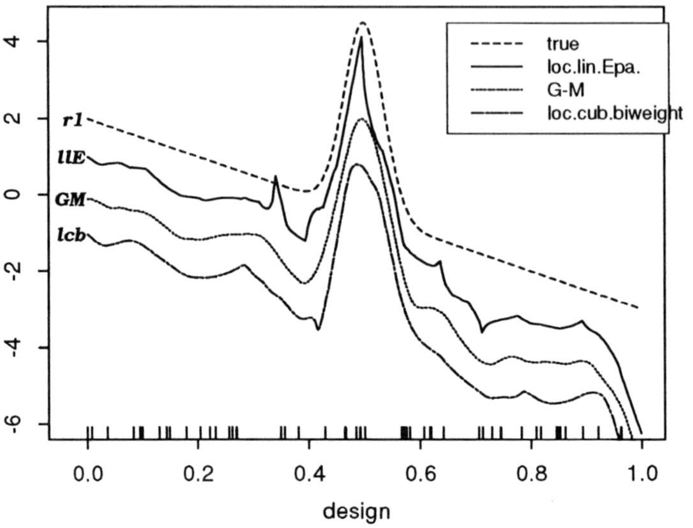

Figure 4: Comparison of fits on simulated data using realization-wise optimal bandwidth: Dashed: true function, solid line: local linear fit using Epanechnikoff weights, dotted: kernel fit, dash-dotted: local cubic fit with biweights.

(i) Local polynomial ridge regression

(ii) a heuristic technique for identifying regions in the design space which lead to a high variance and increase the bandwidth there appropriately.

Both methods look promising but more detailed work is needed. If successful, local polynomials could eventually become the method of choice for fitting regression functions nonparametrically.

References

[1] Brockmann, M., Gasser, Th. and Herrmann, E. (1993): Locally adaptive bandwidth choice for kernel regression estimators. *J. Amer. Statist. Assoc.* **88** 1302–1309.

[2] Chu, C.K. and Marron, J.S. (1991): Choosing a kernel regression estimator. *Statistical Science* **6** 404–433.

[3] Devroye, L.P. (1978): The uniform convergence of nearest neighbor regression function estimators and their application in optimization. *IEEE Trans. Inform. Theory* **IT–24** 142–151.

[4] Donoho, D., Johnstone, I., Kerkyacharian, G. and Picard, D. (1994): Wavelet Shrinkage: Asymptopia? *J. Roy. Statist. Soc.*, to appear.

[5] Fan, J. (1993): Local linear regression smoothers and their minimax efficiencies. *Ann. Statist.* **21** 196–216.

[6] Fan, J. and Gijbels, I. (1992): Variable bandwidth and local linear regression smoothers. *Ann. Statist.* **20** 2008–2036.

[7] Gasser, Th. and Müller, H.–G. (1979): Kernel estimation of regression functions. *Smoothing techniques for curve estimation. Lecture Notes in Mathematics* #757 23–68. New York, Springer.

[8] Gasser, Th. and Müller, H.–G. (1984): Estimating regression functions and their derivatives by the kernel method. *Scand. J. Statist.* **11** 171–185.

[9] Gasser, Th., Müller, H.–G. and Mammitzsch, V. (1985): Kernels for nonparametric curve estimation. *J. Roy. Statist. Soc. B* **47** 238–252.

[10] Gasser, Th., Sroka, L. and Jennen–Steinmetz, Ch. (1986): Residual variance and residual pattern in nonlinear regression. *Biometrika* **73** 625–633.

[11] Gasser, Th., Kneip, A., and Köhler, W. (1991): A flexible and fast method for automatic smoothing. *J. Amer. Statist. Assoc.* **86** 643–652.

[12] Granowski, B.L. and Müller, H.–G. (1989): On the optimality of a class of polynomial kernels. *Statist. and Decisions* **7** 301–312.

[13] Hall, P., Sheather, S.J., Jones, M.C. and Marron, J.S. (1991): On optimal data–based bandwidth selection in kernel density estimation. *Biometrika* **78** 263–269.

[14] Hall, P. and Marron, J.S. (1991): Lower bounds for bandwidth selection in density estimation. *Probab. Th. Rel. Fields* **90** 149–173.

[15] Härdle, W., Hall, P. and Marron, J.S. (1988): How far are automatically chosen regression smoothing parameters from their optimum? (with discussion). *J. Amer. Statist. Assoc.* **83** 86–101.

[16] Härdle, W. (1990): *Applied Nonparametric Regression*. Cambridge Univ. Press, Cambridge.

[17] Hart, J.D. (1991): Kernel regression estimation with time series errors. *J. Roy. Statist. Soc. B* **53** 173–188.

[18] Herrmann, E., Gasser, Th. and Kneip, A. (1992): Choice of bandwidth for kernel regression when residuals are correlated. *Biometrika* **79** 783–796.

[19] Jennen–Steinmetz, Ch. and Gasser, Th. (1988): A unifying approach to nonparametric regression estimation. *J. Amer. Statist. Assoc.* **83** 1084–1089.

[20] Macauley, F.R. (1931): *The Smoothing of Time Series.* National Bureau of Economic Research, New York.

[21] Müller, H.–G. and Stadtmüller, U. (1987): Variable bandwidth kernel estimators of regression curves. *Ann. Statist.* **15** 182–201.

[22] Nadaraya, E.A. (1964): On estimating regression. *Theory Probab. Appl.* **9** 141–142.

[23] Reinsch, C.H. (1967): Smoothing by spline functions. *Numer. Math.* **10** 177–183.

[24] Seifert, B. and Gasser, Th. (1994): *Finite Sample Variance of Local Polynomials: Analysis and Solutions.* Manuscript.

[25] Seifert, B., Brockmann, M., Engel, J. and Gasser, T. (1994): Fast algorithms for nonparametric curve estimation. *J. Comp. Graph. Statist.* **3** 192–213.

[26] Sheather, S. and Jones, M.C. (1991): A reliable data–based bandwidth selection method for kernel density estimation. *J. Roy. Statist. Soc. B* **53** 683–690.

[27] Wang, K. and Gasser, Th. (1994): *Optimal Rates of Convergence for Local Bandwidth Selection.* Manuscript.

[28] Watson, G.S. (1964): Smooth regression analysis. *Sankhyā A* **26** 359–372.

On the Philosophical Foundations of Statistics: Bridges to Huber's work, and recent results

Frank Hampel
ETH Zürich

Abstract

After a few words in honor of P.J. Huber and a brief introduction into a new frequentist paradigm for the foundations of statistics, which uses upper and lower probabilities and the concept of "successful bets", the talk discusses some very recent research results with more general implications: Successful bets for the normal and the exponential shift model, asymptotically successful bets, and a sketch for approximately successful bets in the context of robust statistics.

Key words and phrases: Robust statistics, Huber-estimator; foundations of statistical inference, upper and lower probabilities, successful bets, prediction, parametric models, frequentist interpretation of probabilities, epistemic probabilities, learning process, state of total ignorance; successful bets for normal and exponential shift model, asymptotically successful bets, approximately successful bets in the robustness context.
AMS 1991 subject classifications: Primary 62A99; secondary 62F35.

1 A Personal Introduction

It is a great honor and pleasure for me to speak on the occasion of Peter Huber's 60th birthday. Since my arrival at the University of Zürich in 1968, and until his departure from ETH to Harvard about 10 years later, we had a wonderful scientific relationship, and we managed to keep up our friendship ever since. Our scientific relationship was not of the usual type; we never wrote a joint paper together; but rather we exchanged our ideas and results while pursuing our own lines of research which were often closely parallel, and certainly I was greatly stimulated and motivated in this atmosphere of deep mutual understanding and pursuit of common scientific aims.

It would be easy to become nostalgic about such a common past; and it would take far too much time to try to give an appreciation of Huber's major scientific achievements, even in robust statistics alone. Therefore let me just mention one famous result, the Huber-estimator given by Huber's ψ-function. It was first published exactly 30 years ago (Huber 1964; reprinted, with an introduction, in Kotz and Johnson (eds.) 1992, Vol. II) and is now often simply denoted by the name of its inventor, which can be pronounced and written in many languages and scripts of the world, including Greek, Chinese (e.g., "Double–Tiger Sun"), and Tifinagh. Peter Huber himself can even denote it in cuneiform writing.

Starting from this, let me offer to him as a little "toy" birthday present the idea of a "Huber without corner," more formally given by $\xi(x-S) = \mathrm{E}_\Phi\ \psi_k(x-T_n)$, where T_n is the classical Huber-estimator for n observations with its ψ_k-function, and Φ is the standard normal. Graphically, for $k = 1.5$ and $n = 20$ it differs quantitatively very little from the Huber-function, but for $n = 5$ the difference becomes more noticeable. At least for $n \geq 5$, the highly smooth curves in the corners for different values of n differ almost exactly by affine transformations from each other, presenting an aesthetically pleasing picture. The definition may also be iterated, at least a few times. Its (higher order) Fréchet-differentiability and its smoothness in the strong sense of Mächler (1989, 1994) may be analysed, and its empirical small sample and small sample asymptotic behavior may be studied and compared in their relation to the classical Huber with, for example, Andrews's sine and biweight in their relation to three-part descending M-estimators (Andrews et al. (1972), p. 155, and Hampel et al. (1986), p. 408), but also as a function of the small sample sizes for which it may be most interesting.

Before I come to the main topic of my talk, let me conclude the personal introduction for Peter Huber with three selected citations: with Schiller's famous statement about stupidity (as opposed to ignorance, to be discussed later) from the "Jungfrau von Orleans" 3, 6; with a citation from Ibsen's "Peer Gynt" used by Jeffreys ((1939) 1961, p. 193), and with Satz 7 of Wittgenstein's "Tractatus logico-philosophicus."

2 The Inference Problem and Successful Bets

Huber has not only worked in robust statistics (cf. Huber (1981)), data analysis and computer intensive methods. There are also two interesting though probably rather unknown papers on the philosophical foundations of statistics (Huber (1973, 1976)) based on the usefulness of capacities in robust statistics and pointing towards a generalization of probabilities in inference which few other researchers had become aware of as an important step beyond the classical paradigms of inference. Because of these papers and my present interest in foundations, I'd like to speak about my most recent results in foundations, and in particular I'd like to outline the first connection between my approach to foundations and robust statistics. Before I can do so, I have to give a brief introduction into the inference problem and my approach to it.

One of the most basic problems in the foundations of statistics is the following: Given a parametric model, possibly given an observation from that model, and starting either with the state of total ignorance or with some prior information about the parameter, what can we infer about a future observation from that model? A prototype example are two independent binomials, one observed in the past, one to be observed in the future, with the same parameter p. A variant of the problem (which looks simpler at first, but is really less well defined) is to make inference about the unknown parameter.

Partial solutions to this problem have been given by Bayes, Fisher, Jeffreys, Neyman, the Neobayesians (starting with de Finetti), Dempster and Shafer, among others. None is fully satisfactory or complete. It is worthwhile studying Bayes' (1763) paper, and especially his scholium in detail (cf. Hampel (1994)), in particular since his uniform prior for p was partly blindly believed, partly severely attacked (and both possibly for the wrong reasons). Briefly said, Bayes was not a Bayesian, but a frequentist. Only in his scholium does he deviate from his frequentist framework, but what he seems to say is: Within the paradigm of ordinary probabilities, I see no other solution for starting with the state of total ignorance than the one given (which, by the way, has a deeper interpretation—in terms of maximum entropy for the discrete predictive distribution—than most readers seem to have realized). Another solution was proposed by Fisher (1930, 1956) with his fiducial argument, which has led to much controversy and confusion, but which has at most very limited applicability, even if it is accepted. Jeffreys (1939) tried to propose a canonical Bayesian prior for the state of total ignorance, while de Finetti and the Neobayesians suggested a more or less arbitrary and completely subjective first prior. Dempster (1968) and Shafer (1976) used (formally) upper and lower probabilities in their belief function theory and thus were able to express the state of total ignorance formally in a reasonable way; but all statements are again subjective, and the rules for updating and combining evidence look nice, but arbitrary. (Similar remarks can be applied to fuzzy set theory.) Finally Neyman (1957, 1971) explicitly denied the existence of inductive inference and the inference problem and considered only objective (aleatory, as opposed to epistemic) probabilities with their mathematical properties, and inductive behavior. The resulting arbitrariness of procedures in the Neyman–Pearson theory can perhaps be described by a citation on where to go from here from "Alice in Wonderland" (cf. Jeffreys (1939) 1961, p. 245), while the arbitrariness of subjective priors and methods has been likened to "gauging a fever thermometer by means of the patient's shivers" (van Dantzig (1957)). (By the way, in 1971 Peter Huber jokingly added a "Statistical Priesthood III" to van Dantzig's two papers with this title.)

Since 1988, I am trying to find out what can be achieved for the inference (and decision) problem within a frequentist framework (Hampel (1993a, 1993b)). Some main features and results are: the stress on prediction (as opposed to parameter estimation and testing); the necessity to work with (something like) upper and lower probabilities; the explicit use of epistemic probabilities (referring to our state of knowledge, not to the true unknown state of Nature as exclusively in the Neyman–Pearson theory) which however are intersubjective and not subjective, and which have a frequentist interpretation. The framework is still incomplete (especially the uniqueness problem is not yet solved), but it shows already that it is possible to develop an objective, frequentist solution to the inference problem which contains the Bayesian approach (for aleatory priors) as a special case, which contains Fisher's fiducial probabilities, as far as valid, as a very special case (and bridges the gap between discrete and continuous model distributions), which combines features of the Neyman–Pearson, Dempster–Shafer, likelihood and information theoretic approaches, and which allows an objective learning process, starting with the state of total ignorance, and arriving asymptotically at the true probability model, thereby solving a major problem of artificial intelligence (AI). (Some special solutions turn out to be Choquet capacities of order ∞, however I do not require this axiomatically.) There are also potential drawbacks, depending on one's viewpoint, as in other frequentist theories: thus, the strong likelihood principle is not generally valid, and the solutions depend also on potential observations which have not been observed.

In the following more formal description of this approach, I shall omit all the ramifications leading to information theory, Bayes theory and prior knowledge in between total ignorance and Bayesian priors, the limited validity and proper interpretation of the fiducial argument, confidence intervals and tests, etc., and concentrate on one of the key concepts, namely the concept of "successful bets," and on the most challenging situation, namely starting with "total ignorance" about the parameter (given a parametric model).

Consider a given parametric model $(\Omega, \mathcal{A}, \{P_\theta, \theta \in \Theta\})$ with two random variables X (past observation) and Y (future observation) defined on it. Let M_0, with $0 \leq M_0 \leq 1$, be a function on the subsets of Θ which describes our apriori knowledge on θ and, very loosely speaking, tells us how likely every parameter set may be. For consistency reasons, we require that each M_0 is $\geq$ some proper Bayesian prior distribution π. We shall mainly consider the "state of total ignorance" given by $M_0(B) \equiv 1, \forall\, \emptyset \neq B \subset \Theta$, and $M_0(\emptyset) = 0$, and denoted briefly by "$M_0 \equiv 1$."

We now describe our partial knowledge about the future occurrence of the event $Y \in A$ $(A \neq \emptyset)$, given M_0 and after $X = x$ has been observed, by two equivalent functions:

$$m = m(A|x, M_0) = m(Y \in A|X = x, M_0), \quad 0 \leq m \leq 1$$

(corresponding to Möglichkeit, possibility, plausibility, upper probability, alternating capacity in various systems) and

$$\underline{m} = \underline{m}(A|x, M_0) = \underline{m}(Y \in A|X = x, M_0), \quad 0 \leq \underline{m} \leq 1, \quad \underline{m} \leq m$$

(corresponding to Sicherheit, necessity, belief, lower probability, monotone capacity)

$$\text{with } m(A|x, M_0) + \underline{m}(A^c|x, M_0) \equiv 1.$$

An interpretation of m and $\underline{m}$ is:

$$\underline{m}(A|\ldots) = c \Leftrightarrow \text{ I am willing to bet } c : (1-c) \text{ on } A.$$

The expected gain of my bet with stake 1 is then

$$-c(1-P(A))+(1-c)P(A)=P(A)-c.$$

The smaller c, the larger is my expected gain (or the smaller my expected loss), but I am also trying to offer informative bets with as large a c as I deem reasonable, because a large c means more (true or assumed) knowledge.

Note that I am not necessarily willing to bet $(1-c):c$ on A^c. On the one hand, I could always find subjective Bayesians willing to accept my bet on A, including all those with subjective probability 1 for A^c; on the other hand, I consider a neutral betting partner who accepts all my bets, such as Nature, a "statistician's stooge" (I.J. Good), or my own scientific conscience—the imagined bet is just a measure of my strength of conviction of the occurrence of A.

A natural requirement for the bets I am willing to offer is that their expected gain be ≥ 0, no matter what the possible $\theta's$ (or priors for θ) are. Now we have to consider the more complicated situation that I started with total ignorance, but have made an observation $X=x$. Both m and $\underline{m}$ are then functions of x, for every fixed A. Obviously, if A is highly probable under all values of θ highly likely under x, I should be able to place a strong bet (i.e., with large c) on A. However, with most parametric models, it is always possible that the true θ is highly unlikely under x and gives a very low probability for A, hence my expected gain for my conditional bet, given x, may be strongly negative. In other words, there are usually no nontrivial conditional bets with expected gain ≥ 0 for every x separately. In order to get out of total ignorance and learn (asymptotically the true probability of A), I have to be able to compensate the rare but bad cases with a misleading x by being allowed to average my gain over all values of X (under the true but unknown distribution of X). Somewhat surprisingly, this is indeed possible, and it leads to the following definition (in terms of the equivalent m):

$$m(\text{and } \underline{m}) \text{ leads to " successful bets"}$$
$$\text{iff } \mathrm{E}_\theta\, m(A|X, M_0 \equiv 1)\ \geq P_\theta(Y\in A), \quad \forall\theta,$$

where the expectation is with respect to the distribution of I_A and the full distribution of X. This means, my strategy or set of bets (one for every x) $(1-m(A|X,M_0\equiv 1)) : m(A|X,M_0\equiv 1)$ on A^c has (overall) expected gain ≥ 0, no matter what the true unknown θ is.

For completeness, the general definition is given:

$$m(\text{and } \underline{m}) \text{ leads to " successful bets"}$$
$$\text{iff } \mathrm{E}_\pi\, m(A|X, M_0) \geq \mathrm{E}_\pi\, P_\theta(Y\in A), \quad \forall\pi\leq M_0,$$

where the expectations are (also) with respect to every prior π for θ permitted by M_0.

The operational interpretation is the following: I can embed my conditional bet, given x (where the x has been selected at random by P_θ!), into a sequence of independent such bets (in different experiments!), and under weak conditions on the size of the stakes and of the betting ratios, my average gain will most probably be $\geq -\epsilon$ for a sufficiently large number of bets.

It can be seen qualitatively that "asymptotically" for highly informative X (e.g., X being a vector of independent $X_1, X_2, \ldots, X_n$ which for $n \to \infty$ determines θ a.s.) under mild conditions suitable successful m and $\underline{m}$ converge to $P_\theta(A)$ a.s.: Apart from a small probability ϵ, θ lies in a narrow confidence interval, and given some smoothness of the parametric model, all $P_\theta(A)$ derived from the confidence interval are very close to $P_\theta(A)$. Any m that allows for this latter inaccuracy and the unlikely event of noncoverage (and not much more), leads to successful bets and converges to $P_\theta(A)$.

To see numerically that there are nontrivial successful bets, let us consider the case of two independent binomial distributions $X \sim B(n, \theta)$ and $Y \sim B(k, \theta)$ for $n = k = 2$. For $A = \{Y = 1\}$, and abbreviating $m(A|x, M_0 \equiv 1)$ by $m(x)$ $(x = 0, 1, 2)$ for the moment, we have to find 3 numbers $m(x)$ such that

$$P_\theta(A) = 2\theta(1-\theta) \leq (1-\theta)^2 m(0) + 2\theta(1-\theta)m(1) + \theta^2 m(2), \quad \forall\theta.$$

"Admissible" (minimal) solutions are all triples with $0 \leq m(0) = m(2) = 1 - m(1) \leq 1$. For all other A, the admissible solutions are $m(A|x, M_0 \equiv 1) = I_A(x)$. A particularly interesting and simple successful (though not admissible) solution for all $A \neq \emptyset$ is obtained by computing the probability $m_1(A|x, M_0 \equiv 1) := P_{\hat\theta}(A)$ under the maximum likelihood estimator $\hat\theta = x/n$, and determining the smallest constant c such that $(m_1 + c) \wedge 1$ is successful. In our example $(n = k = 2)$, $c = 1/4$. Besides having other properties, this solution is also a 2-alternating Choquet capacity.

3 New Results

3.1 Successful bets for the normal shift model

One of the prototype continuous models in statistics is the normal shift model. Giving successful bets for the normal prediction problem shows that the approach can also be applied to continuous models. Moreover, in a way the solution bridges the gap between likelihoods and (some generalization of) probabilities (with a frequentist interpretation), something Fisher (1956) apparently might have loved to do (according to an attempt to read between his lines): likelihoods are transformed into frequentist statements, without additional auxiliary observations.

Let $X_1, X_2, \ldots$ (past) and $Y_1, Y_2, \ldots$ (future) be i.i.d. $\sim N(\theta, 1)$. Let $\overline{X}_n = \sum_1^n X_i/n$ and $\overline{Y}_k = \sum_1^k Y_j/k$; let $A \subset R^1$ (measurable) be given. Given $\overline{X}_n$ (and total ignorance regarding θ), how can I bet successfully against $\{\overline{Y}_k \in A\}$? In other words, what is a successful $m(\overline{Y}_k \in A|\overline{X}_n,\ M_\theta \equiv 1)$?

We can find nontrivial solutions for every A, for example by trial and error, but that would be rather cumbersome. Simple general solutions, based on

$m_1(A|\ldots) = P_{\overline{X}_n}(A)$ (the probability under the maximum likelihood estimator), can be found using a two-step Taylor-expansion:

$$\begin{aligned} P_{\overline{X}_n}(\overline{Y}_k \in A) &= P_{\theta+(\overline{X}_n-\theta)}(\overline{Y}_k \in A) \\ &= P_\theta(\overline{Y}_k \in A) + (\overline{X}_n - \theta)\int_A (-k)\varphi'(k^{1/2}(\overline{Y}_k - \theta))d\overline{Y}_k \\ &\quad + \frac{1}{2}(\overline{X}_n - \theta)^2 \int_A k^{3/2}\varphi''(k^{1/2}(\overline{Y}_k - \theta - \vartheta(\overline{X}_n - \theta)))d\overline{Y}_k, \end{aligned}$$

with $0 < \vartheta < 1$, and φ the standard normal density.

The integral of the second term does not depend on $\overline{X}_n$, and the expectation of the first factor in it (with respect to the distribution of $\overline{X}_n$) is zero. The integral of the third term is generously bounded by k in absolute value, and the expected squared term is $1/n$. Hence

$$| \mathrm{E}_\theta^{\overline{X}_n} P_{\overline{X}_n}(\overline{Y}_k \in A) - P_\theta(\overline{Y}_k \in A)| \leq \frac{1}{2}\frac{k}{n}.$$

Therefore we can define successful m's in the following ways:

$$\begin{aligned} m(\overline{Y}_k \in A|\overline{X}_n,\ M_0 \equiv 1) &= P_{\overline{X}_n}(\overline{Y}_k \in A) + \frac{1}{2}\frac{k}{n} \\ &\qquad \text{if } \sup_\theta P_\theta(\overline{Y}_k \in A) \leq 1 - \frac{1}{2}\frac{k}{n} \\ &= \text{(i), (ii), or (iii) otherwise.} \end{aligned}$$

(i) $m \equiv 1$

(ii) $m = \sup_\theta P_\theta(\overline{Y}_k \in A)$

(iii) $m = aP_{\overline{X}_n}(\overline{Y}_k \in A) + b$, with $a = s/c$, $b = 1 - s(1-s)/c$, $s = 1 - \sup_\theta P_\theta(\overline{Y}_k \in A)$, $c = \frac{1}{2}\frac{k}{n}$.

The (somewhat awkward) last three variants ensure that even for sets which under some θ have a high probability, the linear transformation of their estimated probabilities is ≤ 1 and its expectation $\geq$ the true probability. Apart from these exceptional sets, the solution has the same simple structure as the last binomial solution given above.

The constant $\frac{1}{2}\frac{k}{n}$ shows that (at least with this method) prediction (for fixed A) is the more accurate, the longer the past and the shorter the future is, and that a relatively long future cannot be predicted at all nontrivially. This makes sense, once one thinks more about it; in particular, it is not possible to make this kind of predictive claims about fixed (as opposed to random) parameter sets or parameters.

3.2 Successful bets for the exponential shift model

The exponential shift model is a prototype of a class of "nonregular" models where the likelihood suddenly jumps to zero and the maximum likelihood estimator is not asymptotically normal. The fact that even here a solution can be found easily, gives some hope for such practically important problems as variance component models, with their potentially negative unbiased variance estimates. (However, it should be clear that the solutions given here are only for the prediction problem and not for the parameter estimation problem, for various reasons, cf. above.)

Let X and Y be independent, without loss of generality X with density $n\exp(-n(x-\theta))$ for $x \geq \theta$ and 0 otherwise, and Y with density $\exp(-(y-\theta))$ for $y \geq \theta$ and 0 otherwise. Then $P_\theta(Y \in A) = \int I_{\{y \geq \theta\} \cap A} \cdot \exp(-(y-\theta))dy$. For $\Delta := X - \theta \geq 0$ with $\mathrm{E}\,\Delta = \frac{1}{n}$ we obtain

$$\begin{aligned} P_X(Y \in A) &= P_{\theta+\Delta}(Y \in A) \\ &= \int (I_{\{y \geq \theta\} \cap A} - I_{\{\theta \leq y \leq \theta+\Delta\} \cap A}) \exp(-(y-\theta)) \exp(\Delta) dy \\ &\geq P_\theta(Y \in A) - \int I_{\{\theta \leq y \leq \theta+\Delta\} \cap A} \exp(-(y-\theta)) dy \\ &\geq P_\theta(Y \in A) - \Delta \end{aligned}$$

because the first factor in the last integrand is bounded by $I_{\{\theta \leq y \leq \theta+\Delta\}}$ and the second factor by 1. Hence

$$\mathrm{E}_\theta^X P_X(Y \in A) \geq P_\theta(Y \in A) - \frac{1}{n}.$$

Therefore the following m-functions are successful:

$$\begin{aligned} m(Y \in A | X, M_0 \equiv 1) &= P_X(Y \in A) + \frac{1}{n}, \quad \text{if } \sup_\theta P_\theta(Y \in A) \leq 1 - \frac{1}{n} \\ &= \text{(i), (ii), or (iii) otherwise.} \end{aligned}$$

(i) $m \equiv 1$

(ii) $m = \sup_\theta P_\theta(Y \in A) =: 1 - s$

(iii) $m = aP_X(Y \in A) + b$, with $a = ns$ and $b = 1 - ns(1-s)$.

The solutions are qualitatively very similar to the ones for the normal distribution.

3.3 Asymptotically successful bets in the "regular" asymptotic case

The (strict) solution for the normal shift model provides an approximate ("asymptotic") solution for all "regular," asymptotically normal and independent statistics X (based on the past) and Y (to be considered in the future).

Let X and Y be independent and approximately or asymptotically normal,

$$X \dot{\sim} N(\theta, (I_X(\theta))^{-1}),\ Y \dot{\sim} N(\theta, (I_Y(\theta))^{-1}),$$

where $\dot{\sim}$ means "approximately distributed as" and the variances are the actual variances, with

$$I_X(\theta) \cong -\frac{\partial^2}{\partial\theta^2} \log \text{lik}_x(\theta)$$

and

$$I_Y(\theta) \cong -\frac{\partial^2}{\partial\theta^2} \log \text{lik}_y(\theta)$$

evaluated at the maximum likelihood estimator for θ if these functions are sufficiently smooth. Then we can replace k by $-\frac{\partial^2}{\partial\theta^2} \log \text{lik}_y(\theta)$ and n by $-\frac{\partial^2}{\partial\theta^2} \log \text{lik}_x(\theta)$ in the formulas for the normal shift model $\mathcal{N}(\theta, 1)$ in order to get an approximate solution. Of course, one would have to consider regularity conditions, types of convergence and remainder terms in every case or class of cases in order to get a properly "asymptotic" solution for some (imagined) sequence of X's in which the X at hand can be embedded. In practice, the approximation can probably be improved considerably by first using a variance stabilizing transformation.

3.4 Sketch for approximately ("asymptotically") successful bets in the robustness context

So far we have considered strict parametric models and strict or approximate distributions of statistics. But how can we deal with the following situation: " I expect about 10% gross errors, which could be anywhere, in this sort of data" (Hampel (1973), Huber (1973))? More generally, reality shows that parametric models are almost always only approximations and need to be treated as such. There exists now a considerable technology for dealing with approximate parametric models, or full neighborhoods of such models (cf. Huber (1981), Hampel et al. (1986), and the literature therein), and we can try to apply it to our problem. We shall make free use of concepts such as gross-error sensitivity and their practical interpretation, which is described in the literature. Our aim is again to find a useful approximation for a fixed given situation, but there are usually many possible ways to infinity in the background, providing various "asymptotic" results. Since this is a first attempt, no regularity conditions are discussed.

Let a parametric model $\{P_\theta\}$ and a number ϵ $(0 \leq \epsilon \leq 1)$, such as 5 % or 10 % be given, and assume the true underlying distribution is in $P_{\theta,\epsilon}$, where either

$$P_{\theta,\epsilon} = \{P : P = (1-\epsilon)P_\theta + \epsilon H,\ H \text{ arbitrary }\} \quad \text{(gross error model)}$$

or

$$P_{\theta,\epsilon} = \{P : \|P - P_\theta\|_{\text{total variation}} \leq \epsilon\} \quad \text{(total variation model)}.$$

One possible formalism for independent X's and Y's lets

$$X_1,\ldots,X_n \text{ i.i.d.} \sim F_X \in P_{\theta,\epsilon} \quad\text{and}\quad Y_1,\ldots,Y_k \text{ i.i.d.} \sim F_Y \in P_{\theta,\epsilon}$$

(F_X and F_Y need not be the same; moreover, the i.i.d. condition could also be weakened). Let $T(X_1,\ldots,X_n)$ and $S(Y_1,\ldots,Y_k)$ be robust and sufficiently smooth (e.g., Fréchet-differentiable) such that in good approximation under P_θ

$$T \sim N(\theta, \sigma_T^2/n)$$

with gross-error sensitivity γ_T^* and change-of-variance sensitivity κ_T^*, and

$$S \sim N(\theta, \sigma_S^2/k)$$

with γ_S^* and κ_S^*. Let A a (sufficiency "nice") set in the range of S be given, and define

$$H_A(\mu,\sigma^2) := \int_A dN(\mu,\sigma^2).$$

Thus, under P_θ, $P(S\in A) \cong H_A(\theta,\sigma_S^2/k)$, and we can again try to estimate it by $P_T(S\in A)$, obtaining from our previous results (3.1 and 3.3)

$$|\,\mathrm{E}_\theta\, P_T(S\in A) - P_\theta(S\in A)| \lessapprox \frac{1}{2}\frac{\sigma_T^2/n}{\sigma_S^2/k}.$$

Now come the robustness considerations.

Under some other distribution in $P_{\theta,\epsilon}$, the approximate expectation θ of S may change by about $\epsilon\gamma_S^*$, and its variance σ_S^2/k by about $\epsilon\kappa_S^*\sigma_S^2/k$. These changes may be put into $H_A(\theta,\sigma_S^2/k)$, hence H_A may change by $\lessapprox 0.4\epsilon\gamma_S^* k^{\frac{1}{2}}/\sigma_S + \epsilon\kappa_S^*\sigma_S/(8k^{\frac{1}{2}})$.

For T under $P_{\theta,\epsilon}, \theta$ may change by $\epsilon\gamma_T^*$, and σ_T^2/n by $\epsilon\kappa_T^*\sigma_T^2/n$ in $N(\theta,\sigma_T^2/n)$. Putting things together, we obtain (in obvious notation)

$$\begin{aligned}&|\,\mathrm{E}_{\theta,\epsilon}\, P_T(S\in A) - P_{\theta,\epsilon}(S\in A)|\\ &\lessapprox\ 0.4\epsilon(\gamma_T^*+\gamma_S^*)k^{\frac{1}{2}}/\sigma_S + \frac{1}{2}\frac{\sigma_T^2(1+\epsilon\kappa_T^*)/n}{\sigma_S^2(1-\epsilon\kappa_S^*)/k} + \frac{1}{8}\epsilon\kappa_S^*\sigma_S/k^{\frac{1}{2}}.\end{aligned}$$

This formula (whose middle term could perhaps also be simplified) can be looked at in different ways, and again some things can be learned from it. For $n\to\infty$, $k\to\infty$, $k/n\to 0$ the dominant term becomes $0.4\epsilon(\gamma_T^*+\gamma_S^*)k^{\frac{1}{2}}/\sigma_S$ which means that for large k no nontrivial inference is possible! The bias from contamination, together with the increasing concentration and therefore quick change of the distribution of S possibly at a boundary of A prevents an informative prediction. For $k/n\to\infty$, also the second term becomes large. We see, even stronger than in the strict parametric case, that with a long future, no nontrivial prediction for a fixed A is possible. This holds a fortiori for fixed (though not for random) parameters and parameter sets. For moderately small k (and not too small n), solutions analogous to the previous ones can be written down.

Let me conclude my talk with two quotes. One, by R. von Mises, sounds suitable to the genius loci and also fits to Peter Huber: " The leitmotif, the ever recurring melody, is that two things are indispensable in any reasoning, in any description we shape of a segment of reality: to submit to experience and to face the language that is used, with unceasing logical criticism."

The other one, by Artemus Ward, is a nice general warning of blindness, prejudice, and "false clarity," in science as well as in daily life: "It ain't so much the things we don't know that get us in trouble. It's the things we know that ain't so."

References

[1] Andrews, D.F., Bickel, P.J., Hampel, F.R., Huber, P.J., Rogers, W.H. and Tukey, J.W. (1972): *Robust Estimates of Location: Survey and Advances*. Princeton University Press, Princeton, N.J.

[2] Bayes, T. (1763): An essay towards solving a problem in the doctrine of chances. *Philos. Trans. Roy. Soc. London A* **53** 370–418. Reprinted in *Biometrika* **45** 293–315, and in (Pearson, E. S., Kendall, M. G., eds.), (1970): *Studies in the History of Statistics and Probability* 131–153, Griffin, London. Deutsche Übersetzung in (Schneider, I., ed.), (1988): *Die Entwicklung der Wahrscheinlichkeitstheorie von den Anfängen bis 1933* 108–115 und 135–144. Wissenschaftliche Buchgesellschaft, Darmstadt.

[3] Dempster, A.P. (1968): A generalization of Bayesian inference. *J. Roy. Statist. Soc. B* **30** 205–245.

[4] Fisher, R.A. (1930): Inverse probability. *Proc. Camb. Phil. Soc.* **26** 528–535.

[5] Fisher, R.A. (1956): *Statistical Methods and Scientific Inference*. Oliver and Boyd, London. (2nd ed. 1959, reprinted 1967).

[6] Hampel, F. (1973): Robust estimation: A condensed partial survey. *Z. Wahrsch. verw. Gebiete.* **27** 87–104.

[7] Hampel, F. (1993a): Some thoughts about the foundations of statistics. In *New Directions in Statistical Data Analysis and Robustness* (S. Morgenthaler, E. Ronchetti, W. A. Stahel, eds.), 125–137. Birkhäuser Verlag, Basel.

[8] Hampel, F. (1993b): Predictive inference and decisions: Successful bets and enforced fair bets. *Proc. 49th Session of the ISI, Contrib. Papers, Book 1. Firenze (Italy)*, 541–542.

[9] Hampel, F. (1994): *Zur Grundlagendiskussion in der Statistik*. Festvortrag, Univ. Dortmund, to appear.

[10] Hampel, F.R., Ronchetti, E.M., Rousseeuw, P.J. and Stahel, W.A. (1986): *Robust Statistics—The Approach Based on Influence Functions*. John Wiley, New York.

[11] Huber, P.J. (1964): Robust estimation of a location parameter. *Ann. Math. Statist.* **35** 73–101.

[12] Huber, P.J. (1973): The use of Choquet capacities in statistics. *Bull. Internat. Statist. Inst.* **45** (4) 181–191 (with discussion).

[13] Huber, P.J. (1976): Kapazitäten statt Wahrscheinlichkeiten? Gedanken zur Grundlegung der Statistik. *Iber. Deutsch. Math.-Verein.* **78** (2) 81–92.

[14] Huber, P.J. (1981): *Robust Statistics*. John Wiley, New York.

[15] Jeffreys, H. (1939): *Theory of Probability*. Clarendon Press, Oxford. Later editions: 1948, 1961, 1983.

[16] Kotz, S. and Johnson, N.L. (1992): *Breakthroughs in Statistics*. Vol II: Methodology and Distribution. Springer-Verlag, New York.

[17] Mächler, M.B. (1989): *'Parametric' Smoothing Quality in Nonparametric Regression: Shape Control by Penalizing Inflection Points*. Ph.D. Thesis, no. 8920, Swiss Federal Institute of Technology (ETH), Zürich.

[18] Mächler, M.B. (1994): *Estimating Distributions with a Fixed Number of Modes*. Festkolloquium and Workshop in Honor of Peter J. Huber, Univ. of Bayreuth.

[19] Neyman, J. (1957): 'Inductive behaviour' as a basic concept of philosophy of science, *Rev. Internat. Statist. Inst.* **25** 7–22.

[20] Neyman, J. (1971): Foundations of behavioristic statistics. In *Foundations of Statistical Inference* (V. P. Godambe and D. A. Sprott, eds.), Holt, Rinehart and Winston of Canada, Toronto.

[21] Shafer, G. (1976): *A Mathematical Theory of Evidence*. Princeton University Press, Princeton, N.J.

[22] van Dantzig, D. (1957): Statistical priesthood (Savage on personal probabilities). *Statistica Neerlandica* **2** 1–16.

Data Based Prototyping

Thomas M. Huber
Datavision AG, Klosters

Matthias Nagel
ZEG Berlin–Zwickau

Abstract

Data Based Prototyping is a method for bringing massive sets of raw data into an analyzable form and to customize an analysis system around the data. To obtain optimal results, data preparation should be geared towards the types of analyses planned and therefore should be done by experienced data analysts and not programmers. In this paper we outline the key ingredients of Data Based Prototyping and illustrate these with two case studies. The first of the two shows how 1 gigabyte of general hospitalization statistics can be made available for epidemiological studies, while the second example illustrates how data based prototyping can and should be applied to the development of complex information systems.

Key words and phrases: Databased prototyping, dynamical graphics, management systems, exploratory data analysis.
AMS 1991 subject classifications: Primary 62-07; secondary 90-08, 62H25.

1 Why this Paper?

One of the many things that Peter J. Huber has done in his career is develop the data analysis system PC-ISP. The initial system was conceived at Harvard with the assistance of David Donoho and Mathis Thoma. At the time, Peter J. Huber was experimenting with interactive graphical data analysis. He realized that interactive graphics would be useless without a means for data preparation and manipulation. They therefore added the interactive ISP command language as a driver for the powerful dynamic graphics system. The goal was always very clear: to design a general purpose data analysis tool. Because of this requirement they invested much thought and care into the design of a balanced system: if one made the building blocks (commands) too small, the user would waste time programming. If one made them too large, the system would only work in a few special situations.

In 1983, with the advent of the PC, Peter and his wife, Effi Huber–Buser, completely rewrote the system for the new low cost hardware giving it the necessary stability for third party users. In 1987 one of the authors decided to set up his own business to continue the commercialization of PC-ISP.

Unfortunately, the initial venture tried to sell the software as a general purpose tool for exploratory data analysis to statisticians—a major mistake due to shoddy market research. The author had to realize that unlike his father most statisticians were not at all interested in data analysis.

Because of this, the focus was altered from trying to sell the software to actually solving real world problems. The projects were acquired in a somewhat haphazard fashion from many different areas and applications. Eventually, we realized that the common thread among most of these problems was that someone else had failed or refused to pick up the challenge. We also noticed that our way of working contrasted quite markedly to the way others tried to solve similar problems.

Our projects were characterized by a fairly aggressive trial-and-error approach that yielded faster and better results. For marketing purposes we call this "data based prototyping". In earlier days Peter J. Huber and David Donoho used the term "rapid prototyping" to describe this way of working with ISP.

We felt that "rapid prototyping" was too close to the terminology used by the computer scientists. Their prototyping essentially is playing with the hollow user interface—every time one gets close to some substance a message like "function not available" appears.

Data Based Prototyping is the opposite: one works on the substance and solves the "hard" problems before worrying about the user interface. Especially, since the latter is not really subject to much variation given the standards set by the Windows environments. The user interface is supposed to guide the user and hence strongly depends on the intelligence of the underlying software.

The second thing we noticed was that our way of working did not really make a distinction between data analysis projects and projects where the goal was to develop analytical software or an information system. We found that the two types of projects were in fact remarkably similar: In either case one is trying to find a way to handle larger amounts of data. The difference lay in the elaboration of the user interface and safeguards.

The third observation was that we were considerably faster and more flexible than others in data handling. We attribute this to the good design underlying PC-ISP, which unlike other, higher level software tools is geared towards getting the job done (despite the data).

2 The Reality of Data

Data analysis should always orient itself on the reality of the data. A goal which is unproblematic, if one can proceed in a planned and orderly fashion and had the opportunity of influencing the process of data collection from the very beginning (see type A in fig. 1). This is tantamount to forcing the data to conform to a preconceived statistical analysis.

Usually, one encounters significantly more complex data structures requiring a much more differentiated approach to data analysis.

For the purposes of this paper we have grouped the situations known to us into three categories roughly described as follows (of course hybrid forms do occur):

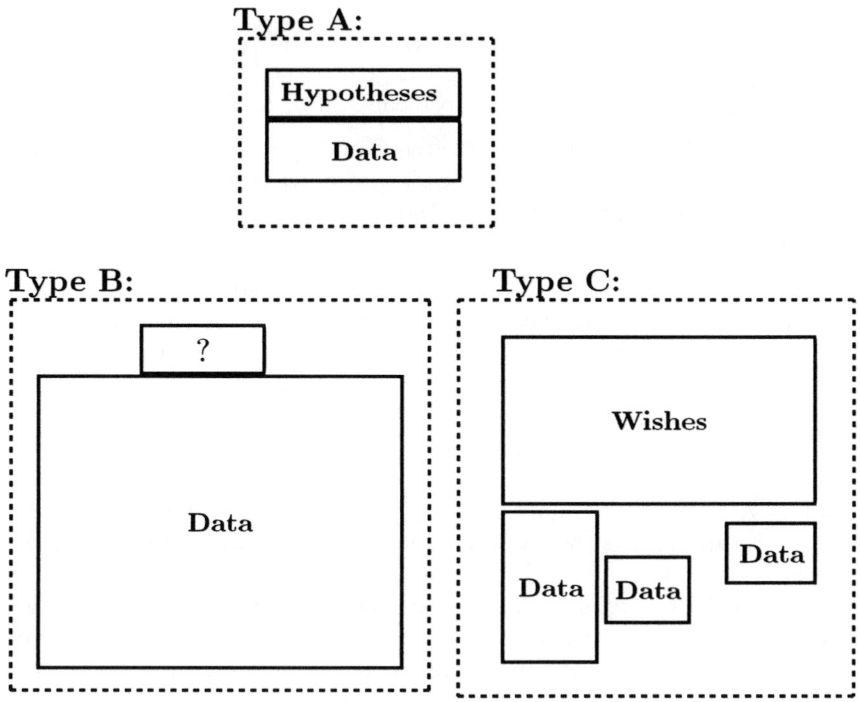

Type A: The "balanced" situation is a result of planned experiments with clearly defined goals or hypothesis-driven surveys with a manageable number of observations and variables. Unfortunately, situations of type A seem to be more the exception rather than the rule. Nevertheless, they are the "standard" case used in the copious statistical literature.

Type B: The "data available" situation In our experience the type B situation is most common. Here, the analyst is confronted with a large data set with only a hazy idea of the kind of analyses to undertake. Usually, the data itself must be investigated for usability, since it may have been gathered at different times and places under perhaps widely varying circumstances. Data sets are fairly large and easily reach a size in the order of 10–100 MBytes. They may be characterized by a large number of variables, observations, or both.

Such situations are often encountered in the analysis of meta-data, where one might want to investigate the data under other aspects than the one they were originally collected for.

Typical for these situations is that analysis steps and strategies are not preconceived or pre-planned but must be formulated in the course of an initial (exploratory) analysis, since the internal structure of the data is unknown.

Type C: The "prospective elysium" situation underlies many large software development projects. The implicit goal is very clear: A system should be developed which solves ALL problems.

The type C situation is most often found in connection with information systems. Typically, little or no data are available, but one is optimistic that they will be available by the time system development is finished. Data collection usually is in progress and can no longer be influenced.

On closer inspection one can see that the customer views the problem mostly as one of data access: Much time is spent on the definition of data structures and eventually one loses track of the real purpose of the system, which is to help in decision making. On top of this we detect a rather common belief that the analysis of the data will ultimately be just as easy as the collection using simple statistical tabulations.

B and C type situations cannot be analyzed directly using fancy statistical techniques. The analyst first needs to get a very thorough understanding of the data and what one wants to do with it. This is a real data exploration, the first stage in Data Based Prototyping. As one gets a clearer picture of the quality of the data and of the types of analyses that are possible, one will start aggregating the data and bringing it into an analyzable form. In many cases the early phases of Data Based Prototyping can be seen as a migration of data from the B or C structure to the A-type.

3 Data Based Prototyping

The goal of data based prototyping is the development of an analysis system, which uses a series of adequate and, for the problem at hand, appropriate algorithms to handle large and/or complex data analysis problems. Short development times are of considerable importance.

This prototype may be developed for one's own use to assist in routine analysis work or may be produced for a third party, where the latter requires a much greater programming effort.

Data based prototyping is an experimental process with a large learning component: One continuously acquires new information about the data, in particular detail aspects and may experiment with heuristic methods and algorithms.

There is no formula for data based prototyping, however, the following components represent the key ingredients of the process:

- Visual data inspection

- Data preparation
- Analysis strategy/hypotheses formulation
- Prototyping and scripts
- Systematic analyses
- Iteration
- Programming of custom system

The list makes it clear that data based prototyping is a form of exploratory data analysis. The individual items shall briefly be explained here:

Visual Data Inspection

Visualization is essential for a first contact with the data. By looking at the data it is possible to detect structural problems: In the simplest case this may be a rapid discovery of extreme outliers or missing data values, which may be due to omissions or erroneous measurements or statements. After the visual inspection one can make a general statement about the quality of the data and can gain an understanding about the types of analyses that will be possible with the data.

Visual data exploration must be fast. Traditional static graphics are rarely of use, since they do not permit interaction, which is absolutely essential for an efficient processing at this step. The ideal tool for this activity is a powerful dynamic graphics system **GEARED TOWARDS ANALYSIS** and not presentation (for example: higher screen resolutions are less useful, since they require more processing time, while not revealing more information). Interaction must be fast and not necessarily "intuitive" in the sense of commonly available windows systems, which require far too many mouse operations to perform even simple tasks making interaction with the data extremely tedious.

Our experience has shown that we routinely deal with data comprising about 10'000 to 20'000 observations (a modest 4×4 scatterplot matrix will result in up to 16 times as may individually addressed points on the screen). At this level the response of the graphics system should still be more or less instantaneous. Beyond this number of observations the direct visualization as multidimensional scatterplots usually is no longer useful and the data must be sampled or aggregated prior to visualization.

Data Preparation

The formal data analysis starts with the preparation of the data. This must be done carefully and with the specifics of the data in mind. Without proper care and foresight one may easily make certain analyses impossible or false, requiring additional data preparation later.

Aside from the elimination of data entry and systematic errors, this step involves the assignment of standardized and unique names for variables and attributes, re-coding data to sensible value ranges, and assigning standardized missing values, etc. Typically, one will also bring the data into an easy to read format, which in our experience is one or more direct (random) access file(s) without header.

The effort needed for a careful data preparation is often underestimated. For this work one is heavily dependent on efficient routines for sorting, selection, searching, matching, and aggregation. These one typically would expect to find in commercial data base programs, however, these programs contain many hidden limitations, are generally very inefficient and hence not the systems of choice for data preparation.

Good data preparation requires an understanding of the types of analyses to be performed. For example, data aggregation should be strong enough to allow efficient processing of the data later, while still yielding the necessary level of detail required by the analysis. This understanding can be gained to some extent via the visual data exploration described above.

Analysis Strategy

After having gained an initial understanding of the data, one can proceed to the formulation of an analysis strategy. This should be done together with the user of the data.

An important goal in this initial ad-hoc analysis is, to make hazy ideas precise, to structure and to modularize the analysis process in order to reduce complexity. Instead of lengthy discussions on whether or not to use a particular model or approach, one should try to apply them directly to the data. Only from the results can one really judge the usefulness of a particular approach and whether it merits further investigation and elaboration. It often is more constructive to try out a questionable method and let the person suggesting it decide to abandon it, than try to argue on a theoretical basis.

This kind of ad-hoc improvisation requires a powerful data analysis language with macro capabilities to adapt algorithms to the needs at hand.

Prototype and Script

The "improvisation" described in the previous section must be done interactively using a suitable command language, in our case the ISP-language. We find it ideally suited for this purpose, since it is a so called "command-response language", where every command produces a usable output, which can be checked for plausibility immediately. Experience shows that even very experienced data analysts are not capable of producing more than a few error-free analysis steps in a row (not due to typos!). Frequent plausibility checks at all intermediate levels are absolutely critical for producing correct final results.

All analysis steps need to be documented in a script file, which can later be edited and re-run. This way it is possible to backtrack from a dead-end path and continue an otherwise promising investigation in a different direction.

The type of work described here is fairly unique to ISP. To our knowledge there are no other systems that allow such an easy interplay between interactive ad-hoc analysis and automated documentation which yields executable batch routines. Menu-systems are of little use, since they provide only very limited automatic documentation functions, which generally are impossible to edit.

Once the successful paths of analysis have been identified, one can start elaborating the scripts by taking their core functions (which now are already debugged!) and adding the necessary ancillary features and safeguards (e.g., remove explicit references or file names, add some exception handling, etc.). Typically, one will also have to extract different functions from the strictly sequential script and make them into separately executable macros. Most importantly, for the end user, one needs to add an easy to use menu-interface, so that he can start working with the prototype at the earliest possible date.

The input of the end user, based on active use of the system, should be carefully reviewed and incorporated as modifications into the current prototype or added as new features to a next version.

Systematic Data Analysis

Once the prototype is made available to the end user, the actual data analysis begins. The user typically needs to process a considerable amount of data on a recurring basis. After a while one will note which methods or algorithms stand the test of time and what features are missing or could be further automated.

The user acquires considerable new understanding of the data and frequently has the possibility of adapting or expanding the scope of data collection. Previously this was only of secondary importance since he did not really have the means of analyzing the data.

Iteration

Data Based Prototyping is an iterative process. After working with the prototype for a while, a new development cycle starts, which needs to incorporate the developed skills and more sophisticated needs of the end user. The extent of the system may grow considerably as additional features get added and more analytical power is made available to the user.

In a final iteration the functionality of the system will probably be shrunk somewhat through the elimination of unneeded features or methods that did not find general acceptance. Some analytical steps can be streamlined through greater automation of certain processes, which are now better understood. Data Based Prototyping produces "lean" systems.

Programming of Custom System

At this stage one has achieved the following:

1. The user is able to analyze his data and is already working with "his" system,

2. The user and the data analysis specialist have an intimate understanding of the particularities and structure of the data, since one has intensively analyzed and prepared them,

3. The system contains exactly those modules and functions that are needed to do the work.

If appropriate, one can now commence the migration from prototype to final custom software. Since structure, content and functionality of the software is clearly defined, one can now concern oneself with optimizing the user interface, and other external features.

This is in fact an extremely important aspect of data-based prototyping, which we have neglected a couple of times and also paid for: **KNOW WHEN TO STOP**. At some point the problem is solved and hence the prototyping phase is finished. Once additions start becoming more cosmetic in nature it usually is time to convert the prototype into a dedicated program.

The tools used for data based prototyping are optimized towards solving the data problems and not fine-tuning the user's aesthetic wishes. The initial speed with which the "hard" part of the problem got solved will slow down to that of a normal programmer and frustration about the lack of progress will set in on all sides. Based on the prototype it is fairly easy to write up a specification for the new system and the development will in fact progress quite rapidly.

Within a windows environment, the often fairly hierarchical structure of the prototype, which helped guide the user, must be abandoned in favor of a more object oriented approach. To achieve a similar functionality, one needs to prepare the data in a somewhat more complex structure. Often one will need several smaller data files linked through elaborate indexing to get acceptable response times.

Our experience in this area is limited to three different implementations. Two of which using VISUAL BASIC under MS-Windows and one under UNIX. The first two were both situations where data was to be made available to a broader audience. This allowed us to keep data preparation and all major calculations off line (under ISP) avoiding the cumbersome programming and slow processing speed under the windows system. Essentially, we took advantage of the flexible windows generator of VISUAL BASIC to create a fast viewing system, while relegating all calculations "off line" to the much better suited ISP-System. A structure like this is only possible, when one has a very clear understanding of what the end user really needs.

The third implementation under UNIX is actually a hybrid system involving a UNIX-version of ISP: Using the Motif and X-resources to create a (massive) front-end user interface we accessISP via a TCP-IP communications link and let ISP do all calculations, many of which are elaborate smoothing and aggregation functions. The advantage of this structure is that the dynamic graphics are also available. (Interactive graphics is something the developers of the X-system have not foreseen, making porting a rather painful process).

4 Examples of Data Based Prototyping

The following examples shall serve to illustrate how Data Based Prototyping can be used to attain fast results. As mentioned earlier, we got most of our projects because others refused to touch them or have failed.

4.1 Examples for Type B-Situations

The "data available" situation is typical in the analysis of meta-data and usually involves large data sets. The following type-B data were investigated by the authors during the previous three years:

- DASKE data conglomerate on pesticide applications (over 1 GB of data) (see Nagel (1994), Nagel et al. (1994) and Nagel and T. Huber (1994))
- Hospitalization data of the former German Democratic Republic (900 MB of data), (see Nagel et al. (1994))
- Survey on occult practices among youths (see T. Huber and Nagel (1994))
- Health and environmental surveys in Germany
- Nitrate contamination of water sources in the former GDR (time series) (see Nagel and T. Huber (1992))
- Relationship of lead contamination in blood and drinking water (see Höring et al. (1992))

The following example illustrates the importance of good data preparation and how analysis sometimes only is possible after such a preparation:

Hospitalization Data of the German Democratic Republic (GDR)

Aside from the usual mortality data, the GDR also collected a number of other routine medical statistics that basically could be used as a basis for epidemiological studies. One of these were the standardized patient records on every hospitalization, which were kept since 1968. Since 1979 these records were stored in computer readable form and contain among other things the following information for each patient and hospitalization:

- patient number

- date of birth, sex, age, domicile
- ward/department administering treatment
- date of admission, length of stay, form of release
- treatment diagnoses (ICD-9 codes)

For the period from 1985 to 1989 this data occupies 900 MB.

Since data in this form and completeness would not be available for Germany in the near future, the data base seemed a logical candidate as a basis for a health reporting system which needs to be established for the five new German states. In an assessment by the federal ministry of health the hospitalization data was deemed representative and valuable, both from an epidemiological as well as economical standpoint.

The preparation of the data had to be done considering both contents and technical limitations. To avoid double counting, one first had to combine all multiple records for the same person, diagnosis, and year (as can occur in a transfer between hospitals). Furthermore, the hospitalization data was combined with mortality data for the same period (1985–1989) to get the complete mortality and morbidity of the former GDR. In a final step all personal ID numbers were eliminated to anonymize the data.

These steps did not really lead to a data reduction and the data still consisted of about 15 million records. Traditional data base systems are unable to handle such quantities efficiently enough for the goal of interactive analysis on a PC. For this reason a technical aggregation was necessary:

All data had to be read and written in a new record-oriented structure, which was devoid of all redundant or unnecessary information and in which the variables were stored as compactly as possible. The real data compression occurred in the aggregation of all cases into one record for each ICD-code, domicile (district), age (5-year age bracket), sex, and year.

The result of this data preparation was essentially a very large contingency table, where each cell was defined by diagnosis x district x age group x sex x year for a theoretical maximum of $1200 \times 216 \times 18 \times 2 \times 5$ cells (the final count actually was 2.3 million cells or records). The entire process of re-coding and aggregation took 508 minutes on a 33 MHz 486 PC using PC-ISP.

For a while one worked with the resulting data file until one felt confident about the types of analyses to be expected. At that point another compression run was performed converting the data to a fully adequate 2 byte word length and eliminating the variable containing the diagnosis number. Instead the data was sorted and accessed via an index file with pointers to the records pertaining to a particular diagnosis.

After these two iterations the original 900 MB of raw data had been compressed to a significantly more manageable 13.9 MB with no significant loss of information for the purpose of analysis. The access times are also quite impressive and definitely allow an interactive analysis of this data: *to extract all records pertaining to a single diagnosis* PC-ISP *takes 0.9 seconds, for two diagnoses 1.04 seconds, and a group of 160 diagnoses 38 seconds (see summary in Table 1).*

Diagnostic group	Nr. of ICD-9 codes	Raw data records	Aggregated data records	Aggregation time (min)	Extract time (s)
Pulmonary	59	1'341'675	157'609	40	31
Injuries, Poison	160	1'317'292	389'039	105	(38) 75
Tumors	69	829'474	212'025	29	35
Cardiovascular (male)	70	1'177'593	109'544	31	20
Endocrine	80	664'835	160'878	13	13
Infections	130	297'259	132'192	5	18

Table 1: Extent of hospitalization data and aggregation/extraction times for a sample of diagnostic groups. Extract times reflect times after initial aggregation (extract times after second aggregation using pointer files are about 40% to 80% lower than these values)

Using these compact and manageable data files it is possible to perform a wide array of analyses. To make the data available to a wider group of people and to simplify routine tasks, a small menu system was developed to help produce various cartographic displays, tables, charts, as well as statistical tests.

These menu-functions serve several purposes: on the one hand an epidemiologist can start using the data and sight them for reliability and usability while at the same time formulating ideas for further study. On the other hand, the data analyst can quickly extract and filter the data to then use them for experimentation with new/different analytical methods.

One can also react quickly to current problems or inquiries. For example, after press reports of physical deformations in people living along the North Sea coast, we were able to quickly check whether similar phenomena occurred in the East German coastal regions.

Finally, since the data offers a complete description of the health and well-being of the population of a country, the data is ideally suited for combination with environmental and other data.

The main focus of the data based prototyping effort here was on the suitable data aggregation that (i) makes the data manageable while (ii) avoiding restrictions on the desired analyses. Without this all subsequent analyses would have become illusory. It should be obvious that the data preparation is part of the analysis and should be performed by an experienced data analyst and not be delegated to a programmer. It is not just an issue of proper choice of data base program!

By using one's head rather than brute force, it is also possible to analyze data that commonly is thought to be way beyond the scope of the available resources, both in terms of hardware and manpower. This data was prepared in a matter of a few days on a normal 486-type PC. We would like to contrast this with an example of similar magnitude (from Statistical Computing and Graphics, see Schall (1993)):

> *The UCLA computing center wanted to make the PUMS (Public Use Microdata Sample) of census data for California available for research use. For this purpose it downloaded the data from tapes onto disk using IBM's DB2 relational data base management system. Access to the data is via the SAS system, which is used for variable selection and subsetting. The hardware used is an IBM ES/9000 mainframe with one gigabyte of memory.*
>
> *Using this impressive hardware it only "... takes less than five minutes to subset the data set for blacks, and merge the 1.5 million person records with almost 600'000 household records to create a final extraction of 300'000 cases."*

Not everybody has a supercomputer available, so that the use of these types of data sets in this fashion is limited to people with access to a heavily sponsored computing center. What is even more staggering is the amount of time required to prepare the data: *"The entire process from designing the database to making new census data available in a production version can take up to nine months, depending on the size and complexity of the data. ..."*. If traditional approaches have such enormous resource requirements, it is not surprising that census data never gets used properly.

4.2 Examples for Type C-Situations

The "prospective elysium" situation is typical in many large software development projects. The analysis system is designed by a committee and very often features must be included for political reasons to avoid conflicts among the committee members. The result is then a monster, that takes ages to develop and, when finally in implementable form, turns out to be not nearly as useable as originally envisioned.

The main problem in these large development projects is that one is trying to specify a new way of working with significantly more and better data, without ever having had a chance of using even a fraction thereof.

The authors have had the chance of observing the following type-C situations, for which they have devised or are devising alternative solutions:

- Road maintenance planning (STREETMAN)
- Radar data analysis (IRAS/AMOR)
- Water quality of East Germany (TISP)

The example below on the STREETMAN project illustrates some of the complexities of large scale data analysis and what leads to the typical phenomenon of overburdening a system right from the beginning. We are currently in the middle of this project and will use the example to show how the focus of the prototyping work will change over time.

Road Maintenance and Quality Management

We are currently just starting the second round of development of a data-based prototype for the analysis of road quality parameters and planning of maintenance activities. The project is called STREETMAN and is being developed in close conjunction with a small engineering firm, that measures road quality parameters and the head of a regional highway authority, who is responsible for a total of about 700 km of highways.

The engineering firm had just finished a major project concerning the measurement of safety parameters for all the German Autobahnen for which they had to deliver over 40'000 pages of charts and tables. They realized that nobody would be able to digest all this material. Therefore, we suggested making an interactive system, which allows the user to quickly zoom into the relevant sections and also perform multivariate analyses of road parameters and then only generate hardcopy for the information really needed.

Aside from our project, there is a development underway, sponsored by the federal government, to create a huge management information system, which is supposed to combine all possible aspects of road maintenance in one software. The specification of data requirements is more or less finished (and spans several volumes). We have already heard rumblings by road maintenance managers, who do not see how they will ever be able to produce the desired data let alone maintain the level of detail on an ongoing basis. These concerns are put off by the committee as being the typical gripes of operational people.

Our initial approach was to produce a system that can get by with a minimum of data. Using PC-ISP we developed a small visualization system, that gives the user the possibility of viewing and sighting the available road parameters using several different graphics displays, essentially an on-line version of the paper output. Since one is usually interested in certain segments of the road, we added some filter functions using the interactive graphics (the data base approach of taking all segments with parameter A above a certain value is of no use here, since one usually also needs to see the segments which lie adjacent to the bad ones to assess maintenance needs). As an additional feature we added the possibility of viewing chosen parameters in a geographic context using a network/node-structure.

One data problem we stumbled over early on, was how to incorporate the measured data into the node structure used (as a coordinate system) to describe the roads: the measurements are made by a measurement vehicle driving along the highway. On an Autobahn, which is not curvy, the deviation of the driven distance to the actual road kilometer will be fairly small and of marginal impact.

All existing programs were geared towards the measurement of the Autobahnen. Now we were looking at overland roads which passed through towns, merged with other streets, sometimes were quite curvy, and could have road construction requiring detours. Suddenly, the measured distance showed substantial deviations from the actual road kilometer, plus the location of a problem area had to be identified much more precisely, since the maintenance manager was only responsible for overland roads and not for those parts that lay within town limits.

The handling of this problem turned out to be quite tricky, requiring three prototype versions, until a reasonably practical solution was arrived at. Again the prototypes were needed for the engineering firm to understand the problem and adapt the documentation standards during the measurement itself.

The ability to easily visualize road quality parameters was already a large step forward for the road maintenance manager. The possibility of multivariate inspection of trouble spots not just on a road by road but also on a larger regional basis was of major assistance in maintenance and upkeep planning. However, the planning process was still manual and there were quite a few uncertainties about optimization and priorization of the necessary activities.

It was then decided to enlarge the functionality of STREETMAN to include some planning and optimization functions. This, however, entails adding considerably more data inputs to the system:

1. construction parameters of the road segment,
2. usage data,
3. special features/complications (bridges, railroad crossings, accident sites, etc.),
4. visual assessments of road condition, and
5. costs for various maintenance and upkeep measures.

Here the problem starts getting interesting. We are given the task of "optimizing" maintenance activities over a network of roads consisting of several thousand 100-meter segments (the unit of aggregation for road parameters). On the data level alone several content problems need to be resolved, since one is trying to combine data of widely varying quality:

Measurements: automatic data collection with a very high resolution of 1 meter or less. These are aggregated to 100 meter segments. We have found mean, standard deviation, and worst value to be most useful.

Visual assessments: subjective estimation of non-measurable quality attributes, such as frequency and size of cracks, general condition of surface and substructure, etc. These assessments are made by different people with varying levels of skill and care. One of the challenges is to standardize these values and try to eliminate possible "assessor biases".

Spot measurements: e.g., traffic volume usually is measured at a few points within the net and must be allocated to the remainder of the net.

Gaps/Omissions: In order to be able to optimize, ALL data must be available for ALL roads. Unfortunately, there will always be several large gaps or omissions in the data. These must be filled with estimates by an expert who is reasonably familiar with the roads. Such a temporary and rough assessment sometimes is unavoidable, since some measurements are destructive and costly and thus must initially be substituted with estimates. As new information becomes available, these estimates must be revised. Experience shows that experts need help in making such estimates and plausibility checks by the data analyst are of major importance.

Only when the input data is complete and accepted as such by the relevant people, can one continue with the planning per se.

We are now in the process of modularizing the planning/optimization process. This is of major importance because of the substantial complexity of the problem as well as for acceptance of the results. Resistance towards computer support is quite large in some circles (today the allocation of the sizeable funds is still very much up to the judgment of the planner and to higher-level bureaucrats in the ministries, since the optimization goal is vaguely defined as "reducing overall cost of traffic").

Based on the available input data we will need to develop algorithms for determining the appropriate action. Here also we find a large discrepancy between theory and practice. There already are some complex theoretical models which are largely ignored by the practitioners who rely almost exclusively on visual assessments. The challenge is to find a practical solution that builds on realistically available data to derive suggested actions. It is important that the practitioner can compare the suggested solution with his solution and that discrepancies are carefully documented by the data analyst in what might be called "knowledge extraction". The system will never be able to find a 100solution. However, it should be able to deliver a good first cut given the available data.

Only after agreement on the type extent of a measure can one proceed with the ranking / optimizing in the context of the network. The clear separation is important to keep earlier assessment steps objective. We already have several different concepts in mind for the ranking procedure. All of which must be tried out and eventually weeded out as practical experience with the system grows.

This level of system development reflects the current level of planning. We believe it will help in and enhance the current method, however, the real step forward will occur when one starts looking into long-range planning. Over time we will have available long-term data on road parameters, road deterioration, and maintenance effects. This promises to become an interesting and non-trivial multivariate data analysis from which modelling parameters will need to be derived.

It should be quite apparent that this is a major project. It is very real in its focus (ultimately, it will have an impact on the deployment of large amounts of money) with very many non-trivial data analytical problems. It is not just a question of specifying data structures and hardware requirements and organizing the data import into the system. That is the easy part—making the data and system work for you is the hard part.

5 Conclusion

With this article we hope to have shown that data based prototyping is in many cases the only way to get at real data. Most data is not available in the pre-packaged form assumed by statistics text-books (and statistics-packages).

In fact, this point should be addressed much more carefully. We are somewhat concerned about how far statistics education is removed from the real world. There are many, very real data analytical problems that are never seen by a statistician. This is largely due to the fact that statisticians in general are not interested in problems where they cannot apply their favorite test or method. We observe how the unwillingness to help gradually eliminates statisticians from modern data analytical processes. Data analysis is falling more and more into the hands of programmers or others who are not trained to do the job, but seem to be more willing to tackle a challenge.

Data based prototyping is mostly data analysis requiring an open minded and **critical evaluation of the data**, a professional attitude towards problem solving ("what can I do with the data?" and not "that's not significant" or simply "you can't do that"), and the appropriate tools.

We have found PC-ISP to be an extremely versatile tool of unmatched power and flexibility. Of course there are many other tools and languages around, which can tackle some aspect of the data based prototyping process. Perhaps they can even do a better job in their special functionality. We have seen a fair number of different software packages and programming languages, none of which could adequately cover the entire range of analytical and data-handling requirements and most of them showed major shortcomings in terms of design and flexibility and hence cannot be considered suitable for data-based prototyping.

We attribute this lack of software to the fact that most people simply are not aware of how superficially their data gets analyzed. Large software projects, such as the typical information systems are firmly in the grasp of programmers who generally view them as data base problems. They are not seen as data analysis problems and hence get tackled in a very ineffective and extremely costly way.

Data Based Prototyping can be a lot of fun, particularly in the early stages where creativity and imagination can get you far. It can also bring its share of frustrations. The people you work with quickly get accustomed to the speed of development and expect everything to go fast. They also do not realize the real value of the prototyping process, which lies in the rapid separation of the usable from the useless (an avoided disaster cannot be seen and hence does not get credit).

We hope to have shown that with Data Based Prototyping there is a (better) way to tackle large data analysis problems. The hardware to handle large amounts of data sits on almost every desktop. What is missing is the software and, to some extent, the mental attitude. Today's over-emphasis on "looks" is dangerous and removes the problem-solving focus. "Good looks" absorb most of the available hardware power, which could be used for data handling (who needs a beautifully styled shovel?). "User friendliness" is another sore point: a good program will be used many years and should not be judged by how well you got along in the first few hours—all those helpful features will later become dead weights.

Large data sets are different and cannot be handled the same way as text book examples. In each of our projects we have worked the data on an ad-hoc basis using our imagination and creativity. Some of these ideas can be systematized but a good portion will remain on an ad-hoc level. This methodological work will only be possible with a solid practical footing—extrapolating traditional methods will not work. Large data sets are available in abundant quantities. People just do not realize that they can be analyzed thoroughly and cost effectively.

References

[1] Höring, H., Englert, N., Krause, C., Meyer, E., Nagel, M. and F. Schiller (1992): *Lead in Drinking Water and Blood. A Study in School Children from Southern Saxonia, Germany*. In Proc. WHO/UNEP GEMS/HEAL Meeting, Bangkok, Thailand, 16.–19. November 1992.

[2] Huber, T. and M. Nagel: Look first, then model: Exploring the data using the interactive graphics of PC-ISP. In *Computational Statistics* (in press).

[3] Nagel, M. (1994): *Graphische Analyse der möglichen PBSM–Belastung des Trinkwassers durch Auswertung von Daten aus landwirtschaftlichen Schlägen (DASKE-Datei)*. In Trinkwasser: Vorkommen von PBSM, Organisch-chemische Stoffe zur Pflanzenbehandlung und Schädlingsbekämpfung einschließlich ihrer toxischen Hauptabbauprodukte im Trinkwasser der neuen Länder. Berichtsband zur Tagung vom 12. Oktober 1993. Landesgruppen Trinkwasser und Fachkommission Soforthilfe Trinkwasser des Bundesministeriums für Gesundheit gemeinsam mit Institut für Wasser-, Boden- und Lufthygiene und Umweltbundesamt, 33–49.

[4] Nagel, M., Clässen, E., Dahlke, D. and L. Heinemann (1994): *Die Analyse von medizinischen Massendaten*. In SoftStat'93 (Faulbaum, F., ed.), 339–346.

[5] Nagel, M., Huber. T. and H. Höring (1991): *Hochinteraktive Graphik und ihre Anwendung auf Daten mit territorialem Bezug*. In SoftStat'91 (Faulbaum, F., ed.), G. Fischer, Stuttgart.

[6] Nagel, M. and T. Huber: *A Graphical Analysis System of Pesticide Contamination—A Successful Application of Data-Based Prototyping*. Submitted to Environmetrics.

[7] Schall, M. (1993): Easy access to census data. In *Statistical Computing & Graphics* **4**, 7–8.

Robust Regression with a Categorical Covariable

Mia Hubert Peter J. Rousseeuw
University of Antwerp

Abstract

A fast algorithm is presented for robust estimation of a linear model with a distributed intercept. This is a regression model in which the data set contains groups with the same slopes but different intercepts, a situation which often occurs in economics. In each group, the algorithm first looks for outliers in (x, y)-space by means of a robust projection method. Then a modified version of the resampling technique is applied to the whole data set, in order to find an approximation to least median of squares or other regression methods with a positive breakdown point. Because of the preliminary projections, the number of subsets may be drastically reduced. Simulations and examples show that the overall computation time is substantially lower than that of the straightforward algorithm. The method is illustrated with a real data set.

Key words and phrases: Algorithms, computation time, distributed intercept, outlier detection, positive-breakdown methods.
AMS subject classifications: 62F35; 62J05.

1 Introduction

We consider the regression model

$$y = \beta_1 x_1 + \beta_2 x_2 + \cdots + \beta_p x_p + f(x_{p+1}) + \varepsilon, \tag{1.1}$$

where the regressors $x_1, \ldots, x_p$ take on real values and ε is normally distributed with mean 0 and variance σ^2. In the usual linear model one puts $x_{p+1} = 1$, hence $f(x_{p+1}) = \beta_{p+1}$ is the intercept term. In this paper, we will consider the more general situation where x_{p+1} is a categorical (or nominal) variable which can take on q different values. Therefore $f(x_{p+1})$ can also take on q values, which we will denote by $\delta_1, \ldots, \delta_q$.

Equivalently, we can say that each of the n observations belongs to one of q groups, according to its value of x_{p+1}. This relation is easily expressed by the introduction of dummy variables. We can therefore rewrite (1.1) as the following model:

$$y_i = \sum_{j=1}^{p} \beta_j x_{ij} + \sum_{k=1}^{q} \delta_k I_{ik} + \varepsilon_i, \qquad i = 1, \ldots, n, \tag{1.2}$$

where

$$\begin{aligned} I_{ik} &= 1 \quad \text{if case } i \text{ belongs to group } k \\ &= 0 \quad \text{otherwise.} \end{aligned}$$

Thus all groups obey a linear structure, with the same slope parameters $\beta_1, \ldots, \beta_p$. On the other hand, the intercept term δ_k (for $k = 1, \ldots, q$) differs between the groups. Therefore we call (1.2) a linear model with a distributed intercept. In simple regression it describes q parallel regression lines. In multiple regression, we will fit q parallel hyperplanes. Note that the usual linear model with a single intercept is still a particular case, with $q = 1$. One could also consider (1.2) as a linear model with $p+q$ explanatory variables, but then the continuous regressors and the dummies would be treated in the same way, which would increase the computational effort for the robust algorithms considered in this paper.

The least squares method (LS) fits the model (1.2) in a nonrobust way. For instance, it is possible to apply the standard calculations by processing the dummy variables in the same manner as the continuous ones, as described by Draper and Smith (1981), Montgomery and Peck (1982), and Chatterjee and Price (1977). We can also carry out an analysis of covariance, which directly assumes x_{p+1} to be a categorical variable. We refer to Montgomery (1991) and Edwards (1985) for more details.

Unfortunately, the least squares method is very sensitive to outliers. Even a small fraction of contamination can influence the regression estimates. In our model, both outliers in the y-direction and in the x-direction can occur. Our aim will be to estimate the $p+q$ coefficients $\beta_1, \ldots, \beta_p$ and $\delta_1, \ldots, \delta_q$ in a robust way. Armstrong and Frome (1977) developed a linear programming algorithm to obtain the least absolute value (L_1) estimate. This approach is already more robust against outliers in the $y-$ direction, but not against outliers in the $x-$ direction which can still tilt the estimated hyperplanes. Rousseeuw and Wagner (1994) proposed an algorithm to approximate the least median of squares (LMS) solution. This algorithm can deal with vertical outliers as well as bad leverage points, but it consumes a large amount of computation time. Here, we will focus on obtaining a faster method.

In the next section we will describe the proposed algorithm. Section 3 studies its behaviour on generated data, whereas Section 4 analyzes a real data set.

2 Description of the Algorithm

The algorithm essentially consists of two parts. For each group, we first detect outliers in $(\mathbf{x}, y)$-space by means of a robust projection method. Then we apply an adapted version of the resampling algorithm. The latter was originally devised for the single-intercept linear model $(q = 1)$, where it is used for computing the least median of squares estimator (Rousseeuw (1984)) given by

$$\underset{T}{\text{minimize}} \underset{i=1,\dots,n}{\text{median}} \quad r_i^2(T), \tag{2.1}$$

where r_i stands for the regression residual.

Let us now consider both parts of the algorithm in turn.

2.1 Projection method

We will denote the different groups by G_k (for $k = 1, \dots, q$) with $n_k = |G_k|$. The index i always runs through the whole data set, whereas j indexes the objects of one group. An observation $z_j = (x_{1j}, x_{2j}, \dots, x_{p+1j}, y_j)$ is written as $(\mathbf{x}_j, y_j)$. Suppose now we are investigating G_k.

Analogous to Donoho and Gasko (1992), we define the *outlyingness* of an observation $z_j = (\mathbf{x}_j, y_j)$ as:

$$u_j = \sup_{\mathbf{v} \neq 0} \frac{|z_j \mathbf{v}^t - L(z_1 \mathbf{v}^t, \dots, z_{n_k} \mathbf{v}^t)|}{S(z_1 \mathbf{v}^t, \dots, z_{n_k} \mathbf{v}^t)} \tag{2.2}$$

where $L(z_1 \mathbf{v}^t, \dots, z_{n_k} \mathbf{v}^t)$ and $S(z_1 \mathbf{v}^t, \dots, z_{n_k} \mathbf{v}^t)$ are the univariate least median of squares (LMS) location and scale estimators applied to the projections of the data on the direction $\mathbf{v}$. These robust estimators correspond with the midpoint and the length of the shortest half (see Rousseeuw and Leroy (1987)). Note that (2.2) measures the outlyingness of the points z_j relative to the group G_k. From a computational point of view, is it necessary to restrict ourselves to a finite set of directions $\mathbf{v}$. Rousseeuw and van Zomeren (1990) propose to take all

$$\mathbf{v}_l = \mathbf{x}_l - M \quad \text{for} \quad l = 1, \dots, n_k,$$

where M is the coordinatewise median of the data (here, of the group G_k). The corresponding u_j are then easy to compute. Note that these u_j are no longer affine invariant (because we do not consider all vectors $\mathbf{v}$), but they are still invariant for permutations of the data, as well as translations and scalar factors.

We calculate the u_j in the $(p+1)$-dimensional space $(\mathbf{x}, y)$ and not only in the space of the explanatory variables. The latter is done by Rousseeuw and van Zomeren (1991) before they apply L_1 regression, which works well because the L_1 method is rather resistant against vertical outliers. But our purpose here is to indicate all possible outliers in order to speed up the second stage of our algorithm, which is based on subsets consisting of clean data.

Having the collection $\{u_j, j = 1, \dots, n_k\}$, we will select a set P_k of points with relatively small u_j. There exists a tradeoff between the dangers of selecting too many points (some of which may be outlying) or too few points. Here, our strategy will be to select approximately $2n_k/3$ points, as described in the next subsection.

2.2 Resampling method

In each group G_k we will select points with small u_j. The complete algorithm is organized as follows.

For $k = 1, \ldots, q$ do:

Step 1. Consider the k-th group G_k and check whether it contains at least $\lceil \frac{3}{2}(p+1) \rceil$ data points. (Otherwise skip this value of k).

Step 2. Carry out the projection method on G_k as described in Section 2.1. Then compute the order statistic $u_{(h)}$ of $\{u_1, \ldots, u_{n_k}\}$ where $h = [\frac{2n_k}{3}]$. Select the set P_k consisting of all points $(\mathbf{x}_j, y_j)$ in G_k for which

$$u_j \leq u_{(h)}.$$

Step 3. Repeat $nrep(k)$ times:

- Draw a subset of $p+1$ points from P_k. (This is possible by Step 1.)
- Compute the hyperplane formed by this $(p+1)$-subset. A linear system has to be solved, which results in the slopes $\hat{\beta}_1, \ldots, \hat{\beta}_p$ and an intercept term.
- For each group a separate intercept is obtained. First calculate the values $d_i = y_i - \hat{\beta}_1 x_{i1} - \cdots - \hat{\beta}_p x_{ip}$ for $i = 1, \ldots, n$. For each group G_l (where $l = 1, \ldots, q$) then apply a robust location estimator to those d_j belonging to G_l, yielding an intercept denoted by $\hat{\delta}_l$.
- For each point i, compute the residual $r_i = y_i - \hat{\beta}_1 x_{i1} - \cdots - \hat{\beta}_p x_{ip} - \hat{\delta}_l$ where $\hat{\delta}_l$ is the intercept of its group. From the entire collection of residuals $\{r_i, i = 1, \ldots, n\}$ we then obtain the objective function

$$\operatorname*{median}_{i=1,\ldots,n} \tag{2.3}$$

Enddo $k = 1, \ldots, q$.

Step 4. Report the estimate $(\hat{\beta}_1, \ldots, \hat{\beta}_p, \hat{\delta}_1, \ldots, \hat{\delta}_q)$ with the smallest objective function (2.3).

This resampling algorithm is related to that in Rousseeuw and Wagner (1994), but now the $(p+1)$-subsets are drawn from P_k and not from G_k, the whole group. The advantage is that $nrep(k)$, the number of replications, may be taken quite small. On the other hand, the intial projections do not need much computation time.

Once the above algorithm has been carried out, it remains possible to assign weights to the observations based on their LMS residuals, and then to perform a reweighted least squares analysis.

In the next section we will give more information about $nrep(k)$, illustrate the method on a generated data set, and compare it with the earlier algorithm.

3 Simulation Results

We have written a Fortran program named PROGROUP, an abbreviation which stands for Program for RObust reGression on gROups Using Projections. It is a revised and extended version of the program PROGRESS (Rousseeuw and Leroy (1987)) for robust multiple regression. PROGROUP fits a distributed intercept model (1.2) by means of the algorithm of Rousseeuw and Wagner (method 1), and by means of our new algorithm with the preliminary projections (method 2). (For completeness, the program also provides a third method in which the slopes as well as the intercepts may vary between the groups. This method corresponds to carrying out a robust regression in each group, but its computation is much faster than with PROGRESS.)

The number of subsamples used by method 1 is the same as in PROGRESS. It was chosen such that the probability of drawing at least one good subsample is near 1. (A "good" subsample consists only of uncontaminated points.) The first column of Table 1 shows this number m for the different values of p. For smaller values of n we draw *all* subsets, hence m equals C_n^{p+1}.

	Method 1		Method 2
p	m	for	m
1	1000	$n > 50$	250
2	1500	$n > 22$	375
3	2000	$n > 17$	500
4	2500	$n > 15$	625
5	3000	$n > 14$	750
6	3500	all n	875
7	4000	all n	1000
8	4500	all n	1125
9	5000	all n	1250
10	5500	all n	1375
≥ 11	6000	all n	1500

Table 1: Number m of subsamples used in PROGROUP for method 1 and 2

The next column of Table 1 lists the number used in the second method. Our simulations have indicated that it is sufficient to consider about one fourth of the original number, since then the objective function value is close to that of method 1.

For each group G_k we compute $nrep(k)$ accordingly, by

$$nrep(k) = \left[m \frac{n_k}{n} \right].$$

For $p \leq 5$ and small n_k we also compute the number of all possible $(p+1)$-subsets of P_k. If this is less than $nrep(k)$, we of course consider all these combinations. (Note that the original PROGRESS contains a "quick" option using a different number of subsamples. The corresponding "quick" option in PROGROUP again takes one fourth of those numbers.)

To illustrate the effectiveness of the new algorithm we will first apply it to a generated data set. We constructed 4 groups of 100 points each, following the model

$$y_i = x_{i1} + x_{i2} + x_{i3} + \sum_{k=1}^{4} \delta_k I_{ik} + \varepsilon_i, \qquad i = 1, \ldots, 400 \tag{3.1}$$

where

$$\begin{array}{lcl} x_{ij} & \sim & \mathcal{N}(0,10) \quad i = 1, \ldots, 400; \, j = 1,2,3 \\ \varepsilon_i & \sim & \mathcal{N}(0,1) \quad \; i = 1, \ldots, 400 \\ \delta_k & = & 10, 20, 30 \text{ or } 40. \end{array}$$

This means that we are dealing with 3 continuous factors and 4 groups, with intercept parameters $10, 20, 30$ and 40. The slope parameters of the hyperplanes are equal to 1. We then included contaminated points by replacing 20% of the x_{i1} by $x_{i1} \sim \mathcal{N}(100, 10)$, yielding bad leverage points.

Table 2 shows the parameter estimates given by both methods. The number of subsamples used, the objective function and the computation time on a Sun workstation and on a 486-PC are also given. The latter are measured in seconds and represent the processing time needed for calculating the coefficients (without drawing the residual plots).

	Method 1 (without proj.)	Method 2 (with proj.)
$\hat{\beta}_1$	1.015	0.986
$\hat{\beta}_2$	0.995	0.976
$\hat{\beta}_3$	0.983	0.995
$\hat{\delta}_1$	9.919	9.893
$\hat{\delta}_2$	20.342	20.076
$\hat{\delta}_3$	30.128	30.039
$\hat{\delta}_4$	40.111	40.292
# subsamples	2000	500
obj. function	1.309	1.307
time (Sun)	19.9 sec.	5.7 sec.
time (PC)	147 sec.	43 sec.

Table 2: Results of PROGROUP on a generated data set

Both methods provide accurate estimates. A complete output shows that they find the same outliers and have simular residual plots. As expected, we see that the new algorithm is executed about four times faster. The actual ratio was 3.5, due to some input/output operations which both algorithms have in common. Because the objective function values are comparable too, we thus may conclude that the new algorithm performs as well as the slower one.

Simular results were obtained in other simulation experiments where 20% of vertical outliers were generated (as in Rousseeuw and Leroy (1987), p. 209), and for other choices of p and q.

4 Case Study

In econometrics, Mincer (1974) has introduced the so-called *earnings function* which corresponds to the model

$$\ln(y) = \beta_1 s + \beta_2 e + \beta_3 e^2 + \delta + \varepsilon, \tag{4.1}$$

where $y =$ personal income, $s =$ years of schooling, and $e =$ years of experience. As it stands, (4.1) is a linear model with a single intercept, but it is now generally assumed that the intercept may vary substantially across sectors (that is, in different industries). For a recent study, see Wagner (1990).

Our data set consists of 296 observations (=persons) belonging to 7 different sectors in Switzerland. (This data set originated from a social survey in 1987 and was analyzed by Rousseeuw and Wagner (1994) with the earlier algorithm, which makes it a good basis of comparison.) We want to estimate the parameters $\beta_1, \ldots, \beta_3, \delta_1, \ldots, \delta_7$ of the model

$$\ln(y_i) = \beta_1 s_i + \beta_2 e_i + \beta_3 e_i^2 + \sum_{k=1}^{7} \delta_k I_{ik} + \varepsilon_i, \quad i = 1, \ldots, 296 \tag{4.2}$$

with $I_{ik} = 1$ if person i works in sector k. Table 3 specifies the sectors.

k	Sector
1	high-tech manufacturing
2	other manufacturing industries
3	building and construction
4	wholesale and retail trade
5	transportation and communication
6	banking and insurance
7	other services

Table 3: List of sectors

We first looked at the distributions of the individual variables, and searched for multicollinearity between the predictor variables. Scatterplots and correlation coefficients did not turn up a linear relationship between any two of them. (Of course, we know that x_2 and x_3 are quadratically related.) We also considered a rotating point cloud graphic of the three regressors, which only gave an indication about the connection between x_2 and x_3, so multicollinearity does not seem to be a problem here.

	Method 1		Method 2	
	estimate	rank	estimate	rank
$\hat{\beta}_1$	0.08561		0.05484	
$\hat{\beta}_2$	0.04778		0.05242	
$\hat{\beta}_3$	-0.00070		-0.00081	
$\hat{\delta}_1$	6.72751	4	6.99306	4
$\hat{\delta}_2$	6.80748	6	7.02210	6
$\hat{\delta}_3$	6.56816	2	6.91435	2
$\hat{\delta}_4$	6.62184	3	6.87065	1
$\hat{\delta}_5$	6.77848	5	7.00435	5
$\hat{\delta}_6$	7.05075	7	7.32903	7
$\hat{\delta}_7$	6.55178	1	6.92371	3
# subsamples	2000		497	
obj. function	0.288		0.293	
time (Sun)	15.2 sec		4.4 sec	
time (PC)	111 sec		30 sec	

Table 4: Regression results on wage differences in Switzerland

Table 4 lists the results of both algorithms on this data set. We see that the objective function values are again close to each other, and that the execution time is reduced about 3.5 times. The coefficients differ slightly, but they all have the same sign and magnitude. Let us also look at the relative ranks of the intercepts, shown in separate columns. Only the groups with ranks 1 and 3 are interchanged, but the other ranks remain. We may conclude that sectors 1, 2, 5 and 6 pay the best salaries, while sectors 3, 4 and 7 score less.

A gaussian $Q - Q$ plot (Figure 1) confirms that the residuals are roughly normally distributed. Only a small number of outliers causes the distribution to be heavier-tailed. Therefore, the normality assumption holds for a majority of the residuals.

Note

The Fortran source code of PROGROUP is available at the email addresses **hubert@wins.uia.ac.be** and **rousse@wins.uia.ac.be**, from the authors.

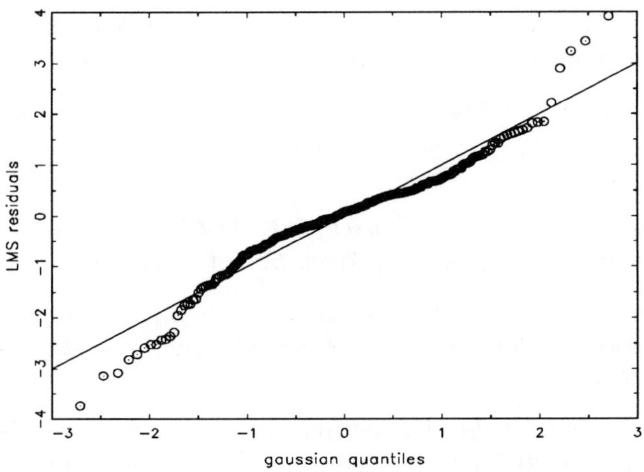

Figure 1: Gaussian $Q - Q$ plot of the residuals

References

[1] Armstrong, R.D. and Frome, E.L. (1977): A special purpose linear programming algorithm for obtaining least absolute value estimators in a linear model with dummy variables. *Communications in Statistics: Simulation and Computation B* **6** 383–398.

[2] Chatterjee, S. and Price, B. (1977): *Regression Analysis by Example*. John Wiley, New York.

[3] Donoho, D.L. and Gasko, M. (1992): Breakdown properties of location estimates based on halfspace depth and projected outlyingness. *Ann. Statist.* **20** 1803–1827.

[4] Draper, N.R. and Smith, H. (1981): *Applied Regression Analysis*. John Wiley, New York.

[5] Edwards, A.L. (1985): *Multiple Regression and the Analysis of Variance and Covariance*. W.H. Freeman and Company, New York.

[6] Mincer, J. (1974): *Schooling, Experience and Earnings*. Columbia University Press, New York.

[7] Montgomery, D.C. (1991): *Design and Analysis of Experiments*. John Wiley, New York.

[8] Montgomery, D.C. and Peck, A.E. (1982): *Introduction to Linear Regression Analysis*. John Wiley, New York.

[9] Rousseeuw, P.J. (1984): Least median of squares regression. *J. Amer. Statist. Assoc.* **79** 871–880.

[10] Rousseeuw, P.J. and Leroy, A.M. (1987): *Robust Regression and Outlier Detection*. John Wiley, New York.

[11] Rousseeuw, P.J. and Wagner, J. (1994): Robust regression with a distributed intercept using least median of squares. *Comput. Statist. & Data Analysis* **17** 65–76.

[12] Rousseeuw, P.J. and van Zomeren, B.C. (1990): Unmasking multivariate outliers and leverage points. *J. Amer. Statist. Assoc.* **85** 633–639.

[13] Rousseeuw, P.J. and van Zomeren, B.C. (1992): A comparison of some quick algorithms for robust regression. *Comput. Statist. & Data Analysis* **15** 107–116.

[14] Wagner, J. (1990): Sektorlohndifferentiale in der Bundesrepublik Deutschland: Empirische Befunde und ökonometrische Untersuchungen zu theoretischen Erklärungen. *Discussion Paper No. 154, Fachbereich Wirtschaftswissenschaften, Universität Hannover*.

Robustness in Discriminant Analysis

Y.S. Kharin *
Belarussian State University

Abstract

The problems of discriminant analysis in $\mathbb{R}^N$ for L classes are considered for the situations, when hypothetical (classical) model of data is distorted. Classification of distortion types is given. Robustness of classical decision rules is evaluated to the distortions of probability density functions of observations to be classified, and robust decision rules are constructed.

Key words and phrases: Discriminant analysis, types of distortions, breakdown point, robust decision rule.
AMS 1991 subject classifications: Primary 62H; secondary 62C.

1 Introduction

The main title of modern statistical data analysis and discriminant analysis, especially, is formulated by P.J. Huber (1981): "Statistical inferences are based only in part upon the observations. An equally important base is formed by prior assumptions about the underlying situation ...". The system of these prior assumptions is called by hypothetical model of data. Unfortunately, our prior assumptions cannot be exact. That is why real data does not correspond to the assumed hypothetical model absolutely: There are some distortions of the model. Because of this fact the classical discriminant decision rules (DR), which are optimal for the hypothetical model, can be non-stable and, in particular, for distorted models lose their optimality. The following topical problems formulated by P.J. Huber have their impact on modern statistics and are investigated in this paper: (A) description and classification of distortion types in discriminant analysis; (B) evaluation of robustness (stability) of classical discriminant decision rules to distortions; (C) evaluation of breakdown points; (D) construction of robust decision rules (RDR), which are not much sensitive to definite distortion-types; (E) construction of special software for robust discriminant analysis.

In Berger (1985, 1990) and Kadane (1978, 1984) the problems (B) and (C) are considered for the cases when prior probabilities and loss functions of general

* These investigations were supported by Belarussian National Found of Science, Grant F40.263

model of statistical decisions theory are distorted. In Kazakos (1982), Lachenbruch (1966), Lawoko (1986), Randless (1978) and Tiku et al. (1986) the problems (B) and (C) are investigated mainly for "contaminations" of multivariate Gaussian p.d.f.. Here and in Kharin (1981, 1983, 1984, 1991, 1992, 1993) we consider the problem (A) and the problems (B)–(E) for different types of distortions of p.d.f. and data.

2 Hypothetical Model of Discriminant Analysis

In $\mathbb{R}^N$, let the random observations $x = (x_j) \in \mathbb{R}^N$ be registered from $L \geq 2$ classes, with prior probabilities $\pi_1, \ldots, \pi_L$ respectively. An observation from the class Ω_i is a random vector $X_i \in \mathbb{R}^N$ with hypothetical conditional p.d.f. $p_i^0(x)$ for $i \in S = \{1, \ldots, L\}$. When the set of densities $p_i^0(\cdot)$ are unknown, they need to be estimated by a so-called training sample $Z \subset \mathbb{R}^{Nn}$ of size n. Usually Z is classified:

$$Z = Z_1 \cup \ldots \cup Z_L,$$

where $Z_i = \{z_{i1}, \ldots, z_{in_i}\}$ is a random sample of size n_i from Ω_i, and the $Z_1, \ldots, Z_L$ are jointly independent. Otherwise, Z is called non-classified training sample. We consider a given loss matrix $W = (w_{ij})$, where $w_{ij} \geq 0$ is the loss incurred if an observation from Ω_i is assigned to Ω_j, where $i, j \in S$. The problem of discriminant analysis is the classification of a random observation $X \in \mathbb{R}^N$ from one of the L classes Ω_i with minimal expected losses.

The randomized form of a DR is defined as a set of L critical functions:

$$\chi = (\chi_i) = \chi(x, Z) = (\chi_i(x, Z)), \qquad i \in S, \; x \in \mathbb{R}^N, \tag{2.1}$$

where $\chi_i(x, Z) \in [0, 1]$ is the probability of the random decision $d = i$ of assigning x to Ω_i; in particular, we have the identity

$$\chi_1(x, Z) + \cdots + \chi_L(x, Z) \equiv 1. \tag{2.2}$$

When $\chi_i(x, Z) \in \{0, 1\}$ for all $i \in S$, the DR (2.1) takes the non-randomized form:

$$d = d(x, Z) = \sum_{i=1}^{L} i \, \mathrm{I}_{V_i}(x), \qquad x \in \mathbb{R}^N, \; d \in S, \tag{2.3}$$

where $\mathrm{I}_A(x)$ is indicator function of a set $A \subset \mathbb{R}^N$ and

$$V_i = V_i(Z) = \{x : \chi_i(x, Z) = 1\} = \{x : d(x, Z) = i\} \subset \mathbb{R}^N$$

is the region in observation space, where the decision $d = i$ is made.

Denoting expectation by $\mathrm{E}\{\cdot\}$, the risk functional—expected loss—is

$$r = r(\chi; \{p_i^0\}) = \sum_{i,j \in S} \pi_i w_{ij} \, \mathrm{E}\Big\{ \int_{\mathbb{R}^N} \chi_j(x, Z) p_i^0(x) \, dx \Big\} \geq 0. \tag{2.4}$$

A Bayesian decision rule (BDR) minimizes the risk (2.4) for given conditional p.d.f. $\{p_i^0(\cdot)\}$, prior probabilities $\{\pi_i(\cdot)\}$, and loss matrix W, and takes the form (2.3), which is

$$d = d_0(x) = \sum_{i \in S} i \prod_{\substack{k \in S \\ k \neq i}} \mathrm{I}(f_{ki}^0(x) \geq 0), \tag{2.5}$$

where $\mathrm{I}(\cdot \geq 0) = \mathrm{I}_{[0,\infty)}(\cdot)$ is the Heavyside unit function, and

$$f_{ki}^0(x) = \sum_{l \in S} c_{lki} p_l^0(x), \qquad c_{lki} = \pi_l(w_{lk} - w_{li}). \tag{2.6}$$

Its risk—minimal possible value—is called Bayesian risk:

$$r_0 = \sum_{i \in S} \pi_i r_{0i}; \qquad r_{0i} = \sum_{j \in S} w_{ij} \int_{V_j^0} p_i^0(x)\, dx. \tag{2.7}$$

When $\{p_i^0(\cdot)\}$ are unknown, the family of plug-in decision rules

$$\hat{d}(x, Z): \mathbb{R}^N \times \mathbb{R}^{Nn} \longrightarrow S$$

is used in practice: A plug-in DR is defined by substituting any consistent statistical estimator $\hat{p}_i(\cdot)$, based on the training sample Z, for true values $p_i^0(\cdot)$ respectively, in the BDR (2.5) and (2.6).

3 Types of Distortions

The types of distortions of the hypothetical model in discriminant analysis can be classified into three subsets:

D.1 Small-sample effects;

D.2 Distortions of models for observations to be classified;

D.3 Distortions of models for training samples.

D.1 includes the distortions caused by the substitution of $\hat{p}_i(\cdot)$ for the true p.d.f. $p_i^0(\cdot)$ in a BDR. There is a well-known result of Glick (1972) that, under weak conditions, the risk r of a plug-in DR $\hat{d}(\cdot)$ converges to r_0 as $n \to \infty$. But for "small" sample sizes n the deviation $r - r_0 > 0$ can be significant for practice and will depend on the type of estimators: parametric (D.1.1) or non-parametric (D.1.2). For the investigation of the difference $r - r_0$ the method of risk asymptotic expansions used by Okamoto (1963), Raudis (1976), Kharin (1981, 1983) and other authors turned out fruitful.

In D.2 the hypothetical p.d.f. $p_i^0(\cdot)$ are distorted: an observation from Ω_i in reality may have any p.d.f.

$$p_i(\cdot) \in \mathcal{P}_i(\varepsilon_{+i}), \qquad i \in S, \tag{3.1}$$

where $\mathcal{P}_i(\varepsilon_{+i})$ is a suitable set of distorted densities for the class Ω_i with distortion radius $\varepsilon_{+i} \geq 0$; if $\varepsilon_{+i} = 0$, then $\mathcal{P}_i(0) = \{p_i^0(\cdot)\}$ is the one-point set consisting of the hypothetical p.d.f.. Distortions of type D.2.1 and D.2.2 occur if, respectively, a parametric or a non-parametric formulation of the set $\mathcal{P}_i(\varepsilon_{+i})$ is employed.

D.2.1 includes 3 main subsets of distortions:

D.2.1.1 Errors in parameters assignment;

D.2.1.2 Finite mixtures of ε-close distributions;

D.2.1.3 Additive distortions of observations.

D.2.2 includes 4 main subsets of distortions (Kharin (1992)):

D.2.2.1 Huber type of distortions;

D.2.2.2 Distortions in L_2 (and χ^2) metric;

D.2.2.3 Distortions in Kolmogorov metric;

D.2.2.4 Random distortions of densities.

In D.3 the hypothetical assumptions about the training samples Z, $Z_1, \ldots, Z_L$ are broken:

D.3.1 Parametric ε-nonhomogeneity of Z_i $(i \in S)$;

D.3.2 Dependence of sample elements;

D.3.3 Misclassification of the training sample;

D.3.4 Outliers in the training sample;

D.3.5 Missing values of feature variables.

It should be noticed that the problem of identification of these distortions becomes essential for applied statistical analysis.

4 Decision Rules Robustness to the Distortions of Observations to be Classified

In accordance with the classification scheme given in the previous section, we investigate here the distortions of the following two types: D.2.2.1 and D.2.2.2 .

Let the conditional p.d.f. $p_i(\cdot)$ be subjected to Huber type (D.2.2.1) distortions (3.1), where

$$\mathcal{P}_i(\varepsilon_{+i}) = \{p_i(\cdot) \mid p_i(x) = (1-\varepsilon_i)p_i^0(x) + \varepsilon_i h_i(x),\ 0 \leq \varepsilon_i \leq \varepsilon_{+i}\}, \tag{4.1}$$

where $h_i(\cdot)$ may be an arbitrary density of "contaminating" distribution and $\varepsilon_{+i} < 1$ is the distortion (contamination) radius $(i \in S)$. This means that the class Ω_i of observed objects actually consists of two subclasses:

$\Omega_i = \Omega_i^0 \cup \Omega_i^h$ such that $\Omega_i^0 \cap \Omega_i^h = \emptyset$, where Ω_i^0 is the most frequently observed (well-studied) subclass, whereas Ω_i^h is the seldom observed (non-studied) subclass. An observation from Ω_i^0 follows the known (or at least, estimated) density $p_i^0(\cdot)$, and an observation from Ω_i^h the unknown density $h_i(\cdot)$. An object from Ω_i belongs to Ω_i^0 with probability $1-\varepsilon_i$, and to Ω_i^h with probability ε_i.

To evaluate the stability of any DR χ given by (2.1) and (2.2) we define the functional of maximum risk:

$$r_+ = r_+(\chi) = \sup\Big\{ r(\chi; \{p_i(\cdot)\}) \;\Big|\; p_i(\cdot) \in \mathcal{P}_i(\varepsilon_{+i}) \Big\}. \tag{4.2}$$

The coefficient of unstability for the BDR (2.5) and (2.6), if $r_0 > 0$, is:

$$\kappa_+ = \kappa_+(\chi^0) = (r_+(\chi^0) - r_0)/r_0 \geq 0. \tag{4.3}$$

For given $\delta > 0$, the δ-admissible distortion radius by definition is:

$$\varepsilon_+(\delta) = \inf\Big\{ \alpha \;\Big|\; \sup_{\{0 \leq \varepsilon_{+i} \leq \alpha\}} \kappa_+(\chi^0) \geq \delta \Big\}, \tag{4.4}$$

and the breakdown point is

$$\varepsilon_+^* = \inf\Big\{ \alpha \;\Big|\; \sup_{\{0 \leq \varepsilon_{+i} \leq \alpha\}} r_+(\chi^0) \geq r^* \Big\}, \tag{4.5}$$

where $r^* = \sum_{i,j=1}^{L} \pi_i w_{ij}/L$ is the value of risk for equiprobable guesses.

Define a RDR $\chi^* = \big(\chi_1^*(x), \ldots, \chi_L^*(x)\big)$ as a DR that minimizes the maximum risk:

$$r_+^* = r_+(\chi^*) = \inf_{\chi} r_+(\chi). \tag{4.6}$$

Introduce the notations: $w_{i+} = \max_{j \in S} w_{ij}$, and

$$\overline{w}_+ = \sum_{i \in S} \pi_i w_{i+}, \qquad q(p_i^0; \chi_j) = \int_{\mathbb{R}^N} \chi_j(x) p_i^0(x)\, dx \in [0,1].$$

Theorem 4.1 *If Huber type distortions of form (3.1) and (4.1) occur, then the maximum risk value of a BDR (2.5) and (2.6) is given by*

$$r_+(\chi^0) = r_0 + \sum_{i \in S} \varepsilon_{+i} \pi_i (w_{i+} - r_{0i}).$$

PROOF The proof is by maximization of the risk functional (2.4) in the equivalent form

$$r(\chi^0; \{p_i\}) = r_0 - \sum_{i \in S} \varepsilon_i \pi_i r_{0i} + \sum_{i \in S} \varepsilon_i \pi_i \sum_{j \in S} w_{ij} q(h_i; \chi_j^0)$$

with respect to contaminating densities $h_i(\cdot)$ and radii ε_i. ////

Corollary 4.2 *For a BDR (2.5) and (2.6), the δ-admissible distortion radius (4.4) and breakdown point (4.5) are*

$$\varepsilon_+(\delta) = \min\Big\{1,\ \delta r_0 \Big/ \sum_{i\in S} \pi_i(w_{i+} - r_{0i})\Big\},$$
$$\varepsilon_+^* = (r^* - r_0)/(\overline{w}_+ - r_0).$$

Corollary 4.3 *For a* $0,1$ *-loss matrix, the risk* r *becomes the classification error probability, and we have*

$$\varepsilon_+(\delta) = \min\{1,\ \delta\, r_0/(1-r_0)\}, \qquad \varepsilon_+^* = 1 - \big(L(1-r_0)\big)^{-1}.$$

Notation: $\delta_{ij} = \mathrm{I}(i=j)$ [Kronecker symbol].

Theorem 4.4 *Under Huber type distortions, for every* $x \in \mathbb{R}^N$ *, a RDR has the following form:*

$$\chi_i^*(x) = \delta_{i,d_*(x)}, \qquad d_*(x) = \arg\min_{k\in S} \sum_{j=1}^{L} \pi_j(1-\varepsilon_{+j})p_j^0(x)w_{jk}.$$

PROOF By solving the minimax problem (4.2) and (4.6); for details, the reader is conferred to Kharin (1992; Theorem 5.2 and proof, p 108). ////

Corollary 4.5 *If all classes* Ω_i *are equidistorted (that is,* $\varepsilon_{+i} = \varepsilon_+ \ \forall i \in S$*) then the notions of RDR and BDR coincide.*

By the method of risk asymptotic expansion with respect to $\varepsilon_+ = \max \varepsilon_{+i}$ (cf. Kharin (1992)) it can be shown that

$$r_+(\chi^*) = r_+(\chi^0) - \sum_{l,m\in S} \rho_{lm}\pi_l\pi_m\varepsilon_{+l}\varepsilon_{+m} + \mathrm{O}(\varepsilon_+^3), \tag{4.7}$$

where ρ_{lm} are expansion coefficients, not dependent on ε_{+l}. The quadratic form in (4.7) is non-negative by definition, so the gain is

$$b = r_+(\chi^0) - r_+(\chi^*) \geq 0.$$

For example, if the hypothetical model is the linear Fisher model with two classes:

$$p_i^0(x) = \varphi_N(x \mid a_i, B), \qquad i \in S = \{1,2\} \tag{4.8}$$

where a_i and B are the hypothetical mean vector and covariance matrix, respectively, then

$$b = \frac{\sqrt{\pi_1(1-\pi_1)}\,(\varepsilon_{+2}-\varepsilon_{+1})^2}{2\sqrt{2\pi}\,\Delta\exp(\Delta^2/8)}$$

with the Mahalanobis interclass distance $\Delta = \sqrt{(a_2-a_1)^\top B^{-1}(a_2-a_1)}$.

Consider now the case D.2.2.2 of distortions (3.1), for $i \in S$, in L_2-metric:

$$\mathcal{P}_i(\varepsilon_{+i}) = \left\{ p_i(\cdot) \,\middle|\, \int_{\mathbb{R}^N} \frac{(p_i(x) - p_i^0(x))^2}{\psi_i(x)}\, dx = \varepsilon_i^2,\ \ 0 \le \varepsilon_i \le \varepsilon_{+i} \right\}, \tag{4.9}$$

where $\psi_i(\cdot) \ge 0$ is a non-negative normed weight function,

$$\int_{\mathbb{R}^N} \psi_i(x)\, dx = 1.$$

If $\psi_i(\cdot) \equiv p_i^0(\cdot)$, then we obtain the ε_{+i}-neighbourhood (4.9) of the hypothetical p.d.f. in χ^2-metric.

Notation: Let $\mathrm{E}_{\psi_i}\{\cdot\}$ denote expectation w.r.t. the density $\psi_i(\cdot)$ and, for $i \in S$, define radii $\varepsilon_i^* \ge 0$ by

$$\varepsilon_i^* = \left| \mathrm{E}_{\psi_i}\Big\{ \Big(\sum_{k\in S} w_{ik}\big(\chi_k(x) - q(\psi_i;\chi_k)\big)\Big)^2 \Big\} \right|^{1/2} \times \left| \Big\{ \inf_x \Big(\frac{\psi_i(x)}{p_i^0(x)} \sum_{l\in S} w_{il}\big(\chi_l(x) - q(\psi_i;\chi_l)\big)\Big) \Big\}^- \right|^{-1}$$

where y^-, for of any number y, denotes the negative part.

Theorem 4.6 *If distortions (3.1) and (4.9) are present with $\varepsilon_{+i} \le \varepsilon_i^*$ $(i \in S)$, then for an arbitrary DR $\chi(\cdot)$ the maximum risk value (4.2) is*

$$r_+(\chi) = r(\chi;\{p_i^0\}) + \sum_{i\in S} \varepsilon_{+i}\pi_i \left| \mathrm{E}_{\psi_i}\Big\{ \Big(\sum_{k\in S} w_{ik}\big(\chi_k(x) - q(\psi_i;\chi_k)\big)\Big)^2 \Big\} \right|^{1/2}.$$

PROOF Maximization in (4.2) under the constraints (4.9) by the method of indefinite Lagrange multipliers. Then maximize w.r.t. $\varepsilon_i \in [0, \varepsilon_{+i}]$. For details, the reader is conferred to Kharin (1992; Theorem 5.4 and proof, pp 115–116).////

Corollary 4.7 *In the case of the χ^2-metric and for radii*

$$\varepsilon_{+i} \le \varepsilon_i^* = \sqrt{\sum_{j\in S} \Big(\frac{w_{ij}}{r_{0i}}\Big)^2 q(p_i^0;\chi_j) - 1} \qquad (i \in S),$$

the coefficient of unstability (4.3) for a BDR is given by

$$\kappa_+(\chi^0) = \sum_{i\in S} \pi_i\varepsilon_{+i} \sqrt{\sum_{j\in S} \Big(\frac{w_{ij} - r_{0i}}{r_0}\Big)^2 q(p_i^0;\chi_j^0)} \ge 0.$$

Corollary 4.8 *For a* $0,1$ *-loss matrix* W *, under the conditions of Corollary 4.7, we have*

$$\varepsilon_i^* = \sqrt{q(p_i^0;\chi_i^0)/(1-q(p_i^0;\chi_i^0))}, \qquad i \in S,$$

$$\kappa_+(\chi^0) = \frac{1}{r_0}\sum_{i\in S}\pi_i\varepsilon_{+i}\sqrt{q(p_i^0;\chi_i^0)(1-q(p_i^0;\chi_i^0))}\,,$$

and breakdown point

$$\varepsilon_+^* = (1-L^{-1}-r_0)\Big(\sum_{i\in S}\pi_i\sqrt{q(p_i^0;\chi_i^0)(1-q(p_i^0;\chi_i^0))}\,\Big)^{-1}.$$

Corollary 4.9 *For a BDR* $\chi^0(\cdot)$ *, the hypothetical Gaussian model (4.8), and a* $0,1$ *-loss matrix, in case* $L=2$ *and if the classes* Ω_1 *and* Ω_2 *are equiprobable* ($\pi_1=\pi_2=\frac{1}{2}$) *and equidistorted* ($\varepsilon_{+1}=\varepsilon_{+2}$) *, then the* δ *-admissible distortion radius and breakdown point are*

$$\varepsilon_+(\delta) = \delta\big(1/\Phi(\Delta/2)-1\big)^{1/2},$$

$$\varepsilon_+^* = (\Phi(\Delta/2)-1/2)\Big(\Phi(\Delta/2)(1-\Phi(\Delta/2))\Big)^{-1/2}.$$

Introduce the notations:

$$a_{kj}(x;\chi) = \sum_{i\in S}\pi_i\varepsilon_{+i}w_{ik}w_{ij}c_i(\chi)\psi_i(x),$$

$$b_j(x;\chi) = \sum_{i\in S}\pi_i w_{ij}\Big(p_i^0(x) - \varepsilon_{+i}c_i(\chi)\psi_i(x)\sum_{k\in S}w_{ik}q(\psi_i;\chi_k)\Big),$$

$$c_i(\chi) = \Big|\mathrm{E}_{\psi_i}\Big\{\Big(\sum_{k\in S}w_{ik}(\chi_k(x)-q(\psi_i;\chi_k))\Big)^2\Big\}\Big|^{1/2}, \qquad i,j,k\in S. \quad (4.10)$$

Theorem 4.10 *Under the assumptions of Theorem 4.6, the set of robust critical functions* $\chi^* = \{\chi_j^*(x)\}$ *are the solution of the following optimization problem:*

$$\int_{\mathbb{R}^N}\Big(\sum_{k,j=1}^{L}a_{kj}(x;\chi)\chi_j(x)\chi_k(x) + \sum_{j=1}^{L}b_j(x;\chi)\chi_j(x)\Big)dx \to \min_{\chi} \qquad (4.11)$$

under the constraints (2.2).

PROOF Putting the "least favourable density" found in Theorem 4.6 into the risk functional (2.4) and using the notations (4.10), we get the minimization problem (4.6) in equivalent form (4.11). ////

The optimization problem (4.11) can be considered as perturbation of the quadratic optimization problem w.r.t. the coefficients $\{a_{kj},\ b_j\}$). To find the RDR χ^* as solution to (4.11), we use the method of successive approximations.

Our results of robustness investigation in discriminant analysis are used in the applied program package ROSTAN (RObust STatistical ANalysis) for IBM PC/AT published by the Department of Mathematical Modeling and Data Analysis in Belarussian State University.

References

[1] Berger, J.O. (1985): *Statistical decision theory and Bayesian analysis.* Springer-Verlag, New York.

[2] Berger, J.O. (1990): Robust Bayesian analysis. *J. Statist. Plann. Inference* **25** 303–328.

[3] Glick, N. (1972): Sample-based classification procedures derived from density estimators. *J. Amer. Statist. Assoc.* **67** 116–122.

[4] Huber, P.J. (1981): *Robust Statistics.* John Wiley, New York.

[5] Kadane, J.B. et al. (1978): Stable decision problems. *Ann. Statist.* **6** 1095–1110.

[6] Kadane, J.B. (ed., 1984): *Robustness of Bayesian Analysis.* North-Holland, New York.

[7] Kazakos, D. (1982): Statistical discrimination using inaccurate models. *IEEE Trans.* **IT-28** 720–728.

[8] Kharin, Y.S. (1981): About statistical classification accuracy at MC-estimators using. *Probability Theory and Its Applications* **26** 866–867.

[9] Kharin, Y.S. (1983): Investigation and optimization of Rosenblatt-Parzen classificator by asymptotic expansions. *Automatics and remote control* 1 91–100.

[10] Kharin, Y.S. (1983): About decision rule robustness under misclassifications presence in training samples. *Automatics and remote control* 11 100–110.

[11] Kharin, Y.S. (1984): Robustness investigation for the decision rules by risk asymptotic expansion method. In *Proceedings of the Third Prague Symposium on Asymptotic Statistics (P. Mandl ed.)*, 309–317. H.X.-Oxford, Elsevier, Amsterdam.

[12] Kharin, Y.S. et al. (1991): Asymptotic robustness of discriminant procedures for dependent and non-homogeneous observations. In *Probability Theory and Mathematical Statistics* 602–610. Proc. of the 5 Vilnius Conf., Utrecht, Netherlands.

[13] Kharin, Y.S. (1992): *Robustness in Statistical Pattern Recognition.* Minsk, Universitetskoje.

[14] Kharin, Y.S. and Zhuk, E.E. (1993): Asymptotic robustness in cluster-analysis for the case of Tukey-Huber distortions. In *Information and Classification* (Opitz, O., Lausen, B., Klar, R., eds.), 31–39. Springer-Verlag, New York.

[15] Lachenbruch, P.A. (1966): Discriminant analysis when the initial samples are misclassified. *Technometrics* **8** 657–662.

[16] Lawoko, C.R.O. et al. (1986): Asymptotic error rates of the W and Z statistics when the training observations are dependent. *Pattern recognition* **18** 467–471.

[17] Okamoto, M. (1963): An asymptotic expansion for the distribution of the linear discriminant function. *Ann. Math. Statist.* **34** 1286–1301.

[18] Randless, R.H. (1978): Generalized linear and quadratic discriminant functions using robust estimates. *J. Amer. Statist. Assoc.* **73** 564–568.

[19] Raudis, S.H. (1976): Sample size finiteness in classification problems. *Statistical control problems* **18** 1-180.

[20] Tiku, M.L. et al. (1986): *Robust Inference*. Marcel Dekker, New York.

M-Estimation and Spatial Quantiles

V. Koltchinskii *
University of Giessen

Abstract

An extension of quantile function (M-quantiles) related in a certain sense to M-parameters of probability distribution defined by convex minimization is considered. Asymtotics of empirical M-quantiles are studied.

Key words and phrases: M-parameter, M-estimator, quantile function, empirical process, Bahadur–Kiefer representation.
AMS 1980 subject classifications: 60F05; 62E20.

1 M-Parameters and Quantiles

Let $(\mathbf{X}, \mathcal{A}, P)$ be a probability space, and let $f : \mathbb{R}^d \times \mathbf{X} \mapsto \mathbb{R}^1$. Suppose that, for any $s \in \mathbb{R}^d$, $f(s, .)$ is P-integrable. Define

$$f_P(s) := \int_{\mathbf{X}} f(s,x)P(dx), \qquad s \in \mathbb{R}^d.$$

If there exists a point s_0 which minimizes f_P on $\mathbb{R}^d$, it is called an *M-parameter* of P with respect to f.

Given a sample $(X_1, \ldots, X_n)$ from the distribution P, let P_n denote the empirical measure based on this sample. An M-parameter of P_n is called an *M-estimator*.

Another way to define M-parameters and M-estimators is to consider a vector-valued function $F : \mathbb{R}^d \times \mathbf{X} \mapsto \mathbb{R}^d$ such that, for any $s \in \mathbb{R}^d$, the section $F(s, .)$ is P-integrable so that one can define a map $F_P : \mathbb{R}^d \mapsto \mathbb{R}^d$ by

$$F_P(s) := \int_{\mathbf{X}} F(s;x)P(dx), \qquad s \in \mathbb{R}^d.$$

If there exists a solution s_0 of the equation $F_P(s) = 0$, it is called an M-parameter of P while a solution of the empirical equation $F_{P_n}(s) = 0$ is called an M-estimator.

* Supported by the Alexander von Humboldt Foundation while on leave from the University of Kiev

Under certain smoothness assumptions one could take $F(s;x) := \nabla f(s;x)$ and get $F_P = \nabla f_P$.

If F_P is a one-to-one map (at least locally from a neighborhood U onto a neighborhood V of 0), then there exists only one M-parameter in U, and it coincides with $F_P^{-1}(0)$.

The asymptotic theory of M-estimators was developed by P.J. Huber (1964, 1967, 1981).

It could be reasonable to study, along with a minimization problem,

$$f_P(s) \to \inf, \qquad s \in \mathbb{R}^d$$

a family of perturbed problems

$$f_P(s) - \langle s, t\rangle \to \inf, \qquad s \in \mathbb{R}^d,$$

where $t \in \mathbb{R}^d$.

Definition 1.1 We call a minimal point of the functional

$$f_{P,t}(s) := f_P(s) - \langle s, t\rangle, \qquad s \in \mathbb{R}^d$$

an (M,t)*-parameter* of P. An (M,t)-parameter of the empirical measure will be called an (M,t)- *estimator* based on a sample $(X_1, \dots, X_n)$.

Note that a function $f_P^* : \mathbb{R}^d \mapsto \mathbb{R}^1$ defined by

$$f_P^*(t) := -\inf_{s\in\mathbb{R}^d} f_{P,t}(s) = \sup_{s\in\mathbb{R}^d} [\langle s, t\rangle - f_P(s)]$$

is known in convex analysis as the *Young–Fenchel conjugate* of f_P (in smooth analysis it coincides with the Legendre transformation of f_P).

Example 1.2 Consider the kernel

$$f(s,x) := 0.5(|s-x| - |x| + s), \qquad s \in \mathbb{R}^1, \ x \in \mathbb{R}^1.$$

Note that $f(s,x)$ is convex with respect to s. Therefore, f_P is a convex function. The distribution function F_P of P is a subgradient of f_P and the quantile function F_P^{-1} is a subgradient of f_P^*. For any $t \in [0,1]$, (M,t)-parameters of P with respect to f coincide with quantiles of P.

It could be shown that for any kernel $f(s,x)$ which is convex with respect to s P-a.s., so that f_P is a convex function, the subdifferential ∂f_P of f_P has some properties of the distribution function and its inverse $(\partial f_P)^{-1} = \partial f_P^*$ has some properties of the quantile function. For any $t \in \mathbb{R}^d$, $(\partial f_P)^{-1}(t)$ is the set of all (M,t)-parameters of P.

To be more specific, let φ be a convex nondecreasing function on $[0,\infty)$, differentiable in $(0,\infty)$, define

$$\Phi(\lambda) := \varphi'(\lambda), \qquad \lambda > 0$$

and choose any $s_0 \in \mathbb{R}^d$. Denoting the Euclidean norm on $\mathbb{R}^d$ by $|.|$, define a kernel

$$f(s;x) := \varphi(|s-x|) - \varphi(|s_0-x|), \qquad s, x \in \mathbb{R}^d.$$

We assume that the functions

$$\varphi(|s-.|) - \varphi(|s_0-.|), \qquad s \in \mathbb{R}^d$$

are P-integrable and define

$$f_P(s) := \int_{\mathbb{R}^d} (\varphi(|s-x|) - \varphi(|s_0-x|))P(dx), \qquad s \in \mathbb{R}^d.$$

Introduce

$$\rho_\varphi := \lim_{\varepsilon \downarrow 0} \frac{\varphi(\varepsilon)-\varphi(0)}{\varepsilon}, \qquad R_\varphi := \lim_{\lambda \uparrow \infty} \frac{\varphi(\lambda)}{\lambda}.$$

A map $F_P : \mathbb{R}^d \mapsto \mathbb{R}^d$,

$$F_P(s) := \int_{\{x \neq s\}} \Phi(|s-x|) \frac{s-x}{|s-x|} P(dx), \qquad s \in \mathbb{R}^d$$

is a subgradient of the convex function f_P. ////

Main properties of the map F_P.

(i) For any P the map F_P is monotone in the sense that

$$\langle s_1 - s_2, F_P(s_1) - F_P(s_2) \rangle \geq 0, \qquad s_1, s_2 \in \mathbb{R}^d.$$

(ii) Let A_P denote the set of all atoms of P, and let $p(s) := P(\{s\})$, $s \in A_P$. Then F_P is continuous in $\mathbb{R}^d \setminus A_P$, and for $s \in A_P$

$$F_P(s+\varepsilon v) \longrightarrow F_P(s) + p(s)\rho_\varphi v, \qquad \varepsilon \to 0, \; |v| = 1.$$

(iii) $F_P(\lambda v) \longrightarrow R_\varphi v, \qquad \lambda \to \infty, \; |v| = 1$.

Suppose now that either φ is strictly convex, or it is strictly increasing and P is not concentrated in a line (under these conditions f_P is strictly convex).

(iv) F_P is a strictly monotone map in the sense that

$$\langle s_1 - s_2, F_P(s_1) - F_P(s_2) \rangle > 0, \qquad s_1, s_2 \in \mathbb{R}^d, \; s_1 \neq s_2.$$

(v) Denoting by $B(s;\delta)$ the open ball and $\bar{B}(s;\delta)$ the closed ball with the center $s \in \mathbb{R}^d$ and radius $\delta > 0$, let

$$U_P := B(0; R_\varphi) \setminus \bigcup_{s \in A_P} \bar{B}(F_P(s); p(s)\rho_\varphi).$$

All balls $\bar{B}(F_P(s); p(s)\rho_\varphi)$ for $s \in A_P$ are mutually disjoint. The map F_P is one-to-one from $\mathbb{R}^d$ onto $U_P \cup F_P(A_P)$ and it is a homeomorphism of $\mathbb{R}^d \setminus A_P$ and U_P. If P is nonatomic, then F_P is a homeomorphism of $\mathbb{R}^d$ and $B(0; R_\varphi)$.

Let us extend the map F_P^{-1} (defined in a natural way on the set U_P) to the whole ball $B(0; R_\varphi)$ assuming that, for $t \in B(0; R_\varphi) \setminus U_P$, $F_P^{-1}(t)$ coincides with the unique $s \in A_P$ such that $t \in \bar{B}(F_P(s); p(s)\rho_\varphi)$. Then

(vi) For any $t \in B(0; R_\varphi)$ the unique (M, t)-parameter of P equals $F_P^{-1}(t)$. In particular, the unique M-parameter of P equals $F_P^{-1}(0)$.

(vii) If P_θ denotes the θ-shift of the distribution P, then

$$F_{P_\theta}(s) = F_P(s - \theta), \qquad s \in \mathbb{R}^d,$$

$$F_{P_\theta}^{-1}(t) = \theta + F_P^{-1}(t), \qquad t \in B(0; R_\varphi).$$

(viii) For any orthogonal transformation O in $\mathbb{R}^d$,

$$F_{P \circ O^{-1}}(s) = OF_P(O^{-1}s), \qquad s \in \mathbb{R}^d,$$

$$F_{P \circ O^{-1}}^{-1}(t) = OF_P^{-1}(O^{-1}t), \qquad t \in B(0; R_\varphi).$$

(ix) Let P be a Borel probability measure symmetric with respect to the center $a \in \mathbb{R}^d$ (it means that for a random vector X with the distribution P $X - a$ and $a - X$ are identically distributed). Then

$$F_P(s) = -F_P(2a - s), \qquad s \in \mathbb{R}^d,$$

$$F_P^{-1}(-t) = 2a - F_P^{-1}(t), \qquad t \in B(0; R_\varphi).$$

(x) Let P be a Borel probability measure in $\mathbb{R}^d$, spherically symmetric with respect to the center $a \in \mathbb{R}^d$. Then there exists an increasing continuous function $\psi_P : (0, \infty) \mapsto (0, \infty)$ such that $\psi_P(0+) = \rho_\varphi$, $\psi_P(\infty) = R_\varphi$, and

$$F_P(s) = \begin{cases} \psi_P(|s - a|)\frac{s-a}{|s-a|} & \text{for } s \neq a \\ 0 & \text{for } s = a, \end{cases}$$

$$F_P^{-1}(t) = \begin{cases} a + \psi_P^{-1}(|t|)\frac{t}{|t|} & \text{for } t \neq 0 \\ a & \text{for } t = 0. \end{cases}$$

(xi) In many cases the map F_P characterizes P, so that $F_P = F_Q$ implies that $P = Q$. In particular, this holds if $\varphi(u) = u^r$ where $r \geq 1$ and $r \neq 2k$, for all $k = 1, 2, \ldots$.

(xii) Suppose that P is nonatomic. Given a probability Q in $\mathbb{R}^d$, denote

$$P_\varepsilon = (1 - \varepsilon)P + \varepsilon Q, \qquad \varepsilon \in [0, 1].$$

If, for some $t \in B(0; R_\varphi)$, $F_P^{-1}(t)$ is not an atom of Q and F_P is continuously differentiable in a neighborhood of the point $F_P^{-1}(t)$ with the nondegenerate derivative $F_P' \circ F_P^{-1}(t)$, then [1], on $\varepsilon \to 0$.

$$\frac{F_{P_\varepsilon}^{-1}(t) - F_P^{-1}(t)}{\varepsilon} \longrightarrow -\mathrm{inv}(F_P' \circ F_P^{-1}(t))(F_P - F_Q) \circ F_P^{-1}(t).$$

It follows that the influence curve of $F_P^{-1}(t)$ is, for $t \neq F_P(x)$,

$$IC(x; P; F_P^{-1}(t)) = -\mathrm{inv}(F_P' \circ F_P^{-1}(t))(F_P - F_{\delta_x}) \circ F_P^{-1}(t).$$

Remark 1.3 Dudley and Koltchinskii (1992) (see also Koltchinskii (1994a,c)) and Chaudhuri (1993) studied a map

$$F_P(s) := \int_{\{x \neq s\}} \frac{s-x}{|s-x|} P(dx), \qquad s \in \mathbb{R}^d$$

(so, they have $\Phi \equiv 1$) and the corresponding spatial quantile function F_P^{-1}. In this case $F_P^{-1}(0)$ coincides with the rather well known L_1- or Haldane's spatial median which is defined as a minimal point of the functional

$$f_P(s) := \int_{\mathbb{R}^d} (|s-x| - |x|) P(dx).$$

Koltchinskii (1994b) studied the case of the function $\varphi(u) = u^r, r > 1$ (or, equivalently, $\Phi(u) = ru^{r-1}$).

We are not going to give here the proofs of the properties (i)–(xii). Some of them (like (i), (iv)–(vi)) are just the properties of subgradients of convex or strictly convex functions (see Rockafellar (1970)). Properties (ii), (iii) and (vii)–(x) follow from the integral representation of F_P. Property (xi) could be obtained as a corollary of the results of Mattner (1992), and property (xii) easily follows from Lemma 3.2 below.

These properties show that F_P could be considered as an extension of the distribution function, and F_P^{-1} as an extension of the quantile function (although we understand quite well that the similarity with the classical distribution and quantile functions is not complete). We call such functions like F_P and F_P^{-1} *M-distribution function* and *M-quantile function* of P, respectively. M-quantile functions could be used, in particular, to extend L- and R-estimators to multivariate case.

The study the asymptotic behavior of the empirical M-quantile process

$$n^{1/2}(F_{P_n}^{-1}(t) - F_P^{-1}(t)), \qquad t \in B(0; R_\varphi).$$

is our main goal. ////

[1] $\mathrm{inv}(A)$ denotes the inverse of a nondegenerate linear operator A.

2 Asymptotics of M-Quantiles.

Let S be an open subset (or the closure of an open subset) of $\mathbb{R}^d$, and let $(L, |.|)$ be a separable Banach space. We define $\ell^\infty(S) = \ell^\infty(S; L)$ to be the normed space of all uniformly bounded functions $Y: S \mapsto L$ with the uniform norm

$$\|Y\|_S := \sup_{s \in S} |Y(s)|.$$

Let $(\Omega, \Sigma, \Pr)$ be a probability space. Speaking about random elements and random functions etc., we do not always suppose them to be Σ-measurable.

We define

$$\Pr^*(A) := \inf\{\Pr(B) : B \in \Sigma, B \supset A\}.$$

$$\mathrm{E}^* \xi := \inf\{\mathrm{E}\, \eta : \eta \Sigma\text{-measurable}, \eta \geq \xi\}.$$

Let $\{\xi_n\}_{n\geq 1}$ be a sequence of random functions $\xi_n: S \mapsto L$. $\{\xi_n\}_{n\geq 1}$ *converges weakly* in the space $\ell^\infty(S)$ to a random function $\xi: S \mapsto L$, or

$$\xi_n \xrightarrow{w} \xi, \qquad n \to \infty,$$

if ξ is uniformly bounded and uniformly continuous on S a.s., and for any bounded continuous functional $\Psi: \ell^\infty(S) \mapsto \mathbb{R}$ we have

$$\mathrm{E}^* \Psi(\xi_n) \longrightarrow \mathrm{E}\, \Psi(\xi), \qquad n \to \infty.$$

$\{\xi_n\}_{n\geq 1}$ *converges weakly locally* to a random function ξ, or

$$\xi_n \xrightarrow{w,loc} \xi, \qquad n \to \infty,$$

if, for any compact subset K of the interior of S, $\{\xi_n\}_{n\geq 1}$ converges weakly to ξ in the space $\ell^\infty(K)$. Let Φ be an increasing continuous function from $(0, \infty)$ into $(0, \infty)$. Suppose that

$$\Phi(\lambda_1 + \lambda_2) \leq C[\Phi(\lambda_1) + \Phi(\lambda_2)], \qquad \lambda_1, \lambda_2 > 0 \tag{2.1}$$

with some positive constant C.

Assume also that for all $c > 0$

$$\frac{\Phi(\lambda + c)}{\Phi(\lambda)} \longrightarrow 1, \qquad \lambda \to \infty.$$

Consider an M-distribution function

$$F_P(s) := \int_{\{x \neq s\}} \Phi(|s - x|) \frac{s - x}{|s - x|} P(dx), \qquad s \in \mathbb{R}^d.$$

Theorem 2.1 *If*

$$\int_{\mathbb{R}^d} \Phi(|x|)P(dx) < \infty, \tag{2.2}$$

then a.s., for any compact $K \subset \mathbb{R}^d$,

$$\|F_{P_n} - F_P\|_K \longrightarrow 0, \qquad n \to \infty.$$

If $\Phi(\infty) < \infty$ *then a.s.*

$$\|F_{P_n} - F_P\|_{\mathbb{R}^d} \longrightarrow 0, \qquad n \to \infty.$$

Let W_P° denote a P-Brownian bridge; that is a centered Gaussian process $W_P^\circ(A)$ indexed by Borel sets $A \subset \mathbb{R}^d$ with the covariance

$$\mathrm{E}\, W_P^\circ(A) W_P^\circ(B) = P(A \cap B) - P(A)P(B).$$

Theorem 2.2 *If* P *is nonatomic and*

$$\int_{\mathbb{R}^d} \Phi^2(|x|)P(dx) < \infty, \tag{2.3}$$

then

$$n^{1/2}(F_{P_n} - F_P) \stackrel{w,loc}{\longrightarrow} \xi, \qquad n \to \infty$$

where

$$\xi(s) = \int_{\{x \neq s\}} \Phi(|s-x|)\frac{s-x}{|s-x|} W_P^\circ(dx), \qquad s \in \mathbb{R}^d.$$

If $\Phi(\infty) < \infty$ *then*

$$n^{1/2}(F_{P_n} - F_P) \stackrel{w}{\longrightarrow} \xi, \qquad n \to \infty.$$

Theorem 2.3 *Suppose that* P *is nonatomic and it is not concentrated in a line. Let* U *be an open subset of* $\mathbb{R}^d$ *and let* $V := F_P(U)$. *If condition (2.3) holds, and if* F_P *is continuously differentiable in* U *with the nondegenerate derivative* $F_P'(s)$ *for* $s \in U$, *then, as* $n \to \infty$,

$$n^{1/2}(F_{P_n}^{-1} - F_P^{-1}) \stackrel{w,loc}{\longrightarrow} \eta \quad \textit{in } V$$

where

$$\eta := -\mathrm{inv}(F_P' \circ F_P^{-1})(\xi \circ F_P^{-1}).$$

Moreover, for any compact $K \subset V$,

$$n^{1/2}\|F_{P_n}^{-1} - F_P^{-1} + \mathrm{inv}(F_P' \circ F_P^{-1})(F_{P_n} - F_P) \circ F_P^{-1}\|_K \longrightarrow 0$$

in (outer) probability, as $n \to \infty$.

Remark 2.4 Under some additional conditions one could get kind of Bahadur–Kiefer representations for M-quantiles. For example, assume that Φ is twice continuously differentiable and concave. Suppose that P is nonatomic, it is not concentrated in a line, condition (2.3) holds and, for some $s_0 \in \mathbb{R}^d$,

$$\int_{\{x \neq s_0\}} \frac{\Phi^2(|s_0 - x|)}{|s_0 - x|^2} P(dx) < \infty. \tag{2.4}$$

Put $t_0 := F_P(s_0)$. Then the sequence of random vectors

$$n[F_{P_n}^{-1}(t_0) - s_0 + \mathrm{inv}(F_P'(s_0))(F_{P_n}(s_0) - t_0)]$$

converges weakly to a nondegenerate distribution in $\mathbb{R}^d$.

If P has a density, which is bounded in a neighborhood of s_0, then condition (2.4) holds if and only if

$$\int_0^1 \lambda^{d-3}\Phi^2(\lambda)d\lambda < \infty.$$

In particular, if $\Phi(0+) > 0$, then this condition is satisfied if $d \geq 3$. In the case when it is not satisfied, one might expect slower rates of the Bahadur–Kiefer approximation. ////

Example 2.5 Suppose that $\Phi \equiv 1$. Then

$$F_P(s) = \int_{\{x \neq s\}} \frac{s - x}{|s - x|} P(dx), \qquad s \in \mathbb{R}^d$$

and F_P^{-1} is the quantile function introduced by Dudley and Koltchinskii (1992) and Chaudhuri (1993). $F_P^{-1}(0)$ is the L_1-estimate or Haldane's median of P.

Let

$$\varepsilon_n := \begin{cases} n^{-1} & \text{for } d > 2 \\ n^{-1}\log^{1/2} n & \text{for } d = 2 \\ n^{-3/4} & \text{for } d = 1. \end{cases}$$

Suppose that P has a uniformly bounded density. Let $s_0 \in \mathbb{R}^d$. Assume that F_P is twice continuously differentiable in a neighborhood of s_0 with the nondegenerate derivative $F_P'(s_0)$. Put $t_0 := F_P(s_0)$. Then the distributions of random vectors

$$\varepsilon_n^{-1}[F_{P_n}^{-1}(t_0) - s_0 + \mathrm{inv}(F_P'(s_0))(F_{P_n}(s_0) - t_0)]$$

converges weakly in $\mathbb{R}^d$ as $n \to \infty$ (Koltchinskii (1994 a, c)).

Given any set K, we put

$$\Delta_n(K) := \|F_{P_n}^{-1} - F_P^{-1} + \mathrm{inv}(F_P' \circ F_P^{-1})(F_{P_n} - F_P) \circ F_P^{-1}\|_K.$$

Suppose that F_P is twice continuously differentiable in an open set $U \subset \mathbb{R}^d$ with the nondegenerate derivative $F'_P(s), s \in U$. Denote $V := F_P(U)$. Then for any compact $K \subset V$ there exists a constant $C > 0$ such that

$$\Pr^*(\{\Delta_n(K) \geq \varepsilon_n u\}) \leq C \exp\{-\frac{u}{C}\},\ u > C \log n$$

(see Koltchinskii (1994a)).

Koltchinskii (1994c) suggested a general technique of getting distributional Bahadur–Kiefer representations for functional inverses and applied this technique to the spatial quantiles introduced by Dudley and Koltchinskii (1992) and Chaudhuri (1993). Arcones and Mason (1992) developed an approach to Bahadur–Kiefer representations (distributional and a.s.) for M-estimators and obtained many remarkable representations for special M-estimators. ////

3 Proofs of the Main Results

PROOF OF THEOREMS 2.1 AND 2.2 We are going to use some facts on empirical processes. Let $\mathcal{F}$ be a class of measurable functions defined on a measurable space $(\mathbf{X}, \mathcal{A})$ (specifically, on $\mathbb{R}^d$ with the Borel σ-algebra).

Let P be a probability measure on $(\mathbf{X}, \mathcal{A})$. We define a probability space $(\Omega, \Sigma, \Pr)$ as

$$(\mathbf{X}^\infty, \mathbf{A}^\infty, P^\infty) \times ([0,1], \mathcal{B}, \lambda),$$

where $\mathcal{B}$ denotes the Borel σ-algebra and λ the Lebesgue measure on $[0,1]$. A sequence $\{X_n\}_{n\geq 1}$ of independent observations with the common distribution P is defined as

$$X_n(\omega) := x_n, \quad \omega = (x_1, x_2, \ldots, u) \in \Omega.$$

Let P_n be the empirical measure based on the sample $(X_1, \ldots, X_n)$. A signed random measure $Z_n := n^{1/2}(P_n - P)$ is called the empirical process.

Given a signed measure ν and measurable function f let $\nu(f) := \int_{\mathbf{X}} f d\nu$. Usually the empirical process Z_n is considered a stochastic process $Z_n(f)$ indexed by functions f.

We denote by $\ell^\infty(\mathcal{F})$ the linear normed space of all uniformly bounded functions on $\mathcal{F}$ with the uniform norm $\|.\|_{\mathcal{F}}$,

$$\|Y\|_{\mathcal{F}} := \sup_{f \in \mathcal{F}} |Y(f)|.$$

Assume that $\mathcal{F} \subset L_1(\mathbf{X}, dP)$. $\mathcal{F}$ is called a *P-Glivenko–Cantelli class* (or, $\mathcal{F} \in GC(P)$) if $\Pr^*$ a.s.

$$\|P_n - P\|_{\mathcal{F}} \longrightarrow 0, \qquad n \to \infty.$$

A *P-Brownian bridge* is a centered Gaussian process $W_P^\circ(f)$ indexed by functions $f \in L_2(\mathbf{X}, dP)$ with the covariance

$$\mathrm{E}\, W_P^\circ(f) W_P^\circ(g) = P(fg) - P(f)P(g).$$

One could also define the values of W_P° on measurable sets $W_P^\circ(A) := W_P^\circ(1_A)$ and write

$$W_P^\circ(f) = \int_{\mathbf{X}} f dW_P^\circ.$$

Assume that $\mathcal{F} \subset L_2(\mathbf{X}, dP)$. Let us introduce a metric ρ_P in $L_2(\mathbf{X}, dP)$ setting

$$\rho_P(f,g) := (P(f-g)^2 - (P(f-g))^2)^{1/2}.$$

$\mathcal{F}$ is called a *P-pregaussian class* if there exists a uniformly bounded and ρ_P uniformly continuous on $\mathcal{F}$ version W_P° of P-Brownian bridge.

$\mathcal{F}$ is called a *P-Donsker class* (or, $\mathcal{F} \in CLT(P)$) if it is P-pregaussian and for any uniformly bounded and $\|.\|_{\mathcal{F}}$ continuous functional $\Psi : \ell^\infty(\mathcal{F}) \mapsto \mathbb{R}$ we have

$$\mathrm{E}^* \Psi(Z_n) \longrightarrow \mathrm{E}\,\Psi(W_P^\circ), \qquad n \to \infty.$$

Let $S \subset \mathbb{R}^d$, and let $\mathcal{F} = \{f(s,.) : s \in S\} \subset L_2(\mathbf{X}, dP)$ be a parametric class of functions. We consider a stochastic process

$$\xi_n(s) := Z_n(f(s,.)), \qquad s \in S.$$

If $\mathcal{F}$ is a P-Donsker class and if $f(s,.)$ is continuous on S with respect to the metric ρ_P, then

$$\xi_n \xrightarrow{w} \xi, \qquad n \to \infty,$$

where

$$\xi(s) := \int_{\mathbf{X}} f(s,x)\, W_P^\circ(dx).$$

A measurable function $E_{\mathcal{F}}$ on $\mathbf{X}$ is called an envelope of $\mathcal{F}$ if

$$|f(x)| \le E_{\mathcal{F}}(x), \qquad x \in \mathbf{X},\ f \in \mathcal{F}.$$

Let $\mathcal{P}(\mathbf{X})$ denote the set of all probability measures on $(\mathbf{X}, \mathcal{A})$. Given any $Q \in \mathcal{P}(\mathbf{X})$ and $1 \le r < \infty$ let $d_{Q,r}$ denote the metric of the space $L_r(\mathbf{X}, dQ)$. Let $H_{d_{Q,r}}(\mathcal{F};\varepsilon)$ denote the ε-entropy of the class $\mathcal{F}$ with respect to the metric $d_{Q,r}$. For $\varepsilon > 0$ we define

$$H_r(\mathcal{F};\varepsilon) = H_{r,E_{\mathcal{F}}}(\mathcal{F};\varepsilon) := \sup_{Q \in \mathcal{P}(\mathbf{X})} H_{d_{Q,r}}(\mathcal{F};\varepsilon(\int_{\mathbf{X}} E_{\mathcal{F}}^r dQ)^{1/r}).$$

Suppose that $\mathcal{F}$ is a P-empirically measurable class of functions (for example, $\mathcal{F} \in NLDM(P)$; see Gine and Zinn (1986)).

We shall use the following lemma (see Pollard (1990) for the statements of such and an even more general kind).

Lemma 3.1 *(i) If*

$$H_1(\mathcal{F};\varepsilon) < \infty, \qquad \varepsilon > 0 \tag{3.1}$$

and

$$\int_{\mathbf{X}} E_{\mathcal{F}} dP < \infty, \tag{3.2}$$

then $\mathcal{F} \in GC(P)$.

(ii) If

$$\int_0^\infty H_2^{1/2}(\mathcal{F};\varepsilon)d\varepsilon < \infty \tag{3.3}$$

and

$$\int_{\mathbf{X}} E_{\mathcal{F}}^2 dP < \infty, \tag{3.4}$$

then $\mathcal{F} \in CLT(P)$.

In particular, conditions (3.1) and (3.3) hold if $\mathcal{F}$ is a Vapnik–Chervonenkis subgraph class, that is, the class of sets

$$\{\{(x,t) : f(x) \geq t \geq 0\} : f \in \mathcal{F}\} \cup \{\{(x,t) : f(x) \leq t \leq 0\} : f \in \mathcal{F}\}$$

is Vapnik–Chervonenkis. In this case we have

$$H_r(\mathcal{F};\varepsilon) = O(\log \varepsilon^{-1}), \qquad \varepsilon \to 0 \tag{3.5}$$

(see Pollard (1984, 1990)).

Define

$$F(s,x) := \begin{cases} \Phi(|s-x|)\frac{s-x}{|s-x|} & \text{for } x \neq s \\ 0 & \text{otherwise .} \end{cases}$$

If $\Phi(\infty) < \infty$ let $K := \mathbb{R}^d$. Otherwise, let K be an arbitrary compact in $\mathbb{R}^d$.

We are going to apply Lemma 3.1 to the following class

$$\mathcal{F} := \mathcal{F}^{(1)} \cup \cdots \cup \mathcal{F}^{(d)}$$

where, $x^{(j)}$ denoting the j-th component of any vector $x \in \mathbb{R}^d$,

$$\mathcal{F}^{(j)} := \{F^{(j)}(s,.) : s \in K\}.$$

If $\Phi(\infty) < \infty$ then $\mathcal{F}$ is a uniformly bounded class of functions (and the function $E_{\mathcal{F}}(x) \equiv \Phi(\infty)$ is its envelope). Otherwise, condition (2.1) implies that there exist constants $C_1, C_2 > 0$ such that the function

$$E_{\mathcal{F}}(x) := C_1 + C_2\Phi(|x|)$$

is an envelope of the class $\mathcal{F}$.

We also define

$$F_1(s,x) := \begin{cases} \Phi(|s-x|) & \text{for } x \neq s \\ 0 & \text{otherwise.} \end{cases}$$

and

$$F_2(s,x) := \begin{cases} E_{\mathcal{F}}(x)\frac{s-x}{|s-x|} & \text{for } x \neq s \\ 0 & \text{otherwise.} \end{cases}$$

Let

$$\mathcal{F}_1 := \{F_1(s,.) : s \in K\}$$
$$\mathcal{F}_2^{(j)} := \{F_2^{(j)}(s,.) : s \in K\}, \qquad 1 \leq j \leq d.$$

We claim that $\mathcal{F}_1$ and $\mathcal{F}_2^{(j)}$, $1 \le j \le d$ are Vapnik–Chervonenkis subgraph classes. As to $\mathcal{F}_1$, see Pollard (1984), Ch. 2, Example 26. As to $\mathcal{F}_2^{(j)}$, consider the set

$$\begin{aligned}
&\{(x,t) : F_2^{(j)}(s,x) \ge t \ge 0\} \\
&= \{(x,t) : s^{(j)} - x^{(j)} \ge \frac{t|s-x|}{E_{\mathcal{F}}(x)},\ x \ne s,\ t \ge 0\} \cup \{(s,0)\} \\
&= \{(x,t) : (s^{(j)} - x^{(j)})^2 - \frac{t^2|s-x|^2}{E_{\mathcal{F}}^2(x)} \ge 0,\ x \ne s,\ t \ge 0\} \cup \{(s,0)\}.
\end{aligned}$$

Let Ψ denote the linear span of functions 1, $x^{(j)}$, $(x^{(j)})^2$, t^2, $t^2\frac{x^{(j)}}{E_{\mathcal{F}}^2(x)}$ and $t^2(\frac{x^{(j)}}{E_{\mathcal{F}}(x)})^2$. Since Ψ is a finite dimensional space of functions, the class

$$\mathcal{C} := \{\{(x,t) : \psi(x,t) \ge 0\} : \psi \in \Psi\}$$

is Vapnik–Chervonenkis (see Pollard (1984)). Clearly, the class

$$\tilde{\mathcal{C}} := \{C \cap \{(x,t) : x \ne s, t \ge 0\} \cup \{(s,0)\} : C \in \mathcal{C}, s \in K\}$$

is Vapnik–Chervonenkis, too.

Since for any $s \in K$

$$\{(x,t) : F_2^{(j)}(s,x) \ge t \ge 0\} \in \tilde{\mathcal{C}},$$

we can conclude that

$$\{\{(x,t) : F_2^{(j)}(s,x) \ge t \ge 0\} :\ s \in K\}$$

is a Vapnik–Chervonenkis class. Similarly, the class

$$\{\{(x,t) : F_2^{(j)}(s,x) \le t \le 0\} :\ s \in K\}$$

is also Vapnik–Chervonenkis. Therefore, the class $\mathcal{F}_2^{(j)}$ is Vapnik–Chervonenkis subgraph.

Clearly, the function $E_{\mathcal{F}}$ is an envelope of $\mathcal{F}_1$ and $\mathcal{F}_2^{(j)}, 1 \le j \le d$.

We use now the following elementary inequality (for all $s_1, s_2 \in K$)

$$\begin{aligned}
|F^{(j)}(s_1,x) - F^{(j)}(s_2,x)| &= |(\Phi(|s_1-x|) - \Phi(|s_2-x|))\frac{s_1^{(j)} - x^{(j)}}{|s_1-x|} \\
&\quad + \Phi(|s_2-x|)(\frac{s_1^{(j)} - x^{(j)}}{|s_1-x|} - \frac{s_2^{(j)} - x^{(j)}}{|s_2-x|})| \\
&\le |\Phi(|s_1-x|) - \Phi(|s_2-x|)| \\
&\quad + E_{\mathcal{F}}(x)|\frac{s_1^{(j)} - x^{(j)}}{|s_1-x|} - \frac{s_2^{(j)} - x^{(j)}}{|s_2-x|}| \\
&\le |F_1(s_1,x) - F_1(s_2,x)| \\
&\quad + |F_2^{(j)}(s_1,x) - F_2^{(j)}(s_2,x)|.
\end{aligned}$$

It follows that for any probability measure Q on $\mathbb{R}^d$ and $1 \le r < \infty$ such that

$$\int_{\mathbb{R}^d} E^r_{\mathcal{F}}(x) Q(dx) < \infty,$$

we have, for all $s_1, s_2 \in K$,

$$d_{Q,r}(F^{(j)}(s_1,.), F^{(j)}(s_2,.)) \le d_{Q,r}(F_1(s_1,.), F_1(s_2,.)) + d_{Q,r}(F_2^{(j)}(s_1,.), F_2^{(j)}(s_2,.))$$

Hence

$$H_{d_{Q,r}}(\mathcal{F}^{(j)};\varepsilon) \le H_{d_{Q,r}}(\mathcal{F}_1;\frac{\varepsilon}{2}) + H_{d_{Q,r}}(\mathcal{F}_2^{(j)};\frac{\varepsilon}{2})$$

and

$$H_r(\mathcal{F}^{(j)};\varepsilon) \le H_r(\mathcal{F}_1;\frac{\varepsilon}{2}) + H_r(\mathcal{F}_2^{(j)};\frac{\varepsilon}{2}). \tag{3.6}$$

Since $\mathcal{F}_1$, $\mathcal{F}_2^{(j)}, 1 \le j \le d$ are Vapnik–Chervonenkis classes, one can conclude from (3.5) and (3.6) that

$$H_r(\mathcal{F}^{(j)};\varepsilon) = O(\log \varepsilon^{-1}), \qquad \varepsilon \to 0.$$

Clearly we also have

$$H_r(\mathcal{F};\varepsilon) = O(\log \varepsilon^{-1}), \qquad \varepsilon \to 0.$$

Since condition (2.2) implies (3.2), we can apply Lemma 3.1 (i) and obtain that $\mathcal{F} \in GC(P)$. The statement of Theorem 2.1 now follows immediately.

Similarly, condition (2.3) implies (3.4), and it follows from Lemma 3.1 (ii) that $\mathcal{F} \in CLT(P)$.

Note that the function $F(s,x)$ is continuous at a point $s = s_0 \in \mathbb{R}^d$ unless $x = s_0$. Since P is nonatomic, it follows that $F(s,x)$ is continuous at any point s_0 for P-almost all x. Using dominated convergence theorem one can prove (under condition (2.3)) that $F(s,.)$ is continuous with respect to s in the metric ρ_P.

Since $\mathcal{F}$ is a P-Donsker class we can conclude that the process

$$\xi_n(s) = n^{1/2}(F_{P_n}(s) - F_P(s)) = Z_n(F(s,.)), \qquad s \in \mathbb{R}^d$$

converges to

$$\xi(s) = \int_{\mathbb{R}^d} F(s,x) W_P^\circ(dx), \qquad s \in \mathbb{R}^d$$

weakly locally if condition (2.3) holds, and even weakly in $\ell^\infty(\mathbb{R}^d)$ if the condition $\Phi(\infty) < \infty$ holds. ////

PROOF OF THEOREM 2.3 We prove the theorem under the condition that $\Phi(\infty) < \infty$. The proof in general case (when Φ might be unbounded, but condition (2.3) holds) is just slightly more complicated.

We are going to use some inequalities for functional inverses (see, Koltchinskii and Dudley (1992) and Koltchinskii (1994a)).

Given an open set $S \subset \mathbb{R}^d$ and a function x from S into a linear normed space $(L, |.|)$ we denote

$$\omega_x(t;\delta) = \omega_x^S(t;\delta) := \sup\{|x(s_1) - x(s_2)| : s_1, s_2 \in B(t;\delta) \cap S\}$$

for $t \in S$ and $\delta > 0$. For any compact $K \subset S$ and $\delta > 0$ let

$$\omega_x(K;\delta) = \omega_x^S(K;\delta) := \sup_{t \in K} \omega_x(t;\delta).$$

Let S, T be open subsets in $\mathbb{R}^d$, and let $G: S \mapsto T$, $H: S \mapsto T$, $H^{(-1)}: T \mapsto S$. Let $I := I_T$ be the identity on T. Denote

$$\delta(H,G) := \|H - G\|_S, \ \theta(H, H^{(-1)}) := \|H \circ H^{(-1)} - I\|_T,$$

$$\gamma(G, H, H^{(-1)}) := \delta(H,G) + \theta(H, H^{(-1)}).$$

We use the following version of Lemma 3.1 in Koltchinskii (1994a), where C denotes some generic positive constant.

Lemma 3.2 *If G is a diffeomorphism of $U \subset S$ and $V \subset T$, then for any compact $K \subset V$ there exists some $\delta_0 = \delta_0(K) > 0$ such that $\gamma(G, H, H^{(-1)}) < \delta_0$ implies that*

$$\begin{aligned} &\|H^{(-1)} - G^{-1} + \mathrm{inv}(G' \circ G^{-1})(H - G) \circ G^{-1}\|_K \\ &\quad \leq C[\omega_{H-G}(G^{-1}(K); C\gamma(G, H, H^{(-1)})) \\ &\qquad + \delta(H,G)\, \omega_{G'}(G^{-1}(K); C\delta(H,G)) \\ &\qquad + \theta(H, H^{(-1)})]. \end{aligned}$$

We apply this lemma to $G := F_P, H := F_{P_n}$ and $H^{(-1)} := F_{P_n}^{-1}$. Take $S = \mathbb{R}^d$ and $T = B(0; R_\varphi)$. Note that since F_P is a one-to-one map from U onto V continuously differentiable at any point $s \in U$ with the nondegenerate derivative $F_P'(s)$, F_P is a diffeomorphism of U and V. Without loss of generality we can and do assume that U (and V) are bounded.

We have

$$\delta(F_{P_n}, F_P) = n^{-1/2}\|\xi_n\|_S. \tag{3.7}$$

It follows from the definition of the inverse $F_{P_n}^{-1}$ that either

$$F_{P_n} \circ F_{P_n}^{-1}(t) = t$$

(namely, if $t \notin \bar{B}(F_{P_n}(X_k); n^{-1}\rho_\varphi)$ for all $k = 1, \ldots, n$) or if

$$t \in \bar{B}(F_{P_n}(X_k); n^{-1}\rho_\varphi),$$

for some $1 \leq k \leq n$, then

$$F_{P_n} \circ F_{P_n}^{-1}(t) = F_{P_n}(X_k).$$

Thus, in any case, we have

$$|F_{P_n} \circ F_{P_n}^{-1}(t) - t| \le n^{-1}\rho_\varphi.$$

and so

$$\theta(F_{P_n}, F_{P_n}^{-1}) \le n^{-1}\rho_\varphi. \tag{3.8}$$

It follows from (3.7) and (3.8) that

$$\gamma(F_P, F_{P_n}, F_{P_n}^{-1}) \le n^{-1/2}\|\xi_n\|_S + \rho_\varphi n^{-1}. \tag{3.9}$$

Let $K \subset V$ be a compact. Since F_P' is continuous in U, we have

$$\delta(F_{P_n}, F_P)\, \omega_{F_P'}(F_P^{-1}(K); C\delta(F_{P_n}, F_P)) = o_p(\delta(F_{P_n}, F_P)), \qquad n \to \infty.$$

Using (3.7) we get

$$\delta(F_{P_n}, F_P)\, \omega_{F_P'}(F_P^{-1}(K); C\delta(F_{P_n}, F_P)) = o_p(n^{-1/2}\|\xi_n\|_S), \qquad n \to \infty.$$

We now use Lemma 3.2. There exists $\delta_0 > 0$ such that the condition

$$n^{-1/2}\|\xi_n\|_S + n^{-1}\rho_\varphi \le \delta_0 \tag{3.10}$$

implies

$$\|F_{P_n}^{-1} - F_P^{-1} + \mathrm{inv}(F_P' \circ F_P^{-1})(F_{P_n} - F_P) \circ F_P^{-1}\|_K \le$$

$$\le Cn^{-1/2}\omega_{\xi_n}(F_P^{-1}(K); \frac{C\|\xi_n\|_S}{n^{1/2}} + Cn^{-1}) + o_p(n^{-1/2}\|\xi_n\|_S) + n^{-1}\rho_\varphi. \tag{3.11}$$

By Theorem 2.2 the sequence $\{\xi_n\}$ converges weakly to a continuous process. It follows that

$$\|\xi_n\|_S = O_p(1), \qquad n \to \infty. \tag{3.12}$$

Asymptotic equicontinuity and (3.12) imply that

$$\omega_{\xi_n}(F_P^{-1}(K); \frac{C\|\xi_n\|_S}{n^{1/2}} + Cn^{-1}) = o_p(1), \qquad n \to \infty. \tag{3.13}$$

It follows from (3.10)–(3.13) that, as $n \to \infty$,

$$\|F_{P_n}^{-1} - F_P^{-1} + \mathrm{inv}(F_P' \circ F_P^{-1})(F_{P_n} - F_P) \circ F_P^{-1}\|_K = o_p(n^{-1/2}).$$

Therefore, the two stochastic processes

$$n^{1/2}(F_{P_n}^{-1} - F_P^{-1}) \quad \text{and} \quad -n^{1/2}\mathrm{inv}(F_P' \circ F_P^{-1})(F_{P_n} - F_P) \circ F_P^{-1}$$

converge weakly locally in V to the same limit (if one of them does). Thus the process

$$-n^{1/2}\mathrm{inv}(F_P' \circ F_P^{-1})(F_{P_n} - F_P) \circ F_P^{-1} = -\mathrm{inv}(F_P' \circ F_P^{-1})(\xi_n \circ F_P^{-1})$$

converges weakly locally in V to the process

$$\eta = -\mathrm{inv}(F_P' \circ F_P^{-1})(\xi \circ F_P^{-1}).$$

as a consequence of Theorem 2.2. ////

References

[1] Arcones, M.A. and Mason D.M. (1992): A general approach to Bahadur-Kiefer representations for M-estimators. Preprint.

[2] Chaudhuri, P. (1993): On a geometric notion of quantiles for multivariate data. Preprint.

[3] Dudley, R.M. and Koltchinskii, V. (1992): The spatial quantiles. Preprint.

[4] Gine, E. and Zinn, J. (1986): Lectures on the central limit theorem for empirical processes. *Lecture Notes in Mathematics* #1221 50–112. Springer, New York.

[5] Huber, P.J. (1964): Robust estimation of a location parameter. *Ann. Math. Statist.* **35** 73–101.

[6] Huber, P.J. (1967): The behavior of maximum likelihood estimates under non-standard conditions. In *Proc. Fifth Berkley Symp. Math. Stat. and Probab.* **1** 221–223. Univ. of California Press, Berkley.

[7] Huber, P.J. (1981): *Robust Statistics*. John Wiley, New York.

[8] Koltchinskii, V. (1994a): Bahadur-Kiefer approximation for spatial quantiles. In *Probability in Banach spaces* (J. Hoffmann-Jorgensen, J.Kuelbs and M.B. Marcus, eds.), 394–408. Birkhäuser, Boston.

[9] Koltchinskii, V. (1994b): Spatial quantiles and their Bahadur-Kiefer representations. In *Asymptotic Statistics*. P. Mandl and M. Huskova, eds.), 361–367. Physica-Verlag, Vienna.

[10] Koltchinskii, V. (1994c): Nonlinear transformations of empirical processes: functional inverses and Bahadur-Kiefer representations. To appear in *Proc. of the 6-th Vilnius Conference on Probability Theory and Mathematical Statistics*.

[11] Koltchinskii, V. and Dudley, R.M. (1992): Differentiability of inverse operators and asymptotic properties of inverse functions. Preprint.

[12] Mattner, L. (1992): Completness of location families, translated moments, and uniqueness of charges. *Probab. Th. Rel. Fields* **92** 137–149.

[13] Pollard, D. (1984): *Convergence of Stochastic Processes*. Springer-Verlag, New York.

[14] Pollard, D. (1990): *Empirical Processes: Theory and Applications*. Institute of Mathematical Statistics, Hayward, California.

[15] Rockafellar, R.T. (1970): *Convex Analysis*. Princeton Univ. Press, Princeton.

Robust Statistical Methods in Interlaboratory Analytical Studies*

Peter Lischer
Swiss Federal Research Station for Agricultural Chemistry and Hygiene of the Environment

Abstract

Interlaboratory analytical study is the general term of an experiment organised by a committee and involving several laboratories to achieve a common goal. Two important types of studies are the method-performance studies and the laboratory-performance studies. The purpose of a method-performance study is to determine the precision and bias characteristics of an analytical test method. A laboratory-performance study ascertains whether the laboratories conform to stated standards in their testing activities. An iterative and a non-iterative method to calculate the estimates in a method-performance study are presented and a new method based on a score-function allows to characterise the performance of laboratories both as groups and individually. This score is a squared Mahalanobis distance with robust estimates of means and covariances. For the latters' determination the specific structure of the interlaboratory-test data is taken into account. Instructive graphical displays support the classification of the laboratories.

Key words and phrases: Interlaboratory studies, robust distance, robust estimation of components of variance, multivariate outlier.
AMS 1991 subject classifications: Primary 62F35; secondary 62J10.

1 Introduction

In order for the results of analytical chemical measurements to be meaningful, procedures must be well developed enough that a reanalysis does not drastically change the conclusions, and well enough specified that different laboratories will achieve similar conclusions for the same sample. This means that there has to be a standard; i.e., a written document that lays down in full details how the test should be carried out. A standardised method has to be robust, that is small variations in the procedure should not produce unexpectedly large changes in the results. (ISO (1987)). Analytical methods and laboratory-competence have

* This paper won the 1995 W. J. Youden Award in Interlaboratory Testing from the American Statistical Association.

to be tested in an interlaboratory study. In such a study, several samples to be analysed are divided, and a part of each sample is sent to each of a number of laboratories. The resulting data are analysed by the referee to yield not only estimates for replication error and for laboratory bias but also the necessary information about the laboratory-performance of all participating laboratories.

2 Method-Performance Studies

2.1 The statistical model

Most method-performance studies consider trials involving n laboratories, each of which analyse a specimen p times (*uniform-level* experiment) or perform a *split-level* test (see below). The procedure is repeated for a number of specimens. Let us first consider uniform level experiments. Then for one particular specimen, every single test result, y_{ij}, $i = 1, 2, \ldots, n$; $j = 1, 2, \ldots, p$, is the sum of three components:

$$y_{ij} = m + b_i + e_{ij} ,$$

where m is the true (or consensus) value, b_i is the laboratory bias with variance σ_L^2 and e_{ij} is the replication error with variance σ_r^2. The b_i and e_{ij} are assumed to be uncorrelated and centred. The parameter σ_r is called repeatability standard deviation and $\sigma_R = (\sigma_L^2 + \sigma_r^2)^{1/2}$ reproducibility standard deviation. Repeatability and reproducibility are the traditional precision parameters in chemistry (ISO (1987)). The repeatability (r) of the method is the value below which the absolute difference between two single analytical results obtained with the same method on identical sample material and under constant conditions as regards laboratory, analyst, apparatus, chemicals and interval of time, is expected (with 95% confidence) to lie. The reproducibility (R) is the value below which the absolute difference between two single analytical results obtained with the same method on identical sample material and under different conditions of laboratory, analyst, apparatus, chemicals and interval of time, is expected (with 95% confidence) to lie. For normally distributed errors we have

$$\begin{aligned} r &= \sqrt{2}\Phi^{-1}(0.975)\,\sigma_r \\ R &= \sqrt{2}\Phi^{-1}(0.975)\sqrt{\sigma_L^2 + \sigma_r^2}, \end{aligned}$$

where $\Phi(z)$ is the cumulative standard normal distribution. The measurements y_{ij} are not all uncorrelated. We have

$$\mathrm{Cov}(y_{ij}, y_{kl}) = \begin{cases} \sigma_L^2 + \sigma_r^2, & \text{if } (i,k) = (k,l) \\ \sigma_L^2, & \text{if } i = k \text{ and } j \neq l \\ 0, & \text{if } i \neq k \end{cases}$$

A drawback of the uniform-level design is that the operator, when testing successively identical samples, may be influenced by the result of his first term. To prevent this an alternative split-level design may be used. In this procedure, instead of using two samples that the operator has been told to be identical, or performing two tests on the same specimen of material, two series of n samples are prepared at slightly different levels $m + \Delta$ and $m - \Delta$ (where Δ is small) and each of the n laboratories receives one sample of series 1 and one sample of series 2 for testing. The values of σ_r and σ_R derived from a split-level experiment are valid for the mean level m.

The aim of a method-performance study consists of finding estimates for the precision parameters σ_r and σ_R which are characteristic of the particular method and not only of the specific study. To achieve this aim, the following conditions must be fulfilled:

a) the participating laboratories and the samples used must be representative of the planned application;

b) the determination of the precision data σ_r and σ_R must be unambiguous and individual extreme results must be taken into account appropriately.

The conditions under a) are relatively easy to meet, although interlaboratory studies often have to be conducted with volunteers instead of randomly selected participants. On the other hand, the conditions in b) raise problems which often have not been satisfactorily solved up to now. The classical analysis of variance supposes normally distributed errors. However, every analytical chemist knows that deviant or suspect results occur much more frequently than the normal distribution would predict. There are many reasons for this; it is enough if just one parameter of the analytical process is not completely under control. Since very few suspect values deviate by an order of magnitude, it is often difficult to decide whether the suspect value should be regarded as valid. In evaluating method-performance studies, extreme results or all results obtained from a suspect laboratory are often eliminated before the analysis of variance is conducted, in order to avoid excessively high values for repeatability and reproducibility and, hence, a bad evaluation of the precision of the method. But any such elimination inevitably entails the risk of underestimating laboratory bias and replication error. An international convention about outlier tests to be used was adopted (Horwitz (1988)). It does not, however, change the unsatisfactory 'either-or' situation, which is typical for all outlier tests: as soon as the conditions for elimination are fulfilled, the value of the desired quantity changes abruptly. Moreover, the proposed Cochran and Grubbs tests are far from the best possible ones; e.g., the Grubbs test cannot even safely reject two distant outliers out of 20 (Hampel (1985)). On the other hand, 30% outliers in method-performance studies are rather the rule than the exception (Horwitz and Albert (1986)). These and other unsatisfactory features of outlier tests led the Swiss Federal Committee for Official Methods in Food Analysis (Lischer (1987), SLB (1989)) and the Analytical Methods Committee of the Royal Society of Chemistry (AMC (1989)) to propose solutions which are similar. Instead of the hitherto usual outlier tests they use robust statistical methods to calculate σ_r

and σ_R. The underlying principle is Huber's proposal 2 (Huber (1964)). In the following we present two methods to estimate the scale parameters σ_r and σ_R, the official SLB-method (SLB (1989)) and an alternative method inspired by Rousseeuw's scale estimator Q_n (Rousseeuw and Croux (1991)).

2.2 The SLB-method for an uniform-level experiment

This method consists of three steps.

1) **Estimation of the laboratory means $m+b_i$ by m_i.** The m_i are the solutions of the following system of equations:

$$\sum_{j=1}^{p} \psi_c\left(\frac{y_{ij}-m_i}{S^*}\right) = 0, \qquad i = 1,2,\ldots,n,$$

where $\psi_c(t) = \max(-c, \min(t,c))$ and

2) **Estimation of σ_r by S_r.** S_r is the solution of

$$\sum_{i=1}^{n}\sum_{j=1}^{p} \psi_c^2\left(\frac{y_{ij}-m_i}{S_r}\right) = n(p-1)\beta,$$

where $\beta = \int \psi_c^2(z)\, d\Phi(z)$.

3) **Estimation of m and σ_L^2.** The solution T of

$$\sum_{i=1}^{n} \psi_c\left(\frac{m_i - T}{\text{MAD}_n}\right) = 0,$$

where $\text{MAD}_n = 1.4826\text{med}_i\{|m_i - \text{med}_i m_i|\}$, is a consistent estimate of m. If T is put in

$$\sum_{i=1}^{n} \psi_c^2\left(\frac{m_i - T}{S}\right) = (n-1)\beta$$

and solved for S, we get a consistent estimate of $\left(\sigma_L^2 + \sigma_r^2/p\right)^{1/2}$. Furthermore, $S_L^2 = S^2 - S_r^2/p$ is a consistent estimate of the interlaboratory variance σ_L^2. If $S^2 - S_r^2/p < 0$, then we put $S_L = 0$.

The two quantities $h_i = (\bar{x}_{i\cdot} - T)/S$ and $k_i = s_i/S_r$, where $\bar{x}_{i\cdot}$ and s_i are mean and within-laboratory standard deviation of laboratory i, will later be used to detect laboratories which have produced unreliable results.

The third step is inspired by Huber's proposal 2, but location T and scale S are calculated separately whereas the proposed algorithm of the Analytical Methods Committee of the Royal Society of Chemistry (AMC (1989)) calculates T and S simultaneously. Reichenbach (1989) compared the two algorithms in a simulation study. He showed that the AMC-algorithm converges slowly and that the breakdown point is lower than 25% for moderate sample sizes. This is too small as 30% outliers are not uncommon in interlaboratory trials. A procedure for evaluating such trials should allow two bad laboratories out of eight. The SLB-procedure with separate determination of location and scale has better convergence properties, a comparable relative efficiency and a breakdown point of $\approx 30\%$.

2.3 An alternative non-iterative method

The scale estimator used in SLB also has some drawbacks. It takes a symmetric view of dispersion, because first a central value is determined and then it attaches equal importance to positive and negative deviations from it, which does not seem to be a natural approach at asymmetric distributions. Further, the finite breakdown points are rather low and the rate of convergence near of the breakdown point can be slow (Reichenbach (1989)). Finally, it is an iterative procedure. Practical chemists, however, prefer explicit formulas. A non-iterative procedure to estimate the precision parameters σ_r and σ_R with 50% breakdown point and a high relative efficiency will be shown below. Basis is the Q_n-estimator proposed by Rousseeuw and Croux (1991). This estimator is given by the 0.25-quantile of the interpoint distances. Let $\mathbf{x} = (x_1, x_2, \ldots, x_n)$ be a set of n observations. Vectors and matrices will be denoted by boldface throughout. Then

$$Q_n(\mathbf{x}) = f_n \, 2.2219 \, \{|x_i - x_j| : i < j\}_{(k)},$$

where $k = \binom{h}{2} \approx \frac{1}{4}\binom{n}{2}$ and $h = \lfloor n/2 \rfloor + 1$ is roughly half the number of observations. The constant f_n is a small-sample correction factor. The scale estimator Q_n does not need any location estimate. Instead of measuring how far away the observations are from the central value, Q_n looks at a typical distance between observations, which is still valid at asymmetric distributions. The Gaussian efficiency of Q_n is 82%.

In the case of a split-level design, Q_n allows immediately to get estimates for σ_r and σ_R. Let $\{(y_{i1}, y_{i2}) : i = 1, 2, \ldots, n\}$ be the results of the experiment, $\mathbf{v} = \{(y_{i1} + y_{i2})/2\}$ and $\mathbf{w} = \{(y_{i1} - y_{i2})/2\}$, so that $\mathrm{Var}[v_i] = \sigma_L^2 + \sigma_r^2/2$ and $\mathrm{Var}[w_i] = \sigma_r^2/2$. Then

$$\hat{\sigma}_r = \sqrt{2}Q_n(\mathbf{w}) \tag{2.1}$$

$$\hat{\sigma}_R = \sqrt{\hat{\sigma}_L^2 + \hat{\sigma}_r^2} = \sqrt{Q_n^2(\mathbf{v}) + Q_n^2(\mathbf{w})} \tag{2.2}$$

For an uniform-level experiment with p replicates, $p \geq 2$ the procedure is similar. Let d_R be the 0.25-quantile of the absolute differences

$$\{\, |y_{ij} - y_{kl}| \;:\; i = 1, 2, \ldots, n-1, j = 1, 2, \ldots, p, \; k = i+1, i+2, \ldots, n, l = 1, 2, \ldots, p \,\}$$

and d_r the 0.25-quantile of the absolute differences

$$\{\, |(y_{ij} - y_{ik}) - (y_{hl} - y_{hm})| \quad : \quad i = 1, 2, \ldots, n-1, j < k, \\ h = i+1, i+2, \ldots, n, l < m \,\}$$

Then

$$\hat{\sigma}_r = 2.2219\, d_r / \sqrt{2}$$

and

$$\hat{\sigma}_R = 2.2219\, d_R .$$

2.4 Graphical display as an aid to analysis

Mandel (1989) presented a procedure for flagging outliers, based on two statistics, called h and k. The h-values are calculated independently for each concentration level. The overall average at that level is subtracted from each cell-average and divided by the standard deviation. It is a measure of where a particular lab's average lies with respect to the consensus value. The k-values are also calculated independently at each level. It is simply the ratio of the within-cell standard deviation to the pooled value over all laboratories at that level. It is evident that this non-robust procedure suffers from the masking effect. But there is a simple remedy, however. Instead of Mandel's h and k we use the two quantities $h_i = (\bar{x}_{i\cdot} - T)/S$ and $k_i = s_i/S_r$ introduced earlier.

We will illustrate this procedure in terms of an interlaboratory study published partially in the SLB (1989). It deals with the photometric determination of nitrate in drinking water at three concentration levels. Every laboratory determined two replicates for each sample (Table 1). The statistical analysis of the data was done with the SLB- and the Q_n-method (Table 2). The estimates do not differ much. The h-values are displayed in Figure 1 and the k-values in Figure 2. h- and k-values with absolute values ≤ 2 are traditionally considered as "satisfactory", between 2 and 3 as "questionable" and with ≥ 3 as "unsatisfactory".

At a glance, we see what is going on: laboratories 1, 11, 12 and 13 got at least one deviant mean value with $|h| > 2$ and laboratory 8 has a high within-laboratory variation for sample 1 and 3 ($|k| > 3$). The organiser of the study has now to find out whether there are any shortcomings in the analytical method.

3 Laboratory-Performance Studies

3.1 Robust distances

This type of interlaboratory analytical study is known as *proficiency testing.* Its aim is to offer the participating laboratories the opportunity to compare their analytical results with those of other laboratories.

	Laboratories							
	1	2	3	4	5	6	7	8
(1)	4.2	3.1	3.2	3.2	3.2	3.2	3.2	3.2
(1)	4.4	3.1	3.2	3.1	3.1	3.3	3.2	2.7
(2)	26.2	26.5	27.0	26.8	26.4	28.8	28.2	26.0
(2)	26.0	26.6	27.2	26.5	26.2	28.0	28.2	25.9
(3)	48.5	44.4	46.4	45.7	44.1	48.8	45.1	45.5
(3)	48.3	44.5	46.6	46.0	45.0	48.5	45.6	49.3

	Laboratories						
	9	10	11	12	13	14	15
(1)	3.0	2.9	3.1	2.6	3.6	3.0	3.1
(1)	3.0	3.1	3.1	2.7	3.5	3.1	3.1
(2)	25.9	26.2	29.6	24.7	29.2	25.1	25.9
(2)	26.0	26.4	30.0	24.1	29.6	25.2	26.0
(3)	43.8	45.0	50.7	45.8	49.0	42.9	44.9
(3)	44.2	45.2	50.6	46.1	50.0	42.9	44.7

Table 1: A trial of determination of nitrate in drinking water [mg/l] for three samples at different concentration levels performed by fifteen laboratories.

	SLB-method			
	$\hat{\mu}$	$\hat{\sigma}_L$	$\hat{\sigma}_r$	$\hat{\sigma}_R$
Sample 1	3.12	0.13	0.08	0.16
Sample 2	26.49	1.45	0.21	1.47
Sample 3	46.18	2.34	0.30	2.36
	Q_n-method			
	$\hat{\mu}$	$\hat{\sigma}_L$	$\hat{\sigma}_r$	$\hat{\sigma}_R$
Sample 1	3.12	0.18	0.13	0.22
Sample 2	26.49	1.31	0.26	1.33
Sample 3	46.18	1.98	0.26	2.00

Table 2: Statistical analysis for nitrate.

Two distinct aims can be formulated (AMC (1992)):

a) to encourage good performance generally, and especially to encourage the use of proper routine quality control measures within individual laboratories; to provide feedback to the laboratories and encourage remedial action where shortcomings in performance are detected;

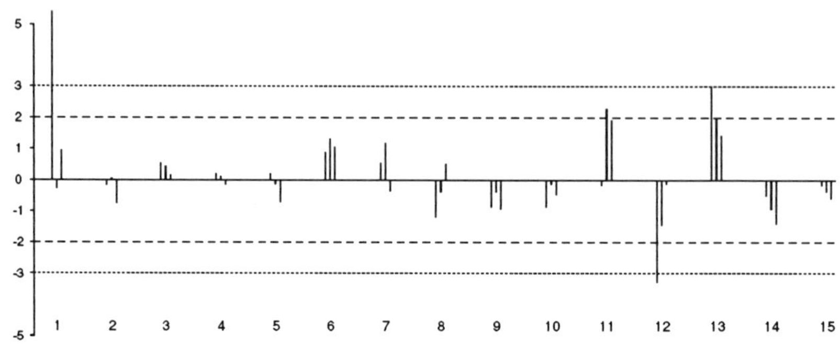

Figure 1: Graph of h-values by laboratories.

b) to provide a rational basis for the selection or licensing of laboratories for a special task and

   likewise to disqualify laboratories for a specific task should their performance fall below a certain standard.

These two aims are somewhat divergent, but the motivation is the same: the identification of laboratories that produce data of unacceptable quality.

Most proficiency testing schemes proceed by comparing the bias estimate $(x - x_{\text{true}})$ with a target value for the standard deviation that forms the criterion of performance. An obvious approach is to form z-scores given by

$$z = \frac{x - x_{\text{true}}}{\sigma},$$

where σ is the target value for the standard deviation. If $\hat{x}$ and $\hat{\sigma}$ are good estimates of x_{true} and the standard deviation σ, respectively, and if the underlying distribution were normal, then z would be approximately normally distributed with a mean zero and a unit standard deviation. Because z is standardised, it can be usefully compared between all analytes, test materials and analytical methods. Values of z obtained from diverse materials and concentration ranges can, therefore, with due caution, be combined to give a composite score for a

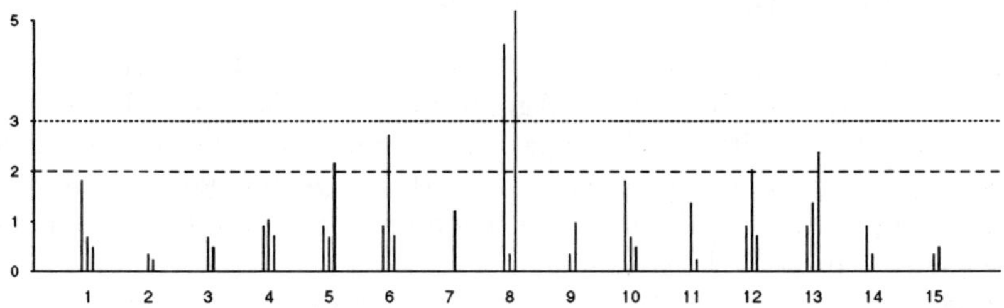

Figure 2: Graph of k-values by laboratories.

laboratory in one round of a proficiency test. AMC (1992) proposes the sum of squared scores as a performance criterion of a laboratory. But this composite score does not take into account the possible correlation structure of the data set.

Suppose we have data of n laboratories, each of which analysed p specimens: $\mathbf{X} = \{\mathbf{x}_1, \mathbf{x}_2, \ldots, \mathbf{x}_n\} = \{(x_{11}, x_{12}, \ldots, x_{1p}), \ldots, (x_{n1}, x_{n2}, \ldots, x_{np})\}$ and we want to diagnose outlying laboratories. The word outlier is applied here to any $\mathbf{x}_i \in \mathbb{R}^p$ that is markedly different from most of the other.

The squared *Mahalanobis distance* for observation $\mathbf{x}_i$ is

$$d_i^2 = (\mathbf{x}_i - \bar{\mathbf{x}})\mathbf{V}^{-1}(\mathbf{x}_i - \bar{\mathbf{x}})^T$$

where $\bar{\mathbf{x}}$ is the mean of the group and $\mathbf{V}$ is the sample covariance matrix. Asymptotically the d_i^2 follow a chi-squared distribution on p degrees of freedom. If $\bar{\mathbf{x}}$ and $\mathbf{V}$ were not estimated but were known population parameters, outlying values of $\mathbf{x}_i$ would yield large values of the squared distance d_i^2. However, the effect of such values on the estimation of $\bar{\mathbf{x}}$ and $\mathbf{V}$ leads to the rapid breakdown of the Mahalanobis distance for the detection of outliers, particularly if several outliers are present. In Rousseeuw and Leroy (1987) it was suggested to calculate robust distances by using the Minimum Volume Ellipsoid estimator. This estimator has a maximal breakdown point but is not very efficient. Further, a rule of thumb of Rousseeuw states that there should be at least five observations per dimension ($n/p \geq 5$). This requirement seems to be unrealistic for laboratory-performance studies as ratios $n/p \approx 2$ are not uncommon. In this case reliable estimates of distances can be got only if the specific structure of the laboratory data is taken into account. This will be discussed below.

3.2 An interlaboratory example

We will illustrate the procedure in terms of a proficiency test carried out at the Swiss Federal Research Station for Agricultural Chemistry and Hygiene of the Environment (FAC) with 23 participating laboratories. The primary goal was a laboratory comparison. A number of dried sewage sludges was analysed for different heavy metals and P and K. In this numerical example the results of the Zn-analyses are presented (Table 3).

Each laboratory L_i, $(i = 1, \ldots, 23)$ had to analyse 10 specimens $S_1, S_2, \ldots, S_{10}$. $S_1, S_2, \ldots, S_5$ were taken from 5 different homogenised samples of dried sludges, $S_6, S_2, \ldots, S_9$ were mixtures: $(S_6 : 50\%S_1 + 50\%S_2; S_7 : 50\%S_3 + 50\%S_2; S_8 : 50\%S_4 + 50\%S_2; S_9 : 50\%S_5 + 50\%S_2)$ and S_{10} was an extracted solution of S_1, produced in a laboratory of the FAC. This specimen was analysed to obtain an estimate of the effect of the sample preparation for different laboratories. The mixtures S_6, S_7, S_8 and S_9 together with S_1, S_3, S_4 and S_5, allowed four alternative determinations of the true content of S_2 (Table 4). Inconsistencies between the direct and the indirect determinations of the true concentration of sample S_2 indicated a bad performance of a laboratory. With the four supplementary results for the true concentration of specimen S_2 it was possible to get a more reliable value of the true content of S_2 and to estimate the scale parameters σ_r and σ_R for this concentration level.

We write the model as

$$x_{ij} = m_j + b_{ij} + e_{ij}\,, \quad i = 1, 2, \ldots, 23; \quad j = 1, 2, \ldots, 10\,,$$

where the m_j are the true (or consensus) values, the b_{ij} are the random laboratory effects assumed to have mean 0 and variance σ_{Lj}^2 and the e_{ij} are the replication errors assumed to have mean 0 and variance σ_{rj}^2. In general σ_{Lj}^2 and σ_{rj}^2 are concentration-dependent but if the same analytical method is used and the concentration levels are not too different the quotient

$$q = \frac{\sigma_{Lj}^2}{\sigma_{Lj}^2 + \sigma_{rj}^2}$$

can be assumed constant $(0 \leq q < 1)$.

Let $z_{ij} = (x_{ij} - \hat{m}_j)/\hat{\sigma}_j$, $i = 1, 2, \ldots, 23$; $j = 1, 2, \ldots, 10$. We have

$$\mathrm{Cov}(z_{ij}, z_{kl}) = \begin{cases} 1\,, & \text{if } (i,k) = (k,l) \\ q\,, & \text{if } i = k \text{ and } j \neq l \\ 0\,, & \text{if } i \neq k \end{cases}$$

Labs	Specimens s_{i1}	s_{i2}	s_{i3}	s_{i4}	s_{i5}	s_{i6}	s_{i7}	s_{i8}	s_{i9}	s_{i10}
1	1531	2945	1264	2068	1150	2252	2109	2526	2096	1499
2	1527	2440	1250	2110	1160	2080	2140	2430	2010	1500
3	1500	2900	1200	1900	1100	2200	1900	2400	2000	1746
4	1490	2830	1210	2060	1090	2220	2060	2440	2030	1350
5	1547	2900	1238	2060	1120	2211	2094	2463	1961	1500
6	1432	—	—	—	1076	2029	—	1813	1368	—
7	1393	2466	1131	1886	1011	2040	1985	2321	1133	1430
8	1579	2954	1188	2240	1229	2313	2172	2642	2160	1524
9	1520	2950	1240	2080	1080	2250	2070	2520	2010	1514
10	1653	3199	1369	2489	1182	2537	2565	2729	2371	1648
11	1577	3030	1264	2113	1129	2296	2191	2676	2129	1600
12	1536	3014	1260	2089	1143	2304	2113	2513	2089	1616
13	1500	3150	1220	2030	1070	1940	1780	2370	1940	1520
14	1512	2779	1252	2089	1120	2245	2042	2517	2000	1485
15	1458	2688	1214	1896	1122	1618	1836	2160	1849	1655
16	1500	3160	1160	2150	1065	2180	2058	2482	2003	1428
17	1525	2865	1245	2125	1120	2215	2180	2600	2065	1495
18	1488	2720	1713	1968	1044	2143	2016	2344	1992	1448
19	1464	2857	1187	2024	1073	2153	2024	2413	1932	1576
20	1600	2400	1300	1900	440	2400	1900	2300	1900	1500
21	1558	3077	1279	2082	1116	2271	2152	2558	2133	1481
22	1595	3271	1344	2241	1202	2420	2280	2670	2134	—
23	1803	2408	1084	2309	1473	2860	2063	2429	1369	1250
$\hat{m}_j$	1527	2882	1238	2075	1114	2222	2071	2479	2003	1507
$\hat{\sigma}_j$	60.0	271.3	62.8	132.5	60.6	153.2	126.3	139.2	128.5	102.5

Table 3: Zn-concentrations [mg/kg] in ten samples of sewage sludges. $\hat{m}_j$ and $\hat{\sigma}_j$ are robust estimates of location m_j and scale $\sigma_j = \left(\sigma_{Lj}^2 + \sigma_{rj}^2\right)^{1/2}$ determined with the SLB-algorithm. We note a relatively high value for $\hat{\sigma}_2$.

The data structure of the z_{ij} is the same as the structure of results of an uniform-level experiment. Therefore, the covariance matrix $\mathbf{S}$ has a very simple form: 1 in the diagonal and q otherwise. With the SLB- or the Q_n-method we get immediately a robust estimate $\hat{q}$ of q resp. $\hat{\mathbf{S}}$ of $\mathbf{S}$ and

$$rd_i^2 = \mathbf{z}_i \hat{\mathbf{S}}^{-1} \mathbf{z}_i^T$$

is the squared robust distance of laboratory i.

The critical value is $\chi^2_{10,0.975} = 20.48$. Laboratories with values of $\mathrm{r}\,d_i^2 > 20.48$ must be considered as unreliable.

In Fig. 3 the z-scores are presented. In the first line below the graphic there are the laboratory codes, in the second the number of analysed specimens, in the third the squared robust distance and in the forth the corresponding p-values. We used the SLB-method to get $\hat{q}$.

Labs	s_{i2}	w_{i1}	w_{i2}	w_{i3}	w_{i4}	$\mathrm{r}\,d_i^2$
1	2945	2973	2954	2984	3042	2.1
2	2440	2633	3030	2750	2860	14.9
3	2900	2900	2600	2900	2900	31.3
4	2830	2950	2910	2820	2970	7.0
5	2900	2875	2950	2866	2802	1.0
6	—	2626	—	—	2550	15.5
7	2466	2687	2839	2756	1255	98.0
8	2954	3047	3156	3044	3091	17.3
9	2950	2980	2900	2960	2940	1.7
10	3199	3421	3761	2969	3560	30.7
11	3030	3015	3118	3239	3129	4.8
12	3014	3072	2966	2937	3035	2.6
13	3150	2380	2340	2710	2810	26.0
14	2779	2978	2832	2945	2880	1.5
15	2688	1778	2458	2424	2576	63.9
16	3160	2860	2956	2814	2941	13.4
17	2865	2905	3115	3075	3010	4.3
18	2720	2798	2319	2720	2940	203.0
19	2857	2842	2861	2802	2791	6.6
20	2400	3200	2500	2700	3360	385.4
21	3077	2984	3025	3034	3150	4.6
22	3271	3245	3216	3099	3066	8.9
23	2408	3917	3042	2549	1265	379.2
$\hat{m}_j$	2882	2924	2925	2874	2928	
$\hat{\sigma}_j$	271.3	241.5	259.1	175.8	231.0	

Table 4: Direct and indirect determination of the Zn-concentration of specimen S_2 ($\mathrm{w}_{i1} = 2 \cdot \mathrm{s}_{i6} - \mathrm{s}_{i1}, \mathrm{w}_{i2} = 2 \cdot \mathrm{s}_{i7} - \mathrm{s}_{i3}, \mathrm{w}_{i3} = 2 \cdot \mathrm{s}_{i8} - \mathrm{s}_{i4}, \mathrm{w}_{i4} = 2 \cdot \mathrm{s}_{i9} - \mathrm{s}_{i5}$) and squared robust distances $\mathrm{r}\,d_i^2$.

3.3 Choice of materials and distribution of samples

A proficiency test must enable the organiser to see whether there is a general improvement in performance in time. But if the same test material is distributed several times, the participants would become aware of the consensus value after the first round and the credibility of the results in successive rounds would be compromised. Therefore the organiser should distribute also mixtures of samples. As we have seen in the above example the true value of sample S_2 could also be determined in an indirect way from samples S_6 and S_1, from S_7 and S_3, from S_8 and S_4 and from S_9 and S_5. We could organise a future proficiency test in the following way: at each round four specimens P_1, P_2, P_3 and P_4 are distributed. P_1 and P_2 are new samples, P_3 and P_4 are mixtures of P_1 and P_2 with P_0, where P_0 is a specimen from an earlier round with a (only for the organiser) known true value. Then the organiser can evaluate an eventual improvement in performance and even estimate the precision parameters σ_r and σ_R at the concentration level of P_0.

At a glance, we see the laboratories which have analytical problems for the

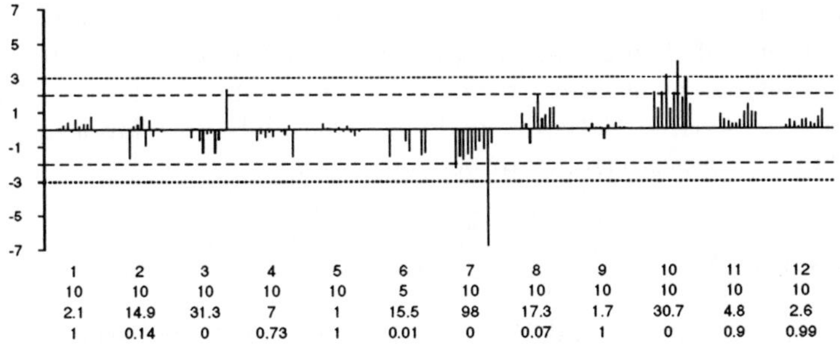

Figure 3: z-scores of Zn-concentrations in sewage sludges and corresponding p-values.

determination of Zn. Similar graphics for the other elements can be done. As the confidentiality of the results is extremely important in this type of laboratory-performance studies, the organiser distributes to the laboratories only graphics which contain exclusively their own results. Instead of the results of one element of all laboratories (Fig. 3), he distributes graphics which show the results of all elements of a particular laboratory.

4 Conclusions

Monitoring the amount of pollutants in soil, water, air, plants, food, etc. is important nowadays. Analytical tests are required to judge contamination. The fascinating thing about analycal chemical measurements is that they can quantify chemical contents objectively. A drawback is that they suffer from a lack of comparability. It is common knowledge amongst those who practise analysis for trade and commerce that analysts can obtain different results on the same material. Obviously it may not be in their interest to expose this fact. It is in the field of public health and environmental monitoring, where determinand concentrations are often small and where slight differences may be significant, that interlaboratory variation has received most attention. The disturbing thing is the suggestion of unreliability and its possible diffusion to the general public as well as to the governments responsible in cases where important decisions must be made on the basis of chemical measurements. However, the situation is not as bad as it seems. If chemists and statisticians collaborate, try to understand each other's problems and use realistic models, representative samples, standardised and robust analytical methods, reference materials, interlaboratory tests and good robust statistics, errors can be controlled.

References

[1] AMC (1989): Analytical Methods Committee. Robust statistics part 2: Interlaboratory trials. *Analyst* **114** 1699–1702.

[2] AMC (1992): Analytical Methods Committee. Proficiency testing of analytical laboratories: Organisation and statistical assessment. *Analyst* **117** 97–117.

[3] Hampel, F.R. (1985): The breakdown points of the mean combined with some rejection rules. *Technometrics* **27** 95–107.

[4] Horwitz, W. Albert, R. (1986): Performance characteristics of methods of analysis used for regulatory purposes. *Paper delivered at the Pittsburgh-Conference and Exposition, Atlantic City*, NJ USA.

[5] Horwitz, W. (1988): Protocol for the design, conduct and interpretation of collaborative atudies. *Pure & Appl. Chem.* **60**(6) 855–864.

[6] Huber P. J. (1964): Robust estimation of a location parameter. *Ann. Math. Statist.* **35** 73–101.

[7] ISO-5725 (1987): *Accuracy (Trueness and Precision) of Test Measurements Part 1: General Principles and Definitions*. International Organisation for Standardisation, Geneva, Switzerland.

[8] Lischer, P. (1987): Robuste Ringversuchsauswertung. *Lebensmittel-Technologie* **20** 167–172.

[9] Mandel, J. (1989): *Interlaboratory Testing and Rejection of Observations.* Proceedings ISO/REMCO 184. International Organisation for Standardisation, Geneva, Switzerland.

[10] Reichenbach, A. (1989): *Robuste Methoden für die Auswertung von Ringversuchen.* Diplomarbeit ETH-Zürich.

[11] Rousseeuw, P.J. and Leroy, A.M. (1987): *Robust Regression and Outlier Detection.* Wiley, New York.

[12] Rousseeuw, P.J. and van Zomeren, B.C. (1990): Unmasking multivariate outliers and leverage points. *J. Amer. Statist. Assoc.* **85** 633–639.

[13] Rousseeuw, P.J. and Croux, C. (1993): Alternatives to the median absolute deviation. *J. Amer. Statist. Assoc.* **88** 1273–1283

[14] SLB (1989): *Schweizerisches Lebensmittelbuch, Kapitel 60: Statistik und Ringversuche.* Eidg. Drucksachen- und Materialzentrale, Bern.

Estimating Distributions with a Fixed Number of Modes

Martin B. Mächler
ETH Zürich

Abstract

A new approach for non- or semi-parametric density estimation allows to specify modes and antimodes. The new smoother is a *Maximum Penalized Likelihood* (MPL) estimate with a novel roughness penalty. It penalizes a relative *change* of curvature which allows considering modes and inflection points. For a given number of modes, the score function, $l' = (\log f)'$ can be represented as $l'(x) = \pm(x - w_1) \cdots (x - w_m) \cdot \exp h_l(x)$, a *semi-parametric* term with parameters w_j (model order m) and nonparametric part $h_l(\cdot)$. The MPL variational problem is equivalent to a differential equation with boundary conditions. The exponential and normal distributions are smoothest limits of the new estimator for zero and one mode, respectively.

Key words and phrases: Nonparametric, semi-parametric, density estimation; smoothing; roughness penalty; maximum penalized likelihood; inflection point; boundary value problem, differential equation, unimodality, multimodality.
AMS 1991 subject classifications: Primary 62G07; secondary 34B10, 41A29, 65D07, 65D10.

1 Motivation and Introduction

The nonparametric estimation of distribution or density functions is well-established with a variety of methods available (see Silverman (1986), Izenman (1991)). Very early on, the number of modes (local maxima) of the unknown density has been considered important (e.g., Cox (1966), Silverman (1981)) and even the number of "bumps" (concave regions between inflection points) has been tried to be kept small (Good and Gaskins (1971), Good and Gaskins (1980)).

Modes are often *the* important attribute of a distribution, since they (informally) lead to conclusions about the underlying population. Tests of unimodality (vs. multimodality) have been developed (see above Hartigan and Hartigan (1985), Minnotte and Scott (1993) for a nice graphical tool). Known examples of such data are (among others)

- Percentages of silica in chondrites (Good and Gaskins (1980)): 3(or more?) modes.

- Old Faithful geyser eruption lengths: 2 or 26 modes?
- Yearly snowfalls in Buffalo: Are there sub-populations (1 or 3 modes)?
- Thickness of Mexican stamps of 1872 (Izenman and Sommer (1988)): 2, 3, 5 or 7 modes?

The present approach is within the framework of *"maximum penalized likelihood estimation"*, MPLE (cf. Thompson and Tapia (1990), ch. 3–5, which "contains" Thompson and Tapia (1978)), using a *novel roughness* penalty. An appealing notion of smoothness is considering *local extrema* and/or *inflection points* of a curve. For nonparametric regression, I *limited the number of inflection points* (Mächler (1989) and Mächler (1993)) as a special case of a very general approach of MPLE. Here, for density estimation, I apply this approach to finding the "best" density function with a given number of local extrema, or almost equivalently, number of modes.[1].

Given independent observations $x_1, x_2, \ldots, x_n$ of $X \sim F$, with "smooth" density $f(x) = \frac{d}{dx}F(x)$ on $[a, b]$ (i.e., $P[a \le X \le b] = 1$ and therefore, $x_i \in [a, b]\, \forall i$), $-\infty \le a < b \le +\infty$, our goal is to estimate f and F such that f has (at most) m^* modes; i.e. typically, $m = (2m^* - 1)_+$ local extrema. We assume that f is twice (continuously) differentiable, and define

$$\begin{aligned} l(x) &:= \log f(x) && \textit{log-density}, \\ \text{hence} \quad l'(x) &= \frac{d}{dx} l(x) = \frac{f'(x)}{f(x)} && \textit{score function}. \end{aligned} \tag{1.1}$$

For differentiable functions f, w is a local extremum if and only if $f'(w) = 0$, or equivalently, $l'(w) = 0$. Denote the m local extrema of f by $w_1, w_2, \ldots, w_m$. From basic calculus we know that l' factors into

$$l'(x) = p_{\mathbf{w}}(x)\,(-1)^j\, e^{h_l(x)}, \quad \text{where} \tag{1.2}$$

$$p_{\mathbf{w}}(x) := (x - w_1)(x - w_2)\cdots(x - w_m), \tag{1.3}$$

and h_l is continuously differentiable, but otherwise "arbitrary" (in the vast class of functions with penalty $\int_a^b h_l'^2\, dt < \infty$). The sign $(-1)^j$ in (1.2), determined by $j = 0$ or 1, is assumed to be given and is almost always negative ($j = 1$), e.g., whenever $b = \infty$ or if there are more modes than antimodes, but $j = 0$, e.g., for a U-shaped density f ($m = 1$).
Note that from (1.2),

$$(\log l'(x))' = (l''/l')(x) = h_l'(x) + \sum_{j=1}^{m} \frac{1}{x - w_j}, \tag{1.4}$$

such that h_l' can be regarded as "l''/l' minus the poles". By specifying a penalty in terms of h_l', we have a semi-parametric model approach, the parameters being the modes and antimodes w_j, $j = 1, \ldots, m$, and nonparametric component h_l.

[1] *"mode"* is used here for maxima x where $f'(x) = 0$. For differentiable densities f, this only excludes maxima at the *end* of the support interval, e.g., $x = 0$ for the exponential distribution (which has zero modes in our definition).

In the remaining sections, I expose the MPLE problem and derive the characterizing Euler–Lagrange differential equation. The implementation is exemplified by a real data application. Properties of the solution and possible extensions are detailed.

2 MPLE — Roughness Penalty

To estimate f (together with F and h_l), we Maximize the Penalized Likelihood criterion, or equivalently,

$$\min_{l \in \mathcal{L}, \mathbf{w}} \left\{ -\sum_{i=1}^{n} l(x_i) + \lambda \int_a^b h_l'^2(t)\, dt \right\}, \tag{2.1}$$

where $\mathcal{L}$ is the set of all functions $l : [a, b] \to \mathbb{R}$ fulfilling (1.2), (1.3) and

$$\int_a^b f(t)\, dt = \int_a^b e^{l(t)}\, dt = 1. \tag{2.2}$$

In the following, the situation for *fixed* $\mathbf{w}$ is considered. The penalty $\int_a^b h_l'^2(t)\, dt$ contains the derivative h_l' which can be expressed as

$$h_l'(x) = \frac{\kappa_L'(x)}{\kappa_L(x)} - \sum_{j=1}^{m} \frac{1}{x - w_j} + l(x) l'(x)\, 3(1 + l^2)^{-1}(x), \tag{2.3}$$

by using (1.4) and $\kappa_L'/\kappa_L = l''/l' - 3ll'(1 + l^2)^{-1}$ where κ_L is the curvature of $L(x) = \int^x l(t)\, dt$. Arguments in Mächler (1989) and Mächler (1993) suggest that a natural roughness penalty for a curve $y = g(x)$ is the *"relative change of curvature"* κ'/κ where $\kappa(x) = g''(x) \left(1 + g'^2(x)\right)^{-3/2}$ is the curvature at x. Eq. (2.3) implies that h_l' approximates the relative change of curvature "apart from the poles" of L.

Using a Lagrange parameter α for the side condition (2.2) and making use of Dirac's δ distribution notation, the problem (2.1) is equivalent to the *variational problem*

$$\min_{l \in \mathcal{L}} \int_a^b \left\{ \alpha\, e^{l(t)} - \sum_{i=1}^{n} \delta(t - x_i) l(x_i) + \lambda h_l'^2(t) \right\} dt. \tag{2.4}$$

3 Differential Equation

Now we apply Theorem 1 of Mächler (1993) (and Mächler (1989)) which gives the Euler–Lagrange ordinary differential equation (=: o.d.e.) and boundary conditions for a vast class of variational problems in a convenient form:
The minimizer f of $\int_a^b \left(S(f(t), t) + \{(d/dt)^k F(t, f^{(\nu)}(t))\}^2\right) dt$ fulfills the o.d.e.

$$F_g \cdot \left(\frac{d}{dt}\right)^{2k} F = \frac{1}{2}(-1)^{\nu+k+1} S_f{}^{[\nu]}(t) \qquad \text{for all } t \in [a, b], \tag{3.1}$$

where $F_g(t,g) := \frac{\partial}{\partial g}F(t,g)$, $S_f{}^{[0]}(x) := \frac{\partial}{\partial f}S(f,x)\Big|_{f=f(x)}$, and $S_f{}^{[j+1]}(x) := \int_a^x S_f{}^{[j]}(t)\,dt$ for $0 \le j < \nu$. The boundary conditions of this very general problem are

$$\textbf{(A)} \qquad S_f{}^{[1]}(b) = S_f{}^{[2]}(b) = \cdots = S_f{}^{[\nu]}(b) = 0, \quad \text{and}$$

$$\textbf{(B)} \qquad (\tfrac{d}{dt})^j F(t) = 0 \qquad \text{for } t \in \{a,b\} \quad \text{and} \quad j \in \{k,\ldots,2k-1\}.$$

For our problem (2.4), $\nu = k = 1$, and $\lambda S(l,x) = \alpha e^l - \sum_{i=1}^n \delta(x-x_i)l(x_i)$. Therefore $\lambda S_f{}^{[1]}(x) = \int_a^x \lambda \frac{\partial}{\partial l}S(l,t)\,dt = \alpha\int_a^x e^{l(t)}\,dt - \sum_{i=1}^n \mathbf{1}_{[x_i\le x]} = \alpha F(x) - nF_n(x)$, where $F_n(x)$ is the empirical distribution function defined as $F_n(x) = n^{-1}\sum_{i=1}^n \mathbf{1}_{[x_i\le x]}$. Further, $F(x,l') = h_l(x) = \log|l'/p_{\mathbf{w}}(x)|$.

The boundary conditions (A) are $0 = \lambda S_f{}^{[1]}(b) = \alpha F(t) - nF_n(b)$ which entails $\alpha = n$ for the Lagrange parameter because $F(b) = F_n(b) = 1$. The conditions (B) are $0 = d/dx\,F = h_l'(x)$ at the boundary $x = a$ or b. The o.d.e. is $F_g\cdot(d/dx)^2F = \frac{1}{2}(-1)^3 S_f{}^{[1]}(x)$, $F_g = \frac{\partial}{\partial g}F(g,x) = \frac{\partial}{\partial g}\log|g/p_{\mathbf{w}}(x)|\Big|_{g=l'} = 1/l'$, $(d/dx)^2F = h_l''$ and $S_f{}^{[1]} = n(F - F_n)$ from above. Therefore, the characterizing Euler–Lagrange o.d.e. for problem (2.1) is

$$h_l''(x) = \frac{l'(x)}{2\lambda}\; n\,(F_n(x) - F(x)), \tag{3.2}$$

with boundary conditions

$$F(a) = 0,\; F(b) = 1, \quad \text{and} \quad h_l'(a) = h_l'(b) = 0, \tag{3.3}$$

where F_n is the empirical distribution function. This 4th order o.d.e. for F can be rewritten as 1st order with four unknown functions (F, l, h_l, h_l'), $F' = \exp(l)$, $l' = (-1)^j p_{\mathbf{w}}\exp(h_l)$, $h_l' = \{h_l\}'$, and h_l'' obeys (3.2). Note that the data only enter through F_n, and the modes (and antimodes) are specified in $p_{\mathbf{w}} = (x - w_1)\cdots(x - w_m)$.

This boundary value o.d.e. is solved numerically to obtain the MPLE for fixed $\mathbf{w} = (w_1, w_2, \ldots, w_M)^\top$. I have used a collocation method o.d.e. solver, "COLNEW" from `netlib`, see Bader and Ascher (1987), and Ascher et al. (1988). Numerical boundary value o.d.e. solvers require an initial "crude" approximate solution to be provided. The *"log-splines"* of Kooperberg and Stone (1992) have been convenient "pre-smoothers" for this. Experience indicates that the locations w_j of modes and antimodes are quite well determined by such a pre-smoother initially. Further, the numerical solution requires the boundaries a and b to be finite. It is quite clear that a limit $a \to \infty$, $b \to +\infty$ exists and is attained to arbitrary precision for finite a and b.

For the final estimate, the minimizer of (2.1), the penalized log-likelihood of the above o.d.e.-solution must be *minimized* over $\mathbf{w}$. This is relatively easy, at least as long as the number m of local extrema is not too large for the problem.

The "Old Faithful geyser eruption lengths" are at least bimodal, and have been used before (e.g., Silverman (1986), Härdle (1990)). Here I use the data set `dat.geyser$x` as supplied in the Statlib (`ftp lib.stat.cmu.edu`) submission `S/haerdle`. These are $n = 272$ observations (at 126 values) of consecutive eruption lengths in minutes (rounded to seconds precision). Figure 1 shows the data and a Gaussian kernel density estimate with window width $h = .3348$ which is default (3.31) from Silverman (1986). The kernel den-

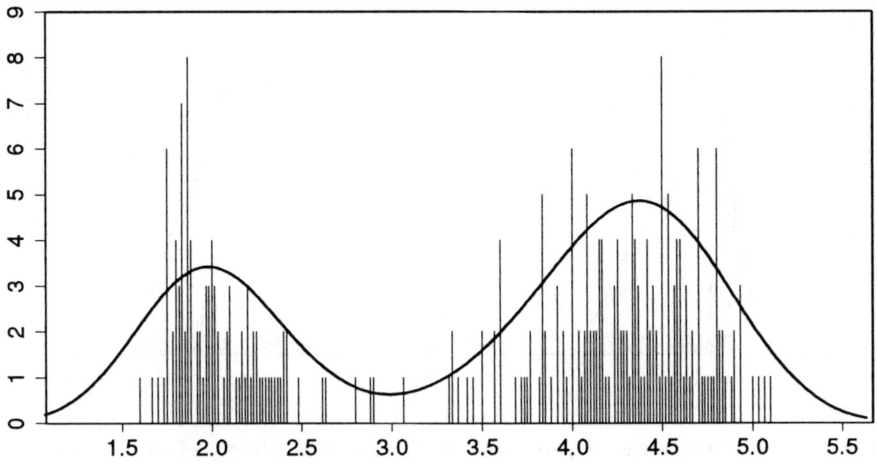

Figure 1: Old Faithful Geyser eruption lengths, $n = 272$; binned data and (Gaussian) kernel density estimate ($\times 10$) with $h = h^* = .3348$.

sity looks perfect showing two modes and no unnecessary bumps. A closer look (e.g., plot $F(x)$ and $F_n(x)$ vs. x) however reveals that the bandwidth chosen is too large. This is no surprise, since the asymptotic for h^* is geared towards unimodal densities (see Silverman (1986), 3.4.2, p.45ff). For goodness of fit, we use distances of the distribution function from the empirical, $\|F - F_n\|$, where $F(x) = \int_{-\infty}^{x} f(t)\,dt$. For simplicity, we use the three scaled L^p norms $\|G\|_1 := r_x^{-1} \int_a^b |G(t)|\,dt$, $\|G\|_2^2 := r_x^{-1} \int_a^b G(t)^2\,dt$, and $\|G\|_\infty := \sup_t |G(t)|$, where $r_x :=$ range$(\{x_i\}) = \max_i(x_i) - \min_i(x_i)$. Note that $\|F - F_n\|_\infty$ is the well known Kolmogorov–Smirnov distance. Further, we use the unit m ("milli"); i.e., thousandths, in order to have more manageable numbers. A further goodness of fit measure is the "average negative log-likelihood"; i.e., $-\frac{1}{n}\sum_{i=1}^{n} l(x_i)$, denoted by ANLL. For the kernel estimate above, ANLL=1.10, and $(L^1, L^2, L^\infty) =$ (31, 33, 76) m. Another very smooth bimodal estimate is the standard Gaussian mixture model, $f(x) = \pi\sigma_1^{-1}\phi((x - \mu_1)/\sigma_1) + (1 - \pi)\sigma_2^{-1}\phi((x - \mu_2)/\sigma_2)$, where ϕ is the standard normal density $\phi(x) = (2\pi)^{-1/2}\exp(-x^2/2)$. For the geyser data, the maximum likelihood estimates for $(\pi,\ \mu_1, \mu_2,\ \sigma_1, \sigma_2)$ are $(0.3484,\ 2.019, 4.273,\ 0.2356, 0.4371)$. The goodness of fit values are ANLL=1.02, and $(L^1, L^2, L^\infty) =$ (14, 18, 49) m which are somewhat better than the above kernel but still not good enough.

More reasonable alternatives are the Gaussian kernel with $h = .1$ and the logspline estimate (with automatic knot choice) which are shown in Figure 2. The estimates of our new approach are much better however. Figure 3 shows

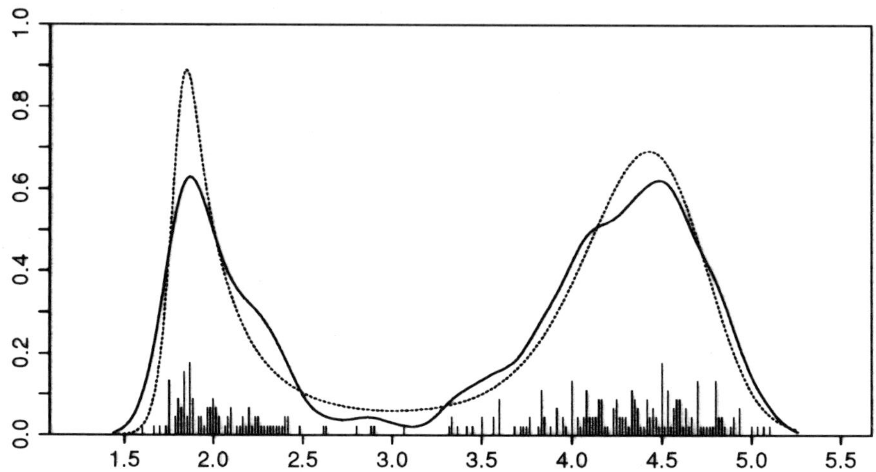

Figure 2: (Gaussian) kernel density with $h = .1$ and logspline (default $\rightsquigarrow$ 6 knots). The kernel density has a third mode which is true even for h as large as almost .13.

$\hat{f}(x)$ for $\lambda = 2^{-2}, 2^{-4}, 2^{-6}$. The images and the goodness-of-fit quantities in

Estimate	ANLL	L^1 [m]	L^2 [m]	L^∞ [m]
kernel $h = .335$	1.100	30.6	33.3	76.1
2 normal mixture	1.016	14.3	18.0	48.6
logspline (6 kn.)	0.957	9.1	11.1	31.9
kernel $h = .1$	0.946	6.3	8.2	29.8
$\lambda = 1/4$	0.948	5.8	8.0	39.3
$\lambda = 2^{-4}$	0.939	5.0	6.7	30.4
$\lambda = 2^{-6}$	0.929	4.7	6.1	22.2

Table 1: Goodness-of-fit values of density/distribution estimates for the geyser data.

Table 1 show that the new densities fit (much) better to the data while being more smooth than the traditional estimates with comparable amount of fit.

More extensive comparisons and simulations will be done in subsequent research, but current experience shows that the new class of density/distribution estimates provides very smooth curves while adapting to the data very well.

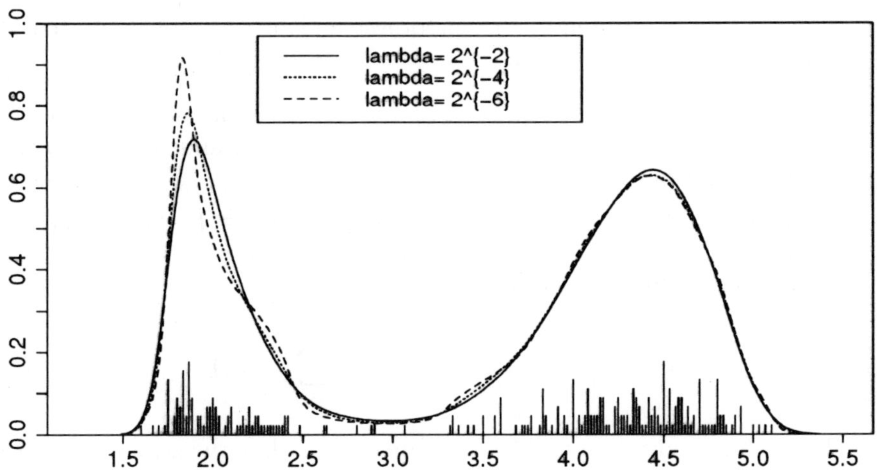

Figure 3: Estimated densities $\hat{f}(x)$ according to the new approach, (1.2), (1.3), (2.1), with two modes ($m = 3$), for $\lambda = 2^{-2}, 2^{-4}, 2^{-6}$. Even the smoothest one ($\lambda = 1/4$) fits better to the data than competitors (s. Table 1).

4 Properties of Solution

By this new approach, we get the distribution function with many derivatives, simultaneously, since F, f, $l' =$ score of f, l'' are continuously differentiable. f''' and l''' are still "cad-lag" (right-continuous, limit from left) being continuous functions of h_l'' and therefore of the empirical distribution F_n.

The prescribed number of modes is the main smoothing parameter. The (extra) smoothing parameter λ is of less importance, since for the "roughest" situation, $\lambda \to 0$ f is still "smooth", whereas for nonparametric estimators, f would converge to the sum of δ-spikes at x_i.

"Most smooth" solution $\lambda \to \infty$. The MPL criterion is solved in the limit $\lambda \to \infty$ by making the penalty $\int h_l'^2(t)\ dt$ equal to zero, and hence, $h_l' \equiv 0$. In other words, the MPL method for $\lambda \to \infty$ consists of doing ordinary *maximum likelihood* within the class of functions satisfying $h_l'(x) \equiv 0\ \forall x \in [a, b]$. Therefore, h_l is constant, say h_0, and $l'(x) = c p_{\mathbf{w}}(x)$ (where the constant $c = (-1)^j e^{h_0}$. Consequently, l is a *polynomial* of degree $m + 1$. We consider the following situations:

- "Zeromodal" case: $m^* = m = 0$: We have $l'(x) \equiv \pm c$ and therefore $f(x) \propto e^{\pm c}$. For $[a, b] = [0, \infty]$, this is the well-known exponential distribution.

- Unimodal case: $m^* = m = 1$. If for the sign, we exclude the rare case $j = 0$, we have $l'(x) = -c(x - w_1)$, and $f(x) \propto \exp(-c(x - w_1)^2/2)$; i.e, for $[a, b] = \mathbb{R}$, f is the normal density with mean w_1 (and ML estimate $w_1 = \bar{x}_n$). Hence, the smoothest limit consists of parametrically estimating a normal density.

- General case: For simplicity, we assume $j = 1$; i.e., a minus sign in (1.2). We have $l'(x) = -c(x - w_1)(x - w_2) \cdots (x - w_m)$, and therefore $l(x) = l_0 - cS_{\mathbf{w}}(x)$, where $S_{\mathbf{w}}(x) := x^m - (\sum w_j)x^{m-1} + (\sum_{i<j} w_i w_j)x^{m-2} - \ldots + (-1)^m \prod w_j$. How is l_0 determined? Let $I(c, \mathbf{w}) := \int_a^b \exp(-cS_{\mathbf{w}}(t))\, dt$. Then, $1 = \int f(x)dx = e^{l_0} I(c, \mathbf{w})$, whence $l_0 = -\log I(c, \mathbf{w})$. Now, maximizing the likelihood means

$$\min_{c,\mathbf{w}} \quad n \log I(c, \mathbf{w}) + c \sum_{i=1}^{n} S_{\mathbf{w}}(x_i) \tag{4.1}$$

  This ML minimization problem is quite simpler than the general MPL problem (2.1).

5 Extensions

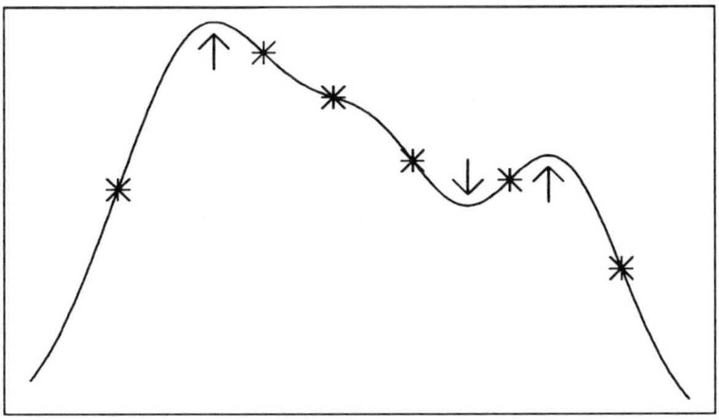

Figure 4: Bumps and Dips

Bumps and Dips. As in Good and Gaskins (1980), one may want to consider bumps and dips (maximal concave and convex regions) of the density. As there is at most one mode in a bump interval, we can effectively control the local extrema via inflection points.

With this approach, we can require even more "qualitative smoothness", and the MPL criterion approximately penalizes relative change of curvature of the log density l. By factoring $l''(x)$ instead of $l'(x)$, we get a very similar o.d.e. boundary value problem with an additional boundary condition ensuring $\int_a^b t \cdot f(t)\, dt = \frac{1}{n}\sum_{i=1}^{n} x_i$; i.e., an exact first moment property. The generalization of considering $l^{(\nu)}$ for $\nu \geq 1$ is straightforward.

A conceptual problem is that for $\nu \geq 2$, the zeros of $f^{(\nu)}$ and $l^{(\nu)}$ no longer coincide. For $\nu = 2$, $l'' = (f''f - f'^2)f^{-2}$, we have at least $f'' \leq 0 \Longrightarrow l'' \leq 0$.

Semi-Parametric Mixture Models. Semi-parametric density models can be envisaged. For instance, a mixture of two,

$$f(x) = pf_1(x) + (1-p)f_2(x), \tag{5.1}$$

where f_1 and f_2 are required to be unimodal or even "unibumpal" (with only one bump) and will be estimated by MPLE ($\nu = 2$) as above. In the limit, for $\lambda_{1,2} \to \infty$, this approach would comprise traditional parametric (Gaussian) mixture models.

References

[1] Ascher, U.M., Mattheij, R. M.M. and Russell, R.D. (1988): *Numerical Solution of Boundary Value Problems for Ordinary Differential Equations.* Prentice–Hall series in computational mathematics. Prentice Hall, Englewood Cliffs, New Jersey.

[2] Bader, G. and Ascher, U. (1987): A new basis implementation for a mixed order boundary value ODE solver. *SIAM Journal on Scientific and Statistical Computing* **8** 483–500.

[3] Cox, D.R. (1966): Notes on the analysis of mixed frequency distributions. *J. Math. Statist. Psychol.* **19** 39–47.

[4] Good, I.J. and Gaskins, R.A. (1971): Non-parametric roughness penalties for probability densities. *Biometrika* **58** 255–277.

[5] Good, I.J. and Gaskins, R.A. (1980): Density estimation and bump-hunting by the penalized likelihood method exemplified by scattering and meteorite data. *J. Amer. Statist. Assoc.* **75** 42–73. Comments & Rejoinder.

[6] Härdle, W. (1990): *Smoothing Techniques With Implementation in S.* Springer-Verlag, Berlin.

[7] Hartigan, J.A. and Hartigan, P.M. (1985): The dip test of unimodality. *Ann. Statist.* **13** 70–84.

[8] Izenman, A.J. (1991): Recent developments in nonparametric density estimation. *J. Amer. Statist. Assoc.* **86(413)** 205–224.

[9] Izenman, A.J. and Sommer, C.J. (1988). Philatelic mixtures and multimodal densities. *J. Amer. Statist. Assoc.* **83** 941–953.

[10] Kooperberg, C. and Stone, C.J. (1992): Logspline density estimation for censored data. *Journal of Computational and Graphical Statistics* **1(4)** 301–328.

[11] Mächler, M. (1989): *'Parametric' Smoothing Quality in Nonparametric Regression: Shape Control by Penalizing Inflection Points.* Ph.D. Thesis, no 8920, ETH Zürich, Switzerland.

[12] Mächler, M. (1995): Very smooth nonparametric curve estimation by penalizing change of curvature. *Ann. Statist* **23**(5), to appear.

[13] Minnotte, M.C. and Scott, D.W. (1993): The mode tree: A tool for visualization of nonparametric density features. *J. of Computational and Graphical Statistics* **2(1)** 51–68.

[14] Silverman, B.W. (1981): Using kernel density estimates to investigate multimodality. *J. Roy. Statist. Soc. B* **43** 97–99.

[15] Silverman, B.W. (1986): *Density Estimation for Statistics and Data Analysis.* Chapman & Hall, London.

[16] Tapia, R.A. and Thompson, J.R. (1978): *Nonparametric Probability Density Estimation.* John Hopkins Univ. Press, Baltimore.

[17] Thompson, J.R. and Tapia, R.A. (1990): *Nonparametric Function Estimation, Modeling, and Simulation.* SIAM, Philadelphia.

Implementing M-Estimators of the Gamma Distribution

A. Marazzi C. Ruffieux

Institut Universitaire de Médicine, Lausanne

Abstract

Several types of M-estimators for multiparameter models have been proposed in Hampel et al. (1986). This paper considers 6 options and specializes them to the Gamma distribution, with the purpose of estimating its mean. It discusses their implementation; i.e., the numerical computation of the estimates, the determination of their tuning parameters, and the evaluation of measures that characterize their distribution (e.g., bias, variance, breakdown point, ...). One of the options is selected for practical use.

Key words and phrases: M-estimates, Gamma distribution.
AMS 1991 subject classifications: Primary 62F35; secondary 65U05.

1 Introduction

We consider the gamma distribution defined by its density

$$f_{\sigma,\alpha}(y) = \frac{1}{\sigma\Gamma(\alpha)}\left(\frac{y}{\sigma}\right)^{\alpha-1}\exp\left(-\left(\frac{y}{\sigma}\right)\right), \qquad y>0,\ \sigma>0,\ \alpha>0, \tag{1.1}$$

where Γ is the Gamma function. When $\sigma = 1$, we write f_α in place of $f_{\sigma,\alpha}$. The cumulative distribution function is denoted by $F_{\sigma,\alpha}$ or by F_α.

The pair (σ, α) is the usual parameter vector; α is the shape parameter, and σ is the scale parameter. However, we often prefer to work with the pair (τ, α), where $\tau = \ln(\sigma)$, or with (ν, α), where $\nu = \ln(\alpha\sigma)$. Often it is convenient to use the abbreviation $\theta = (\sigma, \alpha)^\top$, or $\theta = (\tau, \alpha)^\top$, or $\theta = (\nu, \alpha)^\top$, where θ is a column vector. Our main objective is to estimate the function $\alpha\sigma$ of the parameters; i.e. the mean $\mathrm{E}(Y)$ of a Gamma distributed random variable Y. Thus, in the (ν, α) parametrization, α may be considered a nuisance parameter.

In classical statistics, one assumes that the observations y_i are distributed exactly according to the density f_θ, for a certain unknown value of θ, and estimates this value using the available data. One of the most widely used procedures is maximum likelihood, with the estimate defined as the value $\hat\theta$ which maximizes $\Pi_{i=1}^n f_\theta(y_i)$, or equivalently as that value $\hat\theta$ which minimizes the sum $-\sum \ln f_\theta(y_i)$. Equating to zero the derivatives of this function, one obtains

$$\sum_{i=1}^{n} s(y_i, \theta) = 0, \tag{1.2}$$

where $s(y,\theta) = (s_1(y,\theta), s_2(y,\theta))^\top$ denotes the vector of likelihood scores,

$$s(y,\theta) = \frac{\partial}{\partial\theta} \ln f_\theta(y) = \left(\frac{\partial}{\partial\theta_1} \ln f_\theta(y), \frac{\partial}{\partial\theta_2} \ln f_\theta(y)\right)^\top. \tag{1.3}$$

For the Gamma distribution in the (τ, α) parametrization, the score functions are:

$$s_1(y,\theta) = \left[\frac{y}{\sigma} - \alpha\right], \qquad s_2(y,\theta) = \ln\left(\frac{y}{\sigma}\right) - \dot{\Gamma}(\alpha), \tag{1.4}$$

where $\dot{\Gamma}$ denotes the Digamma function, $\dot{\Gamma}(\alpha) = d\ln\Gamma(\alpha)/d\alpha$, and $\sigma = \exp(\tau)$. The score functions in the other parametrizations are easily obtained observing that, if $\vartheta = g(\theta)$ is a 1-1 parameter transformation (e.g., $\vartheta = (\sigma, \alpha) = (\exp(\tau), \alpha)$ or $\vartheta = (\nu, \alpha) = (\ln(\alpha) + \tau, \alpha)$) with Jacobian matrix $J(\theta) = \partial g/\partial\theta$, the new score functions are $J(\theta)^{-\mathrm{T}} s(y,\theta)$. Note that, in the (τ, α) and the (ν, α) parametrization, the score functions depend on σ via $z = y/\sigma$, and, therefore, the estimation problem is scale invariant. Unfortunately, the problem is not shape invariant.

In robustness theory one realizes that the model f_θ is a mathematical abstraction which is only an idealized approximation of reality. It is therefore necessary to use procedures that behave fairly well under deviations from the assumed model. Such procedures are called robust. One of the best known classes of robust estimators are M-estimators. In the next sections, we consider some of the M-estimators described in Hampel et al. (1986, Chapter 4) and specialize them to the parameters of the Gamma distribution. These estimators are defined as solutions $\hat{\theta}$ of an equation

$$\sum_{i=1}^{n} \psi(y_i, \theta) = 0, \tag{1.5}$$

where ψ denotes a given vector function of (y, θ) with values in $\mathbb{R}^2$. Equation (1.5) is a generalization of the maximum likelihood equation (1.2). The function ψ is a composition of the score functions s and Huber's function h_a with tuning parameter a, which is defined as

$$\begin{aligned} h_a(z) &= \max(-a, \min(z, a)), && \text{when } z \in \mathbb{R}^1, \\ h_a(z) &= z\min(1, a/||z||), && \text{when } z \in \mathbb{R}^2, \end{aligned}$$

where $||z||$ denotes the Euclidean norm of z. Many theoretical results concerning the asymptotic and the robustness properties of M-estimators are available in the literature.

We discuss the numerical computation of M-estimators $\hat{\theta}$ defined by (1.5) for various choices of the function ψ and a few parametrizations of interest. Then, we evaluate and compare numerically some of the measures that characterize the distribution of the corresponding estimates of $\mathrm{E}(Y)$. This includes bias and breakdown point, and estimates of the maximum asymptotic bias and variance over ε-contamination neighborhoods of the model. We aim to select an estimation procedure for $\mathrm{E}(Y)$ with known robustness and asymptotic properties to be used in practice; thus, ease of computation is a major criterion of choice. The selected procedures have been included into the library ROBETH (Marazzi, 1993) and interfaced to S. Programs are available from the authors.

Remark 1.1 Many of the techniques described in this paper can be applied (with obvious changes) to the estimation of the parameters of the Weibull distribution with density

$$f_{\sigma,\alpha}(y) = \frac{\alpha}{\sigma}\left(\frac{y}{\sigma}\right)^{\alpha-1}\exp(-(y/\sigma)^{\alpha}), \qquad y>0,\ \sigma>0,\ \alpha>0. \tag{1.6}$$

Note however that, if Y is distributed according to $f_{\sigma,\alpha}$, the density of $T=\ln Y$ is

$$f_{\tau,v}(t) = \frac{1}{v}\exp\left[\left(\frac{t-\tau}{v}\right) - e^{(t-\tau)/v}\right], \qquad -\infty<t<\infty, \tag{1.7}$$

where $\tau=\ln\sigma$ and $v=1/\alpha$. In this parametrization, τ is a location, v is a scale parameter, and the corresponding estimation problem is location and scale invariant. This is an important simplification compared to the Gamma distribution. ////

2 Estimators

Let $s(\theta,y)$ denote the vector of score functions (1.4) for $\theta=(\tau,\alpha)$. For a fixed value of the parameter b (see below), define $\hat\theta=(\hat\tau,\hat\alpha)^{\top}$ as a solution of

$$\sum_{i=1}^{n}\psi_b\left[A_b(\theta)(s(y_i,\theta)-c_b(\theta))\right]=0. \tag{2.1}$$

We consider two options for the function $\psi_b(z)$, $z=(z_1,z_2)$:

(i) The *shrinking norm estimator* with

$$\psi_b(z)=h_b(z), \qquad b\in\mathbb{R}^1,$$

(ii) The *shrinking component estimator* with

$$\psi_b(z)=(h_{b_1}(z_1),h_{b_2}(z_2))^{\top}, \qquad b=(b_1,b_2)\in\mathbb{R}^2.$$

The parameter b, called *tuning parameter*, is fixed by the user (see Section 6). $A_b(\theta)$ is a 2×2 nonsingular lower triangular matrix; its inverse may be considered as the transposed Jacobian of a parameter transformation $\vartheta=g(\theta)$.

We consider the following options:

(a) The (τ, α) *parametrization* with

$$A_b(\theta) = \mathbb{I} \text{ (unit matrix)};$$

(b) The (ν, α) *parametrization* with

$$A_b(\theta) = \begin{pmatrix} 1 & 0 \\ -\alpha^{-1} & 1 \end{pmatrix};$$

(c) The *standard parametrization* implicitly defined by

$$\int \psi_b \left[A_b(\theta)(s(y,\theta) - c_b(\theta))\right] \psi_b \left[A_b(\theta)(s(y,\theta) - c_b(\theta))\right]^\top f_\theta(y) dy = \mathbb{I}. \quad (2.2)$$

Finally, the function $c_b(\theta)$ is defined by:

$$\int \psi_b \left[A_b(\theta)(s(y,\theta) - c_b(\theta))\right] f_\theta(y) dy = 0. \quad (2.3)$$

Remark 2.1 The special case $A = \mathbb{I}$ of the shrinking component estimator is a simple generalization of the maximum likelihood equations, where the likelihood scores are truncated and "re-centered" by c_b. The "correction" c_b makes the M-estimate asymptotically unbiased if the observations are distributed according to the model (Fisher consistency). Note that all estimators reduce to the maximum likelihood estimator for $b = b_1 = b_2 = \infty$. ////

Remark 2.2 Option (c) defines the *standardized* estimators of Hampel et al. (1986, Chapter 4), where further classes of M-estimators are proposed. In that book, the shrinking norm estimator is called the *optimal B_s-robust estimator*, and the shrinking component estimator is called the *optimal B_s^p-robust estimator for partitioned parameters*. Note however that, despite the name, no optimality result is known for the second proposed estimator, whereas a weak optimality property is known for the optimal B_s-robust case (Hampel et al., 1986, p.244). Both estimators truncate the re-centered score functions of the standard parameter vector $\vartheta = g(\theta)$. Due to the form of the asymptotic covariance matrix of $g(\hat{\theta})$ (Sections 3 and 5), the affine transformation A_b defined by equation (2.2) makes the standard parameters approximately orthogonal. An application of the shrinking norm estimator of the parameters of the Gamma distribution is described in Victoria–Feser (1993). ////

Remark 2.3 Previous experience (Willimann, 1980) suggests that the unstandardized estimators of Hampel et al. (1986) are computationally less reliable than the standardized ones. We do not consider them in this paper. ////

Remark 2.4 Thanks to the scale invariance, the transformation $A_b(\theta)$ and the correction $c_b(\theta)$ defined by (2.2) and (2.3) do not depend on τ. Therefore, we will write $A_b(\alpha)$ and $c_b(\alpha)$ in place of $A_b(\theta)$, $c_b(\theta)$. In the case of the Weibull distribution (1.7), the dependence of on both the parameters τ and v is trivial. ////

3 Evaluation Criteria

Let $\mathcal{Y}$ denote the sample space, $\Theta = \{\theta\}$ the parameter space (of dimension p), and let $y_1, y_2, \ldots, y_n$ denote n observations of the random variable Y. We consider the unknown distribution F of Y, the model distributions F_θ ($\theta \in \Theta$), the empirical distribution F_n, and the distribution Δ_y which puts probability 1 at the point y. The densities of F and F_θ are denoted by f and f_θ. As F_θ is only an idealized approximation of F it is useful to assume that F belongs to a neighborhood $\mathcal{P}$ of one of the F_θ, say, F_{θ^*}. We use the ε-*contamination model*:

$$F \in \mathcal{P}_\varepsilon = \{G \mid G = (1-\varepsilon)F_{\theta^*} + \varepsilon H,\ H \text{ arbitrary}\}, \tag{3.1}$$

where ε is a given percentage of *gross errors*, e.g., $\varepsilon = 5\%$.

In order to evaluate and compare estimates of $\mathrm{E}(Y)$, for the Gamma model, we compute asymptotic bias and variances, estimates of the maximum asymptotic bias over $\mathcal{P}_\varepsilon$, estimates of the maximum asymptotic variances over $\mathcal{P}_\varepsilon$, and asymptotic relative efficiencies with respect to the maximum likelihood estimate. We seek for estimates with small bias and variance over some set of challenge distributions. In particular, we seek for estimates with a small maximum asymptotic bias, a small maximum asymptotic variance and a large asymptotic efficiency with respect to the maximum likelihood estimate at the model F_θ. This section summarizes the definitions of these quantities (see Hampel et al., 1986, for more details). The computational aspects are discussed in Section 5.

Let $\hat\theta(F_n)$ be a p-dimensional M-estimate, where $\hat\theta(F)$ is a solution of

$$\int \psi(y, \hat\theta(F))dF(y) = 0, \tag{3.2}$$

and ψ is a user defined vector valued function $\psi : \mathcal{Y}\times\Theta \to \mathbb{R}^p$. In particular, we consider the cases introduced in Section 2. We use the abbreviation $\hat\theta$ in place of $\hat\theta(F_n)$. Under certain regularity conditions, $\hat\theta$ is asymptotically normally distributed, with asymptotic covariance matrix

$$V(\hat\theta, F) = M(\psi, F)^{-1}Q(\psi, F)M(\psi, F)^{-\top}, \tag{3.3}$$

where

$$Q(\psi, F) = \int \psi(y, \hat\theta(F))\psi(y, \hat\theta(F))^\top dF(y), \tag{3.4}$$

$$M(\psi, F) = -\int \left[\frac{\partial}{\partial\theta}\psi(y,\theta)\right]_{\theta=\hat\theta(F)} dF(y). \tag{3.5}$$

In applications, $V(\hat{\theta}, F_n)/n$ or $V(\hat{\theta}, F_{\hat{\theta}})/n$ are often used as approximations for the covariance matrix of $\hat{\theta}$. If $\hat{\theta}$ is Fisher consistent at F_θ; i.e., if it satisfies $\int \psi(x,\theta)dF_\theta(x) = 0$ for all θ (see equation (2.3)), one obtains:

$$M(\psi, F_\theta) = \int \psi(y,\theta)s(y,\theta)dF_\theta(y), \tag{3.6}$$

where $s(y,\theta)$ is the vector of likelihood scores. This simplifies the computation of $V(T, F_\theta)$.

Let ξ be a one-dimensional parameter defined as a function $\xi = h(\theta)$; e.g., $\xi = \alpha \exp(\tau)$. Let $\hat{\xi}$ be an estimator of ξ, and $V(\hat{\xi}, F)$ its asymptotic variance. We assume that $\hat{\xi}$ satisfies $\hat{\xi}(F_\theta) = \xi$ for all θ, and let $\xi^* = h(\theta^*)$. The *asymptotic bias* of $\hat{\xi}$ under a point contamination $\varepsilon\Delta_y$ of F_{θ^*} is

$$B(\hat{\xi}, \xi^*, \varepsilon, y) = |\hat{\xi}((1-\varepsilon)F_{\theta^*} + \varepsilon\Delta_y) - \xi^*|. \tag{3.7}$$

The *maximum asymptotic bias* of $\hat{\xi}$ over $\mathcal{P}_\varepsilon$ is

$$\mathcal{B}(\hat{\xi}, \xi^*, \varepsilon) = \sup_{G \in \mathcal{P}_\varepsilon} |\hat{\xi}(G) - \xi^*|. \tag{3.8}$$

The *maximum asymptotic variance* of $\hat{\xi}$ over $\mathcal{P}_{\varepsilon^*}$ is

$$\mathcal{V}(\hat{\xi}, \xi^*, \varepsilon) = \sup_{G \in \mathcal{P}_\varepsilon} V(\hat{\xi}, G). \tag{3.9}$$

Let $\check{\theta}$ denote the maximum likelihood estimate of θ and let $\check{\xi} = h(\check{\theta})$. The *asymptotic relative efficiency* of $\hat{\xi}$ with respect to $\check{\xi}$ at the distribution F is

$$\mathrm{ARE}(\hat{\xi}, \check{\xi}, F) = V(\check{\xi}, F)/V(\hat{\xi}, F). \tag{3.10}$$

4 Computation of the Estimates

General suggestions for simultaneously solving (2.1)–(2.3) by means of iterative algorithms can be found in Hampel et al. (1986, Section 4.6b). Given initial values for (τ, α), one computes c_b from (2.3), A_b from (2.2) and improved values for (τ, α) from (2.1). This scheme is usually memory-efficient but typically quite slow, especially when used within a resampling scheme. Therefore, we use the following two step procedure:

(a) For fixed b, solve (2.2) and (2.3) with respect to A_b and c_b for a discrete set of α-values and store the solutions. More precisely, choose an interval $[\alpha_1, \alpha_2]$ and an integer k (e.g., $k = 100$), and tabulate the solutions $A_b(\alpha)$ and $c_b(\alpha)$ of (2.2) and (2.3) for $\alpha = \alpha_1 + (i-1)(\alpha_2 - \alpha_1)/(k-1)$, $i = 1, \ldots, k$.

(b) Solve (2.1) with respect to θ, using a linear interpolation of the tables obtained in step (a) in order to compute $A_b(\alpha)$ and $c_b(\alpha)$ for the required values of α.

In practice, it is necessary to fix α_1 and α_2 so that $[\alpha_1, \alpha_2]$ contains the solution $\hat{\alpha}$ of (2.1). However, the preliminary computation of step (a) substantially accelerates step (b). In the following, we discuss both steps in more detail.

4.1 Computation of the vector c_b

The computation of c_b for given b, α and $A_b(\alpha)$, can be performed by means of the usual algorithms for M-estimates of location.

4.2 Computation of the matrix A_b

For the shrinking norm case, the problem of computing the matrix A_b, for given b, α, and a given value c of $c_b(\alpha)$, can be formulated as the problem of solving for A the equation

$$\text{ave}\{u(|z|)zz^\top\} = \mathbb{I}, \tag{4.1}$$

where $z = (z_1, z_2)^\top = Ax$, $x = s(\theta, y) - c$, $A = A_b(\alpha)$, $\theta = (\tau = 0, \alpha)$, $\text{ave}\{.\} = \int \{.\} f_\alpha(y) dy$ and $u(|z|) = \min(1, b/|z|)^2$. A fixed point algorithm for solving (4.1) is described by Huber (1981, Chapter 8), who also gives a convergence proof.

For the shrinking component case, one has to solve for A:

$$\text{ave}\left\{ \begin{pmatrix} u_1(|z_1|) & 0 \\ 0 & u_2(|z_2|) \end{pmatrix} zz^\top \begin{pmatrix} u_1(|z_1|) & 0 \\ 0 & u_2(|z_2|) \end{pmatrix} \right\} = \mathbb{I}, \tag{4.2}$$

where $u_1(|z_1|) = \min(1, b_1/|z_1|)$, and $u_2(|z_2|) = \min(1, b_2/|z_2|)$. A simple iterative algorithm is obtained as follows. Let $A^{(0)}$ be an initial and $A^{(1)}$ an improved value of A. Replace the arguments of u_1 and u_2 by $|z_1^{(0)}|$ and $|z_2^{(0)}|$, where $z^{(0)} = A^{(0)}x$, and the middle product $zz^\top$ by $A^{(1)}x[A^{(1)}x]^\top$. Then solve for $A^{(1)}$. One obtains:

$$\begin{aligned} a_{11}^{(1)} &= \text{ave}\{u_1(|z_1^{(0)}|)^2 x_1^2\}^{-2}, \\ a_{22}^{(1)} &= \text{ave}\{u_2(|z_2^{(0)}|)^2 (\kappa x_1 + x_2)^2\}^{-2}, \\ a_{21}^{(1)} &= \kappa a_{22}^{(1)}, \end{aligned} \tag{4.3}$$

where $\kappa = -\text{ave}\{u_1(|z_1^{(0)}|)u_2(|z_2^{(0)}|)x_1x_2\}/\text{ave}\{u_1(|z_1^{(0)}|)u_2(|z_2^{(0)}|)x_1^2\}$. A convergence proof of this algorithm is still lacking, but the procedure works well in practice.

Figure 1 shows the elements of the matrix A_b for varying α. They have been obtained by simultaneously solving (2.2) and (2.3) for the shrinking norm and the shrinking component cases. The precision of the two algorithms is similar (but the vertical scales on the diagrams of a_{21} are different).

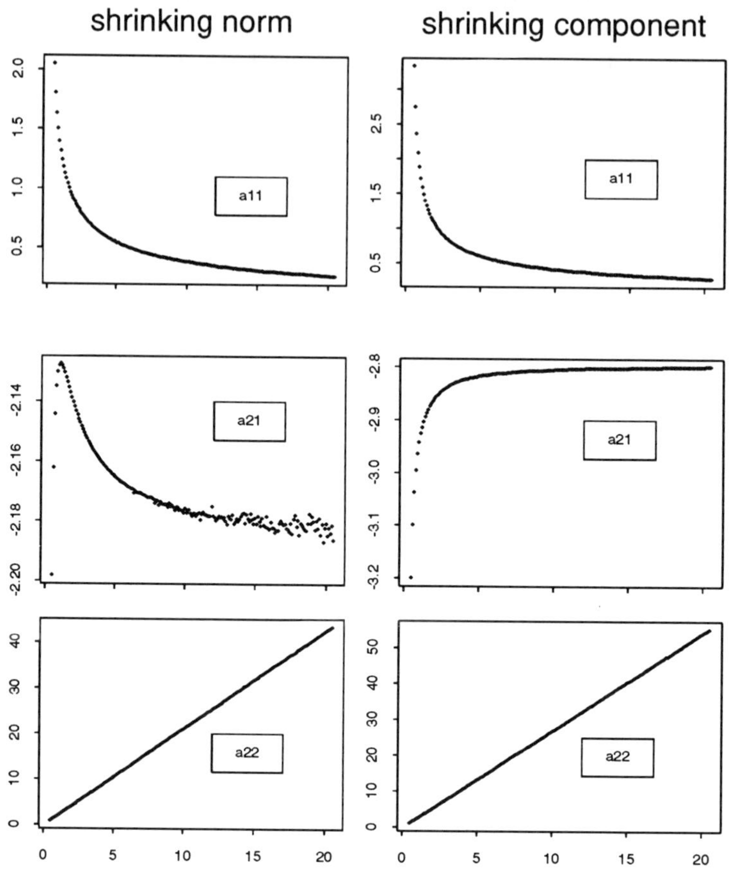

Figure 1: Elements of the matrix $A_b(\alpha)$ for varying α (shrinking norm $b = 2.6$, shrinking component $b_1 = 1.5, b_2 = 1.7$)

4.3 Computation of the parameter estimates

For $\theta = (\tau, \alpha)$, $\sigma = \exp(\tau)$, and $j = 1, 2$ define

$$S_{pj}(\sigma, \alpha) = \frac{1}{n} \sum \psi_{b_j}[\tilde{s}_j(y_i, \theta)], \tag{4.4}$$

$$S_{nj}(\sigma, \alpha) = \frac{1}{n} \sum \tilde{s}_j(y_i, \theta) \min[1, b/|\tilde{s}(y_i, \theta)|], \tag{4.5}$$

where

$$\begin{aligned} \tilde{s}_1(y, \theta) &= a_{11}(\alpha)[y/\sigma - \alpha - c_1(\alpha)], \\ \tilde{s}_2(y, \theta) &= a_{21}(\alpha)[y/\sigma - \alpha - c_1(\alpha)] + a_{22}(\alpha)[\ln(y/\sigma) - \dot{\Gamma}(\alpha) - c_2(\alpha)]. \end{aligned}$$

The components of $\tilde{s}(\theta, y)$ are the recentered score functions of the standard parameters. The shrinking component estimate is the solution of $S_{pj}(\sigma,\alpha) = 0$, $j = 1,2$, and the shrinking norm estimate is the solution of $S_{nj}(\sigma,\alpha) = 0$, $j = 1,2$.

The diagrams in Figure 2 show the functions S_{pj} and S_{nj} for varying σ (some fixed values of α and artificial data sets have been used). For the shrinking

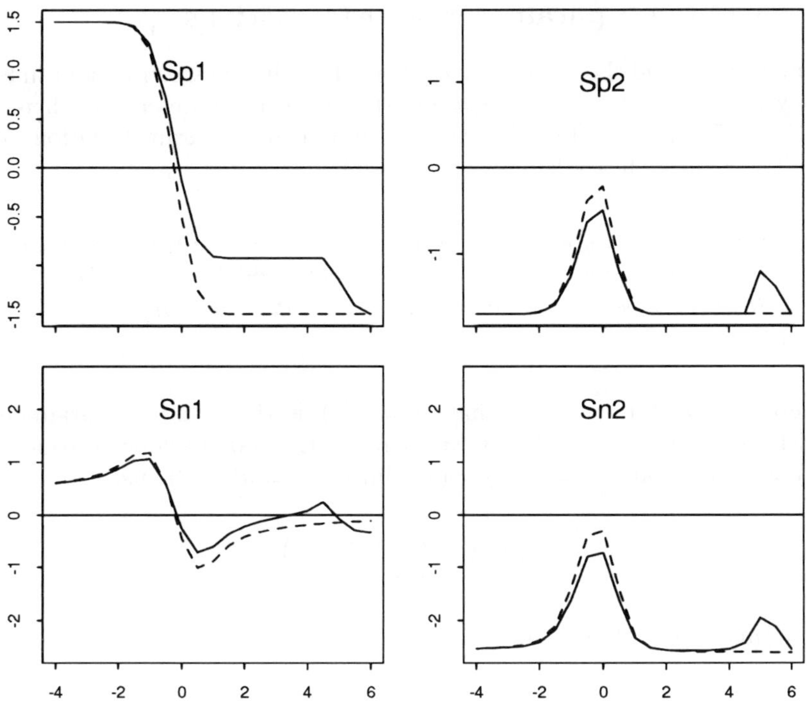

Figure 2: Functions $S_{p1}, S_{p2}, S_{n1}, S_{n2}$ for varying σ (log scale) on artificial data sets

component case, the function S_{p1} is monotone in σ and changes sign. This guarantees the existence of a zero. Thus, the following simple idea works:

(a) Use a regula-falsi method to invert S_{p1} with respect to σ over the interval $[\alpha_1, \alpha_2]$, for which tables of $c_b(\alpha)$ and $A_b(\alpha)$ are available. In other words, for any desired $\alpha \in [\alpha_1, \alpha_2]$ obtain an highly accurate solution $\bar{\sigma}(\alpha)$ of $S_{p1}(\sigma,\alpha) = 0$.

(b) For $\alpha = \alpha_1 + (i-1)(\alpha_2 - \alpha_1)/(k-1)$, $i = 1,\ldots,k$, compute $S_{p2}(\bar{\sigma}(\alpha),\alpha)$. Any time this function changes sign, use a bisection method to solve $S_{p2}(\bar{\sigma}(\alpha),\alpha) = 0$.

This procedure usually finds the solutions of the system $S_{pj}(\sigma,\alpha) = 0$. It can be complemented with a plot of $S_{p2}(\bar{\sigma}(\alpha), \alpha)$, in order to circumscribe the solutions of the second equation, if necessary. On the contrary, the behaviour of S_{nj} is quite complex for $j = 1$ and 2. It suggests that both these equations may have multiple solutions, and solving their system is not as straightforward as in the previous case. In our implementation, we replace the regula-falsi method with a (blind) bisection method in order to (hopefully) find the smallest zero of S_{n1}.

4.4 A note on parameter transformations

The matrix A_b and the correction c_b depend on the parametrization; in Section 2, they have been defined for $\theta = (\tau, \alpha)$. We want to determine them, when equations (2.1),(2.2) and (2.3) are solved in a different parametrization. We use the following elementary lemma (chain rule).

Lemma 4.1 *Let $\vartheta = g(\theta)$ and $\theta = h(\upsilon)$ (i.e., $\upsilon = h^{-1}(\theta)$) define two one-to-one parameter transformations. Let $J(\theta) = \partial g(\theta)/\partial\theta$ be the Jacobian of g, and $C(\upsilon) = \partial h(\upsilon)/\partial\upsilon$ be the Jacobian of h. Then, the Jacobian of $\vartheta = g(h(\upsilon))$ is $J(h(\upsilon))C(\upsilon)$.*

Suppose now, for example, that $\vartheta = g(\theta)$ is the standard parameter vector defined by (2.2) and that $\upsilon = (\nu, \alpha)$ is the new parametrization. Then $A_b(\theta) = J(\theta)^{-\top}$, and $\theta = (\tau, \alpha) = (\nu - \ln(\alpha), \alpha) = h(\upsilon)$. Thus,

$$C(\upsilon) = \begin{pmatrix} 1 & -\alpha^{-1} \\ 0 & 1 \end{pmatrix},$$

and A_b must be replaced with

$$J(h(\upsilon))^{-\top} C(\upsilon)^{-\top} = A_b(\theta) \begin{pmatrix} 1 & 0 \\ \alpha^{-1} & 1 \end{pmatrix}.$$

It is easy to see that c_b must be replaced with $C(\upsilon)^{-\top} c_b(\theta)$. As a corollary, the estimates of the standard parameters ϑ obtained from (2.1)–(2.3), using the (τ, α) or the (ν, α) parametrization, coincide.

5 Computation of Asymptotic Quantities

Let $\hat{\theta}(F_n)$ be an estimate of θ, as defined in Sections 2 and 3. Moreover, let $\xi = h(\theta)$ be a one-dimensional function of θ, e.g., $\xi = \alpha\exp(\tau) = \mathrm{E}(Y)$. We estimate ξ with $h(\hat{\theta})$. In order to evaluate the quantities $\mathcal{B}$ and $\mathcal{V}$, defined in (3.8) and (3.9), we use the influence function and the change of variance function of $\hat{\theta}$ and $h(\hat{\theta})$ (Hampel et al., 1986).

5.1 Influence and change of variance functions of a multidimensional M-estimate

The *influence function* of $\hat{\theta}$ at the distribution F is defined as the p-vector

$$\mathrm{IF}(y;\hat{\theta},F)=\left[\frac{\partial}{\partial t}\hat{\theta}((1-t)F+t\Delta_y)\right]_{t=0}=\lim_{t\to 0}\frac{\hat{\theta}((1-t)F+t\Delta_y)-\hat{\theta}(F)}{t}\qquad(5.1)$$

in those $y\in\mathcal{Y}$ where this limit exists. One obtains

$$\mathrm{IF}(y;\hat{\theta},F)=M(\psi,F)^{-1}\psi(y,\hat{\theta}(F)).\qquad(5.2)$$

The *change of variance function* of $\hat{\theta}$ at the distribution F is defined as the $p\times p$-matrix

$$\mathrm{CVF}(y;\hat{\theta},F)=\left[\frac{\partial}{\partial t}V(\hat{\theta},(1-t)F+t\Delta_y)\right]_{t=0},\qquad(5.3)$$

for all y where this expression exists, where $V(\hat{\theta},F)$ denotes the asymptotic covariance matrix of $\hat{\theta}$ given by (3.3). The elements (h,k) of M and Q are

$$m_{hk}(\psi,F)\;=\;-\int\left[\frac{\partial\psi_h(y,\theta)}{\partial\theta_k}\right]_{\theta=\hat{\theta}(F)}dF(y),\qquad(5.4)$$

$$q_{hk}(\psi,F)\;=\;\int\psi_h(y,\hat{\theta}(F))\psi_k(y,\hat{\theta}(F))dF(y).\qquad(5.5)$$

Thus,

$$\left[\frac{\partial}{\partial t}m_{hk}(\psi,(1-t)F+t\Delta_y))\right]_{t=0}=-m_{hk}(\psi,F)-\left[\frac{\partial\psi_h(y,\theta)}{\partial\theta_k}\right]_{\theta=\hat{\theta}(F)}$$
$$-\,\mathrm{IF}(y;\hat{\theta},F)^{\top}\int\left[\nabla\left(\frac{\partial\psi_h(y,\theta)}{\partial\theta_k}\right)\right]_{\theta=\hat{\theta}(F)}dF(y),\qquad(5.6)$$

where $\nabla(\partial\psi_h/\partial\theta_k)$ denotes the gradient of $\partial\psi_h/\partial\theta_k$, and

$$\left[\frac{\partial}{\partial t}q_{hk}(\psi,(1-t)F+t\Delta_y))\right]_{t=0}=-q_{hk}(\psi,F)+\psi_h(y,\hat{\theta}(F))\psi_k(y,\hat{\theta}(F))$$
$$+\,\mathrm{IF}(y;\hat{\theta},F)^{\top}\int[\psi_h(y,\theta)\nabla\psi_k(y,\theta)+\psi_k(y,\theta)\nabla\psi_h(y,\theta)]_{\theta=\hat{\theta}(F)}\,dF(y),\qquad(5.7)$$

where $\nabla\psi_h$ and $\nabla\psi_k$ denote the gradients of ψ_h and ψ_k. Using these expressions we evaluate

$$\frac{\partial}{\partial t}V=\frac{\partial}{\partial t}\left(M^{-1}QM^{-\top}\right)=$$
$$\left(\frac{\partial}{\partial t}M^{-1}\right)QM^{-\top}+M^{-1}\left(\frac{\partial}{\partial t}Q\right)M^{-\top}+M^{-1}Q\left(\frac{\partial}{\partial t}M^{-\top}\right),\qquad(5.8)$$

where

$$\frac{\partial}{\partial t}M^{-1}=-M^{-1}\left(\frac{\partial}{\partial t}M\right)M^{-1},\quad\text{and}\quad\frac{\partial}{\partial t}M^{-\top}=\left(\frac{\partial}{\partial t}M^{-1}\right)^{\top}.\qquad(5.9)$$

5.2 Influence and change of variance functions of functions of estimates

Let $\vartheta = g(\theta)$ be a differentiable one-to-one parameter transformation with non-singular Jacobian $J(\theta) = \partial g(\theta)/\partial\theta$, and let $\hat{\vartheta} = g(\hat{\theta})$ be an estimate of ϑ. In this section, $\hat{\theta} = \hat{\theta}(F)$. One obtains:

$$\begin{aligned}
\mathrm{IF}(y;\hat{\vartheta},F) &= J(\hat{\theta})^\top \mathrm{IF}(y;\hat{\theta},F), &(5.10)\\
V(\hat{\vartheta},F) &= J(\hat{\theta})^\top V(\hat{\theta},F)J(\hat{\theta}), &(5.11)\\
\mathrm{CVF}(y;\hat{\vartheta},F) &= \mathrm{IF}(y;J(\hat{\theta}),F)^\top \cdot V(\hat{\theta},F)\cdot J(\hat{\theta})\\
&\quad + J(\hat{\theta})^\top \cdot \mathrm{CVF}(y;\hat{\theta},F)\cdot J(\hat{\theta})\\
&\quad + J(\hat{\theta})^\top \cdot V(\hat{\theta},F)\cdot \mathrm{IF}(y;J(\hat{\theta}),F). &(5.12)
\end{aligned}$$

We obtain similar expressions for $\hat{\xi} = h(\hat{\theta})$, where h is a one-dimensional function of θ. It suffices to replace $\hat{\vartheta}$ by $\hat{\xi}$, and $J(\theta)$ with $\nabla h(\theta)$ (the gradient of $h(\theta)$) in (5.9)–(5.12).

5.3 Influence and change of variance function of the mean estimate

Assume now that $\hat{\theta} = (\hat{\tau},\hat{\alpha})$, $\mathrm{IF}(y;\hat{\theta},F)$, $V(\hat{\theta},F)$ and $\mathrm{CVF}(y;\hat{\theta},F)$ are available. Let $\vartheta = g(\theta) = (\exp(\tau),\alpha)^\top$, and $\xi = h(\theta) = \sigma\alpha$, with $\sigma = \exp(\tau)$. Let $\hat{\vartheta} = (\hat{\sigma},\hat{\alpha})^\top = (\exp(\hat{\tau}),\hat{\alpha})^\top$, and $\hat{\xi} = \hat{\sigma}\hat{\alpha}$. In this section, $\hat{\alpha} = \hat{\alpha}(F)$, $\hat{\sigma} = \hat{\sigma}(F)$, and $\hat{\tau} = \hat{\tau}(F)$. Then,

$$J(\theta) = \begin{pmatrix} \exp(\tau) & 0 \\ 0 & 1 \end{pmatrix}.$$

Thus, $\mathrm{IF}(y;\hat{\alpha},F)$ remains unchanged, and

$$\begin{aligned}
\mathrm{IF}(y;\hat{\sigma},F) &= \exp(\hat{\tau})\cdot \mathrm{IF}(y;\hat{\tau},F),\\
V(\hat{\vartheta},F) &= \begin{pmatrix} \exp(2\hat{\tau})v_{\tau\tau} & \exp(\hat{\tau})v_{\tau\alpha} \\ \exp(\hat{\tau})v_{\alpha\tau} & v_{\alpha\alpha} \end{pmatrix},
\end{aligned}$$

where $v_{..}$ denote the elements of $V(\hat{\theta},F)$. The function $\mathrm{CVF}(y;\hat{\vartheta},F)$ follows from (5.12), with

$$\mathrm{IF}(y;J(\hat{\theta}),F) = \begin{pmatrix} \mathrm{IF}(y;\hat{\sigma},F) & 0 \\ 0 & 0 \end{pmatrix}.$$

Finally,

$$\begin{aligned}
\mathrm{IF}(y;\hat{\xi},F) &= \hat{\alpha}\mathrm{IF}(y;\hat{\sigma},F) + \hat{\sigma}\mathrm{IF}(y;\hat{\alpha},F),\\
V(\hat{\xi},F) &= (\hat{\alpha},\hat{\sigma})\cdot V(\hat{\vartheta},F)\cdot(\hat{\alpha},\hat{\sigma})^\top,\\
\mathrm{CVF}(y;\hat{\xi},F) &= (\mathrm{IF}(y;\hat{\alpha},F),\mathrm{IF}(y;\hat{\sigma},F))\cdot V(\hat{\vartheta},F)\cdot(\hat{\alpha},\hat{\sigma})^\top\\
&\quad +(\hat{\alpha},\hat{\sigma})\cdot \mathrm{CVF}(y;\hat{\vartheta},F)\cdot(\hat{\alpha},\hat{\sigma})^\top\\
&\quad +(\hat{\alpha},\hat{\sigma})\cdot V(\hat{\vartheta},F)\cdot(\mathrm{IF}(y;\hat{\alpha},F),\mathrm{IF}(y;\hat{\sigma},F))^\top.
\end{aligned}$$

Figure 3 shows the influence function of the standardized shrinking norm and shrinking component estimates of $\mathrm{E}(Y)$.

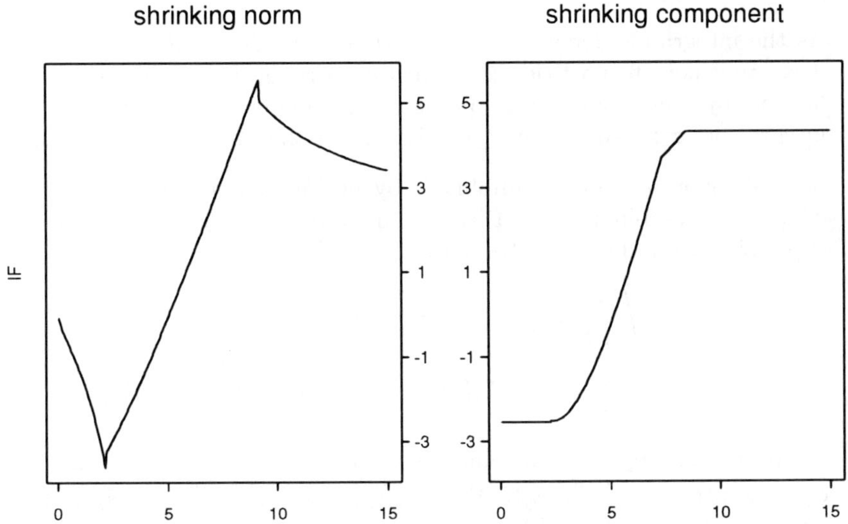

Figure 3: Influence functions of the standardized shrinking norm ($\alpha = 5$, $b = 2.6$) and shrinking component estimates ($\alpha = 5$, $b_1 = 1.5$, $b_2 = 1.7$)

5.4 Maximum bias and variance

The *maximum asymptotic bias* $\mathcal{B}(\hat{\xi}, \xi^*, \varepsilon)$ is approximated by

$$\hat{\mathcal{B}}(\hat{\xi}, \xi^*, \varepsilon) = \varepsilon \cdot \sup_y \mathrm{IF}(y; \hat{\xi}, F_{\theta^*}), \tag{5.13}$$

where $\gamma^*(\hat{\xi}, F) = \sup_y \mathrm{IF}(y; \hat{\xi}, F)$ is the *gross-error sensitivity* of $\hat{\xi}$ at the distribution F. The *maximum asymptotic variance* $\mathcal{V}(\hat{\xi}, \xi^*, \varepsilon)$ is approximated by

$$\hat{\mathcal{V}}(\hat{\xi}, \xi^*, \varepsilon) = V(\hat{\xi}, F_{\theta^*}) \exp\left(\varepsilon \cdot \sup_y \left[\mathrm{CVF}(y; \hat{\xi}, F_{\theta^*}) / V(\hat{\xi}, F_{\theta^*})\right]\right), \tag{5.14}$$

where $\kappa^*(\hat{\xi}, F) = \sup_y \left[\mathrm{CVF}(y; \hat{\xi}, F) / V(\hat{\xi}, F)\right]$ is the *change-of-variance sensitivity* of $\hat{\xi}$ at the distribution F (Hampel et al., 1986).

5.5 Notes on the numerical evaluation of integrals

Several integrals must be evaluated in order to compute the matrices Q and M (according to (3.5) and (3.6)), their derivatives (according to (5.6) and (5.7)), as well as the averages (4.3), and those needed for the determination of $c_b(\alpha)$. We use subroutines for numerical integration taken form the library QUADPACK (Piessens et al., 1980) and pay attention to the following points, in order to maximize accuracy.

1. As the integrands depend on ψ-functions with discontinuous derivatives, it is convenient to partition the integration range $[0,\infty)$ into subintervals limited by the discontinuity points. The upper limit of the numerical integration is set such that $f_\alpha(y)$ is smaller than a specified tolerance.

2. Several integrals can be simplified, by noting that the second derivative of ψ is a combination of Dirac δ-functions. For example, one of the expressions found in (5.6) becomes:

$$\int \frac{\partial^2}{\partial \tau^2}\psi_{b_1}(\tilde{s}_1(y,\theta)f_\alpha(y)dy =$$
$$a_{11}(\alpha)e^{-\tau}y^2 f_\alpha(y)\Big|_{y_2}^{y_1} + a_{11}(\alpha)e^{-\tau}\int_{y_1}^{y_2} yf_\alpha(y)dy, \tag{5.15}$$

where y_1 and y_2 are the solutions of $\tilde{s}_1(y,\theta) = \pm b_1$, There are similar opportunities for $\int \frac{\partial^2}{\partial\tau^2}\psi_{b_2}(\tilde{s}_2))$, $\int \frac{\partial^2}{\partial\alpha\partial\tau}\psi_{b_1}(\tilde{s}_1))$, $\int \frac{\partial^2}{\partial\alpha\partial\tau}\psi_{b_2}(\tilde{s}_2))$. ($\tilde{s}_1$, $\tilde{s}_2$ are defined in Section 4.3).

3. In several cases, integration by parts avoids computation of the second derivative of $A_b(\alpha)$. For example, another integral in (5.6) is:

$$\int \tfrac{\partial^2}{\partial\alpha^2}\psi_{b_1}(\tilde{s}_1(y,\theta))f_\alpha(y)dy =$$
$$\int [-2\tfrac{\partial}{\partial\alpha}\psi_{b_1}(\tilde{s}_1(y,\theta))s_2(y,\theta) + \psi_{b_1}(\tilde{s}_1(y,\theta))\ddot{\Gamma}(\alpha)$$
$$+\psi_{b_1}(\tilde{s}_1(y,\theta))s_2(y,\theta)^2]f_\alpha(y)dy, \tag{5.16}$$

where $\ddot{\Gamma}$ is the second derivative of Γ. The first derivative of $A_b(\alpha)$ can be approximated with the slope of the linear interpolation described at the beginning of this section.

4. A few series developments and continued fractions are useful to evaluate:

$$K_1(t,\alpha) = \int_0^t f_\alpha(y)dy, \tag{5.17}$$
$$K_2(t,\alpha) = \int_0^t \ln(y)f_\alpha(y)dy, \tag{5.18}$$
$$K_3(t,\alpha) = \int_0^t y^k f_\alpha(y)dy. \tag{5.19}$$

We use (Abramowitz and Stegun, 1965; Bhattacharjee, 1970):

$$K_1(t,\alpha) = \sum_0^\infty t f_{\alpha+n}(t)/(\alpha+n) \tag{5.20}$$

$$= \frac{e^{-t}t^{\alpha}}{\Gamma(\alpha)}\left[1+\sum_{n=1}^{\infty}\frac{t^n}{(\alpha+1)(\alpha+2)\ldots(\alpha+n)}\right] \quad (5.21)$$

$$= 1-\frac{e^{-t}t^{\alpha}}{\Gamma(\alpha+1)}\left[\frac{1}{x+}\frac{1-\alpha}{1+}\frac{1}{x+}\frac{2-\alpha}{1+}\frac{2}{x+}\cdots\right], \quad (5.22)$$

$$K_2(t,\alpha) = \ln(t)K_1(t,\alpha)-\int_0^t\frac{1}{y}K_1(y,\alpha)dy \quad (5.23)$$

$$= \ln(t)K_1(t,\alpha)-\left[\sum_{j=1}^{\infty}f_{\alpha+j}(t)\sum_{i=0}^{j-1}\frac{1}{\alpha+i}\right], \quad (5.24)$$

$$K_3(t,\alpha) = \frac{\Gamma(k+\alpha)}{\Gamma(\alpha)}K_1(t,k+\alpha). \quad (5.25)$$

The developments are terminated when the contribution to the development is not greater than the value of a specified accuracy.

6 Tuning Parameters

The most common rule for determining the tuning parameter of an M-estimator is to require that the asymptotic relative efficiency with respect to the maximum likelihood estimator, at the model, equals a given value; e.g., 95%. It is simple to apply this rule to the shrinking norm estimators of $\mathrm{E}(Y)$ because b is a single real number. A few values of b and the corresponding relative efficiencies for the standardized shrinking norm estimator of $\mathrm{E}(Y)$ are given in Table 1.

For the shrinking component estimators there are several pairs of values of (b_1, b_2) whith the same relative asymptotic efficiency, and the rule does not determine $b=(b_1,b_2)$ uniquely. As a remedy, we minimize the approximation $\hat{\mathcal{V}}(\hat{\xi},\sigma^*\alpha^*,\varepsilon)$ of the maximum asymptotic variance, as a function of (b_1,b_2), where α^* and σ^* are rough preliminary estimates of α and σ. The sensitivity of the estimate to contamination depends then on ε and on the optimal value $b(\varepsilon)$ of b: the higher the value of ε, the less is the estimate sensitive to contamination. The choice of ε and $b(\varepsilon)$ is made on the ground of collateral information about the frequency of outliers, or by requiring that the asymptotic relative efficiency of the estimate defined by $\psi_{b(\varepsilon)}$ equals a given value; e.g., 95%. Table 2 gives a few optimal values $(b_1(\varepsilon), b_2(\varepsilon))$ for varying ε, using the standard parametrization with $\alpha^*=1,5,10$ ($\sigma^*=1$). It also gives the corresponding values of $V(\hat{\xi},F_\alpha)$, $\mathrm{ARE}(\hat{\xi},\check{\xi})$, and $\hat{\mathcal{V}}(\hat{\xi},\alpha^*,\varepsilon)$.

7 Empirical Results

Table 3 and 4 summarize a few Monte Carlo experiments for comparing the shrinking component and the shrinking norm estimates in the standard para-

α^*	b	$\text{ARE}(\hat{\xi}, \check{\xi}, F_{\alpha^*})$
1	4.4	0.974
	3.7	0.952
	3.2	0.923
	3.0	0.906
	2.7	0.872
	2.6	0.857
5	3.8	0.975
	3.2	0.952
	2.8	0.923
	2.6	0.901
	2.4	0.873
	2.3	0.855
10	3.7	0.975
	3.1	0.950
	2.8	0.927
	2.6	0.906
	2.4	0.878
	2.2	0.841

Table 1: $\text{ARE}(\hat{\xi}, \check{\xi}, F_{\alpha^*})$ denotes the asymptotic relative efficiency of the standardized shrinking norm estimate $\hat{\xi}$ of $\text{E}(Y)$ with respect to the maximum likelihood estimate $\check{\xi}$ at F_{α^*}. b is the tuning parameter.

metrization. Samples of size 200 were generated from

$$F = (1-\varepsilon)F_\alpha + \varepsilon H, \tag{7.1}$$

using various values of ε and α. The tuning parameters were chosen such that both estimates have the same asymptotic relative efficiency (i.e. the same asymptotic variance) at the model. In Table 3, $H = F_{\sigma=2,\alpha}$. In Table 4, $H = \Delta_{y_0}$, where

$$y_0 \approx \text{argmax}_y B(\hat{\xi}, \alpha, \varepsilon, y). \tag{7.2}$$

Note that y_0 is different for the shrinking component and the shrinking norm cases, and that, due to the form of the influence function, it is difficult to determine y_0 with good accuracy in the shrinking norm case. $\hat{\xi}(F)$ was computed using $(1-\varepsilon)500$ percentile points of F_α and 500ε data points concentrated on y_0. $\bar{\xi}$ is the arithmetic mean of the values of $\hat{\xi} = \exp(\hat{\tau})\hat{\alpha}$ over 200 experiments; $\text{se}(\bar{\xi})$ is the usual estimate of the standar error of the mean, and $\text{sd}(\bar{\xi}) = (V(\hat{\xi}, F_\alpha)/200)^{1/2}$ is the standard error obtained from the asymptotic variance. (Therefore, it is the same for the shrinking norm and the shrinking

α^*	ε	b_1	b_2	$V(\hat{\xi}, \alpha^*)$	ARE$(\hat{\xi}, \check{\xi}, F_{\alpha^*})$	$\hat{\mathcal{V}}(\hat{\xi}, \alpha^*, \varepsilon)$
1	0.01	2.7	2.5	1.058	0.945	1.23
	0.02	2.3	2.1	1.093	0.915	1.41
	0.05	1.7	1.7	1.192	0.839	1.95
	0.07	1.3	1.5	1.313	0.762	2.32
	0.10	1.3	1.3	1.363	0.734	2.96
	0.15	1.3	1.3	1.363	0.734	4.35
	0.20	1.1	1.1	1.611	0.621	6.16
5	0.01	2.1	2.7	5.155	0.970	5.62
	0.02	1.7	2.3	5.303	9.943	6.10
	0.05	1.5	1.7	5.523	0.905	7.38
	0.07	1.5	1.7	5.523	0.905	8.29
	0.10	1.3	1.3	5.978	0.836	9.66
	0.15	1.3	1.3	5.978	0.836	12.28
	0.20	1.3	1.3	5.978	0.836	15.61
10	0.01	2.1	2.7	10.207	0.980	11.03
	0.02	1.7	2.3	10.486	0.954	11.81
	0.05	1.5	1.7	10.889	0.918	13.94
	0.07	1.5	1.7	10.889	0.918	15.39
	0.10	1.3	1.3	11.734	0.852	17.63
	0.15	1.3	1.3	11.734	0.852	21.62
	0.20	1.3	1.3	11.734	0.852	26.50

Table 2: Values of b_1 and b_2 that minimize the approximation $\hat{\mathcal{V}}(\hat{\xi}, \alpha^*, \varepsilon)$ of the maximum asymptotic variance of the standardized shrinking component estimate $\hat{\xi}$ of E(Y) over an ε contamination neighborhood $\mathcal{P}_\varepsilon$ of F_{α^*}. $V(\hat{\xi}, F_{\alpha^*})$ denotes the asymptotic variance of $\hat{\xi}$ at F_{α^*}, ARE$(\hat{\xi}, \check{\xi}, F_{\alpha^*})$ the asymptotic relative efficiency with respect to the maximum likelihood estimate $\check{\xi}$ at F_{α^*}.

component cases, and does not depend on ε.) sd and se also estimate the standard error of $\hat{\xi}$. The values of sd$(\bar{\xi})$ appears to be systematically smaller than those of se$(\bar{\xi})$.

According to the results in Table 3, the shrinking component estimate performs as well as the shrinking norm estimate under a moderate contamination. However, the results of Table 4, suggest that the shrinking component estimate has a smaller bias than the shrinking norm estimate, while maintaining a higher precision, under a very unfavorable contamination.

Table 5 gives values of $B(\hat{\xi}, \alpha, \varepsilon, y_0)$, values of the estimate $\hat{\mathcal{B}}(\hat{\xi}, \alpha, \varepsilon)$ of the maximum asymptotic bias, and values of the estimate $(\hat{\mathcal{V}}(\hat{\xi}, \alpha, \varepsilon)/200)^{1/2}$ of the maximum asymptotic standard error of $\hat{\xi}$. These standard errors can be compared with those based on the asymptotic variance at the model; i.e., with

			Shrinking component			Shrinking norm		
ε	α	sd($\bar{\xi}$)	b_1, b_2	$\bar{\xi}$	se($\bar{\xi}$)	b	$\bar{\xi}$	se($\bar{\xi}$)
0.00	1	0.077	1.7,1.7	1.002	0.078	2.4	0.998	0.077
	5	0.166	1.5,1.7	4.997	0.171	2.6	5.004	0.177
	10	0.234	1.5,1.7	9.890	0.242	2.6	9.984	0.241
0.05	1	0.077	1.7,1.7	1.039	0.078	2.4	1.038	0.077
	5	0.166	1.5,1.7	5.149	0.183	2.6	5.139	0.186
	10	0.234	1.5,1.7	10.289	0.252	2.6	10.256	0.256
0.10	1	0.082	1.3,1.3	1.069	0.098	2.0	1.067	0.095
	5	0.172	1.3,1.3	5.341	0.204	2.2	5.327	0.210
	10	0.242	1.3,1.3	10.506	0.284	2.2	10.451	0.285

Table 3: For each value of α, 200 estimates $\hat{\xi} = \exp(\hat{\tau})\hat{\alpha}$ are computed using samples of size 200 from $(1-\varepsilon)F_\alpha + \varepsilon F_{\sigma=2,\alpha}$. $\bar{\xi}$ is the mean estimate over the 200 samples, se($\bar{\xi}$) the standard error of the mean, and $\text{sd}^2(\bar{\xi}) = (V(\hat{\xi}, F_\alpha)/200)$.

			Shrinking component			Shrinking norm		
ε	α	ARE	b_1, b_2	$\bar{\xi}$	se($\bar{\xi}$)	b	$\bar{\xi}$	se($\bar{\xi}$)
0.05	1	0.84	1.7,1.7	1.143	0.091	2.4	1.146	0.095
	5	0.90	1.5,1.7	5.233	0.198	2.6	5.298	0.195
	10	0.91	1.5,1.7	10.316	0.258	2.6	10.372	0.299
0.10	1	0.73	1.3,1.3	1.260	0.131	2.0	1.298	0.124
	5	0.81	1.3,1.3	5.508	0.215	2.2	5.583	0.231
	10	0.85	1.3,1.3	10.630	0.288	2.2	10.752	0.306

Table 4: For each value of α, 200 estimates $\hat{\xi} = \exp(\hat{\tau})\hat{\alpha}$ are computed using samples of size 200 from $(1-\varepsilon)F_\alpha + \varepsilon\Delta_{y_0}$, where y_0 is very unfavorable for bias. $\bar{\xi}$ is the mean estimate over the 200 samples, and se($\bar{\xi}$) the standard error of the mean.

column sd($\bar{\xi}$) in Table 3. $(\hat{\mathcal{V}}(\hat{\xi}, \alpha, \varepsilon)/200)^{1/2}$ slightly overestimates the standard error se($\bar{\xi}$) given in Table 4, but is closer to se($\bar{\xi}$) than sd($\bar{\xi}$).

Figure 4 shows empirical bias curves $(\varepsilon, B(\hat{\xi}, \alpha, \varepsilon, y_0))$ of the shrinking component estimates with options (a), (b), and (c) for the matrix A_b, for $\alpha = 5$. We obtain similar curves for other values of α. The standard parametrization clearly produces the best bias curves. They suggest that the breakdown point of the corresponding estimate is surprisingly high.

In conclusion, the empirical results of this section and the remarks on the computational aspects point out the standardized shrinking component estimator as the best procedure among those considered. Figure 5, shows the histogram of 2385 lengths of stay of patients that where hospitalized in Switzerland during 1988 for "vaginal delivery"; the two densities of the Gamma distribution drawn on the same figure have been determined by means of the maximum likelihood

		Shrinking component			Shrinking norm		
ε	α	B	$\hat{\mathcal{B}}$	$(\hat{\mathcal{V}}/200)^{1/2}$	B	$\hat{\mathcal{B}}$	$(\hat{\mathcal{V}}/200)^{1/2}$
0.05	1	0.137	0.118	0.098	0.148	0.156	0.105
	5	0.243	0.215	0.192	0.288	0.280	0.211
	10	0.322	0.288	0.264	0.378	0.369	0.290
0.10	1	0.256	0.207	0.121	0.304	0.299	0.136
	5	0.477	0.388	0.219	0.605	0.530	0.263
	10	0.630	0.519	0.296	0.791	0.704	0.359

Table 5: $B = B(\hat{\xi}, \alpha, \varepsilon, y_0)$ is the maximum bias of $\hat{\xi}$ over a point mass contamination $\varepsilon\Delta_y$, for varying y. $\hat{\mathcal{B}} = \hat{\mathcal{B}}(\hat{\xi}, \alpha, \varepsilon)$ is the estimate of the maximum asymptotic bias over an ε-contamination neighborhood of F_α based on the influence function. $\hat{\mathcal{V}} = \hat{\mathcal{V}}(\hat{\xi}, \alpha, \varepsilon)$ is the estimate of the maximum asymptotic variance over an ε-contamination neighborhood of F_α based on the change of variance function.

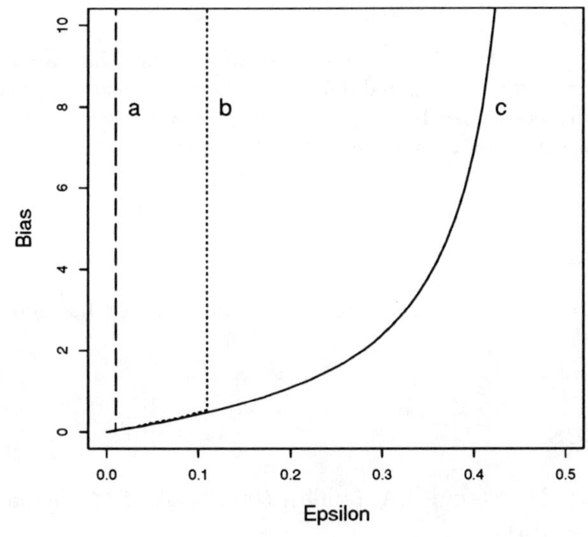

Figure 4: Bias curves of the shrinking component estimates

and the recommended procedure.

The maximum likelihood estimate was $\hat{\alpha} = 5.31$, $\hat{\sigma} = 1.46$ (mean $= 7.81$), and the shrinking component estimate was $\hat{\alpha} = 19.9$, $\hat{\sigma} = 0.38$ (mean $= 7.59$).

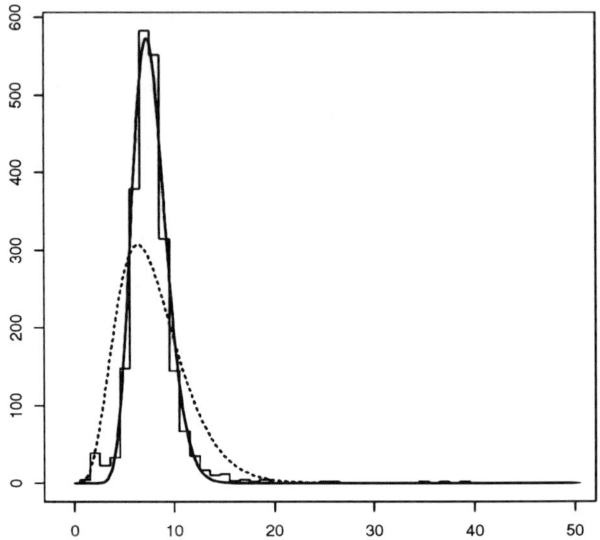

Figure 5: Histogram of 2395 lengths of stay of patients that where hospitalized in Switzerland during 1988 for "vaginal delivery"; the two densities of the Gamma distribution have been determined by means of the maximum likelihood (dotted line) and the standardized shrinking component estimate (solid line)

Acknowledgment

The authors are grateful to Alex Randriamiharisoa for his generous help with programming.

References

[1] Abramowitz M., Stegun I.A. (1965): *Handbook of Mathematical Functions.* Dover, New York.

[2] Bhattacharjee G.P. (1970): Algorithm AS 32: The incomplete gamma integral. *Appl. Statist.* **19** 285–287.

[3] Hampel F.R., Ronchetti E., Rousseeuw P.J., Stahel W.A. (1986): *Robust Statistics—The Approach Based on Influence Functions.* Wiley, New York.

[4] Huber P.J. (1981): *Robust Statistics.* Wiley, New York.

[5] Marazzi A. (1993): *Algorithms, Routines, and S Functions for Robust Statistics.* Wadsworth & Brooks/Cole, Pacific Grove.

[6] Piessens R., de Doncker-Kapenga E., Überhuber C.W., Kahaner D.K. (1980): QUADPACK: *A Subroutine Library for Automatic Integration.* Springer Verlag, Berlin.

[7] Victoria–Feser M.P. (1993): *Robust Estimation of Personal Income Distribution Models.* DARP-4, October 1993. Suntory-Toyota International Centre for Economics and Related Disciplines, London School of Economics, Houghton Street, London WC2A 2AE.

[8] Willimann B. (1980): *Optimale robuste Schätzungen für die Parameter der Gamma-Verteilung.* Diplomarbeit. Fachgruppe für Statistik, ETH Zürich, CH-8057 Zürich.

Constrained M-Estimation for Regression

Beatriz Mendes David E. Tyler
Rutgers University, New Brunswick

Abstract

When using redescending M-estimates of regression, one must choose not only an estimate of scale, but since the redescending M-estimating equations may admit multiple solutions, of which all of them may not be a desired solution, one must also have a method for choosing a desirable solution to the estimating equations. We introduce here a new approach for properly scaling redescending M-estimating equations and for obtaining high breakdown point solutions to the equations by the introduction of the constrained M-estimates of regression, or the CM-estimates of regression for short. Unlike the S-estimates of regression, the CM-estimates of regression can be tuned to obtain good local robustness properties while maintaining a breakdown point of 1/2.

Keywords and phrases: Breakdown point, M-estimates, robust estimation, S-estimates.
AMS 1991 subject classifications: Primary 62F35; secondary 62J05.

1 Introduction

Consider the usual linear regression model

$$\mathbf{y} = \mathbf{X}\boldsymbol{\beta} + \mathbf{e} \tag{1.1}$$

where $\mathbf{y} = (\mathrm{y}_1, \cdots, \mathrm{y}_n)' \in \mathbb{R}^n$ is the response vector, $\mathbf{X}$ is an $(n \times p)$ design matrix with known coefficients x_{ij}, $\boldsymbol{\beta} \in \mathbb{R}^p$ is a p-dimensional vector of unknown parameters, and $\mathbf{e} = (\mathrm{e}_1, \cdots, \mathrm{e}_n)' \in \mathbb{R}^n$ is the error vector. Alternatively, (1.1) may be expressed as

$$\mathrm{y}_i = \boldsymbol{x}_i' \boldsymbol{\beta} + \mathrm{e}_i, \qquad i = 1, \cdots, n \tag{1.2}$$

where $\boldsymbol{x}_i$ is a p-dimensional vector composed of the entries of the ith row of $\mathbf{X}$.

In this paper, we study the redescending M-estimates of $\boldsymbol{\beta}$, or more specifically solutions $\boldsymbol{\beta}_n \in \mathbb{R}^p$, to the p simultaneous equations

$$\mathrm{ave}_i \Big\{ \psi\Big(\frac{\mathrm{y}_i - \boldsymbol{x}_i' \boldsymbol{\beta}}{\mathrm{s}_n}\Big) \boldsymbol{x}_i \Big\} = \mathbf{o} \tag{1.3}$$

where ave_i stands for the arithmetic average over $i = 1, \cdots, n$, s_n is taken to be some estimate of scale and $\psi(r)$ is taken to be some bounded function with $\psi(-r) = -\psi(r)$, $\psi(r) \geq 0$ for $r \geq 0$, and $\psi(r) \to 0$ as $r \to \infty$. For example, the ψ function may be taken to be one of Tukey's biweight ψ functions, which are defined by $\psi_k(r) = \psi_0(\frac{r}{k})$ where

$$\psi_0(r) = \begin{cases} r(1-r^2)^2, & \text{if } |r| \leq 1 \\ 0, & \text{if } |r| > 1 \end{cases} \tag{1.4}$$

and with k being a tuning constant. The estimate of scale s_n is usually normalized so that it is a consistent estimate of ν when the errors e_i are independent normal $\mathcal{N}(0, \nu^2)$. We assume that this is so throughout the paper, and refer to such s_n as a normalized scale statistic.

When using redescending M-estimates of regression, one must choose not only an estimate of scale, but since (1.3) may admit multiple solutions, of which all of them may not be a desired solution, one must also have a method for choosing a desirable solution to (1.3). In the univariate location problem, it is common to let s_n be a multiple of the median absolute deviation about the median (M.A.D.) and to let the solution to (1.3) which is closest to the median be the desired solution (Hampel (1974), Hampel et al. (1986)). These problems are more difficult to resolve in the regression setting since no straightforward analog to the median and M.A.D. exist. The problem of multiple solutions to (1.3) can be avoided if one uses a monotonic ψ function, as originally proposed by Huber (1973). However, as noted in Maronna et al. (1979), this results in an estimate with a zero breakdown point. Redescending M-estimates of regression, on the other hand, can have breakdown points close to 0.5 if the ψ function, the scale statistic s_n and the desired solution to (1.3) are chosen appropriately.

The S-estimates of regression and the MM-estimates of regression proposed by Rousseeuw and Yohai (1984) and Yohai (1987) respectively, represent two classes of estimates which correspond to obtaining high breakdown point solutions to redescending M-estimating equations of the form (1.3). A drawback to the S-estimates of regression is that the S-estimates with higher breakdown points tend to have relatively low efficiencies whenever the error term has a normal distribution. The MM-estimates have high breakdown points and can be tuned to have high efficiencies at the normal model. The MM-estimates are multi-stage estimates with the first stage being based on some initial high breakdown point estimate of the regression parameters. The S-estimates and the MM-estimates are briefly reviewed in the next section.

We introduce here another approach for obtaining high breakdown point solutions to redescending M-estimating equations by the introduction of the constrained M-estimates of regression, or the CM-estimates of regression for short. The CM-estimates of regression and scale are defined respectively to be the solutions $\boldsymbol{\beta}_n$ and σ_n which minimize the objective function

$$L_n(\boldsymbol{\beta}, \sigma) = \text{ave}_i\left\{\rho\left(\frac{y_i - \boldsymbol{x}_i'\boldsymbol{\beta}}{\sigma}\right)\right\} + \log(\sigma) \tag{1.5}$$

over all $\boldsymbol{\beta} \in \mathbb{R}^p$ and $\sigma > 0$ s.t.

$$\text{ave}_i\Big\{\rho(\frac{\mathrm{y}_i - \boldsymbol{x}_i'\boldsymbol{\beta}}{\sigma})\Big\} \le \varepsilon\rho(\infty) \tag{1.6}$$

with $0 < \varepsilon < 1$, and with the function $\rho(r)$ being symmetric, nondecreasing for $r \ge 0$, and bounded. The CM-estimates are regression and affine equivariant as defined for example in Rousseeuw and Leroy (1987), and we show in Section 4 that the breakdown point of the CM-estimates is approximately $\min(\varepsilon, 1-\varepsilon)$, or approximately 0.5 when $\varepsilon = 0.5$. The concept of constrained M-estimation is also used by Kent and Tyler (1993) in relationship to the multivariate location and scatter problem.

When ρ is differentiable, we note in Section 2 that the CM-estimate of regression satisfies the estimating equations

$$\text{ave}_i\Big\{\psi(\frac{\mathrm{y}_i - \boldsymbol{x}_i'\boldsymbol{\beta}}{\sigma})\,\boldsymbol{x}_i\Big\} = \mathbf{o} \tag{1.7}$$

with σ being set equal to σ_n, the CM-estimate of scale, and with $\psi(r) = \rho'(r)$. The CM-estimate of scale is not in general a consistent estimate of ν when the error terms e_i are independent normal $\mathcal{N}(0, \nu^2)$. We show in Section 5 that this can be corrected by defining the normalized CM-estimate of scale to be $\mathrm{s}_n = \frac{\sigma_n}{\lambda_0}$, where $\lambda_0 > 0$ is the value which minimizes the function

$$l(\lambda) = \mathrm{E}[\rho(\tfrac{\mathrm{Z}}{\lambda})] + \log(\lambda) \tag{1.8}$$

s.t. $\lambda \ge \lambda_{\mathrm{L}}$, with λ_{L} being the solution to

$$\mathrm{E}[\rho(\tfrac{\mathrm{Z}}{\lambda})] = \varepsilon\rho(\infty) \tag{1.9}$$

and with Z having the standard normal distribution. Thus, for a given ψ function in (1.3), if the ρ function is chosen so that $\rho'(r)$ is equal to $\psi(\frac{r}{\lambda_0})$ then the CM-estimate of regression satisfies the estimating equation (1.3), with s_n being the normalized CM-estimate of scale.

To illustrate the above concepts, suppose we wish to use a CM-estimate which corresponds to a redescending M-estimate with a high breakdown point and a biweight ψ function with tuning constant k; i.e. ψ_k as defined prior to (1.4). To do so, let $\varepsilon = 0.5$ and define

$$\rho_0(r) = \frac{1}{6}\begin{cases} 3r^2 - 3r^4 + r^6, & \text{for } |r| \le 1 \\ 1, & \text{for } |r| \ge 1, \end{cases} \tag{1.10}$$

and note that $\rho_0{}'(r) = \psi_0(r)$. Considering classes of ρ functions of the form $\rho(r) = \rho_0(\frac{r}{m})$ is not fruitful, since they all define the same CM-estimate of regression. The tuning constant m in this case affects the CM-estimate of scale only up to a constant multiple, and so all such CM-estimates of scale are the same when normalized. Instead, we consider ρ functions of the form $\rho_c(r) = c\rho_0(r)$.

When using $\rho_c(r)$, the value of λ_L does not depend on the tuning constant c since c cancels out on both sides of (1.9). For $\varepsilon = 0.5$, $\lambda_L = 1.5476$. The value of λ_0 itself, however, does depend on c and in order to obtain the biweight ψ function with tuning constant k, we need to choose c so that $\lambda_0 = k$. For $k \geq 1.5476$, we show in Section 6 that this can be done by choosing c such that

$$c^{-1} = \mathrm{E}[\frac{\mathrm{Z}}{k}\psi_0(\frac{\mathrm{Z}}{k})] \tag{1.11}$$

where Z has the standard normal distribution. Equation (1.11) arises from finding the critical points of (1.8) when using ρ_c as the ρ function and setting $\lambda = k$. In the univariate location problem, when s_n is taken to be the M.A.D., in order to obtain a reasonably efficient estimate of location at the normal model and at other models it is often recommended that a tuning constant between 6 and 9 be used for the biweight ψ function (for example see Hoagin, Mosteller and Tukey (1983)). Since a consistent estimate of the standard deviation at the normal model is 1.483 M.A.D., this translates to choosing the tuning constant k to be between 4.045 and 6.069 when s_n is a normalized scale statistc. In general, for the biweight regression M-estimates, a tuning constant of k between 4.045 and 6.069 corresponds to choosing c to be between 3.957 and 7.237. (The asymptotic efficiencies are respectively 0.91 and 0.98.)

The paper is organized as follows. We first give some additional motivation and interpretations of the CM-estimates in Section 2. The functional version of the CM-estimates and conditions for the existence and uniqueness of the CM-estimates and CM-functionals are given in Section 3. In Section 4, the finite sample breakdown points of the CM-estimates are derived. Section 5 treats the influence functions of the CM-functionals and the asymptotic normality of the CM-estimates. In Section 6, we consider general classes of ρ functions of the form $\rho_c(r) = c\rho_0(r)$ and propose a procedure for choosing the tuning constant c in order to achieve desirable robustness and efficiency properties. As previously noted, one such class of ρ functions are those associated with the biweight ψ functions. We show in Section 6 that it is usually possible in general to tune a CM-estimate so that it improves upon an S-estimate with the same breakdown point in terms of both asymptotic efficiency and gross error sensitivity.

2 Background and Motivation

M-estimates. Given an auxilary estimate of scale s_n, an alternative definition of the M-estimates of regression is as the solution $\boldsymbol{\beta}_n$ which minimizes

$$\mathrm{ave}_i\Big\{\rho(\frac{\mathrm{y}_i - \boldsymbol{x}_i'\boldsymbol{\beta}}{\mathrm{s}_n})\Big\} \tag{2.1}$$

over all $\boldsymbol{\beta} \in \mathbb{R}^p$ for some function ρ. If $\rho'(r)$ is proportional to $\psi(r)$, then the critical points of (2.1) satisfy (1.3) and so the minimizer of (2.1) represents a natural choice for the desired solution to (1.3). This approach does not necessarily guarantee that the desired solution represents a high breakdown point solution. The breakdown point depends not only on the breakdown point of the scale estimate, but also on other properties of the scale estimate as well as properties of the ρ function. Furthermore, Huber (1984) shows that for the simple case of location with fixed scale, the breakdown point of redescending M-estimates of location can also depend upon the structure of the good data.

One way to define simultaneous M-estimates of regression and scale is as the solutions $\boldsymbol{\beta}_n$ and σ_n respectively which minimize the objective function $\mathrm{L}_n(\boldsymbol{\beta},\sigma)$ given by (1.5). Maximum likelihood estimates of regression and scale are of this form, but they are associated with unbounded ρ functions. Now, the critical points of $\mathrm{L}_n(\boldsymbol{\beta},\sigma)$ satisfy the estimating equations (1.7) together with

$$\mathrm{ave}_i\Big\{\chi\big(\frac{\mathrm{y}_i - \boldsymbol{x}_i'\boldsymbol{\beta}}{\sigma}\big)\Big\} = 0 \tag{2.2}$$

where $\chi(r) = r\psi(r) - 1$. Thus, a natural solution to the redescending M-estimating equation (1.7) would be that which corresponds to the problem of minimizing $\mathrm{L}_n(\boldsymbol{\beta},\sigma)$. When ρ is bounded, however, it is not meaningful to minimize $\mathrm{L}_n(\boldsymbol{\beta},\sigma)$ unconditionally since $\mathrm{L}_n(\boldsymbol{\beta},\sigma) \to -\infty$ as $\sigma \to 0$. We need to constrain σ away from 0. If we use the obvious constraint $\sigma > \sigma_0$ for some fixed $\sigma_0 > 0$, then the resulting estimate of regression would not be regression and affine equivariant. The constraint (1.6) on the other hand bounds σ away from 0 in a manner which result in an estimate of regression which is regression and affine equivariant. If strict inequality is obtained in (1.6), then the CM-estimates are the local minimum of $\mathrm{L}_n(\boldsymbol{\beta},\sigma)$ and so the simultaneous estimating equations (1.7) and (2.2) hold. If equality is obtained in (1.6), then by using a Lagrange multiplier, it follows that the CM-estimates satisfy the simultaneous estimating equations (1.7) and (1.6). In either case, (1.7) holds when ρ is differentiable.

S-estimates. The formulation of the CM-estimates of regression is somewhat similar to the formulation of the S-estimates of regression. The S-estimates of regression and scale are defined as those values $\boldsymbol{\beta}_n$ and σ_n respectively which result in the minimization of σ over all $\boldsymbol{\beta} \in \mathbb{R}^p$ and $\sigma \in \mathbb{R}^+$ such that constraint (1.6) is satisfied. For $\varepsilon \leq 0.5$, the breakdown point of the S-estimate of regression is approximately ε, see Rousseeuw and Yohai (1984). For the S-estimates, equality is always obtained in (1.6) and so the S-estimate is the same as the CM-estimate whenever equality is obtained in (1.6) for the CM-estimate. The S-estimates satisfy the simultaneous estimating equations (1.7) and (1.6), and in this sense the S-estimates can be viewed as M-estimates. All solutions to the estimating equations though need not correspond to an S-estimate since a global minimum rather than just critical points is being sought. The same statement applies to the CM-estimates.

Analogous to the CM-estimates of scale, the S-estimate of scale is not in general a consistent estimate of ν when the error terms e_i are independent normal $\mathcal{N}(0,\nu^2)$. This can again be corrected by defining the normalized S-estimate of scale to be $\mathrm{s}_n = \frac{\sigma_n}{\lambda_{\mathrm{L}}}$, where λ_{L} is defined as in (1.9). For a given ψ function in (1.3), if the ρ function is chosen so that $\rho'(r)$ is equal to $\psi(\frac{r}{\lambda_{\mathrm{L}}})$ then the

S-estimate of regression satisfies the estimating equation (1.3), with s_n being the normalized S-estimate of scale. The important thing to note here is that the λ_L is a decreasing function of ε since $\rho(r)$ is assumed to be nondecreasing for $r \geq 0$. Large values of λ_L usually give higher efficiency when the errors are Gaussian, and so in order to choose an S-estimate with some desired efficiency, the breakdown point must usually be lowered. For example, using the biweight ρ function given by (1.10), $\varepsilon = 0.5$ implies $k = 1.5476$, which gives a normal asymptotic efficiency of 0.287, whereas to obtain a normal asymptotic efficiency of say 0.95, the value of k should be 4.6904 and this implies a breakdown point of just 0.12.

One interesting way to view the CM-estimates is as adaptive S-estimates. If one had information concerning the maximum proportion of bad observations in a data set, then one could use an S-estimate with ε chosen accordingly in order to maintain some Gaussian efficiency while still protecting against breakdown. The data set itself contains such information and the CM-estimates can be viewed as taking advantage of this information in the sense that the realized CM-estimate of regression with $\varepsilon = 0.5$ is the same as an S-estimate of regression with ε equal to the realized value of

$$\text{ave}_i\Big\{\rho(\frac{y_i - \boldsymbol{x}_i'\boldsymbol{\beta}_n}{\sigma_n})\Big\}\Big/\rho(\infty) \tag{2.3}$$

from the CM-estimate. When sampling from a normal distribution, (2.3) tends to be substantially less than 0.5.

It is helpful to relate the above discussion to the least median of squares (LMS) estimate of Rousseeuw (1984). In simple linear regression, the LMS regression line corresponds to the middle line of the narrowest strip containing half of the observations. The LMS estimate is an S-estimate with $\varepsilon = 0.5$ and

$$\rho(r) = \begin{cases} c, & \text{for } |r| \geq r_0 \\ 0, & \text{for } |r| < r_0, \end{cases} \tag{2.4}$$

where the value of r_0 simply imposes a scaling factor on the S-estimate of scale. The constant $c > 0$ has no effect in defining the LMS estimate since it cancels out in constraint (1.6). The CM-estimate of regression defined using the ρ-function (2.4) and with $\varepsilon = 0.5$ is the middle line associated with the strip which minimizes

$$c(1 - \pi) + \log(w), \tag{2.5}$$

over all strips containing at least one half of the data, where π is the proportion of the data within the strip and w is the width of the strip. Here, c now act as a tuning constant. This CM-estimate is an adaptive version of the LMS estimate in the following sense. If one enlarges the strip containing one half of the data, then both the width of the strip and the proportion of the data within the strip increase. If nearly one half of the data is contaminated, then a relatively small increase in the proportion results in a substantial increase in the width. On the other hand, if the data contain relatively little contamination then the increase in the width relative to the increase in the proportion is modest. The CM-estimate discriminates between these two cases by balancing the proportion π and the width w. In the first case, the CM-estimate and the LMS estimate will essentially result in the same statistics. In the second case the CM-estimate will result in statistics which are more stable than the LMS estimate.

MM-estimates. The MM-estimates introduced by Yohai (1987) are multistage estimates which essentially adjust S-estimates with high breakdown points so that they have high efficiencies at the normal model while mantaining the high breakdown points. To define an MM-estimate of regression, one begins with two bounded, symmetric ρ-functions which are nondecreasing on $\mathbb{R}^+$ so that $\rho_1(r) \leq \rho_2(r)$ and $\rho_1(\infty) = \rho_2(\infty)$. Initially, an S-estimate of regression and scale is obtained using ρ_1 and $\varepsilon = 0.5$. The resulting normalized S-estimate of scale is then used with a redescending M-estimating equation of form (1.3) with the ψ-function taken to be $\psi_2(r) = \rho_2'(r)$. The MM-estimate of regression can then be defined as a solution to (1.3) which minimizes (2.1), where ρ_2 is used as the ρ-function in (2.1). The first stage using ρ_1 and $\varepsilon = 0.5$ determines the breakdown point of the MM-estimate of regression, and the second stage using ρ_2 determines its efficiency. For example, one can choose $\rho_1(r) = \rho_0(\frac{r}{k_1})$ and $\rho_2(r) = \rho_0(\frac{r}{k_2})$, with $k_1 \leq k_2$ and where ρ_0 is given by (1.10). For simplicity, we refer to the resulting MM-estimate as a biweighted MM-estimate. As noted earlier, the initial S-estimate with $\varepsilon = 0.5$ corresponds to using $k_1 = 1.5476$. A normal asymptotic efficiency of 0.95 can be obtained by choosing $k_2 = 4.6904$. Yohai (1987) notes however that higher values of k_2 tend to result in estimates which are less robust away from the Gaussian error model.

The local properties, such as the asymptotic distributions and influence functions, depend only on the M-estimating equations and not on the complete definition of the CM- or MM-estimates. So, for example, a biweighted CM-estimate with say say $c = 3.27$ and corresponding $k = 3.42$ is asymptotically equivalent at the Gaussian error model to a biweighted MM-estimate with $k_2 = 3.42$. Moreover, both estimates have the same breakdown point. The computational effort, however, needed for the MM-estimate is approximately twice that of the CM-estimate due to the multiple stages in the definition of the MM-estimates. A more important difference between the CM- and MM-estimates is their behavior away from the Gaussian error model. The CM-estimates tend to suffer less from the lack of robustness away from the Gaussian error model than the MM-estimates do. This is discussed further in Section 6.

3 Existence of the CM-Estimates and CM-Functionals

To begin, we discuss conditions under which the CM-estimates exist. We assume throughout the paper that ρ satisfies the following conditon. Stronger conditions are needed in later sections to get asymptotic properties of the CM-estimates.

Condition 3.1 *For $r \geq 0$, $\rho(r)$ is nondecreasing. Furthermore, $\rho(-r) = \rho(r)$, $0 = \rho(0) < \rho(\infty) < \infty$ and $\rho(r)$ is continuous at zero.*

Results on existence can be made more general by first introducing the notion of CM-functionals of regression and scale. The CM-functionals are also needed in later sections when discussing the asymptotic properties of the CM-estimates. Consider the random vector $(\mathbf{x}, \mathrm{y})$ with $\mathbf{x} \in \mathbb{R}^p$ and $\mathrm{y} \in \mathbb{R}$, and let F represent its joint distribution function. Given a fixed value $0 < \varepsilon < 1$, the CM-functionals for regression $\boldsymbol{\beta}(\mathrm{F})$ and scale $\sigma(\mathrm{F})$ are defined to be the solutions to the problem of minimizing

$$\mathrm{L}(\boldsymbol{\beta}, \sigma) = \int \rho(\frac{y - \boldsymbol{x}'\boldsymbol{\beta}}{\sigma})\mathrm{dF}(\boldsymbol{x}, y) + \log(\sigma) \tag{3.1}$$

over all $\boldsymbol{\beta} \in \mathbb{R}^p$ and $\sigma \in \mathbb{R}^+$, s.t.

$$\int \rho(\frac{y - \boldsymbol{x}'\boldsymbol{\beta}}{\sigma})\mathrm{dF}(\boldsymbol{x}, y) \leq \varepsilon\rho(\infty). \tag{3.2}$$

The next theorem establishes the existence of the CM-functionals, provided the following condition on $(\mathbf{x}, \mathrm{y})$ holds.

Condition 3.2
(a) $\forall \boldsymbol{\beta} \in \mathbb{R}^p$, $P\{\mathrm{y} = \mathbf{x}'\boldsymbol{\beta}\} < 1 - \varepsilon$.
(b) $\forall \boldsymbol{\beta} \in \mathbb{R}^p$, $\boldsymbol{\beta} \neq \mathbf{o}$, $P\{\mathbf{x}'\boldsymbol{\beta} = 0\} < 1 - \varepsilon$.

Theorem 3.1 *Under Conditions 3.1 and 3.2 there exists a $(\boldsymbol{\beta}_0, \sigma_0) \in \mathbb{R}^p \times \mathbb{R}^+$ which minimizes $L(\boldsymbol{\beta}, \sigma)$ s.t. (3.2). Furthermore, the set of all such $(\boldsymbol{\beta}_0, \sigma_0)$ is bounded away from the boundary set of $\mathbb{R}^p \times \mathbb{R}^+$.*

PROOF Let Θ denote $\mathbb{R}^p \times \mathbb{R}^+$ and $\partial\Theta$ denote the boundary set of Θ. Condition 3.1 on ρ implies that if $(\boldsymbol{\beta}, \sigma) \in \Theta$ then $\mathrm{L}(\boldsymbol{\beta}, \sigma)$ is finite. Moreover, for any pair $(\boldsymbol{\beta}, \sigma) \in \Theta$ we can find an $a \in \mathbb{R}^+$, a sufficiently large such that the pair $(\boldsymbol{\beta}, a\sigma) \in \Theta$ satisfies constraint (3.2). Thus, to complete the proof of existence and to show that the set of all solutions are bounded away from the boundary of Θ we only need to show that if a sequence of solutions $(\boldsymbol{\beta}_k, \sigma_k)$ converge to $\partial\Theta$, then either $\mathrm{L}(\boldsymbol{\beta}_k, \sigma_k) \to \infty$ or the constraint (3.2) is not met for large k.

If constraint (3.2) holds for some $(\boldsymbol{\beta}_k, \sigma_k)$, then for any $s_0 > 0$

$$\varepsilon\rho(\infty) \geq \int_{\mathrm{A}_k} \rho(\frac{y - \boldsymbol{x}'\boldsymbol{\beta}_k}{\sigma_k})\mathrm{dF}(\boldsymbol{x}, y) \geq \rho(s_0)\,\mathrm{Pr}\Big\{\frac{|\mathrm{y} - \mathbf{x}'\boldsymbol{\beta}_k|}{\sigma_k} \geq s_0\Big\}$$

where $\mathrm{A}_k = \{\frac{|\mathrm{y} - \mathbf{x}'\boldsymbol{\beta}_k|}{\sigma_k} \geq s_0\}$. Thus, if constraint (3.2) holds, then for any $\delta > 0$, we can choose $0 < s < \infty$ so that $\rho(s) \geq \frac{\varepsilon}{\varepsilon+\delta}\rho(\infty)$, and hence

$$\mathrm{Pr}\{|\mathrm{y} - \mathbf{x}'\boldsymbol{\beta}_k| < \sigma_k s\} \geq 1 - \varepsilon - \delta \tag{3.3}$$

(a) If $\sigma_k \to \infty$, then $\mathrm{L}(\boldsymbol{\beta}_k, \sigma_k) \to \infty$, and so σ_k is bounded from above.
(b) Suppose $\|\boldsymbol{\beta}_k\| \to \infty$. Express $\boldsymbol{\beta}_k = \|\boldsymbol{\beta}_k\|\boldsymbol{\gamma}_k$. Without loss of generality we can assume $\boldsymbol{\gamma}_k \to \boldsymbol{\gamma}$, since if not, then we can replace $\boldsymbol{\beta}_k$ with a

subsequence which satisfies this assumption. We then have

$$\Pr\left\{\frac{|\mathrm{y}-\mathbf{x}'\boldsymbol{\beta}_k|}{\sigma_k} < s\right\} \leq \Pr\{\mathbf{x}'\boldsymbol{\gamma}=0\} + \Pr\left\{\frac{|\mathrm{y}-\mathbf{x}'\boldsymbol{\beta}_k|}{\sigma_k} < s, |\mathbf{x}'\boldsymbol{\gamma}| \neq 0\right\}$$

with the last term $\to 0$, since σ_k is bounded. By Condition 3.2 (b), this contradicts (3.3).

(c) Suppose $\sigma_k \to 0$. Since we know $\|\boldsymbol{\beta}_k\|$ is bounded, we can assume w.l.o.g. $\boldsymbol{\beta}_k \to \boldsymbol{\beta}$. Thus

$$\Pr\{|\mathrm{y}-\mathbf{x}'\boldsymbol{\beta}_k| < \sigma_k s\} \leq \Pr\{\mathrm{y}=\mathbf{x}'\boldsymbol{\beta}\} + \Pr\{|\mathrm{y}-\mathbf{x}'\boldsymbol{\beta}_k| < \sigma_k s,\ \mathrm{y} \neq \mathbf{x}'\boldsymbol{\beta}\}$$

with the last term $\to 0$. By Condition 3.2 (a), this contradicts (3.3). ////

If F_n represents the empirical distribution of $\{(\boldsymbol{x}_i, y_i),\ i=1,\cdots,n\}$, then the CM-functionals at F_n correspond to the CM-estimates for the data set. For $n \geq p$, a sample is said to be in *general position for regression* if only one hyperplane passes through a given subset of size p and if no two such hyperplanes are the same for different subsets of size p. This is equivalent to having $C_{1,n}=p$ and $C_{2,n}=p-1$ where

$$C_{1,n} = \max_{\beta\in\mathbb{R}^p} \#\{i \mid y_i = \boldsymbol{x}_i'\boldsymbol{\beta},\ 1 \leq i \leq n\}, \tag{3.4}$$

$$C_{2,n} = \max_{\beta\in\mathbb{R}^p, \beta\neq\mathbf{0}} \#\{i \mid \boldsymbol{x}_i'\boldsymbol{\beta}=0,\ 1 \leq i \leq n\}. \tag{3.5}$$

If the sample is in general position for regression, then Theorem 3.1 implies the CM-estimates $(\boldsymbol{\beta}_n, \sigma_n)$ exist for $n > \frac{p}{1-\varepsilon}$. In practice, it is feasible that $C_{1,n}=p$, but if some design points are duplicated then $C_{2,n} > p-1$. In such cases, Theorem 3.1 implies the CM-estimates $(\boldsymbol{\beta}_n, \sigma_n)$ exist for $n > \frac{C_{2,n}}{(1-\varepsilon)}$.

Condition 3.2 (a) is not only a sufficient condition for the existence of the CM-functionals, but the next theorem shows that it is also necessary. This theorem also plays a role in the next section in obtaining the breakdown point of the CM-estimates. We note from the proof that non-existence of the CM-functionals essentially means that $\sigma(\mathrm{F})=0$.

Theorem 3.2 *Under Condition 3.1, if Condition 3.2(a) does not hold, then there does not exist* $(\boldsymbol{\beta}_0, \sigma_0) \in \mathbb{R}^p \times \mathbb{R}^+$ *which minimizes* $L(\boldsymbol{\beta}, \sigma)$ *s.t. (3.2).*

PROOF Suppose Condition 3.2(a) does not hold, then there exists a $\boldsymbol{\beta} \in \mathbb{R}^p$ such that for all $\sigma \in \mathbb{R}^+$

$$\begin{aligned}\int \rho\left(\frac{y-\boldsymbol{x}'\boldsymbol{\beta}}{\sigma}\right)d\mathrm{F}(\boldsymbol{x}, y) &= \int_{y\neq x'\beta} \rho\left(\frac{y-\boldsymbol{x}'\boldsymbol{\beta}}{\sigma}\right)d\mathrm{F}(\boldsymbol{x}, y)\\ &\leq \rho(\infty)\Pr\{\mathrm{y}\neq\mathbf{x}'\boldsymbol{\beta}\} \leq \rho(\infty)\,\varepsilon.\end{aligned}$$

That is, there exists a $\boldsymbol{\beta}$ such that constraint (3.2) holds for all $\sigma \in \mathbb{R}^+$. However, $\mathrm{L}(\boldsymbol{\beta}, \sigma) \to -\infty$ as $\sigma \to 0$. ////

As with the S-functionals, establishing uniqueness of the CM-functionals requires stronger conditions on F. The conditions on F below do not apply to the empirical distribution function F_n and we leave the difficult problem of establishing the uniqueness of the CM-estimates open.

Condition 3.3 *There exists* $\boldsymbol{\beta}_0 \in \mathbb{R}^p$ *such that* $e = y - \mathbf{x}'\boldsymbol{\beta}_0$ *is independent of* $\mathbf{x}$. *Furthermore, the random variable e has an absolutely continuous distribution in* $\mathbb{R}$ *with a density* $g(r)$ *which is bounded, symmetric about 0, decreasing in* $|r|$, *and strictly decreasing in some neighborhood of 0.*

If Condition 3.3 holds, then one would expect the CM-functional of regression $\boldsymbol{\beta}(\mathrm{F})$ to be unique and equal to $\boldsymbol{\beta}_0$. Theorem 3.3 shows this to be true, provided we place the additional conditions on the ρ-function.

Condition 3.4 *The function* $\rho(r)$ *is continuous and there is some* $r_0 \in (0, \infty)$ *such that* $\rho(r) = \rho(\infty)$ *for* $r \geq r_0$.

Theorem 3.3 *Under Conditions 3.1, 3.2, 3.3 and 3.4, there exists a solution* $(\boldsymbol{\beta}(F), \sigma(F))$ *to the problem of minimizing (3.1) s.t. (3.2). Furthermore, any solution must be of form* $(\boldsymbol{\beta}(F), \sigma(F)) = (\boldsymbol{\beta}_0, \sigma_0)$ *for some* $0 < \sigma_0 < \infty$.

PROOF Let $\varepsilon_0 = \mathrm{E}[\rho(\frac{|y - \boldsymbol{x}'\boldsymbol{\beta}(\mathrm{F})|}{\sigma(\mathrm{F})})]$ and note by definition that $(\boldsymbol{\beta}(\mathrm{F}), \sigma(\mathrm{F}))$ minimizes σ s.t. $\mathrm{E}[\rho(\frac{|y - \boldsymbol{x}'\boldsymbol{\beta}|}{\sigma})] = \varepsilon_0 \rho(\infty)$. That is, $(\boldsymbol{\beta}(\mathrm{F}), \sigma(\mathrm{F}))$ corresponds to the S-functional for regression and scale. The theorem follows by noting that the conditions of the theorem are sufficient for showing that the S-functional is uniquely defined with $\boldsymbol{\beta}(\mathrm{F}) = \boldsymbol{\beta}_0$; see, e.g., Theorem 2 in Davies (1990). ////

Theorem 3.3 only states that the CM-functional for regression is uniquely defined under the stated conditions. The CM-estimate of scale is also uniquely defined under the conditions of the theorem provided there exists a unique value σ_0 which minimizes

$$l(\sigma; g) = \mathrm{E}[\rho(\tfrac{\mathrm{e}}{\sigma})] + \log\sigma \qquad (3.6)$$

s.t. $\sigma \geq \sigma_{\mathrm{L}}$, where σ_{L} is the solution to

$$\mathrm{E}[\rho(\tfrac{\mathrm{e}}{\sigma})] = \varepsilon\rho(\infty). \qquad (3.7)$$

This follows from the uniqueness of $\boldsymbol{\beta}(\mathrm{F})$ and the definition of CM-functional.

4 Finite Sample Breakdown Point

We use here the finite sample replacement breakdown point concept introduced by Donoho and Huber (1983). For a regression data set $\mathbf{Y}_n = \{(\boldsymbol{x}_i, y_i);\ 1 \leq i \leq n\}$, let $\hat{\boldsymbol{\beta}}(\mathbf{Y}_n)$ represent an estimator of the regression coefficients $\boldsymbol{\beta}$, and let $\hat{\sigma}(\mathbf{Y}_n)$ represent a scale statistic. Suppose that m arbitrary *bad* values $\mathbf{Y}'_m$, $m \leq n$, replace m original data points of $\mathbf{Y}_n$, producing a corrupted sample $\mathbf{Z}_n$. The data set $\mathbf{Z}_n$ contains a fraction of $\varepsilon_m = \frac{m}{n}$ contaminated values. For a given ε_m, the statistic $(\hat{\boldsymbol{\beta}}(\cdot), \hat{\sigma}(\cdot))$ is said to breakdown at $\mathbf{Y}_n$ under ε_m contamination if either $\sup_{\mathbf{Z}_n} \|\hat{\boldsymbol{\beta}}(\mathbf{Z}_n)\| = \infty$, $\sup_{\mathbf{Z}_n} \|\hat{\sigma}(\mathbf{Z}_n)\| = \infty$, $\inf_{\mathbf{Z}_n} \|\hat{\sigma}(\mathbf{Z}_n)\| = 0$, or if $(\hat{\boldsymbol{\beta}}(\mathbf{Z}_n), \hat{\sigma}(\mathbf{Z}_n))$ is nonexistent for some $\mathbf{Z}_n$. The finite sample replacement breakdown point of $(\hat{\boldsymbol{\beta}}(\cdot), \hat{\sigma}(\cdot))$ at $\mathbf{Y}_n$ is then defined to be ε_n^*, the minimum of all ε_m causing breakdown.

Theorem 4.1 below gives upper and lower bounds for the finite sample replacement breakdown point of the CM-estimates at any arbitrary data set $\mathbf{Y}_n$. The values of $C_{1,n}$ and $C_{2,n}$ are defined respectively by (3.4) and (3.5) in Section 3. For the case of $\mathbf{Y}_n$ being in general position, i.e. $C_{1,n} = p$ and $C_{2,n} = p - 1$, Theorem 4.1 gives

$$\min\{\frac{\lceil n\varepsilon \rceil}{n}, \frac{\lceil n(1-\varepsilon) - p \rceil}{n}\} \le \varepsilon_n{}^* \le \min\{\frac{\lfloor n\varepsilon \rfloor + 1}{n}, \frac{\lceil n(1-\varepsilon) - p \rceil}{n}\} \quad (4.1)$$

where $\lceil x \rceil$ (ceiling function) is the smallest integer greater than or equal to x, and $\lfloor x \rfloor$ (floor function) is the largest integer less than or equal to x. If $n\varepsilon \notin \mathbb{N}$, the upper and lower bounds in (4.1) coincide.

For Y_n in general position Rousseeuw (1984) gives an upper bound for the finite replacement breakdown point for any regression equivariant estimator of

$$\frac{1}{n}\Big(\lfloor \frac{n-p}{2} \rfloor + 1\Big). \quad (4.2)$$

This upper bound is strict if one considers only the regression statistic $\hat{\boldsymbol{\beta}}(\cdot)$. Both Siegel's repeated median (Siegel, 1982) and a variant of the LMS achieve this bound. By choosing $\varepsilon = \frac{n-p}{2n}$ in (4.1) the CM-estimators also achieve this upper bound, provided $\frac{n-p}{2}$ is not an integer. If $\frac{n-p}{2} \in \mathbb{N}$ the breakdown point is $(\frac{n-p}{2} + \frac{1}{2})/n$ for $2\varepsilon = (n-p)/n$. We now state the results for general $\mathbf{Y}_n$.

Theorem 4.1 *If ρ satisfies Condition 3.1, $0 < \varepsilon < 1$, and, for the data set $\mathbf{Y}_n$, $C_n = \min(C_{1,n}, C_{2,n}) < n(1-\varepsilon)$, then the finite sample replacement breakdown point at $\mathbf{Y}_n$ of the regression scale CM-estimate satisfies*

$$\min\{\frac{\lceil n\varepsilon \rceil}{n}, \frac{\lceil n(1-\varepsilon) - C_n \rceil}{n}\} \le \varepsilon_n{}^* \le \min\{\frac{\lfloor n\varepsilon \rfloor + 1}{n}, \frac{\lceil n(1-\varepsilon) - C_{1,n} \rceil}{n}\}.$$

PROOF (a) *Upper bound.* Suppose $m > n\varepsilon$. Consider the sequence of contaminated data sets obtained by replacing $\{(\boldsymbol{x}_j, y_j); 1 \le j \le m\}$ by $\{(\boldsymbol{x}_j, y_{j,k}); 1 \le j \le m\}$ with $|y_{j,k}| \to \infty$ as $k \to \infty$ for $1 \le j \le m$, and let $(\boldsymbol{\beta}_k, \sigma_k)$ represent the corresponding CM-estimates of the contaminated data sets. If breakdown does not occur under ε_m-contamination, then there exist constants a, v_1, v_2 such that $\|\boldsymbol{\beta}_k\| < a < \infty$ and $0 < v_1 < \sigma_k < v_2 < \infty$. This implies that $|y_{j,k} - \boldsymbol{\beta}_k' x_j|/\sigma_k \to \infty$ as $k \to \infty$, for $j = 1, 2, \cdots, m$ and so for any $\delta > 0$ there exists a large enough k so that $\mathrm{ave}_i\ \{\rho(\frac{|y_{i,k} - \boldsymbol{\beta}_k' x_i|}{\sigma_k})\} > \frac{m}{n}\rho(\infty) - \delta$, which contradicts (1.6), and so $\varepsilon_n{}^* \le (\lfloor n\varepsilon \rfloor + 1)/n$.

Now suppose $m \geq n(1-\varepsilon) - C_{1,n}$. By definition, for some $\boldsymbol{\beta} \in \mathbb{R}^p$ there exists $C_{1,n}$ data points in $\mathbf{Y}_n$ so that $y_i = \boldsymbol{\beta}' \boldsymbol{x}_i$, say $1 \leq i \leq C_{1,n}$ without loss of generality. If we replace m other data points in $\mathbf{Y}_n$ by $(y_1, \boldsymbol{x}_1)$, then $\mathbf{Z}_n$ has $C_{1,n}+m$ data points for which $y_i = \boldsymbol{\beta}' \boldsymbol{x}_i$. However, $\frac{C_{1,n}+m}{n} \geq (1-\varepsilon)$ and so by Theorem 3.2 $(\hat{\boldsymbol{\beta}}(\mathbf{Z}_n), \hat{\sigma}(\mathbf{Z}_n))$ does not exist. Thus, ${\varepsilon_n}^* \leq \frac{\lceil n(1-\varepsilon)-C_{1,n}\rceil}{n}$.

(b) *Lower bound.* Suppose $m < \min\{\lceil n\varepsilon\rceil, \lceil n(1-\varepsilon) - C_n\rceil\}$. For any ε_m-contaminated samples $\mathbf{Z}_n = \{(\boldsymbol{x}_i^*, y_i^*);\ 1 \leq i \leq n\}$, Theorem 3.1 ensures the existence of the CM-estimates $(\hat{\boldsymbol{\beta}}(\mathbf{Z}_n), \hat{\sigma}(\mathbf{Z}_n))$. Our goal is to show that for all possible ε_m-contaminated samples $\mathbf{Z}_n$, $\hat{\sigma}(\mathbf{Z}_n)$ is bounded above and below, and $\|\hat{\boldsymbol{\beta}}(\mathbf{Z}_n)\|$ is bounded above. This will imply ${\varepsilon_n}^* \geq \frac{\min\{\lceil n\varepsilon\rceil, \lceil n(1-\varepsilon)-C_n\rceil\}}{n}$.

We first show that $\hat{\sigma}(\mathbf{Z}_n)$ is bounded above over all possible $\mathbf{Z}_n$. Since $\rho(r)$ is continuous at zero, for any $\delta > 0$ there exists a large enough $\lambda < \infty$ so that $\rho(\frac{y_i}{\lambda}) < \delta$ for all $(\boldsymbol{x}_i, y_i) \in \mathbf{Y}_n$, $1 \leq i \leq n$. This implies for any $\delta > 0$, there exists a large enough $\lambda < \infty$ so that for any possible $\mathbf{Z}_n$, $\text{ave}_i\ \{\rho(\frac{y_i^*}{\lambda})\} \leq \delta + \frac{m}{n}\rho(\infty)$. Since $m < n\varepsilon$, there exists a λ, $0 < \lambda < \infty$, so that $(\mathbf{o}, \lambda)$ satisfies constraint (1.6) for any possible $\mathbf{Z}_n$. A contradiction arises if $\hat{\sigma}(\mathbf{Z}_n)$ is not bounded above since for some $\mathbf{Z}_n$ we would have $\mathrm{L}_n(\mathbf{o}, \lambda) < \mathrm{L}_n(\hat{\boldsymbol{\beta}}(\mathbf{Z}_n), \hat{\sigma}(\mathbf{Z}_n))$.

Next we show that $\|\hat{\boldsymbol{\beta}}(\mathbf{Z}_n)\|$ is bounded above over all possible $\mathbf{Z}_n$. For $r \geq 0$, define

$$k(r; \mathbf{Z}_n) = \#\Big\{i \ \Big|\ \frac{|y_i^* - \hat{\boldsymbol{\beta}}'(\mathbf{Z}_n)\boldsymbol{x}_i^*|}{\hat{\sigma}(\mathbf{Z}_n)} \leq r\Big\}.$$

By constraint (1.6), $\frac{n-k(r;\mathbf{Z}_n)}{n}\ \rho(r) \leq \varepsilon\rho(\infty)$ and so for large enough r, say r_u, $k(r; \mathbf{Z}_n) \geq \lceil n(1-\varepsilon)\rceil$ for any $\mathbf{Z}_n$. Since $m < \lceil n(1-\varepsilon) - C_n\rceil$, at least $C_n + 1$ values for which $|y_i^* - \hat{\boldsymbol{\beta}}'(\mathbf{Z}_n)\boldsymbol{x}_i^*| \leq r_u\hat{\sigma}(\mathbf{Z}_n)$ must be from the original data set $\mathbf{Y}_n$, and so

$$\#\{i \ \big|\ |y_i - \hat{\boldsymbol{\beta}}'(\mathbf{Z}_n)\boldsymbol{x}_i| \leq r_u\hat{\sigma}(\mathbf{Z}_n)\} \geq C_n + 1. \tag{4.3}$$

Since $\hat{\sigma}(\mathbf{Z}_n)$ is bounded, $\hat{\boldsymbol{\beta}}(\mathbf{Z}_n)$ must be bounded. Otherwise, there must exist $\boldsymbol{\beta} \neq \mathbf{o}$ so that $\#\{i \mid \boldsymbol{\beta}'\boldsymbol{x}_i = 0\} \geq C_n + 1 \geq C_{2,n} + 1$, which contradicts the definition of $C_{2,n}$.

Finally, we note that (4.3) also implies $\hat{\sigma}(\mathbf{Z}_n)$ is bounded away from zero. Otherwise, there exists a $\boldsymbol{\beta}$ such that $\#\{i \mid y_i = \boldsymbol{\beta}'\boldsymbol{x}_i\} \geq C_n + 1 \geq C_{1,n} + 1$, which contradicts the definition of $C_{1,n}$ ////

5 Asymptotics

In this section, we study the consistency and asymptotic normality of the CM-estimates as well as the influence functions of the CM-functionals. As in Section 3, we let F represent the joint distribution of the $(p+1)$-dimensional random vector $(\boldsymbol{x}, y)$. Throughout this section we assume that Condition 3.2 holds and so the existence of the CM-functional $(\boldsymbol{\beta}(\mathrm{F}), \sigma(\mathrm{F}))$ is assured by Theorem 3.1. We also need $(\boldsymbol{\beta}(\mathrm{F}), \sigma(\mathrm{F}))$ to be uniquely defined, but rather than assuming the conditions used to prove the uniqueness of the CM-functionals in Section 3, we assume the following general uniqueness condition.

Condition 5.1 *The CM-functional* $(\boldsymbol{\beta}(F), \sigma(F))$ *is uniquely defined at* F.

Given uniqueness, we show in Theorem 5.1 that the CM-functionals are weakly continuous and from this the strong consistency of the CM-estimates follow. To establish weak continuity we first strengthen the conditions on ρ.

Condition 5.2 *For* $r \geq 0$, $\rho(r)$ *is bounded, continuous and strictly increasing for* $\rho(r) < \rho(\infty)$. *Furthermore,* $\rho(0) = 0$ *and* $\rho(-r) = \rho(r)$.

Theorem 5.1 *Let* F_k *be a sequence of distribution functions on* $\mathbb{R}^{p+1}$, *such that* $\mathrm{F}_k \to \mathrm{F}$ *weakly. Under Conditions 3.2, 5.1 and 5.2,* $(\boldsymbol{\beta}(\mathrm{F}_k), \sigma(\mathrm{F}_k))$ *exists for large* k *and, for any sequence of solutions* $(\boldsymbol{\beta}(\mathrm{F}_k), \sigma(\mathrm{F}_k))$ *satisfying the definition of the CM-functionals at* F_k, $(\boldsymbol{\beta}(\mathrm{F}_k), \sigma(\mathrm{F}_k)) \longrightarrow (\boldsymbol{\beta}(\mathrm{F}), \sigma(\mathrm{F}))$.

Before proving Theorem 5.1 we prove the following lemma.

Lemma 5.2 *Let* $(\mathbf{x}_k, \mathrm{y}_k)$ *be a sequence of random vectors in* $\mathbb{R}^{p+1}$, *such that* $(\mathbf{x}_k, \mathrm{y}_k) \xrightarrow{\mathrm{d}} (\mathbf{x}, \mathrm{y})$. *Denoting* $\boldsymbol{\theta}_k = (\boldsymbol{\beta}_k, \sigma_k) \in \Theta$ *let* $\boldsymbol{\theta}_k \to \boldsymbol{\theta} = (\boldsymbol{\beta}, \sigma) \in \Theta$. *Then for* $h: \mathbb{R} \to \mathbb{R}$ *bounded and continuous,*

$$\mathrm{E}\Big\{h\Big(\frac{\mathrm{y}_k - \mathbf{x}_k' \boldsymbol{\beta}_k}{\sigma_k}\Big)\Big\} \longrightarrow \mathrm{E}\Big\{h\Big(\frac{\mathrm{y} - \mathbf{x}' \boldsymbol{\beta}}{\sigma}\Big)\Big\}.$$

Proof of Lemma Let $\mathrm{d}_k = \dfrac{\mathrm{y}_k - \mathbf{x}_k' \boldsymbol{\beta}_k}{\sigma_k}$ and $\mathrm{d} = \dfrac{\mathrm{y} - \mathbf{x}' \boldsymbol{\beta}}{\sigma}$, and so $\mathrm{d}_k \to \mathrm{d}$ in distribution. This implies that $\mathrm{E}[h(\mathrm{d}_k)] \longrightarrow \mathrm{E}[h(\mathrm{d})]$ for any bounded continuous function h. ////

Proof of Theorem 5.1 Since $\limsup \mathrm{P}\{\mathrm{y}_k = \mathbf{x}_k' \boldsymbol{\beta}\} \leq \mathrm{P}\{\mathrm{y} = \mathbf{x}' \boldsymbol{\beta}\}$ and $\limsup \mathrm{P}\{\mathbf{x}_k' \boldsymbol{\beta} = 0\} \leq \mathrm{P}\{\mathbf{x}' \boldsymbol{\beta} = 0\}$, we note that Condition 3.2 also holds for F_k for large k, say $k \geq K$, and therefore $\boldsymbol{\theta}_k = (\boldsymbol{\beta}(\mathrm{F}_k), \sigma(\mathrm{F}_k))$ exists for $k \geq K$. Furthermore, the set $\{\boldsymbol{\theta}_k, k \geq K\}$ is contained in a compact subset of Θ, because otherwise there would exist a subsequence of $\boldsymbol{\theta}_k$ diverging to $\partial\Theta$. Using arguments analogous used in proof of Theorem 3.1, it can be shown that such a subsequence is not possible. Thus, to show that $\boldsymbol{\theta}_k$ converges to $(\boldsymbol{\beta}(\mathrm{F}), \sigma(\mathrm{F}))$, it is sufficient to show that: $\boldsymbol{\theta}_k \to \boldsymbol{\theta}_o = (\boldsymbol{\beta}_o, \sigma_o) \in \Theta$ implies that $\boldsymbol{\theta}_o = \boldsymbol{\theta}(\mathrm{F}) = (\boldsymbol{\beta}(\mathrm{F}), \sigma(\mathrm{F}))$.
Define

$$\mathrm{c}(\boldsymbol{\beta}, \sigma) = \int \rho\Big(\frac{y - \boldsymbol{\beta}' \boldsymbol{x}}{\sigma}\Big) \mathrm{dF}(\boldsymbol{x}, y) \Big/ \rho(\infty), \tag{5.1}$$

and let $\mathrm{c}_k(\boldsymbol{\beta}, \sigma)$ and $\mathrm{L}_{(k)}(\boldsymbol{\beta}, \sigma)$ be defined by replacing F by F_k in (5.1) and (3.1) respectively. Note that $\boldsymbol{\theta}_k$ minimizes $\mathrm{L}_{(k)}(\boldsymbol{\theta})$ under the constraint $\mathrm{c}_k(\boldsymbol{\theta}) \leq \varepsilon$, and $\boldsymbol{\theta}(\mathrm{F})$ mimimizes $\mathrm{L}(\boldsymbol{\theta})$ under the constraint $\mathrm{c}(\boldsymbol{\theta}) \leq \varepsilon$.

To complete the proof, assume $\boldsymbol{\theta}_k \longrightarrow \boldsymbol{\theta}_0 \neq \boldsymbol{\theta}(\mathrm{F})$, and let us show that a contradiction arises. Since $\boldsymbol{\theta}(\mathrm{F})$ is unique, then either $\mathrm{L}(\boldsymbol{\theta}_0) > \mathrm{L}(\boldsymbol{\theta}(\mathrm{F}))$ or

$c(\boldsymbol{\theta}_0) > \varepsilon$. By Lemma 5.2, $c_k(\boldsymbol{\theta}_0) \to c(\boldsymbol{\theta}_0)$ and so since $c_k(\boldsymbol{\theta}_0) \leq \varepsilon$ it is not possible for $c(\boldsymbol{\theta}_0) > \varepsilon$. Thus, if $\boldsymbol{\theta}_k \to \boldsymbol{\theta}_0 \neq \boldsymbol{\theta}(F)$, then $L(\boldsymbol{\theta}_0) > L(\boldsymbol{\theta}(F))$. To show this cannot be, the remainder of the proof is divided into two cases.

CASE *A* : $c(\boldsymbol{\theta}(F)) < \varepsilon$. By Lemma 5.2 $L_{(k)}(\boldsymbol{\theta}_k) \to L(\boldsymbol{\theta}_0)$, $L_{(k)}(\boldsymbol{\theta}(F)) \to L(\boldsymbol{\theta}(F))$, and $c_k(\boldsymbol{\theta}(F)) \to c(\boldsymbol{\theta}(F)) < \varepsilon$. If $L(\boldsymbol{\theta}_0) > L(\boldsymbol{\theta}(F))$, then for k sufficiently large $L_{(k)}(\boldsymbol{\theta}_k) > L_{(k)}(\boldsymbol{\theta}(F))$ and also $c_k(\boldsymbol{\theta}(F)) < \varepsilon$, which contradicts the definition of the CM-estimate at F_k.

CASE *B* : $c(\boldsymbol{\theta}(F)) = \varepsilon$. For $\eta \neq 0$, let $\boldsymbol{\theta}_\eta(F) = (\boldsymbol{\beta}(F), (1+\eta)\sigma(F))$. Condition 5.2 implies that for any $\eta > 0$, $c(\boldsymbol{\theta}_\eta(F)) < \varepsilon$ unless $y - \mathbf{x}'\boldsymbol{\beta}(F) = 0$ or $|y - \mathbf{x}'\boldsymbol{\beta}(F)| > r_0\sigma(F)$ with probability 1, where r_0 is such that $\rho(r) = \rho(\infty)$ for $r \geq r_0$. However, this latter condition cannot hold since otherwise, by Condition 3.2, $c(\boldsymbol{\theta}(F)) = P\{y - \mathbf{x}'\boldsymbol{\beta}(F) \neq 0\} > \varepsilon$. Thus, $c(\boldsymbol{\theta}_\eta(F)) < \varepsilon$ for $\eta > 0$. Now, by Lemma 5.2, we have $L(\boldsymbol{\theta}_\eta(F)) \to L(\boldsymbol{\theta}(F))$ as $\eta \to 0$. If $L(\boldsymbol{\theta}_0) > L(\boldsymbol{\theta}(F))$, then we can thus choose η sufficiently small so that $L(\boldsymbol{\theta}_0) > L(\boldsymbol{\theta}_\eta(F))$. The proof then proceeds as in case A, but with $\boldsymbol{\theta}_\eta(F)$ replacing $\boldsymbol{\theta}(F)$. ////

Strong consistency of the CM-estimators follows from Theorem 5.1 since the empirical distribution F_n of the sample $\{(\boldsymbol{x}_1, y_1), \cdots, (\boldsymbol{x}_n, y_n)\}$ converges almost surely to the underlying distribution function of $(\mathbf{x}, y)$. Thus, as $n \to \infty$,

$$(\boldsymbol{\beta}_n, \sigma_n) \xrightarrow{\text{a.s.}} (\boldsymbol{\beta}(F), \sigma(F))$$

for any sequence $(\boldsymbol{\beta}_n, \sigma_n)$ satisfying the definition of the CM-estimators.

Given weak continuity and consistency of the CM-estimators the influence functions and asymptotic distributions follow by using standard expansion arguments on the estimating equations. For simplicity, we restrict our attention to the CM-estimates of regression $\boldsymbol{\beta}_n$ and the functionals $\boldsymbol{\beta}(F)$. For a discussion of the joint local properties of the CM-estimates of regression and scale see Mendes (1994). Similar to (1.7), when ρ is differentiable the CM-functional $\boldsymbol{\beta}(F)$ satisfies the implicit equations

$$\int \psi\left(\frac{y - \boldsymbol{x}'\boldsymbol{\beta}}{\sigma_0}\right) \boldsymbol{x}\, dF(\boldsymbol{x}, y) = \mathbf{o} \tag{5.2}$$

with σ_0 being set equal to the CM-functional of scale $\sigma(F)$, and again $\psi(r) = \rho'(r)$. To insure that expansions of the estimating equations are valid, we again strengthen the conditions on $\rho(r)$ and the conditions on F.

Condition 5.3 *The function $\rho(r)$ is even with $\rho(0) = 0$, and $\rho(r)$ has a continuous second derivative. Furthermore, there exists an r_0, $0 < r_0 < \infty$, such that $\rho(r)$ is strictly increasing for $0 \leq r < r_0$ and $\rho(r) = \rho(\infty)$ for $r \geq r_0$.*

Condition 5.4 *Assume Condition 3.3 holds and let $z = \dfrac{y - \mathbf{x}'\boldsymbol{\beta}(F)}{\sigma(F)}$. Furthermore,*

(a) $A(\psi, F) = E[\psi^2(z)]$ exists,
(b) $B(\psi, F) = E[\psi'(z)]$ exists and is not equal to 0, and
(c) $V(F) = E[\mathbf{x}\mathbf{x}']$ exists and is nonsingular.

Theorem 5.3 *Let $(\mathrm{x}_i, \mathrm{y}_i), i = 1, 2, \cdots, n$, be i.i.d. with distribution F, and assume Conditions 5.1, 5.3 and 5.4 hold. If $(\boldsymbol{\beta}_n, \sigma_n)$ is a sequence of estimates satisfying the definition of CM-estimates, then*

$$\sqrt{n}(\boldsymbol{\beta}_n - \boldsymbol{\beta}_0) \xrightarrow{d} \mathcal{N}_p\Big(\mathbf{o}, \sigma_0^2 \frac{A(\psi, F)}{B^2(\psi, F)} V(F)^{-1}\Big) \tag{5.3}$$

where $A(\psi, F)$, $B(\psi, F)$, $V(F)$ are as in Condition 5.4, $\sigma_0 = \sigma(F)$ $\boldsymbol{\beta}_0 = \boldsymbol{\beta}(F)$.

Theorem 5.3 follows directly from Theorem 4.1 in Yohai (1987). This latter theorem gives the asymptotic normality for general M-estimates of regression of the form (3). Similarly as in Yohai (1987), we have the following result for the influence function of the CM-functional for regression. The influence function of the regression CM-functional at F as a function of $(\boldsymbol{x}_0, y_0) \in \mathbb{R}^{p+1}$ is defined by

$$\mathrm{IF}_{\beta}(\boldsymbol{x}_0, y_0; \mathrm{F}) = \lim_{\varepsilon \to 0+} \frac{\boldsymbol{\beta}((1-\varepsilon)\mathrm{F} + \varepsilon\delta_0) - \boldsymbol{\beta}(\mathrm{F})}{\varepsilon},$$

where δ_0 represents the point mass distribution at $(\boldsymbol{x}_0, y_0)$.

Theorem 5.4 *Assume Condition 5.3 holds for $\rho(r)$ and Conditions 5.1 and 5.4 hold for F. The influence function of the regression CM-functional at F as a function of $(\boldsymbol{x}_0, y_0)$ is given by*

$$IF_{\beta}(\boldsymbol{x}_0, y_0; F) = \frac{\sigma_0 \psi\big(\frac{y_0 - \boldsymbol{x}_0' \boldsymbol{\beta}_0}{\sigma_0}\big)}{B(\psi, F)} V(F)^{-1} \boldsymbol{x}_0 \tag{5.4}$$

where $B(\psi, F)$, $V(F)$ are as in Condition 5.4, and $\sigma_0 = \sigma(F)$, $\boldsymbol{\beta}_0 = \boldsymbol{\beta}(F)$.

Since the expression for the asymptotic covariance matrix in (5.3) and the expression for the influence function (5.4) arise simply from the M-estimating equations for regression, these expressions can also be applied to the S-estimates and MM-estimates of regression, see e.g. Yohai (1987). For the S-estimates, we set $\psi(r) = \rho'(r)$ and $\sigma_0 = \sigma_{\mathrm{L}}$ in (5.3) and (5.4). For the M-estimates defined in Section 2 with $\rho_1(r) = \rho(\frac{r}{k_1})$ and $\rho_2(r) = \rho(\frac{r}{k_2})$ we set $\psi(r) = \rho'(r)$ and $\sigma_0 = \frac{k_2}{k_1}\sigma_{\mathrm{L}}$. The values of σ_0 used for the S-estimates and MM-estimates arise from relating the S-functionals and MM-functionals of regression to implicit equations of form (5.2).

As in Hampel et al. (1986), Section 6.3, the influence function of the regression functional in Theorem 5.4 can be expressed as a product of two factors. Specifically,

$$\mathrm{IF}_{\beta}(\boldsymbol{x}_0, y_0; \mathrm{F}) = \mathrm{IR}_{\beta}(e_0; \mathrm{F})\, \mathrm{IP}_{\beta}(\boldsymbol{x}_0; \mathrm{F})$$

where

$$\mathrm{IR}_{\beta}(e_0; \mathrm{F}) = \frac{\sigma_0 \psi(\frac{e_0}{\sigma_0})}{\mathrm{B}(\psi, \mathrm{F})}, \tag{5.5}$$

and

$$\mathrm{IP}_{\beta}(\boldsymbol{x}_0; \mathrm{F}) = \mathrm{V}(\mathrm{F})^{-1} \boldsymbol{x}_0, \tag{5.6}$$

where $e_0 = y_0 - \boldsymbol{x}_0' \boldsymbol{\beta}_0$. The scalar-valued $\mathrm{IR}_\beta(e_0; \mathrm{F})$ is usually called the *influence of the residuals* and does not depend on the distribution of $\mathbf{x}$. The vector-valued $\mathrm{IP}_\beta(\boldsymbol{x}_0; \mathrm{F})$ is usually called the *influence of position in the factor space*, and does not depend on the distribution of $\mathrm{y} - \mathbf{x}'\boldsymbol{\beta}_0$. As with S- and MM-estimates, although the influence of position in the factor space is unbounded, extreme leverage points do not cause the CM-estimates to breakdown since we know they have high breakdown points. Moreover, the IP_β does not depend upon the choice of the ρ-function for the CM-estimates. This observation, together with the observation that IR_β is bounded, suggests that the supremum of $|\mathrm{IR}_\beta|$ together with the asymptotic efficiencies of the CM-estimate can be used to help suggest a reasonable ρ-function for defining the CM-estimates. We address this problem in the next section.

6 Tuning the CM-Estimates

The breakdown point of the CM-estimates depends upon the choice of ε in (1.6) but not upon the choice of the ρ-function, provided ρ satisfies Condition 3.1. The local properties of the CM-estimates of regression, in particular its influence function and its asymptotic variance-covariance matrix, depend upon the choice of the ρ-function. One way of comparing the local robustness of CM-estimates based upon different ρ-functions is to compare their asymptotic relative efficiency over a range of distributions of interest. Another way is to compare the behavior of the corresponding influence functions along with their asymptotic relative efficiencies at a specific model of interest, usually the normal model. We use the latter approach in this section in developing a procedure for tuning the CM-estimates so that they possess reasonably local robustness properties.

Hereafter, we assume the conditions of Theorem 5.3 and Theorem 5.4 hold. Although the asymptotic variance-covariance matrix of $\boldsymbol{\beta}_n$ is matrix-valued, we note from (5.3) that the asymptotic relative efficiency of $\boldsymbol{\beta}_n$ for varying choices of ρ can be compared by simply comparing the corresponding values of the scalars

$$\alpha(\psi, \mathrm{F}) = \frac{\sigma^2(\mathrm{F})\mathrm{A}(\psi, \mathrm{F})}{\mathrm{B}^2(\psi, \mathrm{F})}. \tag{6.1}$$

From the discussion after Theorem 5.4, we recall that comparisons of influence functions of the CM-functionals of regression for varying choices of ρ can be reduced to comparisons of the real-valued influence of the residuals functions. For comparison purposes, we then use the *residual gross error sensitivities* defined by $\gamma_r^*(\psi, \mathrm{F}) = \sup|\mathrm{IR}_\beta(\boldsymbol{x}_0, y_0; \mathrm{F})|$, where the supremum is over all $(\boldsymbol{x}_0, y_0)$. From (5.5), we have

$$\gamma_r^*(\psi, \mathrm{F}) = \frac{\sigma(\mathrm{F})K(\psi)}{\mathrm{B}(\psi, \mathrm{F})} \tag{6.2}$$

where $K(\psi) = \sup_r |\psi(r)|$. Both $\alpha(\psi, \mathrm{F})$ and $\gamma_r^*(\psi, \mathrm{F})$ depend on F only through the distribution of the error term $\mathrm{e} = \mathrm{y} - \mathbf{x}'\boldsymbol{\beta}(\mathrm{F})$, and neither depends on p, the dimension of the factor space. For a scale family of error distribution with densities $g_\nu(r) = \nu^{-1} g_0(\frac{r}{\nu})$ for $\nu > 0$, we have $\sigma(\mathrm{F}_\nu) = \nu\sigma(\mathrm{F}_0)$, $\mathrm{A}(\psi, \mathrm{F}_\nu) = \mathrm{A}(\psi, \mathrm{F}_0)$ and $\mathrm{B}(\psi, \mathrm{F}_\nu) = \mathrm{B}(\psi, \mathrm{F}_0)$, where the density of the error distribution associated with F_ν and F_0 are g_ν and g_0 respectively. We then have $\alpha(\psi, \mathrm{F}_\nu) = \nu\alpha(\psi, \mathrm{F}_0)$ and $\gamma_r^*(\psi, \mathrm{F}_\nu) = \nu\gamma_r^*(\psi, \mathrm{F}_0)$. Thus, for a given scale family model for the error distribution, we view $\rho(r)$ as a good ρ-function, regardless of p and the distribution of $\mathbf{x}$, if both $\alpha(\psi, \mathrm{F}_0)$ and $\gamma_r^*(\psi, \mathrm{F}_0)$ are relatively small.

The broad problem of choosing a ρ-function from the class of all ρ-functions is not addressed here. Instead, given a ρ-function, a family indexed by a multiplicative tuning constant $c > 0$ can be generated by multiplying $\rho(r)$ by c to obtain $\rho_c(r) = c\,\rho(r)$. For a given error distribution and a given family $\{\rho_c(s), c > 0\}$ one could plot γ_r^* and α as a function of c, and then choose the tuning constant c so that γ_r^* and α are both relatively small.

A plot of γ_r^* and α can be simplified by noting that they depend on the tuning constant c only through the value of $\sigma_0 = \sigma(\mathrm{F})$, and in addition, the value of σ_L defined as the solution to (3.7) does not depend on c. Since σ_0 is not an explicit function of c, we can more easily plot γ_r^* and α as a function of $\sigma_0 \geq \sigma_\mathrm{L}$ and determine a desirable value of σ_0. Given σ_0, we can then find the value of c which produces this value of σ_0.

For $\sigma_0 \geq \sigma_\mathrm{L}$, the corresponding value of c is such that

$$c^{-1} = \mathrm{E}[\psi(\tfrac{\mathrm{e}}{\sigma_0})(\tfrac{\mathrm{e}}{\sigma_0})] \tag{6.3}$$

where again $\psi(r) = \rho'(r)$. This follows since whenever $\sigma_0 > \sigma_\mathrm{L}$, σ_0 is a critical point of $l_c(\sigma; g) = c\,\mathrm{E}[\rho(\frac{\mathrm{e}}{\sigma})] + \log(\sigma)$, and therefore must satisfy $c\,\mathrm{E}[\psi(\frac{\mathrm{e}}{\sigma_0})\frac{\mathrm{e}}{\sigma_0}] = 1$. It is easy to verify from the definition of σ_0, see (3.6) and (3.7), that σ_0 is an increasing function of c, and so to obtain $\sigma_0 = \sigma_\mathrm{L}$ we can choose any c such that $1/c \geq \mathrm{E}[\psi(\frac{\mathrm{e}}{\sigma_\mathrm{L}})(\frac{\mathrm{e}}{\sigma_\mathrm{L}})]$.

The formulas for γ_r^* and α depend upon the value of ε only through σ_L, say $\sigma_\mathrm{L}(\varepsilon)$, which is a decreasing function of ε. As noted in Section 5, the formulas for γ_r^* and α with $\sigma_0 = \sigma_\mathrm{L}(\varepsilon)$ apply to the S-estimates defined using the same value of ε and the same ρ-function. The S-estimates can be tuned by either choosing σ_0 or ε, but once one is specified, the other is determined. If one specifies ε for an S-estimate, then $\sigma_0 = \sigma_\mathrm{L}(\varepsilon)$, and if one specifies σ_0 for an S-estimate then

$$\varepsilon = \frac{\mathrm{E}[\rho(\frac{\mathrm{e}}{\sigma_0})]}{\rho(\infty)}. \tag{6.4}$$

Thus, in contrast to the CM-estimates, the breakdown point and the local properties of the S-estimates are heavily dependent.

To illustrate these ideas, we recall the class of biweighted CM-estimates discussed in Section 1. The corresponding ρ-functions are also used by Rousseeuw and Yohai (1984) in investigating the performance of the regression S-estimates. Hereafter, we set $\varepsilon = 0.5$ and unless stated otherwise we assume the error distribution is Normal$(0, \nu^2)$. As previously noted, it is sufficient to consider the special case $\nu = 1$, and so we assume the error distribution is standard normal. Reverting back to the notation of Section 1, we note that $\sigma_0 = \lambda_0 = k$ and $\sigma_L = \lambda_L = 1.5476$ for this special case.

The values of γ_r^* and α can be expressed in terms of k and linear combinations of cumulative χ-square probabilities. These expressions can be found in Mendes (1994). Figure 1 gives a plot of γ_r^* via the dotted line and a plot of eff $= \frac{1}{\alpha}$ via the solid line as functions of k. At the normal model the value of $\frac{1}{\alpha}$ represents the asymptotic efficiency of the CM-estimate of regression relative to the least squares estimate of regression. Along the top of Figure 1 are the corresponding values of the tuning constant c for a given k. The values of the residual gross sensitivity and asymptotic efficiency at $k = 1.5476$ also correspond to those of the biweighted S-estimates of regression with limiting breakdown point $\varepsilon = 0.5$. As we have anticipated, it is possible to substantially improve upon both the asymptotic relative efficiency and the residual gross error sensitivity of the biweighted S-estimator by using a biweighted CM-estimate. The residual gross error sensitivity is minimized by choosing $k = 3.420$, which corresponds to choosing the tuning constant $c = 3.2707$. For the CM-estimate with $k = 3.420$ we obtain $\gamma_r^* = 1.6421$ and eff $= 0.8467$, whereas for the S-estimate we obtain $\gamma_r^* = 2.8472$ and eff $= 0.287$. It is interesting to note that there is not too much loss in the residual gross error sensitivity if we increase the value of k above 3.420. On the other hand, by doing so we gain in efficiency. If we choose $k = 5.848$ we obtain an asymptotic efficiency of 98 % and a residual gross error sensitivity of 1.9988, which is still considerably smaller than the value $\gamma_r^* = 2.8472$ for the S-estimator. Table 1 summarizes these figures. It is

c	k	eff	γ_r^*
2.568	1.548	0.287	2.8472
3.271	3.420	0.847	1.6421
3.602	3.742	0.886	1.6531
3.901	4.000	0.91	1.6749
4.835	4.690	0.95	1.7705
6.807	5.848	0.98	1.9988

Table 1: Possible choices for k

possible to tune the S-estimates to have a value of k higher than 1.5476, but at the expense of decreasing the limiting breakdown point ε^*. Table 2 gives the values of ε^* computed using (6.3) for different choices of k for the biweighted S-estimates, along with their asymptotic efficiency and residual gross error sensitivities. The results for the asymptotic efficiencies and the breakdown points for the S-estimates correspond to the values given in Table 3 of Rousseeuw and Yohai (1984). Figure 1 can also be interpreted as plots of γ_r^* and eff $= \frac{1}{\alpha}$ as a function of $k = k_2$ under the Gaussian error model for the biweighted MM-estimates discussed in Section 2 based upon an initial S-estimate with $k_1 = 1.5476$. This

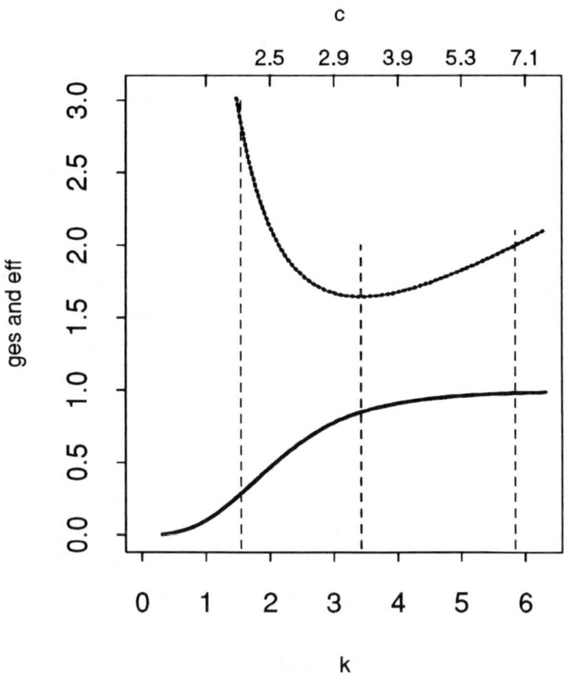

Figure 1. The dotted line represents the residual gross error sensitivity, and the solid line the asymptotic efficiency for the biweighted CM-estimates under the Gaussian model. Vertical lines are drawn at k = 1.548, k = 3.42, and k = 5.848.

follows from the remarks made after Theorem 5.4 and by noting that at the Gaussian model $\sigma_{\rm L} = k_1$ and so $\sigma_0 = k_2$ in (5.3) and (5.4). Thus, given a CM-estimate of regression there always exist an MM-estimate of regression with the same breakdown point and the same values for γ_r^* and eff $= 1/\alpha$ at the Gaussian error model. For two such estimates, however, the values of $\sigma_0 = \sigma(\mathrm{F})$ are not the same when the error distribution is not normal and so the two estimates have differing values of α and γ_r^* outside the Gaussian error model. For example, a biweighted CM-estimate with $c = 3.2707$ (and so $k = 3.420$) and an MM-estimate with $k_2 = 3.420$ both produce the values $\gamma_r^* = 1.6421$ and eff $= 0.847$ at the normal error model. However, at a Cauchy error model, the biweighted CM-estimate with $c = 3.2707$ yields $\sigma_0 = 2.41$ and this gives

S-estimator	ε^*	eff	γ_r^*
$k = 1.548$	0.50	0.287	2.84
$k = 1.988$	0.40	0.462	2.14
$k = 2.560$	0.30	0.661	1.77
$k = 3.420$	0.20	0.847	1.64
$k = 5.182$	0.10	0.966	1.86

Table 2: Asymptotic Efficiency Versus Breakdown Point

$\gamma_r^* = 2.62$ and eff $= \frac{2}{\alpha} = 0.87$. For this case, the biweighted MM-estimate with $k_2 = 3.420$ yields $\sigma_0 = 5.326$ and this gives $\gamma_r^* = 2.85$ and eff $= \frac{2}{\alpha} = 0.81$. At Cauchy model the value of $\frac{2}{\alpha}$ represents the asymptotic efficiency of the CM- or MM-estimate of regression relative to the maximum likelihood estimate at the Cauchy error model.

Hereafter we assume the error distribution is Cauchy. Figure 2 gives a plot of γ_r^* via the dotted line and a plot of eff via the solid line as functions of σ_0. The value of σ_L is now 2.41 and so the values of γ_r^* and eff at 2.41 correspond to those of the biweighted S-estimate with limiting breakdown point of 0.5. The value of σ_0 for the biweighted MM-estimate is $\sigma_0 = \frac{2.41\ k_2}{1.5476} = 1.5578\ k_2$, whereas the value of σ_0 for the biweighted CM-estimate depends on c implicitly, see (3.6) and (3.7). Table 3 displays some values of γ_r^* and eff for the biweighted CM- and MM-estimates which are asymptotically equivalent at the normal error model, i.e. k for the CM-estimate and k_2 for the MM-estimate are taken to be the same. In general, σ_0 tends to be smaller for the CM-estimates than it is for the MM-estimates and overall the CM-estimates behave better than the MM-estimates at the Cauchy error model. Summarizing, we see that properly tuned the CM-estimates behave more like the S-estimates at the Cauchy error model and more like the MM-estimates at the Gaussian error model. Finally, after

CM				*MM*			
k	σ_0	γ_r^*	eff	k_2	σ_0	γ_r^*	eff
1.548	2.41	2.62	0.87	1.548	2.41	2.62	0.87
3.420	2.41	2.62	0.87	3.420	5.33	2.85	0.81
3.742	3.00	2.55	0.90	3.742	5.83	2.96	0.78
4.000	4.00	2.62	0.89	4.000	6.23	3.04	0.76
4.690	6.25	3.05	0.76	4.690	7.30	3.31	0.70
5.848	9.11	3.77	0.61	5.848	9.11	3.77	0.61

Table 3: γ_r^* and eff at the Cauchy distribution

examining Figures 1 and 2, in order to protect against heavy tails, to maintain

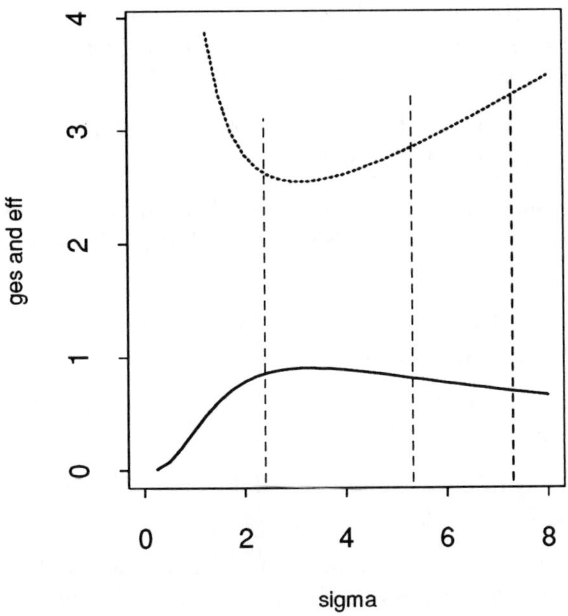

Figure 2. The dotted line represents the residual gross error sensitivity, and the solid line the asymptotic efficiency for the biweighted CM-estimates under the Cauchy model. Vertical lines are drawn at k = 2.41, k = 5.33, and k = 7.30.

reasonable asymptotic efficiencies (85%–91%) at the normal error model, and low residual gross error sensitivities, we recommend tuning the biweighted CM-estimates using c between 3.27 and 3.90. This is summarized in Table 4.

	NORMAL	CAUCHY
eff	0.85 — 0.91	0.87 $\uparrow$ 0.90 $\downarrow$ 0.89
γ_r^*	1.64 — 1.67	2.62 $\downarrow$ 2.55 $\uparrow$ 2.62

Table 4: Asymptotic results for the CM-functionals when c ranges from 3.27 to 3.90

References

[1] Davies, P.L. (1990): The asymptotics of S-estimators in the linear regression model. *Ann. Statist.* **18** 1651–1292.

[2] Donoho, D.L. and Huber, P.J. (1983): The notion of breakdown point. In *Festschrift for Erich L. Lehmann* (P.J. Bickel, K.A. Doksum, J.L. Hodges Jr., eds.), 157–184. Wadsworth, Belmont, Calif.

[3] Hampel, F.R. (1974): The influence curve and its role in robust estimation. *J. Amer. Statist. Assoc.* **69** 383–393.

[4] Hampel, F.R., Ronchetti, E.M., Rousseeuw, P.J. and Stahel, W.A. (1986): *Robust Statistics—The Approach Based On Influence Functions.* John Wiley, New York.

[5] Hoagin, D.C., Mosteller, F. and Tukey, J.W. (editors) (1983): *Understanding Robust and Exploratory Data Analysis.* J. Willey & Sons, Inc.

[6] Huber, P.J. (1973): Robust regression: Asymptotics, conjectures and Monte Carlo. *Ann. Statist.* **1** 799–821.

[7] Huber, P.J. (1984): Finite sample breakdown of M- and P-estimators. *Ann. Statist.* **12**, No. 1, 119–126.

[8] Maronna,R., Bustos,O.H. and Yohai, V.J. (1979): Bias- and efficiency- robustness of general M-estimators for regression with random carriers. In *Smoothing Techniques for Curve Estimation* (T. Gasser and M. Rosenblatt, eds.), Springer, New York.

[9] Mendes, B. (1994): *Constrained M-estimation for Linear Regression Models.* Ph.D. Thesis, Rutgers Univ.

[10] Rousseeuw, P.J. (1984): Least median of squares regression. *J. Amer. Statist. Assoc.* **79** 871–880.

[11] Rousseeuw, P.J. and Leroy, A.M. (1987): *Robust Regression and Outlier Detection.* John Wiley, New York.

[12] Rousseeuw, P.J. and Yohai, V. (1984): Robust regression by means of S-estimators. *Robust and Nonlinear time Series Analysis. Lecture Notes in Statist.* #26 256–272. Springer-Verlag, New York.

[13] Siegel, A.F. (1982): Robust regression using repeated medians. *Biometrika* **69** 242–244.

[14] Kent, J. and Tyler, D.E. (1993): *Constrained M-estimation for Multivariate Location and Covariance Matrices.* Unpublished manuscript.

[15] Yohai, V.J. (1987): High breakdown point and high efficiency robust estimates for regression. *Ann. Statist.* **15** 642–656.

Inference for the Direction of the Larger of Two Eigenvectors: The Case of Circular Elongation

Stephan Morgenthaler *
ETH Lausanne

John W. Tukey †
Princeton University, New Jersey

Abstract

The paper discusses interval estimation of the dominant eigenvector for two-dimensional data. We derive adjustments for both the classical asymptotic formula of Anderson (1963) and two jackknife estimators and exhibit the finite sample behavior of the resulting interval (wedge) estimators in a variety of situations, including nonGaussian and non-circular parent distributions.

Key words and phrases: Interval estimation, eigendirections, jackknife, finite sample adjustments, bivariate, nonGaussian bivariate distributions, nonelliptical bivariate distributions, cross-t, circ-t, exx-t.
AMS subject classifications: 62H25, 62H10, 62G35.

1 Introduction

The simplest truly multivariate thing seems to be the *direction* of eigenvectors—for principal components (and for what can be reduced to principal components). The simplest instance is the direction of the larger-of-two eigenvectors (larger = larger eigenvalue). Asymptotic results were given by Anderson in 1963 in the case of sampling from a normal population. His results were extended to non-normal populations, in particular the elliptical case, by various authors. Tyler (1981) gives broadly applicable general results. Jackknife technology is applicable, but is of course also, hopefully to a lesser extent, asymptotic. What is needed is a technique that gives confidence angles—confidence wedges—with close to nominal coverage over as wide a range of parameters and distributions as is reasonably feasible. This almost certainly implies some finite-sample (and finite eigenvalue-ratio) adjustments for either conventional asymptotic or resampling techniques, and will prove to require radGaussianizing techniques as well. By

* Supported by the grant 20-37150-93 from the Swiss National Science Foundation
† Supported by a grant from the Army Research Office

this last term we mean methods that transform a bivariate point cloud towards the Gaussian idol.

The practical problem requires consideration of:

- observed sample size
- observed ratio of eigenvalues
- assumed shape of distribution

and its treatment involves:

- choice of calculation for finite angles
- choice of classical asymptotic or one of several jackknives

and, usually,

- choice of radGaussianization technique,

as well as

- utilization of an appropriate finiteness correction, to allow for finiteness of both sample size and eigenvalue ratio.

We proceed next to the recommended answers and after expounding them—and providing some structure for shapes, jackknives, and radGaussianization—we turn, separately, to each of several choices of confidence intervals to establish and explain the facts underlying our recommendations. With the usual exceptions

- very large Gaussian samples for classical asymptotic intervals
- very large samples from most shapes for jackknife intervals

any available facts must come from simulations of one sort or another, either pure experimental sampling or carefully swindled Monte Carlo.

Some preliminary results were reported, as illustrative of general principles, in Tukey (1993).

PART A: SHAPES, JACKKNIFES, RADGAUSSIANIZATIONS

2 Shapes

We need to look at a variety of bivariate distribution shapes if we expect our results to be widely useful. It seems unlikely, based on what we know about univariate situations, that irregularity is likely to be a major cause of problems, especially since we are not likely to be concerned with sample sizes below 50 (or below 100). (Someone may sometime want to look at relatively coarse lattice distibutions, where even posing the problem correctly will be a major issue.) This leaves skewness and elongation as the major challenges from one-dimensional causes, and multipole structure (see below) as the most obvious two-dimensional consideration. We shall set skewness and multipolarity aside, so far as the present paper goes, partially in view of its possible cure by reexpression, partially because we find it unclear how best to model it, and partially because we have little experience with it in practical examples. Accordingly we shall focus on living with elongation of our bivariate distributions.

Bivariate Gaussians differ from circular bivariate Gaussians by only the aspect (part of which) we are to study—stretch in some one direction compared to a direction at right angles. We are directly concerned only with the direction, but we will need to utilize the apparent amount of stretch in our calculations.

Circular-t_ν (circ-t_ν) distributions can be generated by dividing circularly distributed 2-dimensional Gaussian vectors (origin = center of distribution) by a chosen multiple of a χ_ν distribution.

Cross-t_ν distributions arise when the u and v components of a circular bivariate Gaussian are divided by (matched multiples of) separate, independent χ_ν distributions.

Exx-t_ν distributions arise when $u+v$ and $u-v$ are similarly separately divided. The result is a 45° rotation of a cross-t_ν distribution.

In general terms circular-t_ν shows elongation, equal in all directions, while preserving circular symmetry. Cross-t_ν and exx-t_ν—which we need to distinguish since we plan to implement non-unity ratios of eigenvalues by expanding the first component of the bivariate vectors—show both elongation and tetrapolarity. They are included in this study in order to demonstrate the extent to which non-ellipticity challenges classical multivariate methods. Even the definition of eigendirections is in such situations far from clear.

More experience with two-dimensional distributions might lead us to more diverse classes of nonGaussian distributions. However, those just described or hinted at seem to cover both elongation and tetrapolarity quite well —and to go far beyond what is usually considered. Based on one-dimensional experience, $\nu = 6$ corresponds to rather mild elongation (rather like the logistic or double exponential, at least moderately far in the tails), while $\nu = 3$ corresponds to strong (but not catastrophic) elongation. ($\nu = 1$ of course produces the unrealistically peaked and very stretched-tailed Cauchy distribution.)

An alternative set of elongated distributions can be obtained by the slash-ν family, in which the divisor for the circular Gaussian vectors would follow, independently of its numerator, $\nu x^{\nu-1}dx$ over [0,1]. In the extreme tails, t_ν and slash-ν, which we will often condense to h_ν, behave similarly. Circular-, cross-, and exx- versions of h_ν are definable in exact parallelism to their definitions for t_ν. The t-derived distributions are more nearly familiar, but the slash-derived ones are more realistic, in both one and two dimensions (especially for small ν, where t_ν is unduly peaked in the middle).

3 Jackknives

We deal here with a class of resampling procedures that involve resamples of smaller size than samples. We prefer them to "bootstrap" procedures which preserve sample size by sampling with replacement, often involving samples with $\geq 60\%$ of their elements involved in artificial ties. There are a variety of jackknife procedures, distinguished mainly by the ratio of size of resamples to size of sample, which will almost certainly lie between $1/2$ and $1 - 1/n$, and by the degree of overlap among resamples.

The original—delete-1—jackknife uses as resamples the n results of deleting one element from the available sample of n (Quenouille (1956) and Tukey (1958)).

The delete-2% jackknife uses 50 resamples that each omit one of 50 nearly equal parts of the original sample. This is included because it limits recalculations, since the number of resamples stays at 50, no matter how large the sample size may be.

The delete-2/7 jackknife uses as resamples all unions of 5 parts out of 7 nearly equal parts into which the original sample has been (randomly) divided. This is mentioned here because some theoretical studies, e.g., Wu (1985), have asymptotic results that suggest using resample sizes in the range

$$0.65 \leq \frac{\text{resample size}}{\text{sample size}} \leq 0.75\,.$$

Since $\binom{7}{2} = 21$, we will have 21 resamples, each of 5 pieces, for a total of 105 $= 21 \cdot 5$ pieces. By symmetry, each piece will appear in 15 resamples. If 21 resamples are not thought to be enough, we might do an additional independent random separation into sevenths, thus generating 21 more resamples. (Each such group is probably worth six degrees of freedom.) Another option would be to "delete 3/10" leaving aside 3 of 10 parts in 120 different ways.

The multihalver jackknife uses resamples of size 1/2 the actual sample size, using a reasonable number of intersecting halvings. It is desirable to have intersecting halvings approach orthogonality wherever possible and convenient to have them reasonably symmetrically related. One convenient approach pairs the elements and plans to put one element from each pair in each and every half. If $n \leq 2^{k+1}$, so that there are $\leq 2^k$ pairs, we can find 2^k sequences of 2^k $+$ and $-$ elements, each pair of which is mutally orthogonal (the columns of the design matrix for a full analysis of a 2^k factorial experiment). If we use the first $n/2$ of the 2^k elements of any such sequence to determine the assignment of the 2 elements of the corresponding pair to the halves corresponding to the sequence, we will get a reasonably satisfactory set of 2^k halvings. (Sequences from Plackett–Burman designs allow us more detailed balance—exact for n a multiple of 4—if desired.) Thus for n=13, for example, we can use the 8 columns of a 2^3 factorial design to create 8 halves. This design matrix is:

+	+	+	+	+	+	+	+
+	−	+	−	+	−	+	−
+	+	−	−	+	+	−	−
+	−	−	+	+	−	−	+
+	+	+	+	−	−	−	−
+	−	+	−	−	+	−	+
+	+	−	−	−	−	+	+
+	−	−	+	−	+	+	−

If the pairing of the observations is (y_1, y_2), (y_3, y_4), (y_5, y_6), (y_7, y_8), (y_9, y_{10}), (y_{11}, y_{12}) and $(y_{13}, -)$, then the corresponding halves are as follows:

y_1	y_1	y_1	y_1	y_1	y_1	y_1	y_1
y_3	y_4	y_3	y_4	y_3	y_4	y_3	y_4
y_5	y_5	y_6	y_6	y_5	y_5	y_6	y_6
y_7	y_8	y_8	y_7	y_7	y_8	y_8	y_7
y_9	y_9	y_9	y_9	y_{10}	y_{10}	y_{10}	y_{10}
y_{11}	y_{12}	y_{11}	y_{12}	y_{12}	y_{11}	y_{12}	y_{11}
y_{13}	y_{13}	−	−	−	−	y_{13}	y_{13}

To keep the computations in check, one can use pairs of blocks of data values with a block size larger than 1.

The multihalver approach is particularly useful in dealing with narrowly estimated quantities, where a few elements of the original samples play an unusually important role in determining the value of the quantity being studied. (Order statistics (including medians) and outliers are two kinds of typical examples.)

Whatever the symmetrical delete-structure adopted, the general pattern of jackknife calculations is as follows, with $q =$ fraction deleted, so that $1/q$ is n, 50, 7/2 or 2, respectively:

$$\begin{aligned}
\text{pseudovalue} &= \frac{1}{q}(\text{result for all}) - \left(\frac{1}{q} - 1\right)(\text{result for resample}) \\
&= 2(\text{result for all}) - (\text{result for half} \\
&\qquad + \text{result for complementary half})/2 \text{ for } q = 1/2 \\
\text{jackknife estimate} &= \text{mean of pseudovalues} \\
\text{basis for variance} &= \sum(\text{pseudovalue} - \text{jackknife estimate})^2
\end{aligned}$$

$$\text{estimated variance of estimate} = (\text{basis for variance})/(\text{divisor})$$

$$\begin{aligned}
\text{confidence interval} &= \text{jackknife estimate} \\
&\quad \pm t_\nu(1 - \alpha/2) \cdot (\text{estimated variance of estimate})^{1/2},
\end{aligned}$$

where "divisor" and ν = (degrees of freedom) are chosen to get correct results when the response being jackknifed is the mean of the values for the elements involved. (Thus, "divisor" = $n(n-1)$, (50)(49), 210/4, (number of halvings) and $\nu = n-1$, 49, 6, (number of halvings), respectively.) Note that $t_\nu(p)$ denotes the p-quantile of the t_ν-distribution.

One reason for a bouquet of jackknives, rather than just one, is the tension between

- elimination of temptation to look at more resamples, which is most complete for the delete-1 jackknife, and
- good performance against narrow estimation, which is (among these candidates) probably best for the multihalver jackknife.

Thus the extent to which a specific class of problems involves narrow estimation is likely to be important in determining the choice of jackknife.

4 Choices in radGaussianization

The simplest approach to radGaussianization observes that, if we knew the center and typical ellipse (shape only) of the distribution, then regarding each data point as defining a vector from the center, and referring this vector to the ellipse, we would get a collection of squared "distances" which could be compared with working values (nearly medians of order statistics) for chi-square on as many degrees of freedom as the dimension of the problem (in our case 2). Such a comparison between working values and actual distances immediately suggests iteration, back and forth between:

- finding low moments, center, and inverse of covariance matrix for the points at the ends of the shrunken vectors, and
- redoing the shrinking, in view of the new squared radii.

There seems to be no particular problem about the first sort of step, but some thought is needed with regard to the second.

Let $r_i^2 = v_i^\top C^{-1} v_i$ be the squared radius corresponding to vector v_i, and let $r_{i|n}^2$ be the ith largest of these, so that $r_{1|n}^2 \geq r_{2|n}^2 \geq \cdots \geq r_{i|n}^2 \geq \cdots$. The corresponding working values (for dimension 2) satisfy

$$a_{i|n} = F^{-1}\left(1 - \frac{3i-1}{3n+1}\right),$$

where F is the exponential distribution with mean 2. Therefore

$$\exp(-a_{i|n}/2) = \frac{3i-1}{3n+1}$$

so that

$$a_{i|n} = -2 \cdot \ln\left(\frac{3i-1}{3n+1}\right) .$$

We are thus led to look at the ratios

$$\frac{r_{1|n}^2}{a_{1|n}}, \frac{r_{2|n}^2}{a_{2|n}}, \ldots, \frac{r_{i|n}^2}{a_{i|n}}, \ldots$$

which in the face of an elongated distribution can be expected to be, generally, decreasing as i increases (as we move inward).

Our shrinkage is intended to make these ratios more nearly alike, at least so far as any general trend is concerned. What happens to short vectors is less important—because of the role of 2nd order moments—than what happens to long vectors.

One extreme approach would be to make all the ratios (after shrinkage) alike, for example by

$$v_{i|n} \longrightarrow v_{i|n}\left(r_{i|n}^2/a_{i|n}\right)^{-1/2} .$$

This implies that $y_j\left(r_{i|n}^2/a_{i|n}\right)^{-1/2}$ replaces the corresponding observation y_j. Another near extreme would adjust only the outer $\alpha\%$ of the ratios, matching at the boundary, so that, with $j = (\alpha/100)n$

$$v_{i|n} \longrightarrow v_{i|n}\left(r_{i|n}^2/a_{i|n}\right)^{-1/2}\left(r_{j|n}^2/a_{j|n}\right)^{1/2} \quad \text{for } i \le j$$

$$v_{i|n} \longrightarrow v_{i|n} \quad \text{for } i > j .$$

(The first extreme corresponds up to a rescaling to the case $\alpha = 100\%$.)

Iteration could be for constant $\alpha\%$ or for changing $\alpha\%$. It would seem reasonable to limit $\alpha\%$, at least in an initial exploration, to 5%, 10%, 20%, 50% and 100%.

In our application we have to be concerned with the effect of radGaussianization on the quantity that concerns us, the direction of the larger eigenvector. Clearly we are concerned with the sample rather than with the population. So one thing we shall need to look at is the spread of the directions for radGaussianized samples as compared to the spread of directions for raw samples. Any or all of sample size, eigenvalue ratio (for raw vectors), and distribution shape can matter here. (The information about spread of directions (of point estimates) is somewhat different from that provided by average length and non-coverage of confidence intervals.)

PART B: CLASSICAL ASYMPTOTICS

5 The Limiting Form

T.W. Anderson in 1963 developed the classical asymptotic results for eigenvector directions when sampling from a normal population. In our two-dimensional case it reduces to

$$\left[\text{variance(direction)}\right]^{-1} = (\text{sample size})\left(\text{ratio}^{1/2} - \frac{1}{\text{ratio}^{1/2}}\right)^2$$

where, since we are being classically asymptotic, the "direction" corresponds to a small angle, φ, between population and sample directions, for which

$$\sin\varphi \approx \varphi \approx \tan\varphi\,.$$

Since $|\tan^{-1} u| < |u| < |\sin^{-1} u|$ for $0 < |u| < 1$, the more inclusive confidence limits will arise if we apply ± 1.96 (asymptotic variance)$^{1/2}$ to $\sin(\hat{\theta} - \theta)$, as in

$$\sin(\hat{\theta} - \theta_L) = \sin(\theta_H - \hat{\theta}) = \pm\min(1, 1.96(\text{asymptotic variance})^{1/2})$$

whence

$$\theta_L, \theta_H = \hat{\theta} \pm \sin^{-1}\left(\min(1, 1.96(\text{asymptotic variance})^{1/2})\right)\,.$$

(We could be slightly more conservative by using the 2-sided 5% point of the t on a finite number of degrees of freedom, but, since we have, in any event, to make a further allowance for finite sample size, it seems simpler not to do this.)

In the case of sampling from non-normal populations these asymptotic result are no longer valid. In the case of elliptical distributions, for example, the asymptotic variance is inflated by a factor of $(1+\kappa)$, where 3κ is the kurtosis of the marginal distribution of a single component (Tyler (1981), Eq. 7.6). For the circular-t_6 distribution κ is equal to 1, which corresponds to a doubling of the asymptotic variance. When the population has sufficiently heavy tails (for example in the circ-t_3 situation), the classical estimators are of no use because of infinite variances.

We clearly ought first ask: "For bivariate Gaussian samples of size n and eigenvalue ratio = ratio, how well does the classical asymptotic formula perform? Exhibit 1 shows the non-coverage probability in %; Exhibit 2 shows the average half length of confidence interval in radians. Each entry in these tables is based on 1000 pseudo-random bivariate samples generated by the multiplicative congruential generator with modulus 2^{31} and multiplier 69069. Standard Gaussian variables were generated by the method of Box–Muller. For each sample size one set of 1000 bivariate samples with independent components was generated. By multiplying the first component of each sample point by the square root of the population eigenvalue ratio, the samples for the other ratios used in the tables were generated from this one set of samples. For each sample the confidence wedge given above was computed and it was determined, whether the first basis vector, i.e. the dominant population eigenvector, was contained within the wedge.

Exhibit 3 shows the increase in length, as a ratio, that would be required if we could make an adjustment based on the population ratio (of eigenvalues). These numbers were computed from the simulated confidence intervals by enlarging or shrinking the wedges by a factor—fixed for each pair of sample size and population ratio—until exactly 950 of the 1000 covered the first basis vector. Since we can only use the sample ratio, we need to take the dependence of the numbers in Exhibit 3 on the population ratio as suggestive rather than definitive.

	population eigenvalue ratio						
sample size	1.2	1.5	2.0	3.0	5.0	10.0	20.0
50	21.1	12.6	9.3	7.7	7.2	6.9	6.9
100	16.9	8.1	6.7	6.2	6.0	5.9	5.9
200	12.0	6.5	5.7	5.4	5.3	5.3	5.3
400	7.3	5.4	4.9	4.9	4.9	4.9	4.9
800	5.4	4.4	4.3	4.3	4.2	4.2	4.2
1600	5.3	5.0	5.0	4.9	5.0	4.4	5.9

Exhibit 1: Gaussian performance of the classical asymptotic formula: % of non-coverage (target = 5%)

	population eigenvalue ratio						
sample size	1.2	1.5	2.0	3.0	5.0	10.0	20.0
50	0.958	0.755	0.462	0.255	0.159	0.0986	0.0658
100	0.899	0.563	0.296	0.173	0.110	0.0690	0.0461
200	0.773	0.377	0.201	0.121	0.078	0.0487	0.0326
400	0.631	0.253	0.141	0.085	0.055	0.0345	0.0231
800	0.452	0.175	0.099	0.060	0.039	0.0244	0.0163
1600	0.301	0.122	0.070	0.043	0.027	0.0172	0.0115

	population eigenvalue ratio						
sample size	1.2	1.5	2.0	3.0	5.0	10.0	20.0
50	6.77	5.34	3.27	1.80	1.12	0.697	0.465
100	8.99	5.62	2.96	1.73	1.10	0.689	0.461
200	10.93	5.33	2.84	1.71	1.10	0.688	0.461
400	12.62	5.06	2.82	1.71	1.10	0.690	0.462
800	12.79	4.95	2.80	1.71	1.10	0.690	0.462

Exhibit 2: Gaussian performance of the classical asymptotic formula: upper panel = average half length of confidence interval; lower panel = same multiplied by (sample size)$^{1/2}$.

	population eigenvalue ratio						
sample size	1.2	1.5	2.0	3.0	5.0	10.0	20.0
50	1.86	1.38	1.185	1.09	1.07	1.06	1.06
100	1.52	1.14	1.06	1.027	1.02	1.015	1.015
200	1.40	1.07	1.05	1.04	1.04	1.03	1.03
400	1.12	1.02	1.00	1.00	1.00	1.00	1.00
800	1.017	0.96	0.96	0.96	0.94	0.94	0.94

Exhibit 3: Gaussian performance of the classical asymptotic formula: Factor of increase in length of confidence interval required to reach 5% non-coverage.

6 First Correction

A simple adjustment to the effective sample size that should improve the non-coverages, would replace the sample size by

$$n^* = \text{sample size}^* = \frac{\text{sample size}}{\text{factor}^2} = \frac{n}{\text{factor}^2},$$

where the adjustment factors have been given in Exhibit 3. Exhibit 4 shows these adjustments as a function of the product (sample size)(population ratio $-$ 1).

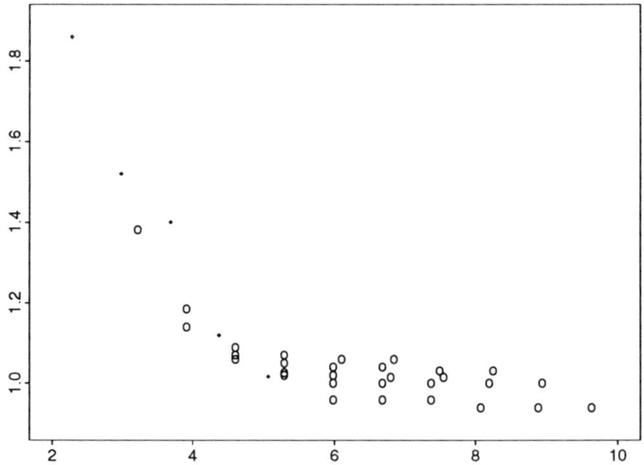

Exhibit 4: The plot shows the values of the blowup factors from exhibit 3 versus the logarithm of the product of (sample size)(population ratio–1). The plot suggests the factor = 1 + 7/((sample size)(population ratio–1)). The points corresponding to the population ratio 1.2 are drawn as asterisks. The fit there could be improved without influencing the fit at the other values of the ratio by using the product (sample size)(population ratio–1) min(2(population ratio–1),1) $^{1/2}$ as the denominator.

Having reached a suggestion of an appropriate adjustment, the next step is to try out such an adjustment—in terms of the observed (sample) ratio or eigenvalues. This means using

$$\text{limits on } \theta = \hat{\theta} \pm 1.96(n^*)^{-1/2}(\text{ratio}^{1/2} - \frac{1}{\text{ratio}^{1/2}})^{-1}$$

where

$$n^* = \frac{n}{((1 + 7/[(\text{sample size})(\text{obs. ratio} - 1)\ (\min\{\,2(\text{obs. ratio} - 1),1\,\})^{1/2}])^2} \ . \tag{6.1}$$

The results are in exhibit 5 for non-coverage %; and in exhibit 6 for average length of confidence interval. Note that the seed values used in simulating the entries of exhibit 5 were different from exhibit 1. Entries within a line are, however, again based on only one set of 1000 bivariate samples.

	population eigenvalue ratio						
sample size	1.2	1.5	2.0	3.0	5.0	10.0	20.0
50	13.6	7.9	6.0	5.6	6.0	6.4	6.6
100	9.6	5.4	4.7	4.7	5.3	5.8	5.8
200	8.1	5.1	5.1	5.1	5.2	5.3	5.3
400	5.4	4.8	4.7	4.9	4.9	4.9	4.9
800	4.6	4.1	4.1	4.1	4.2	4.2	4.2

Exhibit 5: Gaussian performance of the adjusted asymptotic formula: % non-coverage (target = 5%).

	population eigenvalue ratio						
sample size	1.2	1.5	2.0	3.0	5.0	10.0	20.0
50	1.13	0.90	0.55	0.28	0.17	0.10	0.066
100	1.06	0.67	0.33	0.18	0.11	0.070	0.046
200	0.90	0.42	0.21	0.12	0.078	0.049	0.033
400	0.73	0.27	0.14	0.086	0.055	0.035	0.023
800	0.50	0.18	0.10	0.061	0.039	0.024	0.016

Exhibit 6: Gaussian performance of the adjusted asymptotic formula: average half length of confidence interval.

Comparison of exhibit 5 with exhibit 1 shows a marked improvement due to our size/ratio adjustment.

7 NonGaussian Raw Samples

It is clearly now high time that we look at the behavior of the adjusted classical intervals for nonGaussian samples. It seems likely that, instead of a fine grid in [population ratio (of eigenvalues)] and [log sample size], we should at least try out a coarse grid in terms of [log confidence interval half-angle (on average, for Gaussian)] and [log sample size].

Initially, we shall look at a 3×3 grid, with

mean confidence interval half-angles of $\pm 0.2, \pm\ 0.1$ and $\pm\ 0.05$

and

sample size $= 50$, 200 and 800

with the hope that, for most of our exploration we can often work with a well-chosen 2×2 case (possibly extrapolated from the 3×3 on one or both coordinates.) (Notice that the average half-angles are $\pm 11.4°$, $\pm 5.7°$ and $\pm 2.9°$ which seem to cover the overlaps between what we can stand to use and what we are likely to get in practice.) Exhibit 7 shows the population ratios which have been chosen to deliver (approximately) the desired average half-angles.

	half-lengths		
sample size	± 0.2	± 0.1	± 0.05
50	4.0	10.0	33.0
200	2.1	3.7	9.6
800	1.45	2.0	3.7

Exhibit 7: Population ratios to be used to deliver (approximately, for Gaussian parents) the desired confidence interval half-lengths (in radians).

For a first look, we will use the 4 corners of the 3×3 for raw samples from various non-Gaussian distributions. Exhibit 8 has the non-coverages. Exhibit 9 has the average half-angles. The generation of pseudo-random samples from the t-distribution used Gaussian deviates in the numerator and square roots of sums of squared Gaussian deviates in the denominator.

	population shape							
sample size	Gau	circ-t_6	cross-t_6	exx-t_6	circ-t_3	cross-t_3	exx-t_3	circ-t_2
50 [4.0]	5.9	13.8	4.6	15.2	27.6	6.8	38.2	46.6
800 [1.45]	4.1	16.2	6.1	19.9	49.6	12.6	57.0	76.4
50 [33]	6.6	14.8	4.9	16.0	27.9	5.8	39.0	46.0
800 [3.7]	4.1	16.2	6.0	19.9	48.3	7.0	57.0	74.2

Exhibit 8: Non-coverages for 4 corners of the 3×3 and a variety of nonGaussian distributions.

	population shape							
sample size	Gau	circ-t_6	cross-t_6	exx-t_6	circ-t_3	cross-t_3	exx-t_3	circ-t_2
50 [4.0]	0.202	0.205	0.218	0.194	0.235	0.278	0.180	0.210
800 [1.45]	0.197	0.201	0.218	0.186	0.216	0.300	0.167	0.163
50 [33]	0.050	0.050	0.050	0.049	0.052	0.055	0.048	0.053
800 [3.7]	0.050	0.050	0.050	0.049	0.051	0.054	0.048	0.052

Exhibit 9: Average Half-lengths of the confidence intervals in radians

for 4 corners of the 3 $\times$ 3 and a variety of nonGaussian distributions

The worst non-coverage percentages occur for the circular t-distribution on two degrees of freedom. The known asymptotic variance for the circ-t_6 distribution would enlarge our confidence wedge by a factor of $\sqrt{2}$ and would bring the noncoverages for this case under control. Anyone attempting to do principal components for distributions as extreme as these must clearly first robustify the estimation procedure. The adjusted classical interval works reasonably well for the cross-t distributions, where outliers tend to occur along the population eigendirections. If the population ratio becomes small, this feature helps much less, however. The average half-lengths depend only on the behavior of the estimated ratios. They are remarkably stable within each of the four lines of exhibit 9. We will proceed in our further studies with the three nonGaussian situations circ-t_6, exx-t_6, circ-t_3, which give us a range of bad, but not catastrophic behavior for the adjusted classical interval.

8 RadGaussianized Samples

We are now ready—and forced—to look at the effects of radGaussianization both on our problem as a whole and on our working solution for the Gaussian case. We need to look at a variety of radGaussianization and a variety of nonGaussian distributions. To hold the computation down, we will begin with the single, central choice of sample size 200 and population ratio 3.7. Exhibit 10 shows the resulting percent non-coverage for 3-step radGaussianizations with successive cut %'s of 10/10/10, 50/50/50, 100/100/100, 10/20/100, and 5/20/100. The average half-lengths of the confidence intervals in radians lie, in all 24 instances between 0.104 and 0.099. The radGaussianization was implemented as described in Section 4 with the mean of the bivariate sample points as estimate of the center and with C equal to the classical estimate of the covariance matrix. The necessary computations were implemented by way of the singular value decomposition. Note that each column of exhibit 10 is based on a single set of 1000 bivariate samples.

radGau	Gauss	circ-t_6	exx-t_6	circ-t_3
none (raw)	5.1	13.8	18.0	38.5
10/10/10	5.2	9.4	14.1	16.2
50/50/50	5.6	8.3	13.0	15.0
100/100/100	5.4	8.3	12.8	14.9
10/20/100	5.4	8.3	12.5	12.6
5/20/100	5.4	8.2	12.5	12.7
5/20/50/100	5.4	8.1	12.5	12.0

Exhibit 10: Percent non-coverage for samples of size 200 and population ratio 4.0 after radGaussianizing samples from 3 bivariate distributions.

Clearly we have improved the % non-coverage for each of the three challenges. But we need also to ask whether, perhaps, radGaussianization the sample points made the direction of the largest eigenvalue less stable (because of symmetry, which applies to all these examples, there should be no bias). Exhibit 11 shows a quantity not previously of interest, the variance of the (radGaussianized) sample direction. We see that radGaussianizing has reduced the variability, in some cases quite substantially. Thus we need not be surprised that the values in exhibit 10 are close to the target.

radGau	Gauss	circ-t_6	exx-t_6	circ-t_3
none (raw)	0.0026	0.0047	0.0057	0.0237
10/10/10	0.0026	0.0035	0.0044	0.0048
50/50/50	0.0027	0.0034	0.0043	0.0047
100/100/100	0.0027	0.0034	0.0043	0.0048
10/20/100	0.0026	0.0034	0.0043	0.0042
5/20/100	0.0026	0.0034	0.0043	0.0041
5/20/50/100	0.0027	0.0033	0.0042	0.0041

Exhibit 11: Mean squared deviation from population direction for raw and radGaussianized samples (for target ± 0.1 and sample size 200).

8.1 Side Issue

With two directions, one raw and one radGaussianized, we can ask for the best linear combination which will not be the latter unless

$$\text{ave}(\text{raw} - \text{pop'n})^2 \approx \text{ave}(\text{raw} - \text{radGauss})^2 + \text{ave}(\text{radGauss} - \text{pop'n})^2 ,$$

or

$$\text{ave}((\text{radGauss} - \text{raw})\ \text{radGauss}) \approx 0 .$$

Exhibit 12 shows values of ave((raw − radGauss) ^{2}) and of ave((radGauss-raw) radGauss).

radGau	Gauss	circ-t_6	exx-t_6	circ-t_3
none (raw)	0.00000	0.00000	0.00000	0.00000
10/10/10	0.00003	0.00080	0.00091	0.01495
50/50/50	0.00005	0.00087	0.00097	0.01374
100/100/100	0.00005	0.00089	0.00099	0.01350
10/20/100	0.00003	0.00092	0.00101	0.01573
5/20/100	0.00003	0.00094	0.00105	0.01618
5/20/50/100	0.00004	0.00098	0.00109	0.01638

radGau	Gauss	circ-t_6	exx-t_6	circ-t_3
none (raw)	0.00000	0.00000	0.00000	0.00000
10/10/10	0.00001	−0.00021	−0.00035	−0.00198
50/50/50	0.00003	−0.00025	−0.00040	−0.00262
100/100/100	0.00003	−0.00024	−0.00040	−0.00270
10/20/100	0.00002	−0.00023	−0.00039	−0.00187
5/20/100	0.00002	−0.00022	−0.00037	−0.00168
5/20/50/100	0.00002	−0.00021	−0.00036	−0.00161

Exhibit 12: Quantities of interest in determining the best linear combination of the direction based on radGaussianized data and the direction based on the raw data (for target ± 0.1 and sample size 200). upper panel = observed mean squared deviation of radGauss from raw; lower panel = observed covariance between radGauss and the difference (radGauss-raw).

An increase in the corner percentages works best and it seems essential to increase towards 100%. A slower increase such as 10/20/30/50/100 works even better in the circ-t_3 situation.

9 Allowance for (Circular) Elongation

Having done as much as we can to bring down the % of non-coverage by

- adjusting for sample size and sample ratio,

  and

- choosing an effective radGaussianizing pattern

we may still be left with more non-coverage than we like. The next natural step is to drain information from the radGaussianizing process, initially by looking at the % shrinkage for each of the 10% longest vectors and finding the mean of

$$\max\{0, \%\ \text{shrinkage}\} = (\%\ \text{shrinkage})^+$$

for these $n/10$ vectors in the third radGaussianization iteration. We will call the result the "shrink %" and use its average for further adjustment of our confidence angles.

We do not plan to work with

$$\%\ \text{shrinkage}$$

itself for a simple reason. For the Gaussian case among others, the average of this quantity will be negative for about as many samples as it is positive, so that an adjustment proportional to this will contract as many confidence wedges as it expands, with a resultant increase in % non-coverage. By going to the average of the positive % shrinkages we ensure an expansion, by some amount, of each confidence angle and thus, certainly to a decreased non-coverage. If, as a result of such a correction, we make all non-coverages less than 5%, we can make an additional adjustment, not dependant on shrink %, which returns non-coverages closer to 5%.

Exhibit 13 has the non-coverage %'s and the corresponding factors, while exhibit 14 has the mean shrink %'s both for a 2×2 pattern of sample size and population ratio (the corners of our 3×3) and a selection of circularly elongated distributions. The underlying simulations used a fresh set of seed values. As in the previous instances, the entries are based on sets of 1000 pseudo-random samples which were transformations of one set per sample size.

sample size	Gau	circ-t_6	circ-t_3	circ-h_3	circ-t_2	circ-h_2
50 [4]	4.9	8.9	12.2	10.2	15.3	13.6
800 [1.45]	4.5	7.9	12.2	10.0	17.4	17.0
50 [33.0]	5.4	9.3	13.2	11.0	16.3	13.9
800 [3.7]	4.5	7.9	12.2	10.0	17.4	17.0

sample size	Gau	circ-t_6	circ-t_3	circ-h_3	circ-t_2	circ-h_2
50 [4]	1.00	1.20	1.33	1.21	1.45	1.37
800 [1.45]	0.98	1.13	1.33	1.20	1.49	1.47
50 [33.0]	1.02	1.20	1.33	1.22	1.42	1.37
800 [3.7]	0.98	1.14	1.30	1.20	1.44	1.43

Exhibit 13: Non-coverage % (upper panel) and corresponding enlargement factors for selected size-ratio and distribution choices, using the adjustment (6.1) to the sample size and the radGaussianization 5/20/50/100.

sample size	Gau	circ-t_6	circ-t_3	circ-h_3	circ-t_2	circ-h_2
50 [4]	3.24	8.85	16.3	8.4	25.0	16.4
800 [1.45]	0.86	7.85	15.7	6.8	23.9	13.7
50 [33.0]	3.24	8.85	16.3	8.4	25.0	16.4
800 [3.7]	0.86	7.85	15.7	6.8	23.9	13.7

Exhibit 14: Mean shrink %'s for the situations of exhibit 13. These depend on the sample size, but not on the eigenvalue ratio. The numbers shown are the average positive parts of the shrinkages of the 10% longest vectors in the third step of the radGaussianization 5/20/50/100.

We are now ready to look at

$$\frac{(\text{enlargement factor for challenge}) - 1}{(\text{mean shrink \% for challenge})}.$$

These numbers are shown in exhibit 15, which leads us to try an adjustment corresponding to a factor of "0.02" that is to using

$$\text{(previous confidence wedge angle)}(1 + 0.02 \text{ (individual sample shrink \%)}) \tag{9.1}$$

as the new confidence wedge angle.

sample size	Gau	circ-t_6	circ-t_3	circ-h_3	circ-t_2	circ-h_2
50 [4]	0.000	0.023	0.020	0.025	0.018	0.023
800 [1.45]	−0.023	0.017	0.021	0.029	0.021	0.034
50 [33.0]	0.006	0.023	0.020	0.026	0.017	0.023
800 [3.7]	−0.023	0.018	0.019	0.029	0.018	0.031

Exhibit 15: Ratio of change in enlargement factor to shrink % for situations of exhibit 13.

Exhibit 16 has the resulting non-coverages—clearly the adjustment is roughly adequate.

sample size*	Gau	circ-t_6	circ-h_6	circ-t_3	circ-h_3	circ-t_2	circ-h_2
50 [4]	4.6	6.7	5.8	4.7	6.0	4.6	5.7
800 [1.45]	3.7	4.9	5.4	5.1	6.2	5.1	8.6
50 [33.0]	4.9	6.8	6.3	4.8	6.5	4.5	5.9
800 [3.7]	3.7	4.5	5.3	4.4	6.1	4.7	8.4

* population eigenvalue ratio in [ ].

Exhibit 16: % Non-coverage using the adjustment (6.1) to the sample size, radGaussianization 5/20/50/100 and the adjustment (9.1) based on shrink %.

9.1 Further Broadening

We would like to test this confidence wedge in more general circumstances, but prefer to delay this comparison until we have a full panonply of confidence wedges to consider. We note here some plausible combinations of sample size and eigenvalue ratio, namely

- 200 [33] (to yield ± 0.025 radians, = ± 1.4 degrees)
- 200 [4] (to yield ± 0.1 radians, = ± 5.7 degrees)
- 200 [1.5] (to yield ± 0.4 radians, = ± 23 degrees).

PART C: JACKKNIVES

10 Gaussian Performance of the leave-out-one Jackknife

Let $\hat{\theta}_{(i)}$ be the dominant eigendirection if the ith observation is deleted. The general principles from Section 3 lead to the following equation for the ith pseudovalue

$$\text{pseudovalue} = \hat{\theta} + (n-1) \cdot (\hat{\theta} - \hat{\theta}_{(i)}) .$$

Care has to be taken in the evaluation of the difference $(\hat{\theta} - \hat{\theta}_{(i)})$, since this represents a difference between directions. Consequently it has to be computed as $\hat{\theta} - \hat{\theta}_{(i)}$ or $\hat{\theta} - \hat{\theta}_{(i)} + \pi$ or $\hat{\theta} - \hat{\theta}_{(i)} - \pi$ whichever lies between $-\pi/2$ and $\pi/2$. Once the pseudovalues are properly defined, the remaining jackknife quantities are straightforward to compute and lead to the jackknife interval half-length

$$1.96 \cdot \min\left((\text{estimated variance of estimate}, \pi/2)^{1/2}\right) .$$

The non-coverage rates of this interval are shown in exhibit 17. The same sets of bivariate samples were used in computing these numbers as in exhibit 1.

	population eigenvalue ratio						
sample size	1.2	1.5	2.0	3.0	5.0	10.0	20.0
50	29.5	15.6	9.3	7.6	6.7	6.3	6.2
100	21.1	9.9	6.6	5.8	5.8	5.7	5.7
200	16.1	6.9	5.8	5.4	5.3	5.4	5.4
400	8.3	5.4	4.9	4.6	4.5	4.5	4.5
800	6.2	4.3	4.2	4.1	3.9	3.9	3.9

	population eigenvalue ratio						
sample size	1.2	1.5	2.0	3.0	5.0	10.0	20.0
50	0.897	0.716	0.458	0.264	0.166	0.103	0.069
100	0.822	0.530	0.297	0.176	0.112	0.070	0.047
200	0.700	0.363	0.200	0.121	0.078	0.049	0.033
400	0.579	0.250	0.141	0.086	0.055	0.035	0.023
800	0.427	0.175	0.099	0.061	0.039	0.025	0.016

Exhibit 17: Gaussian performance of the leave-out-one jackknife formula; upper panel = % of non-coverage (target = 5%); lower panel = average half length (half angle in radians) of confidence interval (confidence wedge).

Remark 10.1 The behavior of the jackknife interval might turn out to be improved if the directions on the half-circle are re-expressed as real numbers. A simple such re-expression is obtained by taking the pseudo-values

$$(n-1)\cdot \tan\left(\hat{\theta}_{(i)} - \hat{\theta}\right),$$

instead of $\hat{\theta} + (n-1)\cdot(\hat{\theta} - \hat{\theta}_{(i)})$. This corresponds to the projection of the directions onto the tangent to the unit circle at $\hat{\theta}$. A confidence bound b found with these values can be transformed back to the unit circle by

$$b \longrightarrow \hat{\theta} + \arctan(b).$$

The performance of this interval is, however, rather worse than the one based on the less sophisticated jackknife. Exhibit 18 has the numbers.

	population eigenvalue ratio						
sample size	1.2	1.5	2.0	3.0	5.0	10.0	20.0
50	34.5	17.5	10.2	8.0	7.0	6.4	6.2
100	25.0	10.8	7.3	5.9	5.8	5.7	5.7
200	18.5	8.1	6.1	5.5	5.4	5.4	5.4
400	11.7	6.0	4.9	4.7	4.5	4.5	4.5
800	7.2	4.5	4.2	4.2	3.9	3.9	3.9

	population eigenvalue ratio						
sample size	1.2	1.5	2.0	3.0	5.0	10.0	20.0
50	0.732	0.603	0.413	0.255	0.164	0.102	0.068
100	0.674	0.468	0.285	0.174	0.112	0.070	0.047
200	0.591	0.341	0.197	0.120	0.078	0.049	0.033
400	0.501	0.244	0.140	0.086	0.055	0.034	0.023
800	0.389	0.173	0.099	0.060	0.039	0.024	0.016

Exhibit 18: Gaussian performance of the leave-out-one jackknife formula with re-expressed pseudovalues; upper panel = % of non-coverage (target = 5%); lower panel = average half length of confidence interval. (Side track exploration.)

Since the non-coverages in exhibit 17 are quite similar to those in exhibit 1, the same small sample adjustment (6.1) should serve us here as well. Exhibit 19 shows the resulting non-coverage rates.

	population eigenvalue ratio						
sample size	1.2	1.5	2.0	3.0	5.0	10.0	20.0
50	22.7	11.6	6.8	6.2	5.8	5.8	5.8
100	15.9	7.0	5.3	5.4	5.6	5.7	5.7
200	11.8	5.4	5.1	5.0	5.1	5.2	5.3
400	6.6	5.0	4.8	4.5	4.5	4.5	4.5
800	5.0	4.1	4.1	3.9	3.9	3.9	3.9

	population eigenvalue ratio						
sample size	1.2	1.5	2.0	3.0	5.0	10.0	20.0
50	1.056	0.850	0.536	0.287	0.172	0.105	0.069
100	0.974	0.622	0.324	0.182	0.114	0.071	0.047
200	0.823	0.401	0.208	0.123	0.079	0.049	0.033
400	0.663	0.262	0.143	0.086	0.055	0.035	0.023
800	0.470	0.179	0.100	0.061	0.039	0.025	0.016

Exhibit 19: Gaussian performance of the leave-out-one jackknife formula with the small sample adjustment (6.1); upper panel = % of non-coverage (target = 5%); lower panel = average half length of confidence interval.

A different small sample adjustment is called for. Replacing the factor 7 in (6.1) by a factor of 12, for example, does a better job. We will, however, not pursue this issue any further at this moment, since we are likely in any case to require an additional adjustment for elongated tails. ////

11 Gaussian Performance of the Multihalver Jackknife

In Section 3 we briefly discussed the multihalver jackknife. This idea can be applied to the problem of eigendirections with a bit of extra care due to our working with directions rather than with reals. The pseudovalue associated with a half sample h is formally computed as $2\hat{\theta}_{\text{all}} - (\hat{\theta}_{(h)} + \hat{\theta}_{(\neg h)} = \hat{\theta}_{\text{all}} + (\hat{\theta}_{\text{all}} - \hat{\theta}_{(h)} + \hat{\theta}_{\text{all}} - \hat{\theta}_{(\neg h)})/2 = \hat{\theta}_{\text{all}} + (\Delta_{(h)} + \Delta_{(\neg h)})/2$, where $\hat{\theta}_{(h)}$ denotes the eigendirection if the half sample h is deleted and $\hat{\theta}_{(\neg h)}$ is the eigendirection if the complementary half is deleted. The latter two equations exhibit more clearly the bias-correcting nature of the jackknife. In dealing with directions the differences $\Delta_{(h)}$ and $\Delta_{(\neg h)}$ in the second term are evaluated to give a result in the interval from $-\pi/2$ to $\pi/2$, namely

$$\Delta_{(h)} = \begin{cases} \hat{\theta}_{\text{all}} - \hat{\theta}_{(h)} + \pi & \text{if } \hat{\theta}_{\text{all}} - \hat{\theta}_{(h)} < -\pi/2; \\ \hat{\theta}_{\text{all}} - \hat{\theta}_{(h)} & \text{if } -\pi/2 \leq \hat{\theta}_{\text{all}} - \hat{\theta}_{(h)} \leq \pi/2; \\ \hat{\theta}_{\text{all}} - \hat{\theta}_{(h)} - \pi & \text{if } \pi/2 < \hat{\theta}_{\text{all}} - \hat{\theta}_{(h)}. \end{cases}$$

The jackknife variance estimate is based on $\left(\Delta_{(h)} + \Delta_{(\neg h)}\right)^2/4$, where h and $\neg h$ are two halves which combined constitute the full sample. For each choice of h this quantity is an estimate of twice the variance of the eigendirection based on a sample of size $n/2$, so that the mean of $\left(\Delta_{(h)} - \Delta_{(\neg h)}\right)^2/4$ over all different choices of halvings is the jackknife estimate of the variance of the estimated eigendirection.

The generation of half samples that we implemented for our simulations uses sequences of signs of length $2^7 = 128$. Our blocks consist of two halves each of size 1, 1, 1, 2, 4 for samples of size 50, 100, 200, 400 and 800, respectively. Thus there are 25, 50, or 100 blocks. A sequence of signs defines a half sample via inclusion of either the first or the second half of each block, depending on the sign. The blocks were formed randomly by grouping the points in each sample in the order in which they were generated. Exhibit 20 shows the non-coverage rates and the lengths of this interval for Gaussian samples. The same pseudo-random samples as in exhibit 1 were used.

	population eigenvalue ratio						
sample size	1.2	1.5	2.0	3.0	5.0	10.0	20.0
50	34.9	16.3	8.5	7.1	7.1	6.8	6.9
100	25.4	9.3	6.0	6.1	5.8	5.8	5.8
200	17.7	6.6	5.9	5.9	5.7	5.7	5.7
400	9.3	5.9	4.9	5.1	5.0	5.0	5.0
800	6.4	4.6	4.0	4.0	4.0	4.0	4.0

	population eigenvalue ratio						
sample size	1.2	1.5	2.0	3.0	5.0	10.0	20.0
50	0.930	0.706	0.465	0.272	0.167	0.104	0.069
100	0.836	0.537	0.304	0.176	0.112	0.070	0.047
200	0.715	0.375	0.202	0.121	0.078	0.049	0.033
400	0.585	0.253	0.140	0.085	0.055	0.034	0.023
800	0.436	0.175	0.099	0.061	0.039	0.024	0.016

Exhibit 20: Gaussian performance of the multi-halver jackknife for Gaussian samples; upper panel = % of non-coverage (target = 5%); lower panel = average half length of confidence interval.

The behavior of this simpler procedure seems slightly worse than the corresponding leave-out-one jackknife shown in exhibit 17.

12 Performance of the Jackknife Estimates for NonGaussian Samples

For the same situations as in exhibit 8, i.e., the 4 corners and a variety of nonGaussian situations, one observes the following behavior of the leave-out-one jackknife.

	population shape							
sample size	Gau	circ-t_6	cross-t_6	exx-t_6	circ-t_3	cross-t_3	exx-t_3	circ-t_2
50 [4.0]	5.5	7.4	4.4	6.9	9.5	5.3	16.7	16.4
800 [1.45]	3.9	6.4	4.1	6.9	13.6	9.0	19.2	23.7
50 [33]	5.8	8.2	5.3	7.1	10.2	5.1	14.3	14.8
800 [3.7]	3.9	5.8	5.4	4.6	7.0	4.4	8.3	11.7

	population shape							
sample size	Gau	circ-t_6	cross-t_6	exx-t_6	circ-t_3	cross-t_3	exx-t_3	circ-t_2
50 [4.0]	0.218	0.269	0.238	0.272	0.409	0.339	0.377	0.535
800 [1.45]	0.200	0.272	0.227	0.280	0.576	0.450	0.528	0.848
50 [33]	0.053	0.065	0.053	0.068	0.092	0.058	0.102	0.138
800 [3.7]	0.050	0.067	0.050	0.075	0.152	0.058	0.169	0.341

Exhibit 21: Non-coverages of the leave-out-one jackknife interval with correction (6.1) (for the eigenvalue ratio and the sample size) for 4 corners of the 3 × 3 and a variety of nonGaussian distributions; upper panel = % of non-coverage (target = 5%); lower panel = average half length of confidence interval.

The computations leading to these numbers were done with the same sets of samples as in exhibit 8. Applying the half-sample jackknife leads to the following results.

	population shape							
sample size	Gau	circ-t_6	cross-t_6	exx-t_6	circ-t_3	cross-t_3	exx-t_3	circ-t_2
50 [4.0]	5.9	7.1	4.1	9.1	10.3	5.2	20.6	21.3
800 [1.45]	4.4	7.3	3.5	8.1	16.8	7.8	24.2	29.3
50 [33]	6.8	7.9	5.8	9.0	10.0	4.2	18.8	20.2
800 [3.7]	3.9	6.5	5.7	4.9	9.3	3.8	11.9	17.0

	population shape							
sample size	Gau	circ-t_6	cross-t_6	exx-t_6	circ-t_3	cross-t_3	exx-t_3	circ-t_2
50 [4.0]	0.214	0.261	0.246	0.249	0.375	0.375	0.315	0.457
800 [1.45]	0.198	0.270	0.250	0.264	0.530	0.571	0.434	0.752
50 [33]	0.053	0.063	0.053	0.065	0.085	0.063	0.089	0.116
800 [3.7]	0.050	0.066	0.050	0.074	0.141	0.075	0.152	0.289

Exhibit 22: Non-coverages of the half-sample jackknife interval with correction (6.1) (for the eigenvalue ratio and the sample size) for 4 corners of the 3 × 3 and a variety of nonGaussian distributions; upper panel = % of non-coverage (target = 5%); lower panel = average half length of confidence interval.

The two jackknifed intervals behave similarly and are both better than the asymptotic interval shown in exhibit 8. They provide a moderate amount of protection agains circular elongation.

13 RadGaussianized Samples

RadGaussianization with the sequence 5/20/50/100 leads to the following numbers, which are computed by using the same set of samples as in exhibit 13.

sample size	Gau	circ-t_6	circ-h_6	circ-t_3	circ-h_3	circ-t_2	circ-h_2
50 [4]	4.9	8.7	7.3	11.7	9.2	14.5	12.4
800 [1.45]	4.7	8.1	6.3	12.3	10.1	17.9	17.1
50 [33.0]	5.4	9.2	7.7	12.3	10.4	14.4	13.1
800 [3.7]	4.2	7.8	6.3	11.7	9.7	17.5	17.0

Exhibit 23: % Non-coverage of the leave-out-one jackknife using the adjustment (6.1) to the sample size and radGaussianization 5/20/50/100.

The comparison with exhibit 13 shows a slightly better behavior, but not to the point where no further adjustments are needed. If we use the % shrinkage as we did in the case of the asymptotic formula, we obtain the follwing results.

sample size	Gau	circ-t_6	circ-h_6	circ-t_3	circ-h_3	circ-t_2	circ-h_2
50 [4]	4.3	6.0	5.7	4.4	5.4	4.4	4.9
800 [1.45]	4.2	5.1	5.2	5.1	6.7	5.5	8.9
50 [33.0]	4.6	6.5	6.4	4.7	6.0	4.6	5.3
800 [3.7]	3.7	4.9	5.1	4.5	5.9	5.1	8.2

Exhibit 24: % Non-coverage of the leave-out-one jackknife using the adjustment (6.1) to the sample size, radGaussianization 5/20/50/100 and the adjustment (9.1) based on shrink %.

The two following exhibits show the behavior of the multihalver jackknife with radGaussianization.

sample size	Gau	circ-t_6	circ-h_6	circ-t_3	circ-h_3	circ-t_2	circ-h_2
50 [4]	5.3	9.0	7.2	11.4	9.2	14.7	13.3
800 [1.45]	4.4	8.1	6.9	12.4	10.0	17.5	16.5
50 [33.0]	5.7	9.6	7.9	11.5	10.2	14.8	13.7
800 [3.7]	4.5	8.2	6.7	12.1	10.4	17.9	16.9

Exhibit 25: % Non-coverage of the multihalver jackknife using the adjustment (6.1) to the sample size and radGaussianization 5/20/50/100.

sample size	Gau	circ-t_6	circ-h_6	circ-t_3	circ-h_3	circ-t_2	circ-h_2
50 [4]	4.8	6.8	5.8	4.1	5.3	4.6	5.1
800 [1.45]	3.7	4.5	6.0	5.1	6.7	5.8	9.1
50 [33.0]	5.1	7.2	6.3	4.2	6.0	4.5	5.3
800 [3.7]	3.9	4.5	5.4	5.0	6.0	5.3	8.8

Exhibit 26: % Non-coverage of the leave-out-half-repeatedly jackknife using the adjustment (6.1) to the sample size, radGaussianization 5/20/50/100 and the adjustment (9.1) based on shrink %.

14 Reprise of Adjustments

Now that we are doing quite well with all the circular distributions we have tried, it seems time to return to our adjustments and see, if without too great effort, we can do still better.

14.1 The First Adjustment

Clearly we want to begin with (6.1), which we will write in terms of the logarithmic enlargement

$$\frac{1}{2}\log\left(\frac{n}{n^*}\right) = \log\left(1 + \frac{7}{n(\text{observed ratio} - 1)(\min\{2(\text{observed ratio}) - 1), 1\})^{1/2}}\right).$$

If we take exhibits 1 and 5, and convert the % of non-coverage into natural logarithms of a corresponding expansion ratio, we get the results in exhibit 27. These expansion ratios are based on a conversion of non-coverages to their corresponding two-tailed Gaussian quantiles.

A) Unadjusted (calculated from non-coverages interpolated in exhibit 1)

	population eigenvalue ratio					
sample size	1.193	1.42	2.0	3.73	9.72	33
50			156	105	76	72
100			69	50	38	38
200		61	30	17	13	13
400		17	−9	−4	−4	−4
800	17	−21	−32	−32	−37	−37
1600	13	0	0	−4	**	**

B) Adjusted (calculated from non-coverages interpolated in exhibit 5)

	population eigenvalue ratio					
sample size	1.193	1.42	2.0	3.73	9.72	33
50			42	30	57	69
100			−13	−9	34	34
200		4	4	4	13	13
400		−11	−13	−4	−4	−4
800	−19	−42	−42	−37	−37	−32

C) Effect of adjustment (=(B) − (A))

	population eigenvalue ratio					
sample size	1.193	1.42	2.0	3.73	9.72	33
50			− 114	− 75	− 19	− 3
100			− 82	− 59	− 4	− 4
200		− 57	− 26	− 13	0	0
400		− 28	− 9	0	0	0
800	− 31	− 21	− 10	− 5	0	4

D) Needed fraction of (6.1) (= − (A)/(C))

	population eigenvalue ratio					
sample size	1.193	1.42	2.0	3.73	9.72	33
50			137%	140%	400%	2400%
100			84%	85%	950%	1450%
200		107%	115%	130%		
400		61%	neg	neg		
800	55%	neg	neg	neg		

E) $\frac{56}{n}$(values in (C)) − 30 + 13 $\log_2(n/100)$

	population eigenvalue ratio					
sample size	1.193	1.42	2.0	3.73	9.72	33
50			− 171	− 127	− 64	− 46
100			− 76	− 63	− 32	− 32
200		− 93	24	− 21	− 17	− 17
400		− 8	− 5	− 4	− 4	− 4
800	47	8	8	9	9	9

F) Newly adjusted (=(A) − (E))

	population eigenvalue ratio					
sample size	1.193	1.42	2.0	3.73	9.72	33
50			− 15	− 22	12	26
100			− 7	− 13	6	6
200		28	6	− 4	− 4	− 4
400		9	− 9	− 8	− 8	− 8
800	24	− 13	− 24	− 24	− 28	− 28

G) Conversion to % non-coverage

$28 \leftrightarrow 5.7\%$, $26 \leftrightarrow 5.6\%$, $24 \leftrightarrow 5.6\%$, $-28 \leftrightarrow 4.4\%$, $-24 \leftrightarrow 4.5\%$, $-22 \leftrightarrow 4.5\%$.

Exhibit 27: 1000× Natural logs of needed expansion ratios for bivariate Gaussian samples in a variety of population ratio/sample size combinations.

Since it is convenient to match population eigenratios, at least approximately, to sample ones, and to have a consistent set of eigenratios we have gone to the eigenratios that would generate normal round standard errors when (6.1) is not applied. Exhibit 28 does this, expanding exhibit 8.

	half-angles in radians				
sample size	±0.4*	±0.2	±0.1	±0.05	±0.025
25	2.62	5.75	17.6	65	253
50	2.00	3.73	9.72	33	127
100	1.64	2.62	5.75	17.6	65
200	1.42	2.00	3.73	9.72	33
400	1.283	1-64	2.62	5.75	17.6
800	1.193	1.42	2.00	3.73	9.72
1000	1.133	1.283	1.64	2.62	5.75
3200	1.092	1.193	1.42	2.00	3.73

Exhibit 28: Expansion of exhibit 8, after removal of (6.1). Population eigenratios for selected round standard errors at selected round sample sizes

* ±0.4 corresponds to ±27.9°; the next doubling would give ±46° which is unrealistic enough for us to pass it by.

As panel G) of exhibit 27 shows, the predicted non-coverages, which obviously require some checking by resimulation, range from 4.4% to 5.7%, a pleasantly tight distribution for the Gaussian case. Exhibit 29 shows the results of trying the further adjustment, using

$$\frac{1}{2}\log\left(\frac{n}{n^{**}}\right) = \frac{56}{n}\left(\frac{1}{2}\log\left(\frac{n}{n^{*}}\right)\right) + 0.03 - 0.013\log_2\left(\frac{n}{100}\right) \qquad (14.1)$$

which for $n > 800$, requires serious checking, since the last term grows infinitely, if slowly. The improvement of (14.1) over (6.1) can be seen in comparing panel A) with panel B).

A) % Non-coverages when using finite sample correction (14.1)

	population ratio						
sample size	1.193	1.42	2.00	3.73	9.72	33.0	127
13				3.1	5.0	7.2	7.8
50			4.1	3.7	4.0	4.2	4.3
400		5.9	5.3	5.2	5.2	5.2	5.2
800	6.6	5.3	5.0	4.8	4.8	4.8	4.8
3200	5.1	4.7	4.5	4.5	4.5	4.5	4.5

B) % Non-coverages when using finite sample correction (6.1)

	population ratio						
sample size	1.193	1.42	2.00	3.73	9.72	33.0	127
13				9.1	9.7	10.5	10.8
50			5.3	4.7	5.2	5.4	5.4
400		5.4	5.2	5.2	5.2	5.3	5.3
800	4.7	4.1	4.1	4.1	4.3	4.3	4.3
3200	4.2	4.1	4.1	4.2	4.2	4.2	4.2

Exhibit 29: Results of simulation using n^{**} in place of n. The upper panel shows the % non-coverage; the lower panel shows the corresponding numbers for the same sets of samples when using (6.1)

14.2 Allowance for Shrink %

Exhibit 13 is not self consistent. If we accept the non-coverages as correct, the expansion factors based on two-sided Gaussian quantiles, 1.96/(Gau(1-% non-coverage/200)), are calculated to be as follows.

sample size	Gau	circ-t_6	circ-t_3	circ-h_3	circ-t_2	circ-h_2
50 [4]	1.00	1.15	1.27	1.20	1.37	1.31
800 [1.45]	0.98	1.12	1.27	1.19	1.44	1.43
50 [33.0]	1.02	1.17	1.30	1.23	1.40	1.32
800 [3.7]	0.98	1.12	1.27	1.19	1.44	1.43

Exhibit 30: Enlargement factors for selected size/ratio and distribution choices, using the adjustment (6.1) to the sample size and the radGaussianization 5/20/50/100.

If we compare these values with the mean shrink % of exhibit 14, we get the picture in exhibit 31.

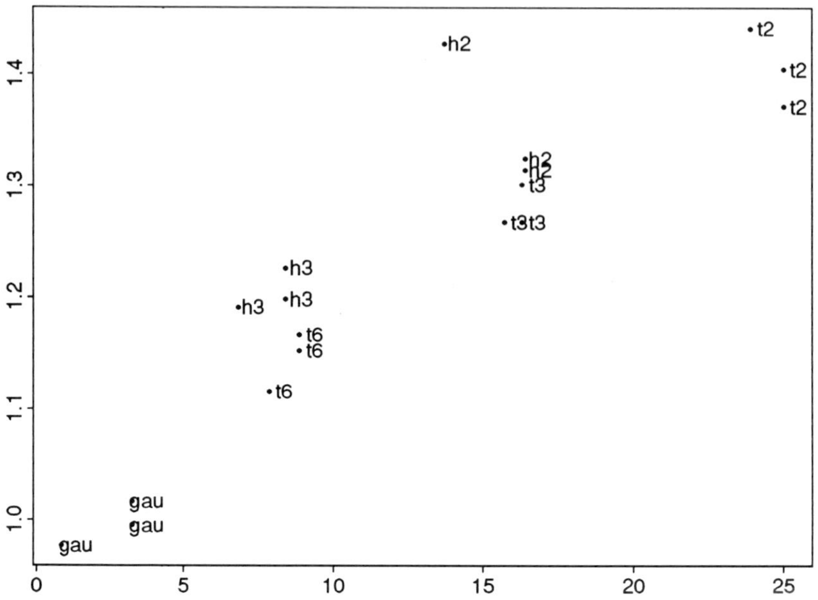

Exhibit 31: The plot shows the values of the blowup factors from exhibit 30 versus the mean shrink % from exhibit 14.

While the straight line is not a bad fit, with the possible exceptions of circ- h_2 for $n = 800$ and circ- t_2 for $n = 50$, the quadratics obviously fit better, and it is evident that behavior for circ- h_j is different from behavior for circ- t_j. There are two possible ways of dealing with the need of circ- h_j for a different dependence of adjustment on shrink %:

- We can say that we think that circ- h_j is unrealistic in the middle—perhaps as much as half as unrealistic as circ- t_j. This would lead us to try to allow for
$$2\,(\text{needed log for h}) - (\text{needed log for t})$$
which might serve in the real world, but would be bad for simulations of circ- h_j—and worse for simulations of circ- t_j, or

- we can try alternative definitions of shrink %, and see if we can find one for which the dependence of needed log on shrink % is more nearly the same for circ- h_j and circ- t_j.

The present shrink %'s are based on a 5/20/50/100 radGaussianization (4 stages) and the average shrinkage of the 10% longest sample vectors in the third stage. Simulations based on n^{**} instead of n^* and with the same four sample size-eigenratio combinations, show that using either the 6% longest or the 20%

longest does not have an appreciable influence on the shrink %. Changing from the third stage to the second stage, however, renders the behavior of the shrink % for circ-h_j and circ-t_j more nearly equal. The adjustment becomes

$$\text{(previous confidence wedge angle)}(1 + 0.015\ \text{(individual sample shrink } \%^*)), \quad (14.2)$$

where the sample shrink $\%^*$ is the average of the 10% longest vectors in the second stage. Exhibit 32 shows the % non-coverage using n^{**} and (14.2) with the same set of samples as we used in exhibit 16.

sample size	Gau	circ-t_6	circ-t_3	circ-h_3	circ-t_2	circ-h_2
50 [4]	5.6	6.0	5.3	3.9	5.0	4.4
800 [1.45]	5.1	5.2	5.3	5.5	6.9	5.1
50 [33.0]	6.0	6.1	5.4	4.1	5.1	4.3
800 [3.7]	5.1	5.0	4.3	5.0	5.7	4.8

Exhibit 32: % Non-coverage using the adjustment (14.1) to the sample size, radGaussianization 5/20/50/100 and the adjustment (14.2) based on shrink $\%^*$.

This alternative adjustment is similar to (9.1). It corrects for samples drawn from a circular circ-h_2 distribution, and is worse for the circular circ-t_2 case.

15 Broadening our Simulation Results

The adjustments (14.1) and (14.2) were developed on a limited number of cases. If we return to the size/ratio pairs of exhibit 7 and resimulate various circular bivariate situations, we obtain the results shown in exhibit 33.

size [ratio]	Gau	circ-t_6	circ-h_6	circ-t_3	circ-h_3	circ-t_2	circ-h_2
50 [4.0]	5.6	6.0	3.4	5.3	3.9	5.0	4.4
50 [10.0]	5.9	6.0	3.8	5.4	4.1	5.1	4.2
50 [33.0]	6.0	6.1	4.0	5.4	4.1	5.1	4.3
200 [2.1]	5.6	4.5	5.2	4.7	4.7	5.6	4.6
200 [3.7]	5.4	4.3	5.0	4.1	4.6	5.0	4.3
200 [9.6]	5.4	4.2	5.0	4.0	4.6	4.9	4.2
800 [1.45]	4.9	6.6	6.0	5.1	5.3	6.6	5.6
800 [2.0]	4.6	6.4	5.5	4.6	4.6	5.8	5.3
800 [3.7]	4.6	6.1	5.3	4.2	4.4	5.6	5.1

size [ratio]	Gau	circ-t_6	circ-h_6	circ-t_3	circ-h_3	circ-t_2	circ-h_2
50 [4.0]	0.217	0.252	0.233	0.288	0.266	0.325	0.309
50 [10.0]	0.108	0.123	0.114	0.139	0.130	0.157	0.150
50 [33.0]	0.054	0.062	0.057	0.069	0.065	0.078	0.075
200 [2.1]	0.196	0.225	0.205	0.258	0.235	0.294	0.277
200 [3.7]	0.104	0.118	0.108	0.135	0.123	0.153	0.143
200 [9.6]	0.052	0.059	0.054	0.067	0.062	0.076	0.072
800 [1.45]	0.193	0.219	0.199	0.255	0.230	0.293	0.273
800 [2.0]	0.100	0.113	0.103	0.130	0.118	0.147	0.139
800 [3.7]	0.050	0.057	0.052	0.065	0.059	0.073	0.069

Exhibit 33: % Non-coverage of Anderson's asymptotic confidence wedge using the adjustment (14.1) to the sample size, radGaussianization 5/20/50/100 and the adjustment (14.2) based on shrink %*. The nomial rate is 5%. The lower panel shows the average half-width of the confidence wedges.

The simulations used sets of 1000 samples with fresh seeds for the pseudo-random number generator. The analogous results for the multihalver jackknife with 128 different halvings are shown in exhibit 34.

size [ratio]	Gau	circ-t_6	circ-h_6	circ-t_3	circ-h_3	circ-t_2	circ-h_2
50 [4.0]	5.5	5.0	3.6	5.1	4.4	4.3	4.2
50 [10.0]	5.8	5.8	3.9	5.4	4.5	4.3	4.4
50 [33.0]	5.9	6.1	4.1	5.1	4.6	4.4	4.6
200 [2.1]	5.4	4.6	5.6	4.7	5.3	5.6	4.4
200 [3.7]	5.1	4.4	5.6	4.4	5.1	5.1	4.3
200 [9.6]	4.8	4.4	5.5	4.4	5.1	4.9	4.2
800 [1.45]	4.8	6.3	5.6	5.3	5.3	7.0	5.7
800 [2.0]	4.8	6.0	5.8	4.5	4.7	6.0	5.5
800 [3.7]	4.8	6.0	5.6	4.4	4.8	5.9	5.5

size [ratio]	Gau	circ-t_6	circ-h_6	circ-t_3	circ-h_3	circ-t_2	circ-h_2
50 [4.0]	0.227	0.263	0.241	0.298	0.275	0.341	0.321
50 [10.0]	0.112	0.128	0.118	0.143	0.134	0.164	0.156
50 [33.0]	0.056	0.064	0.059	0.072	0.067	0.082	0.078
200 [2.1]	0.198	0.227	0.206	0.258	0.236	0.296	0.278
200 [3.7]	0.104	0.119	0.108	0.135	0.123	0.154	0.144
200 [9.6]	0.052	0.060	0.054	0.067	0.062	0.077	0.072
800 [1.45]	0.193	0.219	0.198	0.254	0.229	0.293	0.271
800 [2.0]	0.099	0.113	0.103	0.130	0.118	0.147	0.138
800 [3.7]	0.050	0.057	0.052	0.065	0.059	0.073	0.069

Exhibit 34: % Non-coverage of the halfsample jackknife with 128 orthogonal half samples using the adjustment (14.1) to the sample size, radGaussianization 5/20/50/100 and the adjustment (14.2) based on shrink %*. The data were grouped

into 25 groups of size 2 for n=50, 100 groups of size 2 for n=100, and 100 groups of size 8 for n=800. The nominal rate is 5%. The lower panel shows the average half-width of the confidence wedges.

To appreciate to what extent tetrapolarity still requires our attention, we show some numbers in exhibit 35.

size [ratio]	cross- h_2 Asymptotic	cross- h_2 Jackknife	exx- h_2 Asymptotic	exx- h_2 Jackknife
50 [4.0]	0.5	0.9	12.5	10.3
50 [10.0]	0.5	1.1	12.6	9.7
50 [33.0]	0.5	1.4	12.6	9.7
200 [2.1]	0.2	1.1	14.2	13.0
200 [3.7]	0.1	1.6	13.6	10.0
200 [9.6]	0.1	1.9	13.5	8.7
800 [1.45]	0.4	0.8	22.6	18.1
800 [2.0]	0.0	1.9	21.1	14.2
800 [3.7]	0.2	1.5	20.9	12.9

size [ratio]	cross- h_2 Asymptotic	cross- h_2 Jackknife	exx- h_2 Asymptotic	exx- h_2 Jackknife
50 [4.0]	0.360	0.335	0.277	0.322
50 [10.0]	0.157	0.132	0.141	0.169
50 [33.0]	0.076	0.063	0.071	0.086
200 [2.1]	0.311	0.268	0.252	0.284
200 [3.7]	0.146	0.115	0.138	0.161
200 [9.6]	0.072	0.055	0.070	0.083
800 [1.45]	0.328	0.282	0.244	0.274
800 [2.0]	0.141	0.109	0.132	0.155
800 [3.7]	0.069	0.052	0.067	0.080

Exhibit 35: % Non-coverage of the asymptotic and the halfsample jackknife based on 128 orthogonal half samples. Both methods use the adjustment (14.1) to the sample size, radGaussianization 5/20/50/100 and the adjustment (14.2) based on shrink %*. For the multihalver jackknife, the data were grouped into 25 groups of size 2 for n=50, 100 groups of size 2 for n=100, and 100 groups of size 8 for n=800. The nominal rate is 5%. The lower panel shows the average half-width of the confidence wedges.

The behavior of the confidence wedge is very different in the two cases. The cross leads to smaller than nominal rates, whereas the exx-case has large non-coverages. The jackknife wedge behaves better than the asymptotic formula.

References

[1] Anderson, T.W. (1963): Asymptotic theory for principal component analysis. *Ann. Math. Statist.* **34** 122–148.

[2] Quenouille, M. (1956): Notes on bias in estimation. *Biometrika* **43** 353–360.

[3] Tukey, J.W. (1958): Bias and confidence in not quite large samples (abstract). *Ann. Math. Statist.* **29** 614.

[4] Tukey, J.W. (1993): Major challenges for multiple-response (and multiple-adjustment) analysis. In *Multivariate Analysis: Future Directions* (C.R. Rao, ed.), p. 401–421. North Holland, Amsterdam.

[5] Tyler, D.E. (1981): Asymptotic inference for eigenvectors. *Ann. Statist.* **9** 725–736.

[6] Wu, C.F.J. (1985): *Statistical Methods Based on Data Resampling*. Special invited paper presented at IMS Meeting in Stony Brook.

High Breakdown Point Designs

Christine H. Müller
Freie Universität Berlin

Abstract

A linear model is assumed where the experimental conditions are given by the experimenter so that they are without gross errors. For this situation we regard h-trimmed L_p-estimators which generalize the least median of squares estimator and the least trimmed squares estimators. The breakdown point of these estimators depends strongly on the underlying design and breakdown point maximizing estimators can be derived within these estimators. Also breakdown point maximizing designs can be derived. But these designs are often very different from the classically optimal designs which ensure efficiency also for robust procedures. Therefore we regard compromises providing high breakdown point and high efficiency designs.

Key words and phrases: Linear model, high breakdown point, trimmed L_p-estimator, optimal design.
AMS 1991 subject classification: Primary 62G35; secondary 62K05, 62J05.

1 Breakdown Points for Designed Experiments

We regard a general linear model

$$y = X\beta + z$$

where $y = (y_1, ..., y_N)' \in \mathbb{R}^N$ is the observation vector, $X = (x_1, ..., x_N)' \in \mathcal{X}^N \subset \mathbb{R}^{N\times r}$ the known design matrix, briefly also called the design, $\beta \in \mathbb{R}^r$ the unknown parameter vector and $z = (z_1, ..., z_N)' \in \mathbb{R}^N$ the error vector. We assume that the experimenter can choose freely the experimental conditions x_n, $n = 1, ..., N$, from the experimental region $\mathcal{X}$.

If we have an estimator $\widehat{\beta} : \mathbb{R}^N \times \mathcal{X}^N \rightarrow \mathbb{R}^r$ for the unknown parameter β then we can regard as a robustness measure its breakdown point which is the minimal fraction of outliers which can provide an arbitrarily high bias of the estimator. Originally the breakdown point of an estimator $\widehat{\beta}$ at a finite sample (y, X) was defined by Donoho and Huber (1983) by allowing also outliers in the experimental conditions as

$$\epsilon(\widehat{\beta}, y, X) := \min\{\frac{M}{N};\ M \in \mathbb{N} \quad \text{and} \\ \sup_{(\overline{y},\overline{X})\in\mathcal{Y}_M(y,X)} |\widehat{\beta}(\overline{y},\overline{X}) - \widehat{\beta}(y,X)| = \infty\}, \tag{1.1}$$

where $\mathcal{Y}_M(y, X)$ is the set of all contaminated samples $(\overline{y}, \overline{X}) \in \mathbb{R}^N \times \mathcal{X}^N$ for which $m(1) < ... < m(N-M)$ exist so that $(\overline{y}_{m(i)}, \overline{x}_{m(i)}) = (y_{m(i)}, x_{m(i)})$ for all $i \in \{1, ..., N-M\}$.

In the literature mainly this definition is used by assuming that the experimental conditions are given per random (see for example Rousseeuw and Leroy (1987), Coakley (1991), Mili and Coakley (1993)). But for designed experiments the experimental conditions are given by the experimenter so that outlying experimental conditions can not appear. Therefore here the breakdown point at a given sample (y, X) is defined as in Ellis and Morgenthaler (1992) and Müller (1993) by

$$\epsilon^*(\widehat{\beta}, y, X) := \min\{\frac{M}{N};\ M \in \mathbb{N} \text{ and } \sup_{\overline{y}\in\mathcal{Y}_M(y)} |\widehat{\beta}(\overline{y}, X) - \widehat{\beta}(y, X)| = \infty\}, \tag{1.2}$$

where $\mathcal{Y}_M(y)$ is the set of all contaminated observation vectors $\overline{y} \in \mathbb{R}^N$ for which $m(1) < ... < m(N-M)$ exist so that $\overline{y}_{m(i)} = y_{m(i)}$ for all $i \in \{1, ..., N-M\}$. By regarding the minimum possible breakdown point for an estimator $\widehat{\beta}$ at a given design X, i.e.

$$\epsilon^*(\widehat{\beta}, X) := \min\{\epsilon^*(\widehat{\beta}, y, X);\ y \in \mathbb{R}^N\},$$

we get a breakdown point which is independent of the observation vector y.

2 Breakdown Point Maximizing Estimators

The breakdown point of the classically optimal estimator for β, the least squares estimator (LS estimator) $\widehat{\beta}_{LS}$, is in the original definition of Donoho and Huber (1983) as well as in the definition of (1.2) equal to $\frac{1}{N}$. To improve the breakdown point alternative estimators were proposed. One is the least median of squares estimator (LMS estimator) $\widehat{\beta}_{LMS}$ which minimizes the median of squared residuals instead of the sum of squared residuals, i.e. which is given by

$$\widehat{\beta}_{LMS}(y, X) = \arg\min_{\beta\in\mathbb{R}^r} \operatorname{med}_{n=1,...,N}(y_n - x_n'\beta)^2$$

(see Rousseeuw (1984), Rousseeuw and Leroy (1987)).

But if we have a linear regression model $y_n = \beta_0 + \beta_1 t_n + z_n$ with $t_n \in [-1, 1]$ and $x_n = (1, t_n)'$ and use the classically optimal design (A-, D-, E-optimal) design X^0 with $\frac{N}{2}$ observations at $x = (1, -1)'$ and $\frac{N}{2}$ observations at $x = (1, 1)'$ (see also Section 3) then the breakdown point of the LMS estimator is

$\epsilon^*(\widehat{\beta}_{LMS}, X^0) = \frac{1}{N}$ because the exact fit point satisfies $\epsilon^*(\widehat{\beta}_{LMS}, X^0\beta, X^0) = \frac{1}{N}$. I.e. at X^0 the breakdown point of the LMS estimator coincides with the breakdown point of the LS estimator.

To improve the breakdown point at the classically optimal design X^0 we can regard a generalization of the LMS estimator, namely the h-trimmed L_p-estimators which were regarded in Müller (1993) and which also generalize the least trimmed squares estimators (see also Rousseeuw and Leroy (1987), Coakley (1991), Coakley and Mili (1992), Mili and Coakley (1993)). A h-trimmed L_p-estimator is given by

$$\widehat{\beta}_h(y, X) = \arg \min_{\beta \in \mathbb{R}^r} \sum\nolimits_{n=1}^{h} w_n \, |y_{(n)} - x'_{(n)}\beta|^p$$

where $w_1, \ldots, w_N > 0$, $p > 0$ and $|y_{(1)} - x'_{(1)}\beta|, \ldots, |y_{(N)} - x'_{(N)}\beta|$ are the ordered residuals for a given β, i.e. we have $|y_{(1)} - x'_{(1)}\beta| \leq \ldots \leq |y_{(N)} - x'_{(N)}\beta|$.

As Müller (1993) has shown for h not too large the breakdown point of a h-trimmed L_p-estimator depends only on the trimming value h, on the design X and the sample size N, and it does not depend on p and the weights w_n. For fixed h and N the breakdown point depends on the design X only via $N(X)$ and increases when $N(X)$ decreases. Thereby $N(X)$ is the maximal number of experimental conditions x_n which are lying in a $(r-1)$-dimensional subspace of $\mathbb{R}^r$, i.e.

$$N(X) := \max\{N_V;\ V \text{ is a } (r-1)\text{-dimensional subspace of } \mathbb{R}^r\},$$

where $N_V := \sum_{n=1}^{N} 1_V(x_n)$ (compare also with the quantity $\Delta(Q)$ used in Davies (1993), (1994)). In particular for $N(X) < h \leq \frac{N+N(X)+1}{2}$ the breakdown point of the h-trimmed L_p-estimator is

$$\epsilon^*(\widehat{\beta}_h, X) = \frac{1}{N}(h - N(X)) \tag{2.1}$$

and for $[\frac{N+N(X)+1}{2}] \leq h \leq [\frac{N+N(X)+2}{2}]$ we have

$$\epsilon^*(\widehat{\beta}_h, X) = \frac{1}{N}\left[\frac{N - N(X) + 1}{2}\right],$$

where $[z] := \max\{n \in \mathbb{N};\ n \leq z\}$ (see Müller (1993)).

Every regession equivariant estimator, i.e. every estimator $\widehat{\beta}$ satisfying $\widehat{\beta}(y + X\theta, X) = \widehat{\beta}(y, X) + \theta$ for all $y \in \mathbb{R}^N$ and $\theta \in \mathbb{R}^r$ as the h-trimmed L_p-estimators, has a breakdown point with

$$\epsilon^*(\widehat{\beta}, y, X) \leq \frac{1}{N}\left[\frac{N - N(X) + 1}{2}\right]$$

for all $y \in \mathbb{R}^N$ (see Müller (1993)). Hence the maximum breakdown point at the design X is

$$\epsilon^*(X) := \max\{\epsilon^*(\widehat{\beta}, X);\ \widehat{\beta} \text{ regression equivariant}\} = \frac{1}{N}\left[\frac{N - N(X) + 1}{2}\right]$$

and is attained by h-trimmed L_p-estimators with $[\frac{N+N(X)+1}{2}] \leq h \leq [\frac{N+N(X)+2}{2}]$ so that these h-trimmed L_p-estimators are breakdown point maximizing estimators.

3 Breakdown Point Maximizing Designs

A breakdown maximizing design can be defined as a design X^* which maximizes the maximum breakdown point $\epsilon^*(X)$ within a given class Δ of designs X, i.e.

$$X^* = \arg\max\{\epsilon^*(X);\ X \in \Delta\}.$$

Because the maximum breakdown point is a decreasing function of $N(X)$ we also can define a breakdown point maximizing design as a design X^* which minimizes $N(X)$, i.e.

$$X^* = \arg\min\{N(X);\ X \in \Delta\}$$

(see Müller (1993)). Because of (2.1) the breakdown point maximizing design maximizes not only the breakdown point of the breakdown point maximizing estimator but also the breakdown point of every h-trimmed L_p-estimator with $N(X^*) < h \leq \frac{N+N(X^*)+1}{2}$.

Because we always have $N(X) \geq r-1$ a design X^* with $N(X^*) = r-1$, i.e. a design where the experimental conditions $x_1, ..., x_N$ are in general position (see Rousseeuw and Leroy (1987)), is breakdown point maximizing.

For example for linear regression $N(X)$ is the maximal number of repetitions of a experimental condition x_n so that a design X^* with no repeated experimental conditions x_n satisfies $N(X^*) = 1$ and therefore is breakdown point maximizing. But these breakdown point maximizing designs are very different from the classically optimal design which minimizes some functional ϕ of the covariance matrix $(X'X)^{-1}$ of the least squares estimator as the trace (A-optimality), the determinant (D-optimality) or the maximum eigenvalue (E-optimality) of $(X'X)^{-1}$. The equivalence theorems for A-, D- and E-optimality (see Pukelsheim (1993) pp 221, 212 and 181-182, respectively) provide that in $\Delta = \mathcal{X}^N$ the A-, D- and E-optimal design has $\frac{N}{2}$ repetitions of $x = (1, -1)'$ and $\frac{N}{2}$ repetitions of $x = (1, 1)'$ if $\mathcal{X} = \{1\} \times [-1, 1]$ is the experimental region. Hence for linear regression we have a contradiction between maximum breakdown point and classic optimality.

This contradiction is in particular grave because classically optimal designs provide also high efficiency of some robust procedures. In particular rank-based estimators as regarded in Hössjer (1994) and Huber M-estimators have an asymptotic covariance matrix which coincides with the covariance matrix of the least squares estimator except for some constant which does not depend on the design (see Hössjer (1994), Bickel (1975), Yohai and Maronna (1979)). Also at A-optimal designs the trace of the asymptotic covariance matrix of the optimal robust Hampel-Krasker estimator is minimized (see Müller (1994)). Moreover the classically optimal design for linear regression with $\mathcal{X} = \{1\} \times [-1, 1]$ maximizes the breakdown point of the L_1-estimator (see Ellis and Morgenthaler (1992)).

4 Combining High Breakdown Point and High Efficiency

Estimators which combine high breakdown point and high efficiency can be easily constructed by one-step M-estimators

$$\widehat{\beta}(y, X) = \widehat{\beta}^0(y, X) + \frac{1}{N}\sum\nolimits_{n=1}^{N} \psi(y_n - x_n'\widehat{\beta}^0(y, X), x_n),$$

where the initial estimator $\widehat{\beta}^0$ is a high breakdown point estimator and the score function ψ provides high asymptotic efficiency. If the initial estimator is asymptotically normally distributed with convergence rate of $N^{-1/2}$ as many h-trimmed L_p-estimators (see Hössjer (1994)) then the finite sample breakdown point of $\widehat{\beta}$ coincides with that of the initial estimator $\widehat{\beta}^0$ and the asymptotic covariance matrix is only determined by the score function ψ (see Jurečková and Portnoy (1987)).

To find a design which combines high breakdown point and high efficiency is more difficult. There are situations where even designs exist which are classically optimal as well as breakdown point maximizing (see Müller (1994)) but these situations are rare. More typical is the situation of linear regression where classically optimal designs are very different from the breakdown point maximizing designs. In all these situations constrained problems will be a good compromise to combine high breakdown point and high asymptotic efficiency.

One constrained problem is to minimize a functional ϕ of the covariance matrix $(X'X)^{-1}$ under the side condition that the breakdown point is not too bad; that is,

$$\begin{aligned} X^{**} &= \arg\min\{\phi((X'X)^{-1});\ \epsilon^*(X) \geq B,\ X \in \Delta\} \\ &= \arg\min\{\phi((X'X)^{-1});\ N(X) \leq \overline{B},\ X \in \Delta\}. \end{aligned}$$

For the particular case that $N(X)$ is the maximal number of repetitions as for linear regression this problem was already treated by Wynn (1982), Wierich (1985) and Fedorov (1989) by constructing optimal designs with bounded number of repetitions.

Another constrained problem is to maximize the breakdown point under the side condition that the functional ϕ of the covariance matrix $(X'X)^{-1}$ is not too large; that is,

$$\begin{aligned} X^{**} &= \arg\max\{\epsilon^*(X);\ \phi((X'X)^{-1}) \leq B,\ X \in \Delta\} \\ &= \arg\min\{N(X);\ \phi((X'X)^{-1}) \leq B,\ X \in \Delta\}. \end{aligned}$$

In particular we can set

$$B = \min\{\phi((X'X)^{-1});\ X \in \Delta\}$$

so that we restrict ourselves to ϕ-optimal designs. This of course makes only sense if we have several ϕ-optimal designs. For example for linear regression this makes no sense because the A-, D- and E-optimal design is unique. But there are other problems for which we have several classically optimal designs as the following examples show.

5 Examples

A) Weighing designs for four objects. Here we regard the problem of weighing four objects with a chemical balance. The weights $\beta^1, ..., \beta^4$ of the four objects are unknown and we can use N weighing realizations for determining the weights of the objects. For each weighing realization for each object we have to decide whether it will participate and if it will participate on which side of the balance the object should be put. Hence for the n-th weighing realization we have to choose

$$x_n \in \mathcal{X} = \{-1, 0, 1\}^4$$

where $x_n^i = 0$ means that in the n'th realization the i'th object does not participate while $x_n^i = -1$ and $x_n^i = 1$ means that the i'th object is put on the lefthand side and righthand side of the balance, respectively. Then the result of the n-th weighing realization will be

$$y_n = x_n'\beta + z_n = \beta^1 x_n^1 + ... + \beta^4 x_n^4 + z_n$$

where $\beta = (\beta^1, \beta^2, \beta^3, \beta^4)' \in \mathbb{R}^4$, $x_n = (x_n^1, x_n^2, x_n^3, x_n^4)' \in \mathcal{X}$ and z_n is an error which may be also an outlier if some extraordinary shake happens. Then the equivalence theorems (see as in Section 3 for example Pukelsheim (1993)) provide that every design $X \in \mathcal{X}^N$ satisfying $X'X = N\,E_4$, where E_4 denotes the 4×4-identity matrix, is A-, D- and E-optimal in $\Delta = \mathcal{X}^N$, i.e. classically optimal. In particular if the sample size is $N = 8L$, where L is an integer, then classically optimal designs are $X = (X_1|...|X_1)'$ and $X^* = (X_2|...|X_2)'$, where the matrix $X_1 = (1_4|X_0)' \in \mathbb{R}^{4\times 4}$ is repeated $2L$-times and the matrix $X_2 = ((1_4|X_0)'|(1_4| - X_0)') \in \mathbb{R}^{4\times 8}$ is repeated L times. Thereby we have $1_4 = (1, 1, 1, 1)'$ and

$$X_0 = \begin{pmatrix} 1 & 1 & 1 \\ 1 & -1 & -1 \\ -1 & 1 & -1 \\ -1 & -1 & 1 \end{pmatrix}.$$

The design X has minimum support but the maximum breakdown point at X is approximately the half of that of X^* which is the breakdown point maximizing design within the classically optimal designs in $\mathcal{X}^N$ (see Müller (1994)). Note that X^* is not breakdown point maximizing within all designs of $\mathcal{X}^N$. But if we reduce $\mathcal{X}$ to $\mathcal{X}_0 = \{-1, 1\}^4$ then X^* also is breakdown point maximizing in $\mathcal{X}_0^N$ as Müller (1994) showed.

B) Multiple regression on balls. Another example where several classically optimal designs exist is the problem of multiple regression on balls which also was regarded in Müller (1994). For this problem Müller (1994) showed that classically optimal designs exist which are also breakdown point maximizing within all designs so that in particular they are breakdown point maximizing within the classically optimal designs.

References

[1] Bickel, P.J. (1975): One-step Huber estimates in the linear model. *J. Amer. Statist. Assoc.* **70** 428–434.

[2] Coakley, C.W. (1991): *Breakdown Points under Simple Regression with Replication.* Technical Report No. 91-21, Department of Statistics, Virginia Polytechnic Institute and State University.

[3] Coakley, C.W. and Mili, L. (1993): Exact fit points under simple regression with replication. *Statist. Probab. Lett.* **17** 265–271.

[4] Davies, P.L. (1993): Aspects of robust linear regression. *Ann. Statist.* **21** 1843–1899.

[5] Davies, P.L. (1994): Desirable properties, breakdown and efficiency in the linear regression model. *Statist. Probab. Lett.* **19** 361–370.

[6] Donoho, D.L. and Huber, P.J. (1983): The notion of breakdown point. In *A Festschrift for Erich L. Lehmann* (P.J. Bickel, K.A. Doksum and J.L. Hodges, Jr., eds.), 157–184. Wadsworth, Belmont, Calif.

[7] Ellis, S.P. and Morgenthaler, S. (1992): Leverage and breakdown in L_1 regression. *J. Amer. Statist. Assoc.* **87** 143–148.

[8] Fedorov, V.V. (1989): Optimal design with bounded density: Optimization algorithms of the exchange type. *J. Statist. Plann. Inference* **22** 1–13.

[9] Hössjer, O. (1991): Rank-based estimates in the linear model with high breakdown point. *J. Amer. Statist. Assoc.* **89** 149–158.

[10] Jurečková, J. and Portnoy, S. (1987): Asymptotics for one-step M-estimators in regression with application to combining efficiency and high breakdown point. *Commun. Statist.-Theory Meth.* **16** 2187–2199.

[11] Mili, L. and Coakley, C.W. (1993): *Robust Estimation in Structured Linear Regression.* Technical Report No. 93-13, Dept. of Statistics, Virginia Polytechnic Institute and State University.

[12] Müller, Ch.H. (1993): *Breakdown Points for Designed Experiments.* Preprint No. A-93-29, Freie Universität Berlin, Fachbereich Mathematik und Informatik. *J. Statist. Plann. Inference*, to appear.

[13] Müller, Ch.H. (1994): Optimal designs for robust estimation in conditionally contaminated linear models. *J. Statist. Plann. Inference.* **38** 125-140.

[14] Müller, Ch.H. (1994): *Optimal Breakdown Point Maximizing Designs.* Submitted to the Proceedings of PROBASTAT'94. Tatra Mountains Mathematical Publications, Bratislava.

[15] Pukelsheim, F. (1993): *Optimal Design of Experiments.* Wiley, New York.

[16] Rieder, H. (1994): *Robust Asymptotic Statistics.* Springer, New York.

[17] Rousseuw, P.J. (1984): Least median of squares regression. *J. Amer. Statist. Assoc.* **79** 871–880.

[18] Rousseeuw, P.J. and Leroy, A.M. (1987): *Robust Regression and Outlier Detection.* Wiley, New York.

[19] Wierich, W. (1985): Optimum designs under experimental constraints for a covariate model and an intra-class regression model. *J. Statist. Plann. Inference* **12** 27-40.

[20] Wynn, H. (1982): Optimum submeasures with applications to finite population sampling. In *Statistical Decision Theory and Related Topics III*, Vol. 2, 485–495. Academic Press, New York.

[21] Yohai, V.J. and Maronna, R.A. (1979): Asymptotic behavior of M-estimators for the linear model. *Ann. Statist.* **7** 258–268.

On Bayesian Robustness: An Asymptotic Approach *

Daniel Peña
Universidad Carlos III de Madrid

Ruben H. Zamar
University of British Columbia, Vancouver

Abstract

This paper presents a new asymptotic approach to study the robustness of Bayesian inference to changes on the prior distribution. We study the robustness of the posterior density score function when the uncertainty about the prior distribution has been restated as a problem of uncertainty about the model parametrization. Classical robustness tools, such as the influence function and the maximum bias function, are defined for uniparametric models and calculated for the location case. Possible extensions to other models are also briefly discussed.

Key words and phrases: gross error sensitivity, influence function, maximum bias curve, prior robustness.
AMS 1980 subject classifications: Primary 62F15; secondary 62F35.

1 Introduction

The robustness of Bayesian analysis to possible misspecification of the prior distribution has been extensively studied. Berger (1984,1990) and Kadane (1984) give a detailed review of the subject. See also Walley (1990) and Pericchi and Walley (1991). The main idea in this approach is to consider a whole set of prior distributions $\mathcal{P}$ instead of a single prior distribution Π alone. Then one focuses on a certain measure of interest (e.g., the posterior risk) and calculates its maximum and infimum as Π ranges over $\mathcal{P}$. A widely used class of prior distribution is the ε-contamination class given by

$$\mathcal{P} = \{ (1-\varepsilon)\Pi_0 + \varepsilon Q \ , Q \in \mathcal{Q}\}, \tag{1.1}$$

where $0 < \varepsilon < 1$ is given, and $\mathcal{Q}$ depends on the given setup. This type of contamination family has been used by many researchers including Huber

* Research partially supported by DGICYT grant PB93-0232, Spain, and NSERC and ONR Contract N00014-84-C-0169

(1981), in the classical robustness field, and Box and Tiao (1968) in the Bayesian field. More recently Marazzi (1985), Berger and Berliner (1986), Walley (1990) and Moreno and Cano (1991), among others, used model (1.1) to represent contamination situations.

In general (see for instance Moreno and Cano, 1991) we have that the robustness of the Bayesian inference is analyzed by looking at the infimum and the supremum of

$$\rho(\Pi) = \frac{\int a(\theta) d\Pi(\theta)}{\int b(\theta) d\Pi(\theta)},$$

as Π ranges over $\mathcal{P}$. For instance, if $b(\theta) = f(x|\theta)$, the likelihood density, and $a(\theta) = \theta f(x|\theta)$, $\rho(\Pi)$ becomes the posterior mean. If $a(\theta) = L(\theta, d(x)) f(x|\theta)$ and $b(\theta) = f(x|\theta)$, where $L(\theta, d(x))$ is a loss function for the decision rule $d(x)$, $\rho(\Pi)$ becomes the posterior risk. Finally, if $L(\theta, d(x)) = I_C(\theta)$ where $I_C(\theta)$ is the indicator function for the set C and $b(\theta)$ is as before, then $\rho(\Pi)$ becomes the posterior probability of the set C (a credible set). Although a formal definition of Bayesian robustness is not given, all the literature on this subject shows that the robustness of the Bayesian inference is highly dependent on the observed data.

On the other hand, there is also an extensive literature on classical robustness. A nice presentation of the main topics and ideas underlaying this subject can be found in Hampel et al. (1986). The classical robustness theory basically relies on some asymptotic tools to measure the degree of robustness of a certain procedure, namely: the influence function, the gross-error-sensitivity and the maximum bias function. The asymptotic theory provides a simple setup to study the sensitivity of different statistical procedures to changes on the model's assumptions. One would expect, of course, that the findings from this "asymptotic lab" will be validated by Monte Carlo simulation and practical examples.

No comparable asymptotic developments have occurred, so far, in the Bayesian field. This may be due in part to the fact that a major component of Bayesian procedures, namely the prior distribution, does not seem to survive the limiting process.

In this paper we present an asymptotic approach to study the robustness properties of Bayesian inference under changes in the assumed prior distribution. We will show that the uncertainty about the prior distribution can be recast as a "dual" uncertainty about the model parametrization. The main objective of this paper is to explore this duality. There are two main reasons why we think that this is worthwhile. First, it is well known that some problems which appear to be very difficult or untractable in a given domain (e.g., the time domain) can be easily solved and understood in another (e.g., the frequency domain). Second, and more importantly, we will show that the "parameter" domain allows the use of some asymptotic tools which have proved to be very useful in calssical setup.

The rest of the paper is organized as follows. In Section 2 we derive an asymptotic representation for the changes on the Bayesian posterior density. This representation is obtained by transfering the uncertainty on the prior distribution to a "dual" uncertainty on the parametrization of the model. This is then used to define the Bayesian counterparts for the influence function, the gross-error-sensitivity and the maximum bias curve. This asymptotic approach allows the derivation of general conclusions independently from any given sample situation. In particular, one can state under which conditions and in which sense a given Bayesian procedure will be robust (robustness being measured now by standard robustness tools). In Section 3 we apply the general methodology of Section 2 to the simple location model. Finally, in Section 4, we briefly discuss some possible extensions to the multiple parameter case and to the study of changes on the likelihood part of the model.

2 A New Defintion of Bayesian Robustness

Suppose that a statistician is interested in making inferences on a parameter θ. His prior assumptions are that θ follows a distribution Π_0 and that the sample $X_1, X_2, \ldots, X_n$ are i.i.d. random variables with common distribution $F(x|\theta)$. Suppose that $F(x|\theta)$ is differentiable and has a density $f(x|\theta)$. If Π_0 has also a density, π_0, the posterior distribution for θ under this model is given by

$$\mu_{\Pi_0}(\theta) = k_0 \ \pi_0(\theta) \ f(x_1|\theta) \ldots f(x_n|\theta), \tag{2.1}$$

where $k_0 = (\int f(x_1|\theta) \ldots f(x_n|\theta) \ \pi_0(\theta) \ d\theta)^{-1}$ does not depend on θ.

If the pair $[\Pi_0, F]$ represents the statistician's ideal model only approximately (e.g., Π_0 might represent only a crude approximation for his actual prior beliefs) he may also be interested in knowing how changes on the prior density π_0 and/or the likelihood density f can affect the inferences on θ. That is, what would happen if he used the alternative model $[\Pi, G]$—with prior distribution Π and likelihood distribution G—which is close to $[\Pi_0, F]$ (in some metric) and therefore might also approximately represent his prior beliefs. In this article we will concentrate on the uncertainty about the prior distribution and therefore assume that $F = G$.

The posterior density for θ under (Π, F) is given by

$$\mu_{\Pi}(\theta) = k_1 \ p(\theta) \ f(x_1|\theta) \ldots f(x_n|\theta), \tag{2.2}$$

where $p = \Pi'$, $f = F'$ and k_1 does not depend on θ. The change of variables

$$\tau = T_{\Pi}^{-1}(\theta) = \Pi_0^{-1}(\Pi(\theta)), \qquad \theta = T_{\Pi}(\tau) = \Pi^{-1}(\Pi_0(\tau)) \tag{2.3}$$

where, as usual, $\Pi_0^{-1}(u) = \inf\{t : \Pi_0(t) \geq u\}$, gives the posterior density

$$\mu_{T_{\Pi}}(\tau) = k_1 \ p_0(\tau) \ f(x_1|T_{\Pi}(\tau)) \ldots f(x_n|T_{\Pi}(\tau)), \tag{2.4}$$

Therefore, the uncertainty with regard to the shape of the prior distribution and expressed by the alternative models $[\Pi_0(\theta), F(y|\theta)]$ and $[\Pi(\theta), F(y|\theta)]$, can be recast in terms of the "dual" uncertainty with regard to the *parametrization* of the model and expressed by the alternative parametrizations $[\Pi_0(\theta), F(y|\theta)]$ and $[\Pi_0(\theta), F(y|T_\Pi(\theta))]$.

In practice, of course, the transformation T_Π is unknown and normally taken equal to the identity. Loosely speaking, then, if the absolute difference $|T(\theta) - \theta|$ is small for values of θ which are likely under $\mu_{\Pi_0}(\theta)$, then the posterior inference will be robust (the possible posterior densities will be close). On the other hand, the posterior inference will be non-robust if $|T(\theta) - \theta|$ is large for values of θ which are likely under $\mu_{\Pi_0}(\theta)$.

Example 2.1 This example illustrates the asymmetric roles of Π and Π_0 in determining the behavior of $|T_\Pi(\theta) - \theta|$. In Figure 1 we plot θ versus $|T_\Pi(\theta) - \theta|$ for the cases (i) Π_0 =Normal and Π =Cauchy and (ii) Π_0 =Cauchy and Π =Normal. In the first case, (solid line), $|T(\theta) - \theta|$ quickly becomes very large indicating that the normal prior will usually produce a fairly unstable posterior analysis. In the second case (dashed line), $|T(\theta) - \theta|$ remains fairly small indicating that the Cauchy prior will be more robust. Another simple

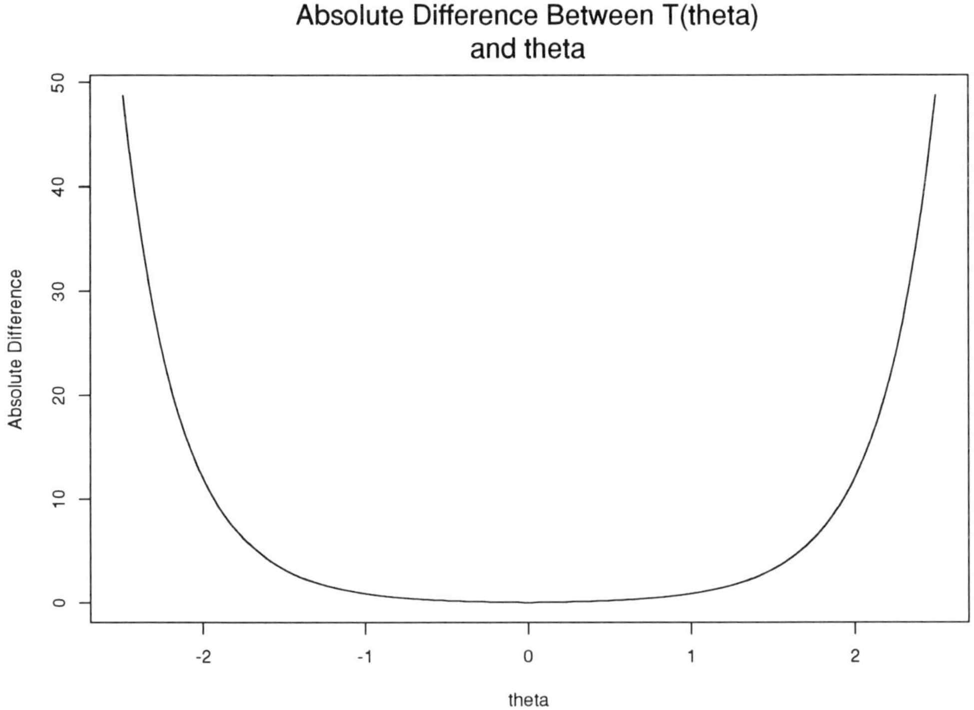

Figure 1:

example is when the prior distributions are $\mathcal{N}(0, 1)$ and $\mathcal{N}(0, \sigma^2)$, with any $\sigma^2 > 1$. The limiting case $\sigma^2 \to \infty$ corresponds to the non-informative prior.

When $\Pi_0 = \mathcal{N}(0,1)$ and $\Pi = \mathcal{N}(0,\sigma^2)$, then

$$|T_\Pi(\theta) - \theta| = |\sigma\Phi^{-1}[\Phi(\theta)] - \theta| = |\theta(\sigma - 1)| \to \infty, \quad \text{as } \sigma \to \infty,$$

for all $\theta \neq 0$. If, on the other hand, $\Pi_0 = \mathcal{N}(0,\sigma^2)$ and $\Pi = \mathcal{N}(0,1)$, then

$$|T_\Pi(\theta) - \theta| = |\Phi^{-1}[\Phi(\theta/\sigma)] - \theta| = |\theta[(1/\sigma) - 1]| \to |\theta|, \quad \text{as } \sigma \to \infty.$$

Therefore, a larger variance will normally lead to a more robust posterior analysis. ////

Example 2.2 This example illustrates how the "prior uncertainty" can be re-expressed in the form of "parametrization uncertainty". Let

$$f(x|\theta) = (1/\theta)\exp(-x/\theta)$$

and suppose that the prior densities $\pi_0(\theta)$ and $\pi(\theta)$ are exponential with means $\log(2)$ and 1, respectively. In this case,

$$\mu_{\Pi_0}(\theta) = k_{\Pi_0} \frac{1}{\log(2)} \exp\{-\theta/\log(2)\}\theta\exp\{-\theta x\}$$

where $\theta = \mathrm{E}[X|\theta]$ represents the conditional mean of X. On the other hand, since

$$T(\theta) = \theta/\log(2),$$

we have that

$$\mu_{T_\Pi}(\tau) = k_\Pi \frac{1}{\log(2)} \exp\{-\theta/\log(2)\}\frac{\theta}{\log(2)} \exp\{-(\theta/\log(2))x\}$$

and the parameter θ now represents the conditional median of X. In this case, the two different exponential prior distributions Π_0 and Π (on the conditional mean of X) gave rise to two different parametrizations of the conditional distribution of X: the mean and the median. ////

The Posterior-Score-Based Discrepancy Function

It is well known that changes on the tails of a density $u(t)$ can be better visualized by looking at the score function,

$$\hat{u}(t) = \frac{d}{dt}\log u(t) = \frac{u'(t)}{u(t)}, \tag{2.5}$$

rather than the density itself. For instance, a plot of the standard normal and Cauchy densities does not reveal the striking tails differences as well as a plot of the the corresponding score functions. Therefore, we will use the posterior score (the score function for the posterior density) to measure the sensitivity to changes on the priors. This approach has also the advantage that the proportionality constants k_0 and k_1 can be safely ignored.

We will pursue an asymptotic approach and compare the posterior asymptotic scores functions before and after the transformation T is applied. In the sequel, we will use θ as a dummy argument for all the posterior score functions involved. More precisely, we will study the limit value, $\Delta(\theta|\Pi, \Pi_0)$, of the discrepancy measure

$$\Delta_n(\theta|\Pi, \Pi_0) = \frac{1}{n}\left[\hat{\mu}_\Pi(\theta) - \hat{\mu}_{T_{\Pi_0}}(\theta)\right] = \frac{1}{n}\sum_1^n \left[\frac{\pi_0(\theta)}{\pi(T(\theta))}\hat{f}(x_i|T(\theta)) - \hat{f}(x_i|\theta)\right],$$

where $\hat{\mu}_\Pi(\theta)$ and $\hat{\mu}_{T_{\Pi_0}}(\theta)$ are given by (2.5), $\hat{f}(x|\theta) = (\partial/\partial t)\log f(x|t)|_{t=\theta}$, and

$$\frac{\pi_0(\theta)}{\pi(T(\theta))} = (\partial/\partial t)T(t)|_{t=\theta}.$$

Suppose now that the "true" mechanism that generates the sequence X_1, $X_2, \ldots$ is the following: given θ_0, the random variables $X_1, X_2, \ldots$ are i.i.d. with common distribution $H_{\theta_0}(x)$. By the Strong Law of Large Numbers, for each θ,

$$\Delta_n(\theta|\Pi, \Pi_0) \to \Delta(\theta|\Pi, \Pi_0), \qquad \text{a.s. } [H_{\theta_0}]$$

where

$$\Delta(\theta|\Pi, \Pi_0) = \mathrm{E}_{H_{\theta_0}}\left\{\frac{\pi_0(\theta)}{\pi(T(\theta))}\hat{f}(x|T(\theta)) - \hat{f}(x|\theta)\right\}. \tag{2.6}$$

For simplicity, we will take $H_{\theta_0}(x) = F(x|\theta_0)$. We will also assume that the prior distribution Π belongs to the family $\mathcal{P}$ given by (1.1), with fixed $0 < \varepsilon < 0.5$ and "contaminating" distribution Q ranging over $\mathcal{Q}$. This last set must be specified in accordance with the situation at hand.

We introduce now the following definitions.

Definition 2.3 *The maximum posterior score change with respect to changes on the prior distribution, $MC(\Pi_0, \theta)$, is defined as*

$$MC(\Pi_0, \theta) = \sup_{\Pi \in \mathcal{P}} |\Delta(\theta|\Pi, \Pi_0)|. \tag{2.7}$$

Observe that $MC(\Pi_0, \theta)$ is a global robustness measure, similar in nature to the maximum bias curve used in classical robustness.

Definition 2.4 *The influence function with respect to changes on the prior distribution is defined as*

$$IF(Q|\Pi_0, \theta) = \lim_{\varepsilon \to 0} \frac{\Delta(\theta|(1-\varepsilon)\Pi_0 + \varepsilon Q, \Pi_0)}{\varepsilon}, \tag{2.8}$$

provided that the limit exists.

Observe that $\mathrm{IF}(Q|\Pi_0, \theta)$ is the directional derivative of $\Delta(\theta|\Pi, \Pi_0)$ in the direction of Q, that is,

$$\mathrm{IF}(Q|\Pi_0, \theta) = \frac{d}{d\varepsilon}\ \Delta(\theta|(1-\varepsilon)\Pi_0 + \varepsilon Q, \Pi_0)\ |_{\varepsilon=0} \tag{2.9}$$

Definition 2.5 *The sensitivity to changes on the prior distribution is defined as*

$$S(\Pi_0, \theta) = \sup_{Q \in \mathcal{P}} |IF(Q|\Pi_0, \theta)|. \tag{2.10}$$

Now we can formally define b-robustness to changes on the prior, where as in Hampel et al. (1986) "b" stands for "bias" (to differentiate this concept from that of variance-robustness).

Definition 2.6 *The Bayesian inference based on Π_0 is said to be b-robust with respect to the prior if the influence function (2.9) is bounded (or equivalently, if $S(\Pi_0, \theta)$ is finite).*

The IF and the S are infinitesimal concepts, similar in nature to Hampel's influence function and gross-error-sensitivity. They give the infinitesimal rate of change of $\Delta(\theta|\Pi, \Pi_0)$ when the assumed model (Π_0, F) is perturbed in the direction of $((1-\varepsilon)\Pi_0 + \varepsilon Q, F)$ (in the case of the IF) and in the worst possible direction (in the case of S).

One would expect that, paralleling results in classical robustness, under regularity conditions

$$S(\Pi_0, \theta) = \frac{\partial}{\partial \varepsilon} MC(\varepsilon|\Pi_0, \theta)|_{\varepsilon=0}.$$

In such case, likewise in the classical setup, the left hand side above would constitute an "easier" way to compute the right hand side.

Let

$$A(\theta) = -\int_{-\infty}^{\infty} \hat{f}(x|\theta) f(x|\theta_0) dx, \tag{2.11}$$

and then we can write

$$\Delta(\theta|\Pi, \Pi_0) = A(\theta) - \frac{\pi_0(\theta)}{\pi(T(\theta))} A(T(\theta)). \tag{2.12}$$

Moreover, let

$$\hat{f}_0'(x|\theta) = \frac{d}{dt}\hat{f}_0(x|t)|_{t=\theta} \quad \text{and} \quad B(\theta) = A'(\theta) = \int_{-\infty}^{\infty} \hat{f}'(x|\theta) f(x|\theta_0) dx, \tag{2.13}$$

where the last equality holds only if differentiation under the integral sign is allowed. Then we have the following result.

Theorem 2.7 *Under regularity assumptions,*

$$IF(Q|\Pi_0,\theta) = \frac{q(\theta)-p_0(\theta)}{\pi_0(\theta)} + \frac{Q(\theta)-\Pi_0(\theta)}{\pi_0(\theta)}\left[B(\theta)-\hat{p}_0(\theta)A(\theta)\right]. \tag{2.14}$$

PROOF Let $h(\varepsilon) = T(\theta) = [(1-\varepsilon)\Pi_0 + \varepsilon Q]^{-1}\Pi_0(\theta)$. Then,

$$\Pi_0(\theta) = (1-\varepsilon)\Pi_0(h(\varepsilon)) + \varepsilon Q(h(\varepsilon)).$$

Differentiating both sides with respect to ε at $\varepsilon = 0$, we obtain,

$$0 = -\Pi_0(h(0)) + \pi_0(h(0))h'(0) + Q(h(0)),$$

Since $h(0) = \theta$ it follows that

$$\frac{d}{d\varepsilon}\, T(\theta) \mid_{\varepsilon=0} \;=\; h'(0) \;=\; \frac{\Pi_0(\theta)-Q(\theta)}{\pi_0(\theta)}.$$

In addition,

$$\begin{aligned}
&\frac{d}{d\varepsilon}\left(\frac{\pi_0(\theta)}{\pi(T(\theta))}\right)\mid_{\varepsilon=0} = \\
&= -\frac{\pi_0(\theta)}{\pi^2(T(\theta))}\Big[-\pi_0(T(\theta)) + (1-\varepsilon)p_0'(T(\theta))\frac{d}{d\varepsilon}\,(T(\theta))\mid_{\varepsilon=0} \\
&\quad + q(T(\theta)) + \varepsilon q'(T(\theta))\frac{d}{d\varepsilon}\,(T(\theta))\mid_{\varepsilon=0}\Big]\mid_{\varepsilon=0} \\
&= -\frac{1}{\pi_0(\theta)}\left[q(T(\theta)) - \pi_0(\theta) + \pi_0'(\theta)\frac{\Pi_0(\theta)-Q(\theta)}{\pi_0(\theta)}\right] \\
&= \frac{p_0(\theta)-q(\theta)}{\pi_0(\theta)} - \hat{\pi}_0(\theta)\frac{\Pi_0(\theta)-Q(\theta)}{\pi_0(\theta)}.
\end{aligned}$$

Therefore, from (3.4)

$$\begin{aligned}
&\mathrm{IF}(Q|\Pi_0,\theta) = \\
&= \frac{d}{d\varepsilon}\Delta(\theta|(1-\varepsilon)\Pi_0 + \varepsilon Q, \Pi_0)|_{\varepsilon=0} = -\frac{d}{d\varepsilon}\left(\frac{\pi_0(\theta)}{\pi(T(\theta))}A(T(\theta))\right)\mid_{\varepsilon=0} \\
&= -\frac{d}{d\varepsilon}\left(\frac{\pi_0(\theta)}{\pi(T(\theta))}\right)\mid_{\varepsilon=0} A(\theta) - \frac{d}{d\varepsilon}\,(A(T(\theta)))\mid_{\varepsilon=0} \\
&= \left[\frac{q(\theta)-p_0(\theta)}{\pi_0(\theta)} - \hat{\pi}_0(\theta)\frac{Q(\theta)-\Pi_0(\theta)}{\pi_0(\theta)}\right]A(\theta) + \frac{Q(\theta)-\Pi_0(\theta)}{\pi_0(\theta)}B(\theta).
\end{aligned}$$

////

Note that this theorem shows that the influence function (2.14) is very large if $\pi_0(\theta)$ is negligibly small when compared with $q(\theta)$. In particular, it is equal to infinity when $\pi_0(\theta) = 0$, and therefore, a π_0 with compact support is automatically non-robust by this criterion. On the other hand, if $\pi_0(\theta) > 0$ for all θ, $\mathrm{IF}(Q|\Pi_0,\theta)$ is bounded (for all Q!) and so, boundedness of (2.14) does not seem to be a good criterion to define the robustness of π_0. This is not necessarily surprising in view of analogous results for regression estimates in the classical robustness setup, where boundedness of the influence function does not necessarily imply robustness of the regression estimate (see for instance Martin et al. (1989)).

Given this limitation, we suggest to use the sensitivity (2.10)—which is given by the supremum of (2.14) when Q ranges over a certain class to determine the robustness merits of two prior distributions over a given interval of values of θ; such interval will be normally decided in accordance with the objectives at hand. In the next section we will compute the influence function for the location model and compare the robustness properties of the normal, double exponential and Cauchy priors, over different parameter ranges.

3 The Location Model

Let us assume that θ is a location parameter and $F(x|\theta) = F_0(x-\theta)$. We make the following assumptions on Π_0 and F_0:

(i) Π_0 has a positive and unimodal density $\pi_0(\theta)$ which is symmetric about a central value θ_0;

(ii) F_0 has a positive and unimodal density f_0 which is symmetric about zero;

(iii) $\psi_0(x) = -f_0'(x)/f_0(x)$ is well defined almost everywhere (set $\psi_0(x) = 0$ where it is not defined); and

(iv) $B(\theta) = A'(\theta) \geq 0, \quad \forall\, \theta \geq 0$.

Observe that ψ_0 is odd and $\psi_0(x) \geq 0$ for $x \geq 0$. Our choice for $\mathcal{Q}$ is given by

$$\mathcal{Q} = \left\{ Q : Q(\theta_0) = \frac{1}{2} \text{ and } q(\theta) \text{ is unimodal and symmetric about } \theta_0 \right\}. \quad (3.1)$$

To simplify the notations we will take $\theta_0 = 0$.

Remark 3.1 By symmetry of Π_0 and Q, $T(-\theta) = -T(\theta)$, for all θ, and $T(\theta) \geq 0$, for $\theta \geq 0$. Since in addition f_0 is symmetric, $A(-\theta) = -A(\theta)$, for all θ, and $A(\theta) \geq 0$ for $\theta \geq 0$. Consequently,

$$\begin{aligned}\Delta(-\theta|\Pi,\Pi_0) &= A(-\theta) - \frac{\pi_0(-\theta)}{\pi(T(-\theta))} A(T(-\theta)) \\ &= -A(\theta) + \frac{\pi_0(\theta)}{\pi(T(\theta))} A(T(\theta)) \\ &= -\Delta(\theta|\Pi,\Pi_0),\end{aligned}$$

and we can concentrate on the case $\theta \geq 0$. ////

Notice that by the assumptions on f_0, $A(t) = -\int_{-\infty}^{\infty} \psi_0(x-\theta) f_0(x)dx > 0$ for all $\theta > 0$. In addition, if $\psi_0'(x) \geq 0$ for all x and differentiation can be performed under the integral sign, then $B(\theta) = A'(\theta) = \int_{-\infty}^{\infty} \psi_0'(x-\theta) f_0(x)dx \geq 0$ for all θ.

If

$$C(\theta) = B(\theta) - \hat{p}_0(\theta)A(\theta),$$

then

$$\mathrm{IF}(Q|\Pi_0, \theta) = \frac{q(\theta) - \pi_0(\theta)}{\pi_0(\theta)} A(\theta) + \frac{Q(\theta) - \Pi_0(\theta)}{\pi_0(\theta)} C(\theta).$$

The functions h_1 and h_2 defined below are needed for the statement and the proof of Theorem 3.2. For any fixed $\theta > 0$ let

$$\begin{aligned} h_1(u,v) &= \left| \frac{u - \pi_0(\theta)}{\pi_0(\theta)} A(\theta) + \frac{v - \Pi_0(\theta)}{\pi_0(\theta)} C(\theta) \right|, \quad \text{and} \\ h_2(\alpha, q) &= \frac{\pi_0(\theta)}{(1-\varepsilon)\pi_0(\alpha) + \varepsilon q(\alpha)} A(\alpha). \end{aligned}$$

Observe that $h_1(q(\theta), Q(\theta)) = |\mathrm{IF}(\Pi|\Pi_0, \theta)|$ and that $A(\theta) - h_2(T(\theta), q) = \Delta(\theta|Q, \Pi_0)$.

Theorem 3.2 *Suppose that $B(\theta) > 0$. Then,*

$$\underline{T}(\theta) = \inf_{Q \in \mathcal{Q}} T(\theta) = \Pi_0^{-1}\left(\frac{\Pi_0(\theta) - \varepsilon}{1 - \varepsilon}\right), \tag{3.2}$$

and, if $\Pi_0(\theta) - .5\varepsilon < 1 - \varepsilon$ (the "regular" case)

$$\overline{T}(\theta) = \sup_{Q \in \mathcal{Q}} T(\theta) = \Pi_0^{-1}\left(\frac{\Pi_0(\theta) - (\varepsilon/2)}{1 - \varepsilon}\right), \tag{3.3}$$

and $\overline{T}(\theta) = \infty$, otherwise. Furthermore, in the regular case,

$$MC(\Pi_0, \theta) = \max\left\{ \frac{A(\overline{T}(\theta))}{(1-\varepsilon)\pi_0(\overline{T}(\theta))}, \frac{A(\underline{T}(\theta))}{(1-\varepsilon)\pi_0(\underline{T}(\theta)) + [\varepsilon/(2\underline{T}(\theta)]} \right\} \tag{3.4}$$

and $MC(\Pi_0, \theta) = \infty$, otherwise. Moreover,

$$S(\Pi_0, \theta) = \max\left\{ h_1(0, .5), \ h_1(0, 1), \ h_1(\frac{1}{2\theta}, 1) \right\}. \tag{3.5}$$

Proof Let $\theta > 0$. For any fixed $q(\theta) = y \geq 0$,

$$|\mathrm{IF}(Q|\Pi_0, \theta)| = h_1(y, Q(\theta))$$

is the absolute value of a linear function of $Q(\theta)$. Since $Q(\theta)$ takes all the values on $[.5 + y\theta, 1]$ as Q ranges over

$$\mathcal{Q}_y = \{ Q \in \mathcal{Q} \ , \ q(\theta) = y \},$$

it follows that

$$\sup_{Q\in\mathcal{Q}_y} |\mathrm{IF}(Q|\Pi_0,\theta)| = \max\{\ h_1(y,.5+\theta y)\ ,\ h_1(y,1)\ \}.$$

Equation (3.5) holds now because $h_1(y,.5+\theta y)$ and $h_1(y,1)$ are the absolute value of linear functions of y which reach their maximum and minimum over $[\theta, 1/2\theta]$ at either $y=0$ or $y=1/(2\theta)$.
To prove (3.4) we first notice that $|\Delta(\theta|Q,\Pi_0)|$ is maximized (in q) when

$$h_2(T(\theta),q) = \frac{\pi_0(\theta)}{(1-\varepsilon)\pi_0(T(\theta))+\varepsilon q(T(\theta))}A(T(\theta))$$

takes either its maximum or its minimum value. Second, since by assumption $B(\theta)\geq 0$, one follows that

$$\frac{\partial}{\partial\alpha}h_2(\alpha,q)\geq 0.$$

So, maximizing or minimizing $h_2(T(\theta),q)$ (for a fixed q) is equivalent to maximizing or minimizing $T(\theta)$. Now, the minimum of $T(\theta)$ (as Q ranges over $\mathcal{Q}$) is given by the smallest value of t for which the equation

$$(1-\varepsilon)\Pi_0(t)+\varepsilon Q(t) = \Pi_0(\theta) \tag{3.6}$$

holds, for some Q. That is,

$$\underline{T}(\theta) = \inf_{Q\in\mathcal{Q}}\{t : (1-\varepsilon)\Pi_0(t)+\varepsilon Q(t) = \Pi_0(\theta)\}.$$

Since,

$$(1-\varepsilon)\Pi_0(t)+\varepsilon Q(t) \leq (1-\varepsilon)\Pi_0(t)+\varepsilon, \quad \forall\, Q\ ,\ \forall\, t,$$

then

$$\underline{T}(\theta) \geq \inf_{Q\in\mathcal{Q}}\{t : (1-\varepsilon)\Pi_0(t)+\varepsilon = \Pi_0(\theta)\} = \Pi_0^{-1}\left(\frac{\Pi_0(\theta)-\varepsilon}{1-\varepsilon}\right).$$

Equation (3.2) follows now because (3.6) is satisfied with

$$t = \Pi_0^{-1}\left(\frac{\Pi_0(\theta)-\varepsilon}{1-\varepsilon}\right),$$

and Q uniform on $[-\underline{T}(\theta)\ ,\ \underline{T}(\theta)]$. Analogously, one notices that the maximum value of $T(\theta)$ is given by the largest value of t for which equation (3.6) holds, for some Q. Since

$$(1-\varepsilon)\Pi_0(t)+\varepsilon Q(t) \geq (1-\varepsilon)\Pi_0(t)+\varepsilon.5, \quad \forall\, Q\ ,\ \forall\, t,$$

we have

$$\overline{T}(\theta) \leq \Pi_0^{-1}\left(\frac{\Pi_0(\theta)-.5\varepsilon}{1-\varepsilon}\right),$$

θ	Normal	Double Exponential	Cauchy
0.00	2.51	2.00	3.14
0.25	2.15	2.13	2.99
0.50	2.02	2.28	3.40
0.75	2.17	2.45	4.20
1.00	3.82	2.65	5.28
1.50	12.37	4.29	9.11
2.00	46.20	7.15	16.39
2.50	206.74	11.22	26.01
3.00	1128.14	16.93	37.97
4.00	63512.33	35.95	68.96
5.00	8286946.00	71.61	108.93
6.00	∞	138.70	158.67
7.00	∞	262.28	217.84

Table 1: The maximum posterior score change for some prior distributions.

provided that $\Pi_0(\theta) - .5\varepsilon \leq 1 - \varepsilon$. Otherwise, it clearly follows that $\overline{T}(\theta) = \infty$. Equation (3.3) follows now because (3.6) is satisfied with

$$t = \Pi_0^{-1}\left(\frac{\Pi_0(\theta) - .5\varepsilon}{1 - \varepsilon}\right),$$

and Q uniform on $[-K\ ,\ K]$, with $K \to \infty$. Finally, $q(\theta) = .5\theta$ and $q(\theta) = 0$ are the largest and smallest possible values for $q(\theta)$, respectively, and therefore (3.4) follows. ////

Table 1 gives the values of $\mathrm{S}(\Pi_0, \theta)$ when Π_0 is the standard normal, double exponential and Cauchy distribution, for several values of θ. In general, one notices that $\mathrm{S}(\Pi_0, \theta)$ is relatively small in places where Π_0 concentrates a relatively large mass. For instance, the double exponential has the smallest sensitivity for $0 \leq \theta \leq 0.25$ and $0.82 \leq \theta \leq 6.42$. The standard normal has the smallest sensitivity (but fairly close to the double exponential) in the range $0.26 \leq \theta \leq 0.81$. Finally the Cauchy dominates the other two when $\theta \geq 6.43$.

The conclusion emerging from these comparisons is the following: other things equal, the double exponential should be preferred over the Normal and Cauchy distributions to achieve a good level of b-robustness.

4 Some Possible Extensions and Concluding Remarks

This paper concentrates on the univariate case, and particularly on the location model. However, the procedures presented in Section 2 can be easily applied to other single parameter situations such as the Binomial, Poisson and exponential models, etc.

The extension to the multiparameter case is also straightforward for the case of independent priors. For general priors, we can define the multivariate analogue of the transformation (2.9) as follows: let $\theta = (\theta_1, \ldots, \theta_k)$ be the vector of parameters and Π_0 and Π be two possible multivariate priors on θ with joint densities π_0 and π, respectively. Let us decompose π_0 as

$$\pi_0(\theta_1, \ldots, \theta_k) = p_{01}(\theta_1)\; p_{02}(\theta_2|\theta_1)\; \ldots\; p_{0k}(\theta_k|\theta_{k-1}, \ldots, \theta_1) \tag{4.1}$$

and analogously for p. Le us call Π_{0i} and Π_i, ($i = 1, \ldots, k$), the corresponding cumulative distribution functions. Then, it can be shown using for instance the Rosemblatt's transformation (see Mardia et al. (1979), pp. 36–37) that the new vector of parameters $\tau = (\tau_1, \ldots, \tau_n)$ defined as

$$\tau_i = \Pi_{0i}^{-1}(\Pi_i(\theta)), \qquad i = 1, \ldots, k, \tag{4.2}$$

has joint distribution Π_0 when θ has joint distribution Π. This result allows the straightforward generalization of our procedures to study the robustness properties of multivariate priors.

The discrepancy measure (2.6) can be easily differentiated with respect to the likelihood part of the Bayesian model by setting

$$f(x|\theta) = (1-\delta)g(x|\theta) + \delta l(x|\theta).$$

In fact, it is not difficult to verify that in this case,

$$\frac{d}{d\delta}\Delta(\theta|\mathcal{D})|_{\delta=0} = \frac{l(x|\theta)}{g(x|\theta)}[\hat{l}(x|\theta) - \hat{g}(x|\theta)].$$

This result can be used together with (2.7) to study the joint sensitivity of the Bayesian inference to changes on both, the prior and likelihood parts of the model.

The asymptotic nature of the present approach carried with it the disadvantage of being an approximation to any given real situation, and its finite sample relevance should be further investigated. On the other hand, it has the advantage of being relatively simple and general, in the sense of allowing the comparison of different Bayesian models independently of the particular sampling situation.

In summary, the present approach seems to offer a way to study the robustness properties of Bayesian procedures. It also allows the use of many tools, initially developed in the classical robustness setup, to the Bayesian field. One would hope, of course, that this connection will lay the ground for further cross-fertilizations between these two important areas of statistical research.

References

[1] Berger, J.O. (1984): The robust Bayesian viewpoint. *In Robustness of Bayesian Analysis* (J. Kadane ed.). North Holland, Amsterdam.

[2] Berger, J.O. (1990): Robust Bayesian analysis: Sensitivity to the prior. *J. Statist. Plann. Inference* **25** 303–328.

[3] Berger, J.O. and Berliner, L.M. (1986): Robust Bayes and empirical Bayes analysis with ε-contaminated priors. *Ann. Statist.* **14** 461–486.

[4] Box, G.E.P. and Tiao, G.C. (1968): A Bayesian approach to some outlier problems. *Biometrika* **75** 651–659.

[5] Hampel, F.R., Ronchetti, E.M., Rousseeuw, P.J. and Stahel, W.A. (1986): *Robust Statistics—The Approach Based on Influece Functions*. Wiley, New York.

[6] Huber, P.J. (1981): *Robust Statistics*. Wiley, New York.

[7] Kadane, J.B. (1984): *Robustness of Bayesian Analysis*. North Holland, Amsterdam.

[8] Marazzi, A. (1985): On constrained minimization of the Bayes risk for the linear model. *Statist. Decisions* **3** 277–296.

[9] Mardia, K.V., Kent, J.T. and Bibby, J.M. (1979): *Multivariate Analysis*. Academic Press, New York.

[10] Martin, R.D., Yohai, V.J. and Zamar, R.H. (1989): Min-max bias robust regression. *Ann. Statist.* **17** 1608–1630.

[11] Moreno, E. and Cano, J.A. (1991): Robust Bayesian analysis with ε-contamination partially known. *J. Roy. Statist. Soc. B* **53** 143–155.

[12] Pericchi, L.R. and Walley, P. (1991): Robust Bayesian credible interval and prior ignorance. *Internat. Statist. Rev.* **58** 1–23.

[13] Rieder, H. (1994): *Robust Asymptotic Statistics*. Springer, New York.

[14] Walley, P. (1990): *Statistical Reasoning with Imprecise Probabilities*. Chapman & Hall, London.

Computational Aspects of Trimmed Single-Link Clustering

Evangelos Tabakis*
Universität Bayreuth

Abstract

The main drawback of the single-link method is the chaining effect. A few observations between clusters can create a chain, i.e. a path of short edges joining the real clusters and thus making their single-link distances small. It has been suggested (Tabakis (1992a)) that a density estimator could help by eliminating those low-density observations creating the problem. Since the amount of trimming necessary is not known, we may have to compute and compare several minimal spanning trees using standard algorithms such as Prim's algorithm. In this paper we are concerned with the computational aspects of the problem. In particular, we prove that trimmed single-link can be applied on n points in $O(n^2)$ time, i.e. its time complexity does not exceed that of the classical single-link method. We also discuss the space requirements of the method.

Key words and phrases: Cluster analysis, single-link, minimal spanning tree, computational complexity.
AMS 1991 subject classifications: Primary 62H30; secondary 62G35, 90C27.

1 Introduction

One of the oldest and most discussed methods in clustering analysis is the single-link (or single linkage) method. There are several way to introduce it and we proceed by the one that links it to the minimal spanning tree.

We remind the reader that a tree is a connected graph with no cycles and that a minimal spanning tree (MST) on the set S of n points with interpoint distances $d(x,y)$, $x,y \in S$ is a tree with vertices S and minimal total length. Having constructed an MST of S, a hierarchy of S-subsets can be obtained. Formally, a hierarchy $\mathcal{H}$ of S-subsets is a class of S-subsets with the following properties:

- $S \in \mathcal{H}$ and $\{x\} \in \mathcal{H}$ for all $x \in S$.

* The author gratefully acknowledges the financial support of the Newton Institute for the Mathematical Sciences in Cambridge where part of this paper was written.

- When $A, B \in \mathcal{H}$ then $A \cap B \in \{\emptyset, A, B\}$

The single-link hierarchy $\mathcal{H}_{SL}$ is then constructed recursively as follows:

- S is in $\mathcal{H}_{SL}$.
- When a set A with at least two points is in $\mathcal{H}_{SL}$, then so are the sets of vertices of the two connected graph components obtained by deleting the longest edge of the MST on A.

A significant advantage of this approach is that it is divisive. In a typical application of the single-link method we would only need to perform a few "cuts" to obtain an adequate clustering. So, the bulk of work is contained in computing the MST. Such a tree can be computed in $O(n^2)$ time by the following algorithm due to Prim:

ALGORITHM 1:

1. Choose any $x_0 \in S$ and let:
   - $V := \{x_0\}$
   - $E := \emptyset$
   - $D(x, V) := d(x, x_0)$ for all $x \in S$
2. While $V \neq S$ do:
   - Choose $x' \in S \setminus V$ that minimizes $D(x, V)$ over $x \notin V$
   - Choose $y \in V$ that attains $d(x', y) = D(x', V)$
   - Add x' to V and (x', y) to E.
   - Update: $D(x, V) := \min\{D(x, V), d(x, x')\}$, $x \notin V$.
3. Define the MST to be the tree with vertices V and edges E.

A proof that the resulting tree is indeed an MST and that it can be obtained in $O(n^2)$ time can be found, for example, in Hu (1982), pp. 27–29.

2 Non-Robustness of the Single-Link

The reason why single-link as described here cannot be used in practice is the so called "chaining" problem. Compare the MST on a sample of 160 points drawn from the mixture:

$$\frac{1}{2}U_{[0,1]\times[0,1]} + \frac{1}{2}U_{[2,3]\times[2,3]}$$

with that on the same 160 points plus 40 points drawn from:

$$N\left(\begin{pmatrix} 1.5 \\ 1.5 \end{pmatrix}, \begin{pmatrix} 1 & 0 \\ 0 & 1 \end{pmatrix}\right)$$

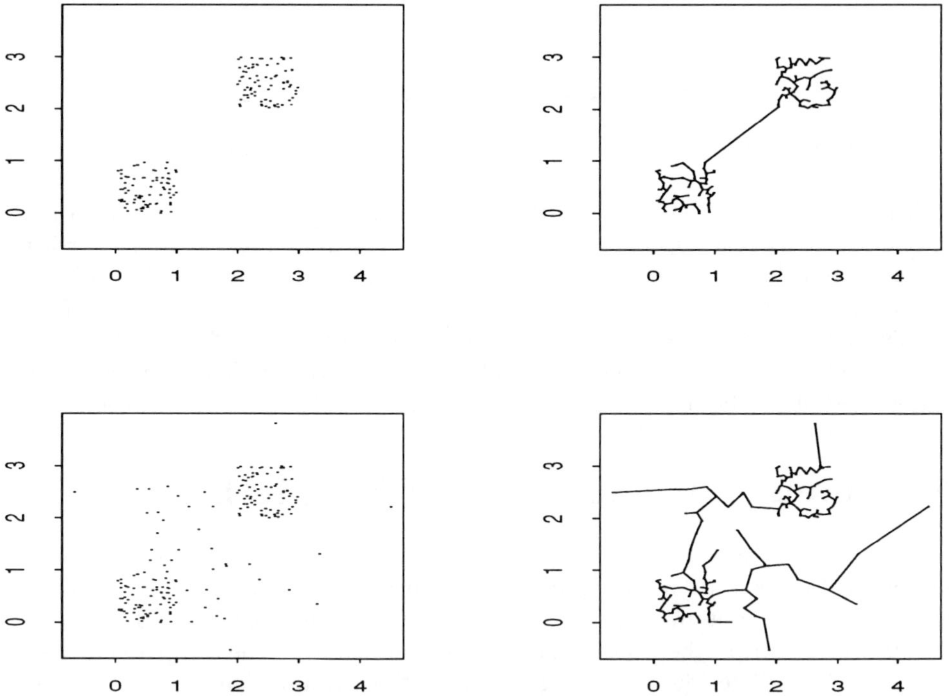

Figure 1: Contamination of a mixture of uniforms

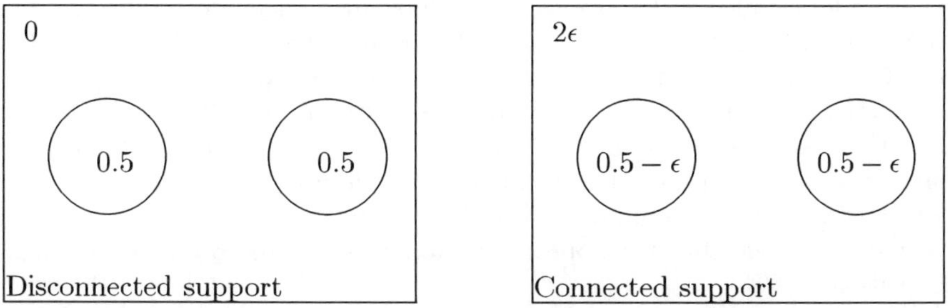

Figure 2: Effect of contamination

(see Figure 1). In both examples, the clustering structure is visually obvious: it consists of two clusters corresponding to the two uniform components of the mixture. For the single-link hierarchy to capture the structure, it would have to produce these two clusters at one of the very first "cuts" of the MST, i.e. after eliminating only a few of the longest edges of the tree. This is certainly what happens in the first sample but not in the second.

Note that the important statistic that guides the clustering process is the longest edge of the tree. Let M_n denote the longest edge of the MST on a set $X_1, X_2, \ldots, X_n$ of iid random vectors with common distribution F. Then we have:

- $\lim_n M_n = 0$ when F has a compact and connected support and
- $\lim_n M_n > 0$ when F has a compact disconnected support.

A proof of this is included in Tabakis (1992b). This obviously creates a robustness problem for M_n since it suffices to deviate from a distribution F with compact and disconnected support by an arbitrary amount to obtain a distribution G whose support is still compact but connected (see Figure 2). This means that the breakdown point of M_n is 0. A rigorous proof of this fact was given in Tabakis (1995). If we now think of the second sample of Figure 1 as a contamination of the first, we have a clear picture of this robustness problem.

3 Robustifying the Single-Link Method

When dealing with real data, one is more likely to come across the contaminated sample of Figure 1 than the "pure" one. Since the statistic M_n is so easily affected by such contamination we will have to make adjustments. Ideally, one would like to "strip" the contaminating observations from the sample and then compute M_n. The idea is similar to the use of trimmed means as location estimators (instead of simply using the non-robust sample mean). In the case of location estimators the type of contamination we mostly want to avoid is that from long-tailed distributions. Therefore, in the sample version we choose to remove the extreme observations and compute the mean of the rest. In our case, the contamination to be avoided is that of sparse points in the point cloud we are trying to cluster. If we assume the existence of a density for the point cloud, then sparse observations would lie in low-density areas and "real" cluster structure would lie in high-density areas. The picture of clusters as islands in a sea of low-density contamination has often been invoked to describe this case. It would then seem that the appropriate action would be to remove low-density observations.

For the moment, let us assume that we have n iid random vectors X_1, $X_2, \ldots, X_n$ distributed according to a known density f, continuous on its support and such that the sets $\{x \in \mathbf{R}^d : f(x) \geq \delta\}$ are compact for any $\delta > 0$. Then, the subsample $S_n(\delta) = \{X_i : f(X_i) \geq \delta\}$ can be assumed to be an iid sample from the density f_δ defined by:

$$f_\delta(x) := \frac{f(x) \cdot 1_{[f \geq \delta]}(x)}{\Pr(f \geq \delta)}.$$

We now define $M_n(\delta)$, the length of the longest edge of the MST on $S_n(\delta)$, and $(A_n(\delta), B_n(\delta))$, the partition of $S_n(\delta)$ resulting from removing that edge from the MST. The existence of a density makes all these objects almost surely unique.

We now hope to be able to distinguish between:

1. Densities f such that f_δ has compact **and** connected support for all $\delta > 0$.

2. Densities f such that f_δ has compact **but** disconnected support at least for some values of $\delta > 0$.

According to the observations in the previous section, $\lim_n M_n(\delta) = 0$ for all $\delta > 0$ in the first case, while there must be some $\delta > 0$ so that $\lim_n M_n(\delta) > 0$ in the second case. We would like to be able to

- decide which class f belongs to and,
- if it belongs to the second class, find a way to produce a sensible clustering.

In Tabakis (1992a), it was argued that if f belongs to the first class, then $M_n(\delta)^d$ decreases like $1/\delta$ and a graphical test of the hypothesis that f is in class A was based on this fact. But, if we decide that f is in class B, how do we produce a clustering of $X_1, X_2, \ldots, X_n$? Keeping with the idea suggested in the beginning of the section, we must throw out a number of low-density observations and work with a subset $S_n(\delta)$ for some δ. Then the single-link method applied to $S_n(\delta)$ will produce clusters $A_n(\delta)$ and $B_n(\delta)$. In the spirit of hierarchical methods, further clustering can be obtained by repeating the procedure on $A_n(\delta)$ and $B_n(\delta)$. But what is the right value of δ?

One can argue that δ must be large enough to allow for a "clear" picture of the structure. But letting it be too large would tend to produce very small clusters containing too few observations to be of any use. A case can be made against the use of $M_n(\delta)$ alone in choosing δ. Choosing δ to maximize $M_n(\delta)$ will result in one of the clusters having only one element. Choosing δ as the point where $M_n(\delta)$ starts to increase (as we let δ increase) may not produce a good separation of clusters. What we really need is to choose δ so as to obtain clusters $A_n(\delta)$ and $B_n(\delta)$ large enough but at the same time separated by a considerable distance $M_n(\delta)$. One way to achieve this is to maximize the separation statistic:

$$T_n(\delta) = \text{card}(A_n(\delta)) \cdot \text{card}(B_n(\delta)) \cdot M_n(\delta).$$

The reader who is familiar with single-link distances will immediately recognize this statistic as the sum of "between" single-link distances for the $(A_n(\delta), B_n(\delta))$ partition. In the one-dimensional case this statistic was proved to be asymptotically equivalent to the average single-link distance on $S_n(\delta)$ and to a U-statistic (Tabakis (1992b)).

4 Trimmed Single-Link Clustering

Let us now return to the real world where f is going to be unknown. However, given a "good" density estimator f_n, the previous definitions can be adjusted by substituting f by f_n, giving computable quantities. To avoid introducing new notation, we will still write $S_n(\delta), M_n(\delta), A_n(\delta), B_n(\delta)$ and $T_n(\delta)$ for the truncated sample, the length of the longest edge, the clusters and the separation statistic resulting from the use of f_n instead of f.

There are two computational tasks to discuss:

- The estimation of the density.
- The computation of the process $T_n(\delta)$.

Combining density estimation with clustering methods is not a new idea. In Kittler (1976), Schaffer et al. (1979) and Kittler (1979), there is a discussion of a mode seeking method that appears to resemble single-link clustering. Another hybrid clustering method using density estimation is described in Wong (1982), Wong and Lane (1983). We will treat the two tasks as separate but the reader might wish to see Rozál and Hartigan (1992) for an effort to combine the two.

The reader is certainly well aware of the variety of density estimation methods. We will only assume that we have a method that will compute a consistent estimate of $f(x)$ (say $f_n(x)$) for a given x in $O(n)$ time. Estimators of the kernel type for example will certainly accomplish the task. Note that we only need to compute $f_n(X_i)$, $i = 1, 2, \ldots, n$ so that the first task will take $O(n^2)$ time.

The real challenge however is the computation of $T_n(\delta)$, $\delta > 0$. First note that the only interesting values of δ are $\delta_i = f_n(X_i)$. It is only at these values that a change in $S_n(\delta)$ takes place. We sort these δ's to obtain $\delta_{(1)} \leq \delta_{(2)} \leq \cdots \leq \delta_{(n)}$. Then $S_n(\delta_{(1)}) \supset S_n(\delta_{(2)}) \supset \ldots \supset S_n(\delta_{(n)})$. As a first step, we reduce the computation of these statistics to the computation of $E_n(\delta_{(i)})$, the set of edges of the MST on $S_n(\delta_{(i)})$. First we need the following:

Lemma 4.1 *There is a constant N_d, depending only on d, such that for any MST in $\mathbb{R}^d$ and for any vertex of that MST, the number of edges adjacent to the vertex is bounded by N_d. Furthermore, $N_d = O(2.74^d)$ as $d \to \infty$.*

PROOF The existence of such a constant is a special case of Lemma 2.4 in Steele (1988). It is based on the fact that no two edges of the MST can form an angle of less than $60°$. Therefore, the number of edges adjacent to any vertex must be bounded by the maximum possible number N_d of edges adjacent to a point in $\mathbb{R}^d$ without any of the edges forming an angle smaller than $60°$. Clearly, N_d only depends on d.

Furthermore, N_d is bounded by the maximum number of $30°$ caps than can be packed on the surface of the unit sphere S^{d-1}. This, in turn, is bounded by the ratio of the surface area A_d^{180} of S^{d-1} to the surface area A_d^{30} of a $30°$ cap. Let $h(x) = \sqrt{1 - x^2}$. Then:

$$\begin{aligned} A_d^{30} &= \int_{\sqrt{3}/2}^{1} h(x)^{d-2} \cdot A_{d-1}^{180} \cdot \sqrt{1 + [h'(x)]^2}\,dx \\ &= A_{d-1}^{180} \int_{\sqrt{3}/2}^{1} h(x)^{d-3}\,dx \end{aligned}$$

and similarly:

$$A_d^{180} = 2A_{d-1}^{180} \int_0^1 h(x)^{d-3} dx.$$

Since $h(x) \leq h(0)$ for $x \in [0,1]$:

$$A_d^{180} \leq 2A_{d-1}^{180} h(0)^{d-3}(1-0) = 2A_{d-1}^{180}.$$

Note that the function h is decreasing in $[\sqrt{3}/2, 1]$. So, for $0 < \epsilon < 1 - \sqrt{3}/2$:

$$\begin{aligned} A_d^{30} &> A_{d-1}^{180} \int_{\sqrt{3}/2}^{1-\epsilon} h(x)^{d-3} dx \\ &\geq A_{d-1}^{180} \left(1 - \epsilon - \frac{\sqrt{3}}{2}\right) [h(1-\epsilon)]^{d-3} \\ &= A_{d-1}^{180} \left(1 - \epsilon - \frac{\sqrt{3}}{2}\right) (2\epsilon - \epsilon^2)^{\frac{d-3}{2}} \\ &> A_{d-1}^{180} \left(1 - \epsilon - \frac{\sqrt{3}}{2}\right) \epsilon^{\frac{d-3}{2}}. \end{aligned}$$

So:

$$N_d < \frac{2}{\left(1 - \epsilon - \frac{\sqrt{3}}{2}\right) \epsilon^{\frac{d-3}{2}}} = \frac{2\epsilon^{3/2}}{\left(1 - \epsilon - \frac{\sqrt{3}}{2}\right)} (\epsilon^{-1/2})^d.$$

Using an ϵ slightly smaller than $1 - \sqrt{3}/2$, eg. $\epsilon = 0.1339$ we get $N_d = O(2.74^d)$ as $d \to \infty$. ////

An MST (S_n, E_n) on $S_n = \{X_1, X_2, \ldots, X_n\}$ can be represented in a variety of ways. We choose the following form: Define an array t of dimension nN_d. Fill the block $t((i-1)N_d + 1)$ to $t(iN_d)$ by the integers j such that $(i,j) \in E_n$ in ascending order. Any remaining values are filled with zeros. By Lemma 4.1, an array of dimension nN_d will be sufficient. As is generally done in the combinatorial optimization literature (see e.g., Papadimitriou and Steiglitz (1982), p. 161), we assume that all integers to be treated by the algorithm can be transcribed in unit time. Under this assumption, we have:

Lemma 4.2 *Given an MST* $TREE_n = (S_n, E_n)$ *on* $S_n = \{X_1, X_2, \ldots, X_n\}$ *with the above described representation and any edge* $e \in E_n$, *we can obtain the two trees* $TREE_n^A(e) = (V_n^A(e), E_n^A(e))$ *and* $TREE_n^B(e) = (V_n^B(e), E_n^B(e))$ *resulting from removing the edge* e *in* $O(n)$ *time and using* $O(2.74^d n)$ *space as* $n, d \to \infty$.

PROOF If $e = (X_i, X_j)$, we search within the blocks $t((i-1)N_d + 1)$ to $t(iN_d)$ and $t((j-1)N_d + 1)$ to $t(jN_d)$ to locate vertices adjacent to i and j. Then we search within the blocks corresponding to these vertices and continue in this fashion until we have exhausted all the vertices in S_n. Every entry in every block corresponds to an edge of the MST. Conversely, every edge appears in exactly two blocks (the ones corresponding to its endpoints). There are exactly $n-1$ edges in the MST and, therefore, there are exactly $2(n-1)$ non-zero entries in the array t. So, it will only take $O(n)$ time to complete the search and transcription (regardless of how large d is).

The space needed to store the array t is $O(nN_d)$. (We know that only $2(n-1)$ entries will be non-zero but we don't know where these entries are going to be.) By Lemma 4.1, this makes the space needed $O(2.74^d n)$. ////

It is now obvious that, once we have the MST on $S_n(\delta_{(k)})$, we will need $O(k)$ time to find $M_n(\delta_{(k)})$ and (because of Lemma 4.2) $O(k)$ time to find the sets $A_n(\delta_{(k)})$ and $B_n(\delta_{(k)})$ and, therefore, compute $T_n(\delta(k))$. Next we give an algorithm that computes the MST on the set $S_n(\delta_{(i)})$, for $1 \le i \le n$. We will present an algorithm that introduces a new vertex w to an MST $\text{TREE}_n = (V_n, E_n)$ of n vertices. Then, the n MSTs on $S_n(\delta_{(i)})$ can be computed by successive use of this algorithm. The existence of a density f assures us that every MST that we will come across will have a unique longest edge. (When working in floating-point arithmetic, this will happen with probability very close to 1). We also agree that an MST with only one vertex has longest edge equal to 0.

ALGORITHM 2

Input:

1. The tree $\text{TREE}_n = (V_n, E_n)$.
2. The vertex $w \notin V_n$.

Output:

The tree $\text{TREE}_{n+1} = (V_{n+1} := V_n \cup \{w\}, E_{n+1})$.

1. LET $d_w := d(w, u_w) := \min_{u \in V_n} d(w,u)$.
   LET e_n be the longest edge of (V_n, E_n).

2. • IF $d_w \ge ||e_n||$ THEN:
     - add (w, u_w) to E_n and w to V_n
     - STOP
   • ELSE compute the trees:

$$\begin{aligned}\text{TREE}_n^A(e_n) &= (V_n^A(e_n), E_n^A(e_n))\\ \text{TREE}_n^B(e_n) &= (V_n^B(e_n), E_n^B(e_n))\end{aligned}$$

   resulting from the removal of e_n.
   LET $d_w^A := d(w, u_w^A) := \min_{u \in V_n^A} d(w,u)$.

LET $d_w^B := d(w, u_w^B) := \min_{u \in V_n^B} d(w, u)$.
IF $d_w^A > d_w^B$ SWAP:

$$\begin{array}{ccc} \text{TREE}_n^A(e_n) & \leftrightarrow & \text{TREE}_n^B(e_n) \\ d_w^A & \leftrightarrow & d_w^B \end{array}$$

IF $d_w^A < ||e_n|| \leq d_w^B$ THEN

- include E_n^B and e_n in the new MST,
- to complete the MST, run the algorithm on $\text{TREE}_n^A(e_n)$ and w.
- STOP

IF $d_w^A \leq d_w^B < ||e_n||$ THEN

- include (w, u_w^A) and (w, u_w^B) in the new MST,
- to complete the MST, run the algorithm on:
  (a) $\text{TREE}_n^A(e_n)$ and w,
  (b) $\text{TREE}_n^B(e_n)$ and w.
- STOP

Theorem 4.3 ALGORITHM 2 *terminates giving a minimal spanning tree on* V_{n+1}.

PROOF The following lemma describes a fundamental property of the MST which will be useful for the proof. It is based on the equivalence of single-link distances and minimal spanning trees as shown in Gower et al. (1969) and a proof of it can be found in Lebart et al. (1984):

Lemma 4.4 *Let* V *be a set of vertices and* d *a distance on* V. *Let* $T = (V, E)$ *be an MST on* V. *For* $u, v \in V$ *and for every path* $\mathbf{p} = (u = w_1, w_2, \ldots, w_n = v)$ *joining* u *and* v $(w_1, w_2, \ldots, w_n \in V)$ *define:*

$$d_{SL}(u, v; \mathbf{p}) = \max_{1 \leq i < n} d(w_i, w_{i+1}).$$

Then, for every edge $e = (u, v) \in E$:

$$d(u, v) = \min\{d_{SL}(u, v; \mathbf{p}), \mathbf{p} \text{ a path joining } u \text{ and } v\}.$$

This is often expressed by saying that the edges of the MST achieve the single-link distance between their endpoints.

We now return to the algorithm. In the first step, the algorithm compares the nearest neighbor edge (w, u_w) to the longest edge e_n of (V_n, E_n). The two possibilities are:

a) The nearest neighbor edge is longer than e_n. Then Lemma 4.4 requires that the nearest neighbor edge becomes the new edge of the MST on V_{n+1}.

The construction is complete because the resulting graph is certainly a tree (connected with no cycles) and the presence of the new edge does not affect the optimality of the old (V_n, E_n) part since every edge of E_n is shorter than the new edge.

b) The nearest neighbor edge is shorter than e_n. Then we compute the distances d_w^A and d_w^B of w from the two connected components $(V_n^A(e_n), E_n^A(e_n))$ and $(V_n^B(e_n), E_n^B(e_n))$ of (V_n, E_n) that we obtain by eliminating e_n. There are two subcases to consider:

b1) $d_w^A < ||e_n|| \leq d_w^B$ (the symmetric case to be treated similarly). Then w must be connected to $(V_n^A(e_n), E_n^A(e_n))$ with the nearest neighbor edge (again by Lemma 4.4). We may have to make more changes in $(V_n^A(e_n), E_n^A(e_n))$ because some of the edges in $E_n^A(e_n)$ may not be optimal any more (in the sense of Lemma 4.4). That is why we have to run the algorithm again using the tree $(V_n^A(e_n), E_n^A(e_n))$ and the new node w as input. However, there is no reason to change anything in $(V_n^B(e_n), E_n^B(e_n))$ which remains an optimal subtree. Even e_n remains in the new tree because we can not improve on bridging $(V_n^A(e_n), E_n^A(e_n))$ and $(V_n^B(e_n), E_n^B(e_n))$ using w. So, we have reduced the problem to introducing w into the smaller tree $(V_n^A(e_n), E_n^A(e_n))$ which has at least one vertex less.

b2) $d_w^A \leq d_w^B < ||e_n||$. Then w must be connected not only to $(V_n^A(e_n), E_n^A(e_n))$ by the nearest neighbor edge, but to $(V_n^B(e_n), E_n^B(e_n))$ by an edge of length d_w^B as well. The edge e_n is no longer the optimal bridge between $(V_n^A(e_n), E_n^A(e_n))$ and $(V_n^B(e_n), E_n^B(e_n))$ since a path using (w, u_w^A) and (w, u_w^B) is going to be better. It must be dropped from the new tree. We now have to recompute the MSTs on both $V_n^A \cup \{w\}$ and $V_n^B \cup \{w\}$. However, both subtrees are smaller than the original by at least one vertex.

The algorithm terminates, since each recursive call on it reduces the number of the remaining edges by at least one. Furthermore, every recursive call (if needed) results in an MST on the subtree considered at that point, provided that the necessary changes in any further subtrees are made. So, recursion will finally give an MST on V_{n+1}. ////

5 Binary Form of an MST

We will need to successively introduce the n observations $X_1, X_2, \ldots, X_n$ in descending order of the density estimator f_n into the tree, each time making sure the tree is an MST and computing the statistic $T_n(\delta_{(i)})$. How long is this going to take?

Since the computation of $\min_{u \in V_n} d(w, u)$ alone will take $O(n)$ time, we need at least $O(n^2)$ time to compute all nearest neighbor distances. Consider now the univariate example shown in Figure 3. We would like to add the new vertex w to the MST on $\{u_1, u_2, \ldots, u_n\}$. Using **ALGORITHM 2**, we will need to recursively call the algorithm on $\{u_2, u_3, \ldots, u_n\}$, then on $\{u_3, \ldots, u_n\}$ and

Figure 3: An example of worst case performance.

continue this way until finally calling the algorithm on $\{u_n\}$. Each recursion will compute the distance of w from subtrees TREE_n^A and TREE_n^B. This gives a total of $\mathrm{O}(n^2)$ time. Does this mean that, in the worst case, we will need $\mathrm{O}(n^3)$ time to build the MSTs on the sets $S_n(\delta_{(i)})$ for $1 \leq i \leq n$?

The answer is no. It suffices to keep track of the tree structure as we build the tree and compute the distances of the new edge w from **all** subtrees in $\mathrm{O}(n)$ time. First we build the MST (V_n, E_n) (eg. the one of Figure 4) in the form of a binary tree B_n with *vertices* $E_n \cup V_n$. The root will be the longest edge e_n of E_n having the longest edges of the two subtrees obtained by removing e_n as *descendants.* The rest is then constructed recursively. The *leaves* of the tree will be the vertices of the MST (edges of length 0). See Figure 5.

Theorem 5.1 *Suppose that we have been given the MST (V_n, E_n) in the form B_n and we want to introduce a new node w. Suppose also that we have computed and saved at every node $e \in E_n$ of B_n the distances*

$$d_w(e) = \min\{d(u,w),\ u \text{ in the subtree whose longest edge is } e\}. \tag{5.1}$$

We also agree that $d_w(v) := d(v,w)$ for all leaves (v) of B_n $(v \in V_n)$. Then the new MST on $V_n \cup \{w\}$ can be constructed and saved as the new binary tree B_{n+1} in $\mathrm{O}(n)$ time using **ALGORITHM 2**.

PROOF The only case that needs some work is the case $d_w^A \leq d_w^B < ||e_n||$ because it is in this case that the root of the B_n must be deleted and substantial changes

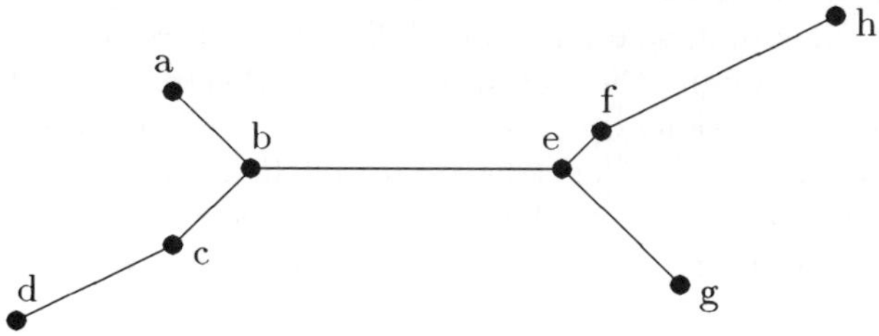

Figure 4: An MST in (V_n, E_n) form

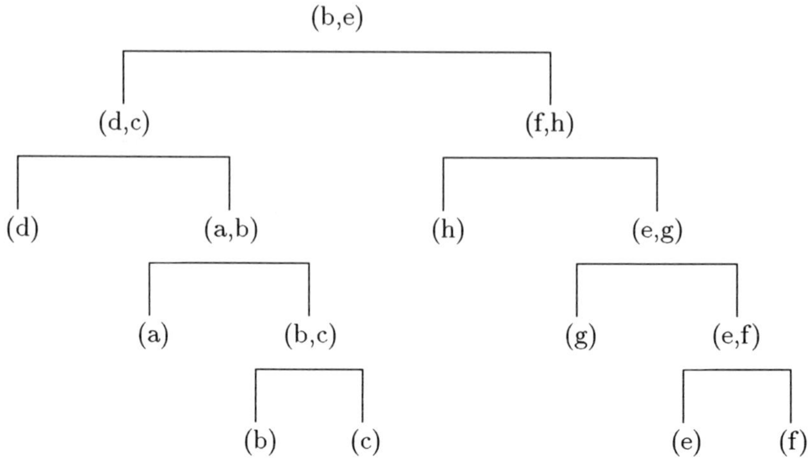

Figure 5: The same MST in B_n form

may be needed. However, it suffices to slightly change this particular step in **ALGORITHM 2**. It should now read:

IF $d_w^A \leq d_w^B < ||e_n||$ THEN

- IF $\max\{||e||,\ e \in E_n^A\} < \max\{||e||,\ e \in E_n^B\}$ SWAP

$$\begin{array}{ccc} \text{TREE}_n^A(e_n) & \leftrightarrow & \text{TREE}_n^B(e_n) \\ d_w^A & \leftrightarrow & d_w^B \end{array}$$

- to complete the MST, run the algorithm on:
  1. $\text{TREE}_n^A(e_n)$ and w to obtain the binary subtree B_{n+1}^A,
  2. $\text{TREE}_n^B(e_n)$ and w to obtain the binary subtree B_{n+1}^B.
- replace the leaf (w) in B_{n+1}^A by the subtree B_{n+1}^B to obtain B_{n+1}.
- STOP

In plain English this means that we first run the algorithm on the subtree with the longest maximal edge (say $\text{TREE}_n^A(e_n)$) and we replace the new leaf (w) by the result of running the algorithm on the other subtree ($\text{TREE}_n^B(e_n)$). Notice that, once the distances $d_w(e)$ are known, the only operations depending on n are the comparisons of $||e||$ (the length of the edge e) and the distances $d_w(e)$. Since there are exactly $n-1$ non-terminal nodes in B_n (the $n-1$ edges in E_n) and each can be used only once, the comparisons will take $O(n)$ time. ////

It remains to show that the computation of $d_w(e)$, $e \in E_n$ and $d_w(v)$, $v \in V_n$ also takes $O(n)$ time. Consider the following algorithm:

ALGORITHM 3

Input:

1. The tree $\mathrm{TREE}_n = (V_n, E_n)$ in the binary form B_n.
2. The vertex $w \notin V_n$.

Output:

The distances $d_w(e)$, $e \in E_n$ and $d_w(v)$, $v \in V_n$ defined in (5.1).

1. Initialize by setting all $d_w(e)$, $e \in E_n$ and $d_w(v)$, $v \in V_n$ equal to 0.
2. Compute distances $d(w, v)$, $v \in V_n$, thus computing the d_w distances for every leaf of B_n.
3. Until a node having no links (neither descendants nor ancestor) is reached:
   (a) If the current node has descendants move to one of them and skip part (b).
   (b) Add the current d_w value of the current node to the current d_w value of its ancestor-node. Then delete the current node and move to the ancestor-node.

In implementing the algorithm, we would not really delete links. We would rather mark them as used. It is not difficult to see now that every node in B_n will be visited twice if it is an edge $e \in E_n$ and once if it is a vertex $v \in V_n$ (i.e. a leaf of B_n). Therefore the whole computation will not exceed $O(n)$ time.

6 Conclusions

We have managed to implement the trimmed single-link method already suggested in Tabakis (1992a) in no more time than the usual single-link method thus gaining in robustness without increasing computational complexity. We based our analysis on the binary representation of the MST where every node represents a subtree pointing to two descendants, the two sub-subtrees obtained by deleting the longest edge of the current subtree. This representation has the additional advantage of allowing us to reapply the trimmed single-link method on subtrees without having to recompute the tree structure.

It is perhaps interesting to add that cases such as the one in Figure 3 have a small probability to occur. In the example of Figure 3, the new vertex w is the last to be added to the tree. If that is the case, it must satisfy:

$$f_n(w) \leq \min\{f_n(u_1), f_n(u_2), f_n(u_3), \ldots, f_n(u_n)\}$$

where f_n is the density estimator used. To the extent that f_n is a good estimator of the true density f, this hardly seems probable, judging from the distances between points.

Having said that, the reader must be warned not to assume that an observation having small estimated density is necessarily going to be more isolated than others and will therefore join the MST by means of a very long edge. In fact, it can be argued that this is exactly the difficulty causing the chaining problem of the single-link method: The statistic d_w in **ALGORITHM 2** is not a good basis for an estimate of the density of w. It would lead to an 1-nearest neighbor density estimator for the density f which is not consistent; k_n-nearest neighbor density estimators must satisfy $\lim_n k_n = \infty$ to be consistent. This fact is also pointed out in Hartigan (1981). Trimmed single-link tries to avoid the problem by imposing the use of a consistent estimator f_n, independent of the MST calculations, and building the MST by introducing nodes in the order suggested by that estimator. In this paper we showed that, since neither the density estimation nor the building of the MST have to exceed $O(n^2)$ time, the overall complexity is still comparable to that of the classical single-link method.

References

[1] Gower, J.C. and Ross, C.J.S. (1969): Minimum spanning trees and single linkage cluster analysis. *J. Roy. Statist. Soc. C* **18** 54–64.

[2] Hartigan, J.A. (1981): Consistency of single linkage for high-density clusters. *J. Amer. Statist. Assoc.* **76** 388–394.

[3] Hu, T.C. (1982): *Combinatorial Algorithms.* Addison-Wesley, Reading.

[4] Kittler, Josef (1976): A locally sensitive method for cluster analysis. *Pattern Recognition* **8** 23–33.

[5] Kittler, Josef (1979): Comments on "Single-link characteristics of a mode-seeking algorithm". *Pattern Recognition* **11** 71–73.

[6] Lebart, L., Morineau, A. and Warwick, K.M. (1984): *Multivariate Descriptive Statistical Analysis: Correspondence Analysis and Related Techniques for Large Matrices.* Wiley.

[7] Papadimitriou, Christos H. and Steiglitz, Kenneth (1982): *Combinatorial Optimization. Algorithms and Complexity.* Prentice-Hall, Englewood Cliff.

[8] Rozál, G.P.M. and Hartigan, J.A. (1994): The MAP Test for Multimodality. *Journal of Classification* **11** 5–36.

[9] Schaffer, E., Dubes, R. and Jain, A. (1979): Single-link characteristics of a mode-seeking clustering algorithm. *Pattern Recognition* **11** 65–70.

[10] Steele, J. Michael (1988): Growth rates of euclidean minimal spanning trees with power weighted edges. *Ann. Prob.* **16** 1767–1787.

[11] Tabakis, Evangelos (1992a): Visualizing Cluster Structure in High Dimensions. *Computing Science and Statistics* **24** 582–586.

[12] Tabakis, Evangelos (1992b): Assessing cluster structure on the real line using single-link distances. Submitted to *Comm. Statist. A*, under revision.

[13] Tabakis, Evangelos (1995): On the length of the longest edge of the minimal spanning tree. In *From Data to Knowledge: Theoretical and Practical Aspects of Classification, Data Analysis and Knowledge Organization* (W. Gaul, D. Pfeifer, eds.), 222–230. Springer-Verlag, Berlin,

[14] Wong, M. Antony (1982): A hybrid clustering method for identifying high-density clusters. *J. Amer. Statist. Assoc.* **77** 841–847.

[15] Wong, M. Antony and Lane, Tom (1983): A kth nearest neighbour clustering procedure. *J. Roy. Statist. Soc. B* **45** 362–368.

Interval Probability on Finite Sample Spaces

Kurt Weichselberger
Ludwig–Maximilians–Universität München

Abstract

In the seventies and early eighties theory of interval probability has gained very much from contributions of Peter Huber. The present article gives an outline of an approach aiming to establish unifying foundations of that theory. A system of axioms for interval probability is described generalizing Kolmogorov's axioms for classical probability and distinguishing two types of interval probability. The first one (*R-probability*) is characterized by employing interval-limits which are not self-contradictory, while the interval-limits of the second (*F-probability*) fit exactly to the set of classical probabilities in accordance to these interval-limits (*structure*). The task of controlling whether an assignment of intervals to each random event of a finite sample space represents at least R-probability or even F-probability is done by means of Linear Programming. The possibility of characterizing R-probability by features of its structure is sketched. A thorough presentation of theory will be given in a forthcoming book by the author.

Key words and phrases: Interval probability, axiomatic foundations of probability.
AMS 1991 subject classifications: Primary 60A05.

1 Introduction

History of interval probability is traced back as far as to Jakob Bernoulli by some authors (e.g., Shafer (1978)), but it was not before 1940, when for the first time the idea of upper and lower limits for probability was systematically introduced by B.O. Koopman (Koopman (1940a), (1940b), (1941)), with respect to epistemic probability. Generally the importance of an approach of this type had not been realized, when Choquet (Choquet (1953)) in the early fifties provided a mathematical tool, the theory of capacities, suitable for handling upper and lower limits instead of exact numbers. From 1964 on it was adapted for use in probability theory by Strassen (1964) and Huber ((1973), (1976), (1981)). In the first line the Huber–Strassen Theorem (Huber and Strassen (1973)) from 1973 was important for making known among probabilists the use of two-monotone capacities instead of classical probability assignments.

While in the following period interest in interval probability focussed on the application of capacities, in his 1981 book Huber (1981,Chapter 10) reports on results concerning a far more general type of interval probability than the one generated by capacities.

Another important contribution of Peter Huber to the theory of interval probability is also contained in this book: Most authors define interval probability by mathematical properties of upper and lower limits such as subadditivity, superadditivity, monotonicity of order two or total monotonicity. Using this type of methods—*outer methods*—it is, however, not possible to find conditions, which are necessary as well as sufficient for interval probability. In contrast Huber (1981, Chapter 10), introduces in his 1981 book interval probability by use of the relation between interval-limits and a certain set of classical probabilities. Such procedures may be called *inner methods*. It is the kind of methods applied to develop a system of axioms for interval probability in the present article.

Between Huber's approach from 1981 and the one chosen here there is one important difference. Huber's basic probabilistic concept is mainly that of expectation, while the approach in the present article uses probability itself as basic concept. Although—at least for finite sample spaces—results concerning probability may be derived from those on expectations by specializing to the outcomes 0 and 1, mathematical treatment in both cases is rather different. Difference between the two approaches may be compared with the difference between the two approaches of Huygens and of Leibniz in the 17th century, the basic concept of Huygens being expectation, that of Leibniz being probability. In the 20th century de Finetti may be taken as representative for the expectation-approach, while Kolmogorov stands for the probability-approach.

In Chapter III it will be shown that results given by Huber and his school about 15 years ago using the expectation-approach can also be achieved in a more direct and practicable way via the probability-approach.

The aim of the present article is to characterize the direct approach to interval probability by the use of a system of axioms which allows to transfer most results of classical probability theory to interval probability. Some basic concepts are described by the author's article for the 2nd Gauß Symposion (Weichselberger (1995a)). A thorough presentation of the theory will be given in the author's forthcoming book "Elementare Grundbegriffe einer allgemeineren Wahrscheinlichkeitsrechnung" (Weichselberger (1995b)).

2 System of Axioms

Mathematically interval probability may be characterized as mapping of a set of random events into a suitable set $\mathcal{Z}$, for which the following relation holds:

$$\{[0;0],[1;1]\} \subseteq \mathcal{Z} \subseteq \mathcal{Z}_0 = \{[L,U] : 0 \leq L \leq U \leq 1\} \tag{2.1}$$

While this understanding incorporates a large number of possible branches of theory, among which two-valued logic as well as classical probability theory can be found, one may restrain to

$$\mathcal{Z} = \mathcal{Z}_0. \tag{2.2}$$

In this case the following set of definitions may be considered as characterizing the system of axioms for interval probability.

Definition 2.1 *Given a sample space* Ω *and a* σ*-field* $\mathcal{A}$ *of random events in* Ω*, a set function* $p(\cdot)$ *defined on* $\mathcal{A}$ *is named K-function, if it obeys the following three axioms:*

I.

$$p(A) \geq 0, \quad \forall A \in \mathcal{A}. \tag{2.3}$$

II.

$$p(\Omega) = 1. \tag{2.4}$$

III. $\forall A_i,\ A_j \in \mathcal{A}$ *with* $A_i \cap A_j = \emptyset$, *if* $i \neq j$:

$$p(\bigcup_{i=1}^{\infty} A_i) = \sum_{i=1}^{\infty} p(A_i). \tag{2.5}$$

Since K-functions have the same properties as classical probabilities, in the following they are named *K-probabilities.*

Definition 2.2 *An interval-valued set function* $P(\cdot)$ *on* $\mathcal{A}$ *is called R-probability, if it obeys the following two axioms.*

IV.

$$P(A) = [L(A); U(A)], \quad \forall A \in \mathcal{A}, \tag{2.6}$$

with

$$0 \leq L(A) \leq U(A) \leq 1, \quad \forall A \in \mathcal{A}. \tag{2.7}$$

V. *The set* $\mathcal{M}$ *of K-functions* $p(\cdot)$ *on* $\mathcal{A}$ *with*

$$L(A) \leq p(A) \leq U(A), \quad \forall A \in \mathcal{A}, \tag{2.8}$$

is not empty.

The name "R-probability" may be related to the word "reasonable". A quadruple consisting of a sample space Ω, a σ-field $\mathcal{A}$ of random events and lower as well as upper interval-limits of a certain assignment representing R-probability on $(\Omega, \mathcal{A})$ will be called an R-probability field $(\Omega; \mathcal{A}; L(\cdot), U(\cdot))$. The set $\mathcal{M}$ of K-probabilities being in accordance with (2.8) is named the *structure* of the R-probability field. Therefore the existence of a non-empty structure is the only condition for any R-probability field.

It is obvious that

$$L(\emptyset) = 0, \quad U(\Omega) = 1 \tag{2.9}$$

are among the necessary conditions for R-probability.

Definition 2.3 *R-probability obeying the following axiom is named F-probability.*

$$VI. \qquad \left.\begin{array}{rcl} \inf\limits_{p \in \mathcal{M}} p(A) & = & L(A) \\ \sup\limits_{p \in \mathcal{M}} p(A) & = & U(A) \end{array}\right\} \forall A \in \mathcal{A}. \tag{2.10}$$

The letter F may be connected with the word "feasible".

In any F-probability field none of the limits $L(\cdot)$ and $U(\cdot)$ are too wide with respect to the structure, while this may be the case for an R-probability field. Furthermore the property of F-probability implies the validity of

$$U(A) = 1 - L(\neg A), \quad \forall A \in \mathcal{A}, \tag{2.11}$$

and of

$$U(\emptyset) = 0, \quad L(\Omega) = 1, \tag{2.12}$$

which by use of the symbol

$$[a] := [a; a] \tag{2.13}$$

together with (2.9) read as

$$P(\emptyset) = [0], \quad P(\Omega) = [1]. \tag{2.14}$$

A triple consisting of a sample space Ω, a σ-field $\mathcal{A}$ of random events and lower interval-limits of any assignment representing F-probability is understood to be an F-probability field $(\Omega; \mathcal{A}; L(\cdot))$.

The given system of axioms may be applied to finite sample spaces and to infinite ones, but in the case of F-probability it proves useful to distinguish continuous F-probability, since Axioms I–VI do not guarantee the continuity of the set functions $L(\cdot)$ and $U(\cdot)$.

While Huber introduces the concept described as R-probability, in the same way Axiom V does (Huber(1981), p 257), he characterizes assignments producing F-probability applying the expression "representable" (Huber(1981), p 255).

In practical use assessments producing R-probability are not self-contradictory and therefore cannot be rejected, but they need not to be perfect, since some of their limits may be too wide with respect to the structure defined by these limits. Should this be true for any R-probability field, it will be called *redundant* in the following.

If otherwise Axiom VI is fulfilled, the resulting F-probability field is uniquely defined by its structure. According to the theory produced by Axioms I to VI all F-probability fields should be understood as perfect generalization of classical probability fields. Correspondingly the property of 2-monotonicity is not necessary for qualifying interval probability fields as perfect generalization of the concept K-probability field. Therefore 2-monotone capacities are understood as one distinguished subgroup of F-probability (denominated *C-probability*) showing favourable mathematical properties but not representing a higher standard of probability assignment.

In practical situations R-probability may be seen as a step in the development of F-probability. The most appealing standpoint concerning the improvement of an R-probability field proposes to pass over to the limits of that F-probability field which is in accordance with the structure of the given field: It is called the *derivable* F-probability field. Its interval- limits are never wider than the original ones and at least some of them are narrower in all cases, in which the original R-probability field is redundant.

This standpoint may be called the *rigid* one. Another standpoint concerning transition from a redundant R-probability field to an F-probability field, the *cautious* standpoint, is described in "Axiomatic foundations ... " (Weichselberger (1995a)).

3 R-Probability and F-Probability

3.1 Checking for R-probability

Let Ω_k be a sample space consisting of singletons $E_1, \ldots, E_k$ and let

$$\mathcal{A} = \mathcal{P}(\Omega_k). \tag{3.1}$$

In this case each element of the power set of Ω_k constitutes a random event. Let further $P(\emptyset) = [0], P(\Omega_k) = [1]$ and

$$0 \leq L(A) \leq U(A) \leq 1, \quad \forall A \in \mathcal{A} \setminus \{\emptyset, \Omega\} =: \mathcal{A}' \tag{3.2}$$

be given. Then the question arises, whether these interval-limits describe R-probability. Due to Axiom V the answer to this question depends on the existence of a non-empty structure in accordance with (3.2). Therefore the set of interval-limits describes R-probability iff at least one K-probability $p(\cdot)$ exists, for which (2.8) comes true.

Since for any K-probability the relations

$$p(A) + p(\neg A) = 1, \quad \forall A \in \mathcal{A}, \tag{3.3}$$

hold, not all of the lower and upper limits must be taken into consideration separately, but for each random event A only the narrower one out of $L(A)$ and $1 - U(\neg A)$.

Definition 3.1 *Let for* $(\Omega_k, \mathcal{A})$ *interval limits* $L(A)$ *and* $U(A)$ *in accordance with (3.2) be given. Then*

$$L'(A) := \max[L(A),\, 1 - U(\neg A)], \quad \forall A \in \mathcal{A}', \tag{3.4}$$

defines consolidated lower interval-limits.

Obviously any K-probability $p(\cdot)$ agrees with relation (2.8), iff

$$p(A) \geq L'(A), \quad \forall A \in \mathcal{A}', \tag{3.5}$$

comes true.

Since the sample space is finite, by solving the following Minimum Problem of Linear Programming it may be decided, whether such K-probability p(.) exists:

$$\left.\begin{aligned} \sum_{E \subseteq A} p(E) &\geq L'(A), \quad \forall A \in \mathcal{A}', \\ \sum_{E \subseteq \Omega} p(E) &= \min! \end{aligned}\right\} \tag{3.6}$$

The number of restrictions in (3.6) is equal to

$$r := |\mathcal{A}'| = |\mathcal{A}| - 2,$$

which because of $\mathcal{A} = \mathcal{P}(\Omega_k)$ becomes

$$r = 2^k - 2.$$

Since the function

$$p_0(E) = 1, \quad \forall E \subseteq \Omega,$$

always produces an initial solution of System (3.6), the set of restrictions is free of contradictions.

By solving Linear Program (3.6), the question concerning the existence of R-probability may be answered: If

$$\min \sum_{\Omega} p(E) > 1, \tag{3.7}$$

it is impossible to find a K-probability $p(\cdot)$ in accordance with the given limits (3.2). Otherwise any solution of (3.6) producing an objective function not larger than 1 makes it possible to construct a linear combination $p(\cdot)$ of initial solution and minimum solution, for which

$$\sum_{\Omega} p(E) = 1 \tag{3.8}$$

comes true. Therefore the existence of a non-empty structure is proven.

For $k = 4$ the set of restrictions and the objective function are shown in the following

Example 3.2 Let $L'(A), A \in \mathcal{P}(\Omega_4)$, be consolidated lower interval-limits. For simplicity write

$$\begin{array}{llll} A_{12} := E_1 \cup E_2, & & A_{123} := E_1 \cup E_2 \cup E_3, & etc. \\ L'_{12} := L'(A_{12}), & & L'_{123} := L'(A_{123}), & etc. \\ p_i := p(E_i), \ i = 1,2,3,4. & & & \end{array}$$

Then the set of 14 restrictions reads as:

$$\begin{array}{ll} p_1 \geq L'_1 & p_2 + p_3 \geq L'_{23} \\ p_2 \geq L'_2 & p_2 + p_4 \geq L'_{24} \\ p_3 \geq L'_3 & p_3 + p_4 \geq L'_{34} \\ p_4 \geq L'_4 & p_1 + p_2 + p_3 \geq L'_{123} \\ p_1 + p_2 \geq L'_{12} & p_1 + p_2 + p_4 \geq L'_{124} \\ p_1 + p_3 \geq L'_{13} & p_1 + p_3 + p_4 \geq L'_{134} \\ p_1 + p_4 \geq L'_{14} & p_2 + p_3 + p_4 \geq L'_{234} \end{array}$$

and the objective reads as

$$p_1 + p_2 + p_3 + p_4 = \min!$$

That this minimum be not larger than 1, constitutes the necessary and sufficient condition for R-probability. ////

Solution of a Minimum Problem of Linear Programming via simplex method nowadays is easily gained using suitable software packages.

System (3.6) may be described as

$$\left.\begin{array}{rcl} \sum\limits_{\Omega} \mathrm{I}_A(E) \cdot p(E) & \geq & L'(A), \quad \forall A \in \mathcal{A}', \\ \sum\limits_{\Omega} p(E) & = & \min! \end{array}\right\} \tag{3.9}$$

or—stating explicitely the non-negativity of $p(\cdot)$, although it is contained in the restrictions:

$$\left.\begin{array}{rcl} \mathcal{B}_{(r,k)} \cdot p_{(k)} & \geq & b_{(r)} \\ p_{(k)} & \geq & 0_{(k)} \\ e_{(k)}^{\top} \cdot p_{(k)} & = & \min! \end{array}\right\} \tag{3.10}$$

with

$$p_{(k)} = \begin{pmatrix} p_1 \\ \vdots \\ p_k \end{pmatrix} \tag{3.11}$$

$$\mathcal{B}_{(r,k)} = (\mathrm{I}_{A_i}(E_j))_{i,j}, \quad 1 \leq i \leq r, \quad 1 \leq j \leq k, \tag{3.12}$$

by numbering the non-trivial random events, i.e. the elements of $\mathcal{A}'$: $A_1, \ldots, A_r$.

$$b_{(r)} = \begin{pmatrix} L'(A_1) \\ \vdots \\ L'(A_r) \end{pmatrix} \tag{3.13}$$

$$e_{(k)}^{\top} = (1, \ldots, 1)\,. \tag{3.14}$$

Due to a famous result of John von Neumann (Kuhn and Tucker (1958; p 117), Dantzig (1963; p 123f)) any Minimum Problem of type (3.10) possesses a dual Maximum Problem with a vector $x_{(r)}$ of non-negative variables:

$$x_{(r)} = \begin{pmatrix} x_1 \\ \vdots \\ x_r \end{pmatrix} =: \begin{pmatrix} x(A_1) \\ \vdots \\ x(A_r) \end{pmatrix}. \tag{3.15}$$

With respect to System (3.10) this is the system:

$$\left.\begin{array}{rcl} \mathcal{B}_{(k,r)}^{\top} \cdot x_{(r)} & \leq & e_{(k)} \\ x_{(r)} & \geq & 0_{(r)} \\ b_{(r)}^{\top} \cdot x_{(r)} & = & \max! \end{array}\right\}, \tag{3.16}$$

producing exactly the same value of the objective function.

Obviously the number of variables in the primal is equal to the number of restrictions in the dual and vice versa.

Therefore the solution of (3.6) may also be achieved by solving the following Maximum Problem:

$$\left.\begin{array}{rcl} \sum\limits_{\mathcal{A}'} \mathrm{I}_A(E) \cdot x(A) & \leq & 1, \quad \forall E \subset \Omega, \\ x(A) & \geq & 0, \quad \forall A \in \mathcal{A}', \\ \sum\limits_{\mathcal{A}'} L'(A) \cdot x(A) & = & \max! \end{array}\right\} \tag{3.17}$$

That the solution of (3.17) must be not larger that 1, is exactly the condition for a non-empty structure as derived in a totally different way by Huber (1981, pp 257 f., Lemma 2.2).

3.2 Checking for F-probability

Let $\mathcal{R}$ be an R-probability field on Ω_k with $P(\emptyset) = [0]$ and $P(\Omega) = [1]$. A necessary condition for $\mathcal{R}$ being an F-probability field has been given by (2.11). If it holds, necessary and sufficient conditions can be achieved by applying Linear Programming again. For each non-trivial random event A_i the validity of condition (2.10) may be controlled directly by solving the Minimum Problems

$$\left.\begin{array}{rcl} \sum\limits_{A} p(E) & \geq & L(A), \quad \forall A \in \mathcal{A}', \\ \sum\limits_{\Omega} p(E) & = & 1 \\ \sum\limits_{A_i} p(E) & = & \min! \end{array}\right\} \quad \forall A_i \in \mathcal{A}. \tag{3.18}$$

If and only if for all elements A_i of $\mathcal{A}'$ the solution of (3.18) is exactly $L(A_i)$, the field $\mathcal{R}$ constitutes an F-probability field.

Therefore answering the question whether a field $\mathcal{R}$ in accordance with (2.11) and (2.14) represents an F-probability field, affords the solution of up to $|\mathcal{A}'|-2$ Minimum Problems of Linear Programming, each of them possessing the same system of $|\mathcal{A}'|-1$ restrictions for k variables, and distinguished by the objective functions. If $\mathcal{A}=\mathcal{P}(\Omega_k)$, the number of Minimum Problems becomes 2^k-2.

Example 3.3 In the case of $k=4$ and $\mathcal{A}=\mathcal{P}(\Omega_4)$ the set of restrictions is given by

$$
\begin{array}{llll}
p_1 & & & \geq L_1 \\
& p_2 & & \geq L_2 \\
& & p_3 & \geq L_3 \\
& & & p_4 \geq L_4 \\
p_1 + p_2 & & & \geq L_{12} \\
p_1 + & & p_3 & \geq L_{13} \\
p_1 + & & & p_4 \geq L_{14}
\end{array}
\qquad
\begin{array}{llll}
& p_2 + p_3 & & \geq L_{23} \\
& p_2 + & & p_4 \geq L_{24} \\
& & p_3 + & p_4 \geq L_{34} \\
p_1 + p_2 + p_3 & & & \geq L_{123} \\
p_1 + p_2 + & & & p_4 \geq L_{124} \\
p_1 + & & p_3 + & p_4 \geq L_{134} \\
& p_2 + p_3 + & & p_4 \geq L_{234}
\end{array}
$$

$$p_1+p_2+p_3+p_4=1$$

If $\mathcal{R}=(\Omega,\mathcal{A},L(\cdot),U(\cdot))$ obeys (2.14) and constitutes an R-probability field, there exists an initial solution in accordance with these restrictions.

Each of the 14 non-trivial random events produces one objective:

$$
\begin{array}{rcl}
p_1 & = & \min! \\
& \vdots & \\
p_2+p_3+p_4 & = & \min!
\end{array}
$$

If and only if each of the 14 solutions of these Minimum Problems equals $L(A)$, the respective lower interval-limit, $\mathcal{R}$ represents an F-probability field. ////

In order to transform the equality in (3.18) in a manner fitting to the inequalities, it may be written as

$$
\left.
\begin{array}{rcr}
\sum_{\Omega} p(E) & \geq & 1 \\
\sum_{\Omega} (-1)\cdot p(E) & \geq & -1
\end{array}
\right\}
\tag{3.19}
$$

In analogy to (3.9) therefore (3.18) can be reformulated:

$$
\left.
\begin{array}{rcl}
\sum_{\Omega} \mathrm{I}_A(E)\cdot p(E) & \geq & L(A), \quad \forall A\in\mathcal{A}', \\
\sum_{\Omega} p(E) & \geq & 1 \\
\sum_{\Omega} (-1)\cdot p(E) & \geq & -1 \\
\sum_{\Omega} \mathrm{I}_{A_i}(E)\cdot p(E) & = & \min!
\end{array}
\right\}
\forall A_i\in\mathcal{A}.
\tag{3.20}
$$

With $\bar{r} := r + 2 = |\mathcal{A}|$ the following matrix describes the coefficients of the restrictions:

$$\bar{\mathcal{B}}_{(\bar{r},k)} := \begin{pmatrix} & & \mathcal{B}_{(r,k)} & \\ 1 & 1 & \dots & 1 \\ -1 & -1 & \dots & -1 \end{pmatrix} \tag{3.21}$$

and the following vector has to be applied instead of $b_{(r)}$:

$$\bar{b}_{(\bar{r})} := \begin{pmatrix} L(A_1) \\ \vdots \\ L(A_r) \\ 1 \\ -1 \end{pmatrix}. \tag{3.22}$$

Completing again by the unnecessary statement of non-negativity one arrives at the following description of the Minimum Problem for random event A_i:

$$\left.\begin{array}{rcl} \bar{\mathcal{B}}_{(\bar{r},k)} \cdot p_{(k)} & \geq & \bar{b}_{(\bar{r})} \\ p_{(k)} & \geq & 0_{(k)} \\ e_{i(k)}^{\top} \cdot p_{(k)} & = & \min! \end{array}\right\} \tag{3.23}$$

with

$$e_{i(k)}^{\top} := (\mathrm{I}_{A_i}(E_1), \mathrm{I}_{A_i}(E_2), \dots, \mathrm{I}_{A_i}(E_k)). \tag{3.24}$$

Again the dual Maximum Problem may be studied. Its vector of variables is now of dimension $\bar{r}$:

$$x_{(\bar{r})} := \begin{pmatrix} x_1 \\ \vdots \\ x_r \\ x_{r+1} \\ x_{r+2} \end{pmatrix}, \tag{3.25}$$

where a correspondence may be seen between x_{r+1} and the event Ω—because of the analogy between the first two lines in (3.20)—and between x_{r+2} and the event $\emptyset$—since the third line in (3.20) may be interpreted as: $p(\emptyset) \geq 0$. Therefore in the following the notation $x_{r+1} =: x_\Omega = x(\Omega)$ and $x_{r+2} =: x_\emptyset = x(\emptyset)$ is applied.

The dual Maximum Problem stemming from the non-trivial random event A_i reads as:

$$\left.\begin{array}{rcl} \bar{\mathcal{B}}_{(k,\bar{r})}^{\top} \cdot x_{(\bar{r})} & \leq & e_{i(k)} \\ x_{(\bar{r})} & \geq & 0_{(r)} \\ \bar{b}_{(\bar{r})}^{\top} \cdot x_{(\bar{r})} & = & \max! \end{array}\right\} \tag{3.26}$$

Applying the definitions given in (3.21), (3.22) and (3.24) this is translated to

$$\left.\begin{array}{rcll}\sum_{\mathcal{A}'} \mathrm{I}_A(E)\cdot x(A)+x(\Omega)-x(\emptyset) & \leq & \mathrm{I}_{A_i}(E), & \forall E\subset\Omega,\\ x(A) & \geq & 0, & \forall A\in\mathcal{A},\\ \sum_{\mathcal{A}'} L(A)\cdot x(A)+x(\Omega)-x(\emptyset) & = & \max! & \end{array}\right\}\ \forall A_i\in\mathcal{A}. \tag{3.27}$$

Systems (3.27) belonging to different random events A_i, differ with respect to their restrictions, all of them possessing the same objective function.

Example 3.4 For $k = 4$ the variables of the Maximum Problem are the components of the vector

$$x^{\top} = (x_1, x_2, x_3, x_4, x_{12}, x_{13}, x_{14}, x_{23}, x_{24}, x_{34}, x_{123}, x_{124}, x_{134}, x_{234}, x_\Omega, x_\emptyset)$$

The systems of restrictions resulting for $A_1 = E_1$ is given by:

$$\begin{array}{l} x_1+x_{12}+x_{13}+x_{14}+x_{123}+x_{124}+x_{134}+x_\Omega-x_\emptyset\leq 1\\ x_2+x_{12}+x_{23}+x_{24}+x_{123}+x_{124}+x_{234}+x_\Omega-x_\emptyset\leq 0\\ x_3+x_{13}++x_{23}++x_{34}+x_{123}+x_{134}+x_{234}+x_\Omega-x_\emptyset\leq 0\\ x_4+x_{14}+x_{24}+x_{34}+x_{124}+x_{134}+x_{234}+x_\Omega-x_\emptyset\leq 0 \end{array}$$

while e.g. for $A_{12} = E_1 \cup E_2$ the vector of boundaries is changed to

$$\begin{pmatrix} \mathrm{I}_{A_{12}}(E_1) & = & 1\\ \mathrm{I}_{A_{12}}(E_2) & = & 1\\ \mathrm{I}_{A_{12}}(E_3) & = & 0\\ \mathrm{I}_{A_{12}}(E_4) & = & 0 \end{pmatrix}.$$

For each of the non-trivial random events the objective is the same, namely

$$\begin{array}{l} L_1x_1 + L_2x_2 + L_3x_3 + L_4x_4 +\\ \quad + L_{12}x_{12} + L_{13}x_{13} + L_{14}x_{14} + L_{23}x_{23} + L_{24}x_{24} + L_{34}x_{34} +\\ \quad + L_{123}x_{123} + L_{124}x_{124} + L_{134}x_{134} + L_{234}x_{234} + x_\Omega - x_\emptyset = \max! \end{array}$$

////

According to the criterion described above, for an F-probability field it is necessary and sufficient, that for each element A_i of $\mathcal{A}'$ the maximum of the objective function be exactly $L(A_i)$.

What was demonstrated by (3.18), (3.20) and (3.23) for lower interval-limits $L(.)$, can as well be applied to upper limits $U(.)$. Then the original—primal—problems are Maximum Problems:

$$\left.\begin{array}{rcll}\sum_{A} p(E) & \leq & U(A), & \forall A\in\mathcal{A}',\\ \sum_{\Omega} p(E) & = & 1 & \\ \sum_{A_i} p(E) & = & \max! & \end{array}\right\}\ \forall A_i\in\mathcal{A}'. \tag{3.28}$$

The necessary and sufficient condition for an F-probability field is now that each solution is equal to $U(A_i)$.

In order to arrive at a representation in analogy to (3.23), the vector of upper limits has to be introduced:

$$\bar{c}_{(\bar{r})} := \begin{pmatrix} U(A_1) \\ \vdots \\ U(A_r) \\ 1 \\ -1 \end{pmatrix}. \tag{3.29}$$

Systems (3.28) then may be described by

$$\left.\begin{array}{rcl} \bar{\mathcal{B}}_{(\bar{r},k)} \cdot p_{(k)} & \leq & \bar{c}_{(\bar{r})} \\ p_{(k)} & \geq & 0_{(k)} \\ e_{i(k)}^{\top} \cdot p_{(k)} & = & \max! \end{array}\right\}. \tag{3.30}$$

The corresponding duals are the following Minimum Problems:

$$\left.\begin{array}{rcl} \bar{\mathcal{B}}_{(k,\bar{r})}^{\top} \cdot y_{(\bar{r})} & \geq & e_{i(k)} \\ y_{(\bar{r})} & \geq & 0_{(\bar{r})} \\ \bar{c}_{(\bar{r})}^{\top} \cdot y_{(\bar{r})} & = & \min! \end{array}\right\}. \tag{3.31}$$

with $y^{\top} = (y_1, y_2, \ldots, y_{r+1}, y_{r+2})$, where $y_{r+1} =: y_{\Omega} = y(\Omega)$ and $y_{r+2} =: y_{\emptyset} = y(\emptyset)$.

If (3.31) are translated in analogy to (3.27) one arrives at

$$\left.\begin{array}{rcll} \sum_{\mathcal{A}'} \mathrm{I}_A(E) \cdot y(A) + y(\Omega) - y(\emptyset) & \geq & \mathrm{I}_{A_i}(E), & \forall E \subset \Omega, \\ y(A) & \geq & 0, & \forall A \in \mathcal{A}, \\ \sum_{\mathcal{A}'} U(A) \cdot y(A) + y(\Omega) - y(\emptyset) & = & \min! & \end{array}\right\} \forall A_i \in \mathcal{A}'. \tag{3.32}$$

The condition that the solutions of (3.32) must be equal to $U(A_i)$, $\forall A_i \in \mathcal{A}'$, is that one given by Huber (1981, p 259) quoting the results of Wolf (1977). Therefore the set of necessary and sufficient conditions for the F-property may be achieved in two very different ways, starting either from the use of expectation or of probability as basic concepts.

3.3 Additional aspects

The use of Linear Programming in order to solve one of Systems (3.18), (3.27), (3.28) or (3.32) secures the practical applicability of the criterion for all relevant sizes of the sample space. Additional advantage is produced since the solutions of System (3.18) or (3.27) represent the lower limits of the derivable F-probability field, while the solutions of System (3.28) or (3.32) represent the upper limits of this field. Therefore, should the given R-probability field prove to be redundant, the proposed methods of control produce immediately the derivable F-probability field.

The following two examples demonstrate the employment of the methods described above in cases, where $k = 3$ or $k = 4$.

Example 3.5 Let an assessment on Ω_3 be intended to represent "probability" be given by:

$$\begin{array}{lcllcl} P(E_1) & = & [0.1; 0.4] & P(E_1 \cup E_2) & = & [0.3; 0.8] \\ P(E_2) & = & [0.3; 0.7] & P(E_1 \cup E_3) & = & [0.3; 0.8] \\ P(E_3) & = & [0.2; 0.8] & P(E_2 \cup E_3) & = & [0.5; 0.8]. \end{array}$$

In order to check whether this defines R-probability, at first one uses (3.4) to gain consolidated interval-limits:

$$\begin{array}{lcllcl} P'(E_1) & = & [0.2; 0.4] & P'(E_1 \cup E_2) & = & [0.3; 0.8] \\ P'(E_2) & = & [0.3; 0.7] & P'(E_1 \cup E_3) & = & [0.3; 0.7] \\ P'(E_3) & = & [0.2; 0.7] & P'(E_2 \cup E_3) & = & [0.6; 0.8]. \end{array}$$

System (3.6) becomes

$$\begin{array}{ccccccl} p_1 & & & & & \geq & 0.2 \\ & & p_2 & & & \geq & 0.3 \\ & & & & p_3 & \geq & 0.2 \\ p_1 & + & p_2 & & & \geq & 0.3 \\ p_1 & & & + & p_3 & \geq & 0.3 \\ & & p_2 & + & p_3 & \geq & 0.6 \\ p_1 & + & p_2 & + & p_3 & = & \min! \ . \end{array}$$

The solution of this Linear Program is $0.8 < 1$: The given assessment produces R-probability.

Since there exist random events A, for which $P(A) \neq P'(A)$, it is evident, that $\mathcal{R} = (\Omega; \mathcal{A}; L(\cdot), U(\cdot))$ is redundant. To answer the question, whether the consolidated R-probability field $\mathcal{R}' = (\Omega_3, \mathcal{P}(\Omega_3); L'(\cdot), U'(\cdot))$ represents an F-probability field, the set of restrictions

$$\begin{array}{ccccccl} p_1 & & & & & \geq & 0.2 \\ & & p_2 & & & \geq & 0.3 \\ & & & & p_3 & \geq & 0.2 \\ p_1 & + & p_2 & & & \geq & 0.3 \\ p_1 & & & + & p_3 & \geq & 0.3 \\ & & p_2 & + & p_3 & \geq & 0.6 \\ p_1 & + & p_2 & + & p_3 & = & 1 \end{array}$$

must be combined with 6 objective functions

$$\begin{array}{rcrclllcl} & & p_1 & = & \min! & \text{Solution:} & 0.2 & & & = & L'(E_1) \\ & & p_2 & = & \min! & \text{Solution:} & 0.3 & & & = & L'(E_2) \\ & & p_3 & = & \min! & \text{Solution:} & 0.2 & & & = & L'(E_3) \\ p_1 & + & p_2 & = & \min! & \text{Solution:} & 0.5 & > & 0.3 & = & L'(E_1 \cup E_2) \\ p_1 & + & p_3 & = & \min! & \text{Solution:} & 0.4 & > & 0.3 & = & L'(E_1 \cup E_3) \\ p_2 & + & p_3 & = & \min! & \text{Solution:} & 0.6 & & & = & L'(E_2 \cup E_3). \end{array}$$

Since the limits $L'(E_1 \cup E_2)$ and $L'(E_1 \cup E_3)$ are not reached by elements of the structure, $\mathcal{R}$ is a redundant R-probability field, not an F-probability field.

Applying the solutions of the Minimum Problems, as far as they are different from the given limits, the derivable F-probability field is produced and thereby through (2.11) all boundaries of the structure $\mathcal{M}$.

$$\begin{array}{rclrcl} P^*(E_1) & = & [0.2; 0.4] & P^*(E_1 \cup E_2) & = & [0.5; 0.8] \\ P^*(E_2) & = & [0.3; 0.6] & P^*(E_1 \cup E_3) & = & [0.4; 0.7] \\ P^*(E_3) & = & [0.2; 0.5] & P^*(E_2 \cup E_3) & = & [0.6; 0.8]. \end{array}$$

In Figure 1 the structure $\mathcal{M}$ of $\mathcal{R}$ is shown. It is represented by a quadrangle.

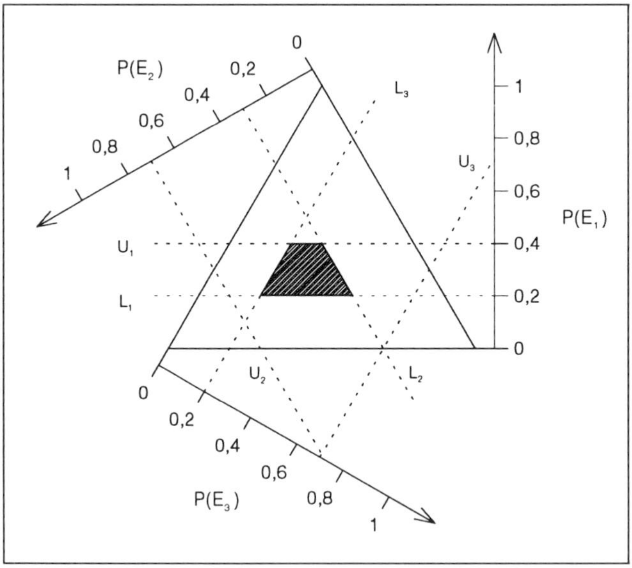

Figure 1: The interval-limits of Example 3.4 and the structure of the R-probability field in barycentric representation

In addition those two consolidated interval-limits are depicted which are not reached by elements of the structure.

From this diagram it becomes evident, that all solutions would remain unchanged, if instead of the full set of restrictions only the following restrictions

$$\begin{array}{ccccccc} p_1 & & & & & \geq & 0.2 \\ & & p_2 & & & \geq & 0.3 \\ & & & & p_3 & \geq & 0.2 \\ & & p_2 & + & p_3 & \geq & 0.6 \\ p_1 & + & p_2 & + & p_3 & = & 1 \end{array}$$

were used. Therefore the information consisting of

$$L'(E_1) = 0.2, \quad L'(E_2) = 0.3, \quad L'(E_3) = 0.2, \quad L'(E_2 \cup E_3) = 0.6$$

is sufficient to determine $\mathcal{M}$. ////

Definition 3.6 *Let* $\mathcal{R} = (\Omega_k; \mathcal{A}; L(\cdot), U(\cdot))$ *and* $\mathcal{M}$ *be its structure. A set* $\mathcal{S}$ *of random events is named* consolidated support *of* $\mathcal{R}$, *if the set of restrictions produced by the corresponding consolidated lower interval-limits is sufficient to determine* $\mathcal{M}$.

In case of Example 3.5 any further reduction of the set of restrictions used above would change at least some of the solutions of the Minimum Problems and thereby produce another $\mathcal{M}$.

Definition 3.7 *Consolidated support* $\mathcal{L}$ *of* $\mathcal{R}$ *is named* shortest consolidated support *of* $\mathcal{R}$, *if for each consolidated support* $\mathcal{S}$ *of* $\mathcal{R}$ *the relation*

$$|\mathcal{S}| \geq |\mathcal{L}|$$

holds. $|\mathcal{L}|$ *is named* complexity $\lambda(\mathcal{R})$ *of the R-probability field* $\mathcal{R}$.

It can be shown that

$$0 \leq \lambda(\mathcal{R}) \leq 2^k - 2$$

describes the narrowest boundaries generally valid for $\lambda(\mathcal{R})$. It should be noted that $\lambda(\mathcal{R}) = 0$ is true if and only if $\mathcal{R}$ represents the *non-selective* F-probability field for $(\Omega_k; \mathcal{A})$ with $L(A) = 0, U(A) = 1, \forall A \in \mathcal{A}'$. On the other hand $\lambda(\mathcal{R}) = 2^k - 2$ is possible only, if $\mathcal{R}$ represents a certain kind of F-probability field.

Example 3.8 Since the R-probability field $\mathcal{R}$ considered in Example 3.5 proves to be redundant, its complexity cannot reach the upper boundary $2^3 - 2 = 6$. Indeed it was shown that $\mathcal{L} = \{E_1, E_2, E_3, E_2 \cup E_3\}$ and therefore $\lambda(\mathcal{R}) = 4$. ////

With respect to its shape, every structure can be seen as polyhedron in $\mathbb{R}_{k-1}$. The number of apices, $\eta(\mathcal{R})$, is equal to the minimum size of a basis allowing the representation of each element of $\mathcal{M}$ as convex linear combination. While for every k a structure consisting of only one element produces $\eta(\mathcal{R}) = 1$ as a minimum, the maximum of $\eta(\mathcal{R})$ for k = 3 is easily seen to be 6, and for $k = 4$ the relation $\eta(\mathcal{R}) = 24$ can be proven.

There exists a rather loose connection between $\lambda(\mathcal{R})$ and $\eta(\mathcal{R})$, and consequently the possibility is given to characterize every R-probability field by means of $\lambda(\mathcal{R})$ and $\eta(\mathcal{R})$. More detailed information concerning these attributes of probability fields are given in the author's "Elementare Grundbegriffe..." (Weichselberger (1995b)).

Finally, an R-probability field for $(\Omega_4, \mathcal{P}(\Omega_4))$ is used to demonstrate the concepts and results of this paragraph. In order to provide for "beautiful" graphical representation it employs rather large distances between the interval-limits $L(\cdot)$ and $U(\cdot)$.

Example 3.9 Let

$$\begin{array}{lcl lcl}
P(E_1) & = & [0.02; 0.35] & P(E_2 \cup E_3) & = & [0.25; 0.75] \\
P(E_2) & = & [0.04; 0.42] & P(E_2 \cup E_4) & = & [0.33; 0.79] \\
P(E_3) & = & [0.08; 0.64] & P(E_3 \cup E_4) & = & [0.25; 0.85] \\
P(E_4) & = & [0.08; 0.71] & P(E_1 \cup E_2 \cup E_3) & = & [0.30; 0.90] \\
P(E_1 \cup E_2) & = & [0.15; 0.72] & P(E_1 \cup E_2 \cup E_4) & = & [0.35; 0.94] \\
P(E_1 \cup E_3) & = & [0.18; 0.67] & P(E_1 \cup E_3 \cup E_4) & = & [0.60; 0.95] \\
P(E_1 \cup E_4) & = & [0.27; 0.65] & P(E_2 \cup E_3 \cup E_4) & = & [0.70; 0.99]
\end{array}$$

be a probability assessment. In order to prove, whether $P(\cdot)$ represents an R-probability field, the consolidated lower interval-limits according to Definition 3.1 are calculated.

$$\begin{array}{lcl lcl}
L'(E_1) & = & 0.02 & L'(E_2 \cup E_3) & = & 0.35 \\
L'(E_2) & = & 0.05 & L'(E_2 \cup E_4) & = & 0.33 \\
L'(E_3) & = & 0.08 & L'(E_3 \cup E_4) & = & 0.28 \\
L'(E_4) & = & 0.10 & L'(E_1 \cup E_2 \cup E_3) & = & 0.30 \\
L'(E_1 \cup E_2) & = & 0.15 & L'(E_1 \cup E_2 \cup E_4) & = & 0.36 \\
L'(E_1 \cup E_3) & = & 0.21 & L'(E_1 \cup E_3 \cup E_4) & = & 0.60 \\
L'(E_1 \cup E_4) & = & 0.27 & L'(E_2 \cup E_3 \cup E_4) & = & 0.70
\end{array}$$

These limits produce the following Minimum Problem:

$$\begin{array}{ll}
p_1 \geq 0.02 & p_2 + p_3 \geq 0.35 \\
p_2 \geq 0.05 & p_2 + p_4 \geq 0.33 \\
p_3 \geq 0.08 & p_3 + p_4 \geq 0.28 \\
p_4 \geq 0.10 & p_1 + p_2 + p_3 \geq 0.30 \\
p_1 + p_2 \geq 0.15 & p_1 + p_2 + p_4 \geq 0.36 \\
p_1 + p_3 \geq 0.21 & p_1 + p_3 + p_4 \geq 0.60 \\
p_1 + p_4 \geq 0.27 & p_2 + p_3 + p_4 \geq 0.70
\end{array}$$

$$p_1 + p_2 + p_3 + p_4 = \min!$$

The solution of this problem is given as $0.725 \leq 1$, proving that the considered assessment represents an R-probability field $\mathcal{R} = (\Omega_4; \mathcal{P}(\Omega_4); L(\cdot), U(\cdot))$. Since it differs from the consolidated field $\mathcal{R}' = (\Omega; \mathcal{P}(\Omega_4); L'(\cdot), U'(\cdot))$, it does not qualify as F-probability field.

To answer the question, whether the field $\mathcal{R}'$ has F-property, 14 Minimum Problems have to be solved, each applying the set of restrictions given above, accomplished by

$$p_1 + p_2 + p_3 + p_4 = 1.$$

The minimum values of the 14 objective functions are found to be

$$\begin{array}{lcllcllcl}
\min\ p_1 & = & 0.02 & \min(p_2+p_3) & = & 0.35 & & \\
\min\ p_2 & = & 0.05 & \min(p_2+p_4) & = & 0.33 & & \\
\min\ p_3 & = & 0.08 & \min(p_3+p_4) & = & 0.30 & > & 0.28 \\
\min\ p_4 & = & 0.10 & \min(p_1+p_2+p_3) & = & 0.37 & > & 0.30 \\
\min(p_1+p_2) & = & 0.15 & \min(p_1+p_2+p_4) & = & 0.375 & > & 0.36 \\
\min(p_1+p_3) & = & 0.21 & \min(p_1+p_3+p_4) & = & 0.60 & & \\
\min(p_1+p_4) & = & 0.27 & \min(p_2+p_3+p_4) & = & 0.70 & &
\end{array}$$

Since not all of the minimum values are identical with the respective $L'(\cdot)$, the R-probability field $\mathcal{R}'$ is shown to be redundant.

The derivable R-probability field is constructed by use of the minimum values applying relation (2.11):

$$\begin{array}{lcllcl}
P^*(E_1) & = & [0.02; 0.30] & P^*(E_2 \cup E_3) & = & [0.35; 0.73] \\
P^*(E_2) & = & [0.05; 0.40] & P^*(E_2 \cup E_4) & = & [0.33; 0.79] \\
P^*(E_3) & = & [0.08; 0.625] & P^*(E_3 \cup E_4) & = & [0.30; 0.85] \\
P^*(E_4) & = & [0.10; 0.63] & P^*(E_1 \cup E_2 \cup E_3) & = & [0.37; 0.90] \\
P^*(E_1 \cup E_2) & = & [0.15; 0.70] & P^*(E_1 \cup E_2 \cup E_4) & = & [0.375; 0.92] \\
P^*(E_1 \cup E_3) & = & [0.21; 0.67] & P^*(E_1 \cup E_3 \cup E_4) & = & [0.60; 0.95] \\
P^*(E_1 \cup E_4) & = & [0.27; 0.65] & P^*(E_2 \cup E_3 \cup E_4) & = & [0.70; 0.98]
\end{array}$$

The structure of the R-probability field in question is depicted in Figure 2.

Since $\mathcal{S} = \{E_1, E_2, E_3, E_4, E_1 \cup E_2, E_1 \cup E_3, E_1 \cup E_4, E_2 \cup E_3, E_2 \cup E_4, E_1 \cup E_2 \cup E_4, E_2 \cup E_3 \cup E_4\}$ represents a consolidated support of $\mathcal{R}$, the complexity of the structure must be at most 11. One counts exactly 11 planes as parts of the surface. This demonstrates that $\mathcal{S}$ represents a shortest consolidated support of $\mathcal{R}$, and $\lambda(\mathcal{R}) = 11$ is true. The number of apices also does not reach the possible maximum value of 24; it is found to be 18. ////

4 Outlook

A few basic concepts and results of the theory of interval probability are presented in this article in order to provide some insight into problems and solutions stemming from the author's method of generalization classical probability theory.

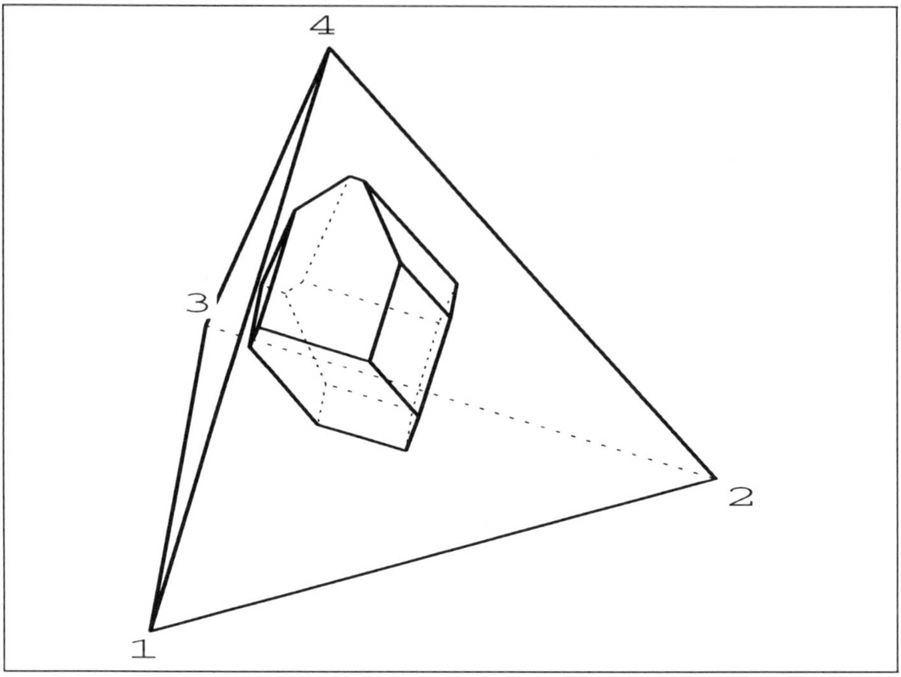

Figure 2: The structure of the R-probability field used in Example 3.8 in barycentric representation seen perspectively

Since Axioms I to VI in theory of interval probability play the role of KOLMOGOROV's Axioms in classical theory, it proves necessary to apply supplementary definitions of conditional probability and of independence in the more general theory too. If some further concepts are adapted in suitable way, it becomes possible to achieve among other results a version of the Weak Law of Large Numbers permitting to state a Frequency Interpretation of interval probability.

While some elements of the resulting theory may also be found in the author's article "Axiomatic foundations ..." Weichselberger (1995a)), a thorough presentation of the theory in its main aspects is given in the forthcoming book "Elementare Grundbegriffe einer allgemeineren Wahrscheinlichkeitsrechnung" (Weichselberger (1995b)).

References

[1] Choquet, G. (1953): Theory of capacities. *Ann. Inst. Fourier* **5** 131–295.

[2] Dantzig, G.B.: *Linear Programming and Extensions*. Princeton University Press, Princeton (New Jersey).

[3] Huber, P.J. (1973): The use of Choquet capacities in statistics. *Bull. Int. Statist. Inst. XLV* **4** 181–188.

[4] Huber, P.J. (1976): Kapazitäten statt Wahrscheinlichkeiten? Gedanken zur Grundlegung der Statistik. *Jber. Deutsch. Math.-Verein.* **78 H.2** 81–92.

[5] Huber, P.J. (1981): *Robust Statistics*. John Wiley, New York.

[6] Huber, P.J. and Strassen, V. (1973): Minimax tests and the Neyman-Pearson lemma for capacities. *Ann. Statist.* **1** 251–263. Correction **2** 223–224, 1974.

[7] Koopman, B.O. (1940a): The bases of probability. *Bull. Amer. Math. Soc.* **46** 763–774. Reprinted in *Studies in Subjective Probability* (Kyburg, H.E. and Smokler, H.E., eds.), John Wiley, New York, 1964.

[8] Koopman, B.O. (1940b): The axioms and algebra of intuitive probability. *Ann. Math.* **41** 269–292.

[9] Koopman, B.O. (1941): Intuitive probability and sequences. *Ann. Math.* **42** 169–187.

[10] Kuhn, H.W. & Tucker, A.W. (1958): John von Neumann's work in the theory of games and mathematical economics. *Bull. Amer. Math. Soc.* **64** 100–122.

[11] Shafer, G. (1978): Non-additive probabilities in the work of Bernoulli and Lambert. *Archive for History of Exact Sciences* **19** 309–370.

[12] Strassen, V. (1964): Meßfehler und Information. *Z. Wahrsch. verw. Gebiete* **2** 273–305.

[13] Weichselberger, K. (1995a): *Axiomatic foundations of the theory of interval probability*. In *Proceedings of the Second Gauß Symposion* (Mammitzsch, V. and Schneeweiß, H., eds.), Vol. B, de Gruyter, Berlin.

[14] Weichselberger, K. (1995b): *Elementare Grundbegriffe einer allgemeineren Wahrscheinlichkeitsrechnung*. Forthcoming.

[15] Wolf, G. (1977): *Obere und untere Wahrscheinlichkeiten*. Ph.D. Thesis, ETH Zürich.

On Consistency of Recursive Multivariate M-Estimators in Linear Models

Yuehua Wu

York University

Abstract

Strong consistency of recursive M-estimators of regression coefficients and scatter parameters in multivariate linear regression models is proved when the designs are nonrandom matrices satisfying $\frac{1}{n}\sum_{i=1}^{n} X_i' X_i \to Q > 0$.

Key words and phrases: M-estimation, recursive algorithm, strong mixing, convergence, and consistency, robust estimation, scatter.
AMS 1991 subject classifications: Primary 62J05, 62F12; secondary 62F35.

1 Introduction

The importance of robust estimation has been recognized in the literature for decades. Since the concepts of robustness and M-estimation were introduced by Huber (1964), they have attracted considerable attention of both theoretical and applied statisticians. For a literature review, the reader is referred to Huber (1973), Dodge (1987) and Bai, Rao and Wu (1991). However, a serious disadvantage of M-estimation, comparing to the Least Squares method, is due to the difficulty of the computation of the M-estimates because almost no M-estimators have closed forms. Thus, there is a great need of developing circular or recursive algorithms to compute the M-estimates or those estimates which are asymptotically equivalent to the M-estimates. First, one may want to refine his estimates from an initial estimate and get an estimate as accurate as the M-estimate after several steps of circular refinements. On the other hand, in applications to prediction, control and target tracking, it is often necessary to re-estimate the parameters as new portions of data are observed. The cases are different in their applications. However, in most cases, the two are asymptotically equivalent, i.e., the limiting properties for one case are true once they are true for the other.

Ruppert (1983) investigated the convergence of the so-called stochastic approximation algorithm. Based on this technique, Englund et al. (1988) proposed a recursive algorithm computing the M-estimates of the location and scale parameters of a univariate population and proved the strong consistency of their recursive estimators. Later, Englund (1989) studied the recursive M-estimators of the location parameters of a multivariate population. Englund (1993) extended this result to the simultaneous recursive M-estimators of the location and scatter parameters of a multivariate population and established the strong consistency of the recursive estimates under certain conditions. On one hand, his result does not cover general linear regression models. On the other hand, more importantly, his main theorem involves an unverifiable condition. Bai and Wu (1993) generalized Englund's algorithm to a recursive M-estimates of regression coefficients and scatter parameters in a general multivariate linear model where the design points are iid (independent and identically distributed) random matrices and succeeded in proving the strong consistency with Englund's unverifiable condition removed. However, it may be of more interest to assume that the design points are known, and the case where the design matrices are random may reduce to a special case of the latter. The aim of this paper is to extend the results of Bai and Wu (1993) to this more general setup.

The organization of this paper is as follows. In Section 2, some notations are introduced and the conditions are stated. In Section 3, some lemmas are given. The main theorems will be established in Section 4.

2 Notations and Conditions

In this paper, we shall consider the following multivariate linear regression model

$$\boldsymbol{y}_i = X_i \boldsymbol{\beta} + \boldsymbol{e}_i\,, \quad i = 1, 2, \ldots \tag{2.1}$$

where X_i, $i = 1, 2, \ldots$ are $m \times p$ matrices, $\boldsymbol{\beta}$ is a p-vector of unknown regression coefficients, and $\boldsymbol{e}_i$, $i = 1, 2, \ldots$ are random m-vectors.

Following Maronna (1976), a class of M-estimators of the regression coefficient and scatter parameters may be defined as the solutions of the following equations:

$$\begin{cases} \sum_{i=1}^n X_i' V^{-1}(\boldsymbol{y}_i - X_i\boldsymbol{\beta}) u_1(\|\boldsymbol{y}_i - X_i\boldsymbol{\beta}\|_V) = \mathbf{0} \\ \\ \sum_{i=1}^n \left[(\boldsymbol{y}_i - X_i\boldsymbol{\beta})(\boldsymbol{y}_i - X_i\boldsymbol{\beta})' u_2(\|\boldsymbol{y}_i - X_i\boldsymbol{\beta}\|_V^2) - V\right] = 0, \end{cases} \tag{2.2}$$

where u_1 and u_2 are suitably chosen functions. In Bai and Wu (1993), the following recursive algorithm has been constructed:

$$\begin{cases} \boldsymbol{\beta}_{n+1} = \boldsymbol{\beta}_n + S_{n+1}^{-1}\, a_n \boldsymbol{h}_1(\boldsymbol{\beta}_n,\ V_n,\ X_{n+1},\ \boldsymbol{y}_{n+1}) \\ V_{n+1} = V_n + (n+1)^{-1} H_2(\boldsymbol{\beta}_n,\ V_n,\ X_{n+1},\ \boldsymbol{y}_{n+1}), \end{cases} \tag{2.3}$$

where

$$\begin{cases} \boldsymbol{h}_1(\boldsymbol{\beta},\, V,\, X,\, \boldsymbol{y}) = X'\tilde{V}^{-1}(\boldsymbol{y} - X\boldsymbol{\beta})u_1(\|\boldsymbol{y} - X\boldsymbol{\beta}\|_{\tilde{V}}) \\ H_2(\boldsymbol{\beta},\, V,\, X,\, \boldsymbol{y}) = (\boldsymbol{y} - X\boldsymbol{\beta})(\boldsymbol{y} - X\boldsymbol{\beta})'u_2(\|\boldsymbol{y} - X\boldsymbol{\beta}\|_{\tilde{V}}^2) - V \\ S_n = \sum_{i=1}^{n} X_i'X_i \\ \|\boldsymbol{y}\|_{\tilde{V}}^2 = \boldsymbol{y}'\tilde{V}^{-1}\boldsymbol{y}, \end{cases} \tag{2.4}$$

$\boldsymbol{\beta}_0$ and $V_0 > 0$ are arbitrary, $\{a_n\}$ satisfies certain conditions and $\tilde{V}$ is a Lipschitz continuous $m \times m$ matrix function of V defined as follows:

> Let λ_i and $\boldsymbol{\alpha}_i$ be the eigenvalues and orthonormal eigenvectors of V respectively. Then $\tilde{V}$ is defined to be the matrix with eigenvalues and eigenvectors $\tilde{\lambda}_i$ and $\boldsymbol{\alpha}_i$, where
>
> $$\tilde{\lambda}_i = \begin{cases} \delta_1 & \text{if} \quad \lambda_i < \delta_1 \\ \lambda_i & \text{if} \quad \delta_1 \le \lambda_i \le \delta_2 \\ \delta_2 & \text{if} \quad \lambda_i > \delta_2, \end{cases}$$
>
> where $0 < \delta_1 < \delta_2 < \infty$ are positive constants which will be further specified when they are needed.

It is not difficult to verify that $\tilde{V}$ is bounded and Lipschitz continuous as a function of V with respect to the Euclidean norm. Then we have

$$0 < \delta_1 \le \lambda_*(\tilde{V}) \le \lambda^*(\tilde{V}) \le \delta_2 \le \infty. \tag{2.5}$$

For later use, define

$$\bar{\boldsymbol{h}}_1 = \bar{\boldsymbol{h}}_1(\boldsymbol{\beta}, V, X) = \mathrm{E}\, \boldsymbol{h}_1(\boldsymbol{\beta}, V, X, \boldsymbol{e}_i), \tag{2.6}$$

and

$$\bar{H}_2 = \bar{H}_2(\boldsymbol{\beta}, V, X) = \mathrm{E}\, H_2(\boldsymbol{\beta}, V, X, \boldsymbol{e}_i). \tag{2.7}$$

To establish our main results, we need the following conditions:

(E) $\{\boldsymbol{e}_i, \quad i = 1, 2, \ldots\}$ is a sequence of i.i.d. p-vectors having a common distribution density

$$|\Sigma|^{-\frac{1}{2}} f(\|\boldsymbol{e}\|_{\Sigma}),$$

where f is decreasing on $[0, \infty)$ and strictly decreasing in a neighborhood of 0.

Remark 2.1 In the literature, some authors assume that the error sequence satisfies a mixing structure (e.g., Englund 1993). Checking their proofs, one may find that the conditions on the mixing structure are only used in proving the convergence of the term $\max|R_{k,\ell}|$ (see (4.2)) by well-known results of limit theorems for mixing sequences in the literature. We do not assume mixing structure for simplicity. In fact, under their mixing structure, our results remain true. ////

(X) There exists $Q > 0$ such that, as $n \to \infty$,

$$\frac{1}{n}\sum_{i=1}^{n} X_i' X_i \longrightarrow Q, \tag{2.8}$$

and for any $\varepsilon > 0$ there is a constant $M = M(\varepsilon) > 0$ and a positive definite matrix Q_M such that, as $n \to \infty$,

$$\frac{1}{n}\sum_{i=1}^{n}(X_i' X_i I_{[||X_i|| \le M]}) \longrightarrow Q_M \tag{2.9}$$

with

$$||Q - Q_M|| < \varepsilon.$$

Remark 2.2 If $\{X_i\}$ is an iid sample of size n drawn from a random matrix with $\mathrm{E}\,\boldsymbol{X}'\boldsymbol{X} = Q > 0$, then (2.8) and (2.9) are true with probability one. In this case, for each $M > 0$, $Q_M = \mathrm{E}\,\boldsymbol{X}'\boldsymbol{X}\, I_{[||\,\boldsymbol{X}\,|| \le M]}$. ////

Let us select a sequence of increasing positive integers $\{n(k)\}$ satisfying

$$\sum_{k=1}^{\infty}(n(k)d_k)^{-1} < \infty, \quad d_k \to 0 \quad \text{as} \quad k \to \infty, \tag{2.10}$$

$$\frac{1}{n(k+1) - n(k)}\sum_{i=n(k)+1}^{n(k+1)} X_i' X_i - Q \longrightarrow 0 \tag{2.11}$$

and

$$\frac{1}{n(k+1) - n(k)}\sum_{i=n(k)+1}^{n(k+1)} X_i' X_i I_{[||X_i|| \le M]} - Q_M \longrightarrow 0, \tag{2.12}$$

where

$$d_k = \sum_{i=n(k)+1}^{n(k+1)} i^{-1}\,. \tag{2.13}$$

In fact, in the sequel, the sequence $\{n(k)\}$ is defined in the following way: Let $\{\alpha(n)\}$ be a monotone sequence of positive numbers satisfying

$$\alpha(1) < 1, \;\; \alpha(n) \to 0, \;\; n\alpha(n) \to \infty \text{ and } \alpha^*(n)/\alpha(n) \to 0,$$

where

$$\alpha^*(n) = \sup_{k\ge n}\{\max[||\frac{1}{m}\sum_{i=1}^{m} X_i'X_i - Q||,\ ||\frac{1}{m}\sum_{i=1}^{m} X_i'X_i I_{[||X_i||\le M]} - Q_M||]\},$$

$||Q - Q_M|| < (1/4)\lambda_*(Q)$ and $\lambda_*(Q)$ denotes the smallest eigenvalue of the symmetric matrix Q. Define $n(1) = 1$ and $n(k+1) = [n(k)/(1-\alpha(n(k)))]+1$, for all $k \ge 1$. Note that the integer sequence $\{n(k)\}$ is strictly increasing. From this, it is easy to see that

$$\frac{n(k+1)}{n(k)} \to 1 \text{ and } \frac{n(k)d_k}{n(k+1)-n(k)} \sim \frac{n(k)\alpha(n(k))}{n(k+1)-n(k)} \to 1. \qquad (2.14)$$

Hence, we have

$$\begin{aligned}
&||\frac{1}{n(k+1)-n(k)}\sum_{i=n(k)+1}^{n(k+1)} X_i'X_i - Q||\\
\le\ & \frac{2}{\alpha(n(k))}\left[||\frac{1}{n(k+1)}\sum_{i=1}^{n(k+1)} X_i'X_i - Q|| + ||\frac{1}{n(k)}\sum_{i=1}^{n(k)} X_i'X_i - Q||\right]\\
\le\ & 4\alpha^{-1}(n(k))\alpha^*(n(k)) \to 0.
\end{aligned}$$

This proves the limit (2.11). The limit (2.12) can be similarly proved. Note that the sequence $\{\alpha(n)\}$ can be chosen with an arbitrarily slow rate, and for all n with $n(k) < n \le n(k+1)$, we may choose $\alpha(n)$ after $n(k)$ is defined. Thus, we may assume that $\alpha(n(k)) > \frac{2+\varepsilon}{k+1}$. Therefore, we have

$$\begin{aligned}
\sum_{k=1}^{\infty}\frac{1}{n(k)d_k} &\le C\sum_{k=1}^{\infty} k\exp\Big(-\sum_{i=1}^{k}\frac{2+\varepsilon}{i}\Big)\\
&\le C\sum_{k=1}^{\infty} k^{-(1+\varepsilon)} < \infty.
\end{aligned}$$

It is obvious that we always have

$$\sum_{k=1}^{\infty} d_k = \infty. \qquad (2.15)$$

(A1) The sequence $\{a_n,\ n = 1, 2, \ldots\}$ is adaptive and $0 < \nu_1 \le a_i \le \nu_2 < \infty$ for some ν_1 and ν_2 and all i.
(A2) The sequence $\{a_n, n = 1, 2, \ldots\}$ satisfies

$$\max_{n(k)<\ell\le n(k+1)} |a_\ell - a_{n(k)}| = \mathrm{o}(1), \quad \text{a.s.} \qquad (2.16)$$

and $0 < \nu_1 \le a_i \le \nu_2 < \infty$ for some ν_1 and ν_2 and all i.

(V) The sequence $\{V_n,\ n = 1, 2, \ldots\}$ of random matrices is adaptive (i.e., V_n is $\mathcal{F}_n$-measurable, where $\mathcal{F}_n$ denotes the sigma field generated by the random vectors $\boldsymbol{y}_i$ for $i = 1, \ldots, n$) and satisfies

$$\max_{n(k)<\ell\le n(k+1)} \|V_\ell - V_{n(k)}\| = o(1), \quad \text{a.s.} \tag{2.17}$$

Condition (V) is not actually a restriction because (2.17) is established in the proof of Theorem 4.2 and the adaptiveness of V_n holds automatically true for the recursive M-estimators defined by (2.3). This condition is made only for the use in Theorem 4.1. In fact, the adaptiveness is not necessary even for Theorem 4.1.

For simplicity, we make the following conventions:

Definition 2.3 *A multivariate function is said to be BLC($\boldsymbol{y}$) if it is bounded and Lipschitz continuous as a function of $\boldsymbol{y}$; a univariate function $v(t)$, $t > 0$, is said to be a BL function if there exists a positive constant M such that*

$$|v(t) - v(s)| \le Mt^{-1}|t - s|, \qquad \textit{for any } t,\ s > 0$$

and

$$|v(t) - v(s)| \le M(ts)^{-1}|t - s|, \qquad \textit{for any } t,\ s > 0\,.$$

(U) $u_1(t)$ and $u_2(t)$ are nonnegative decreasing BL functions such that $tu_j(t)$, $j = 1, 2$, are increasing for $t > 0$ (strictly increasing for all small t). Also, we assume that $tu_2(t) > m$ for some $t > 0$.

Remark 2.4 An M-estimator may also be defined as the solution of a minimization problem associated with a discrepancy function ρ. If ρ is symmetric and convex, and has a continuous derivative φ, then our $u_1(t)$ corresponds to $t^{-1}\varphi(t)$ and $u_2(t)$ corresponds to $u_1(\sqrt{t})/2$, for all $t > 0$. It has been noticed that Huber's celebrated example satisfied this condition. ////

In Lemma A2 of Bai and Wu (1993), it has been proved that the functions $\boldsymbol{y}u_1(\|\boldsymbol{y}\|_{\tilde{V}})$ and $\boldsymbol{y}\,\boldsymbol{y}'u_2(\|\boldsymbol{y}\|^2_{\tilde{V}})$ are BLC$(\boldsymbol{y}, V)$ if Condition (U) is true. Hence we have

$$\begin{aligned}\Big\|(\boldsymbol{y} - X\boldsymbol{\beta}_1)u_1(\|\boldsymbol{y} - X\boldsymbol{\beta}_1\|_{\tilde{V}_1}) &- (\boldsymbol{y} - X\boldsymbol{\beta}_2)u_1(\|\boldsymbol{y} - X\boldsymbol{\beta}_2\|_{\tilde{V}_2})\Big\| \\ &\le C\left(\|X\|\,\|\boldsymbol{\beta}_1 - \boldsymbol{\beta}_2\| + \|\tilde{V}_1 - \tilde{V}_2\|\right),\end{aligned}$$

where $\|V\|^2 = \operatorname{tr}(V'V)$. Consequently we have

$$\begin{aligned}&|\boldsymbol{h}_1(\boldsymbol{\beta}_1, V_1, X, \boldsymbol{e}) - \boldsymbol{h}_1(\boldsymbol{\beta}_2, V_2, X, \boldsymbol{e})| \\ &\quad\le\ C\left(\|X\|^2\|\boldsymbol{\beta}_1 - \boldsymbol{\beta}_2\| + \|X\|\|\tilde{V}_1 - \tilde{V}_2\|\right),\end{aligned} \tag{2.18}$$

and

$$\begin{aligned}&|H_2(\boldsymbol{\beta}_1, V_1, X, \boldsymbol{e}) - H_2(\boldsymbol{\beta}_2, V_2, X, \boldsymbol{e})| \\ &\quad\le\ C\left(\|X\|\|\boldsymbol{\beta}_1 - \boldsymbol{\beta}_2\| + \|\tilde{V}_1 - \tilde{V}_2\|\right),\end{aligned} \tag{2.19}$$

where $\boldsymbol{h}_1$ and H_2 are defined in (2.4).

3 Some Lemmas

Lemma 3.1 *Let $\{z_i \mid i = 1, 2, \ldots\}$ be a martingale difference sequence such that $|1/(n(k+1) - n(k)) \sum_{n(k)+1}^{n(k+1)} \mathrm{E}\, z_i^2| \le C$, where C is a constant, and the constant sequence of $\{a_i\}$ satisfy Conditions (A1) or (A2). Then*

$$\max_{n(k)<\ell\le n(k+1)} \left| \sum_{i=n(k)+1}^{\ell} i^{-1} a_{i-1} z_i \right| = \mathrm{o}(d_k), \quad \text{a.s.} \tag{3.1}$$

PROOF By Condition (A1), $\left\{\sum_{i=n(k)+1}^{\ell} i^{-1} a_{i-1} z_i,\ \ell = n(k)+1, \ldots\right\}$ forms a martingale sequence. Hence, for any $\varepsilon > 0$, by (2.14) and $|1/[n(k+1) - n(k)] \sum_{n(k)+1}^{n(k+1)} \mathrm{E}\, z_i^2| \le C$,

$$P\left\{ \max_{n(k)<\ell\le n(k+1)} d_k^{-1} \left| \sum_{i=n(k)+1}^{\ell} i^{-1} a_{i-1} z_i \right| \ge \varepsilon \right\}$$

$$\le \nu_2^2 \varepsilon^2 d_k^{-2} \sum_{i=n(k)+1}^{n(k+1)} i^{-2}\, \mathrm{E}\, z_i^2 \le C\, (n(k) d_k)^{-1}. \tag{3.2}$$

Then (3.1) follows from (2.10) and (3.2). If Condition (A2) is true, the proof is similar and omitted. ////

Lemma 3.2 *If Condition (X) holds, then for any $\kappa \in (0, 2]$*

$$\sum_{i=n(k)+1}^{n(k+1)} i^{-1} \|X_i\|^{\kappa} = \mathrm{O}(d_k).$$

PROOF By Condition (X), (2.11) and (2.14),

$$\begin{aligned} &\sum_{i=n(k)+1}^{n(k+1)} i^{-1} \|X_i\|^{\kappa} \le \\ &\le \left(\frac{n(k+1) - n(k)}{n(k)}\right) \left(\frac{1}{n(k+1) - n(k)} \sum_{i=n(k)+1}^{n(k+1)} \|X_i\|^{\kappa}\right) \\ &\le C d_k \left(\frac{1}{n(k+1) - n(k)} \sum_{i=n(k)+1}^{n(k+1)} \|X_i\|^2\right)^{\kappa/2} = \mathrm{O}(d_k), \end{aligned}$$

and the lemma is proved. ////

Lemma 3.3 *Suppose that* $\{\boldsymbol{q}_k(.,.)\}$ *is a sequence of uniformly bounded functions from* $R^p \times R^{m^2}$ *to* R^p, $\{d_k\}$ *is a sequence of positive constants,* $\{\boldsymbol{r}_k\}$ *is a sequence of p-vectors and* $\mathcal{S}$ *is a set in* R^{m^2}. *Fix an* $\boldsymbol{\eta}_0$. *For each* k, $\underset{\sim}{V}_k \in \mathcal{S}$. *Define* $\boldsymbol{\eta}_k$ *recursively by*

$$\boldsymbol{\eta}_{k+1} = \boldsymbol{\eta}_k + d_k \boldsymbol{q}_k(\boldsymbol{\eta}_k, \underset{\sim}{V}_k) + \boldsymbol{r}_k .$$

Suppose that

(i) $d_k \to 0$, $\boldsymbol{r}_k = o(d_k)$ *and* $\sum_{k=1}^{\infty} d_k = \infty$.

(ii) *For each small* $\gamma > 0$, *there is a positive constant* $\tau = \tau(\gamma)$ *such that*

$$\sup_k \sup \left\{ (\boldsymbol{x} - \boldsymbol{\eta})'(\boldsymbol{q}_k(\boldsymbol{x}, \underset{\sim}{V}) + \boldsymbol{\varepsilon}) : \|\boldsymbol{x} - \boldsymbol{\eta}\| > \gamma,\ \|\boldsymbol{\varepsilon}\| < \tau,\ \underset{\sim}{V} \in \mathcal{S} \right\} < 0. \tag{3.3}$$

Then $\boldsymbol{\eta}_k \to \boldsymbol{\eta}$, *where* $\boldsymbol{\eta}$ *is defined in* (3.3).

If $\{\boldsymbol{\eta}_k\}$ *is bounded, the conclusion is still true when* (3.3) *is weakened as*

$$\sup_k \sup \left\{ (\boldsymbol{x} - \boldsymbol{\eta})' \boldsymbol{q}_k(\boldsymbol{x}, \underset{\sim}{V}) : \|\boldsymbol{x} - \boldsymbol{\eta}\| > \gamma,\ \underset{\sim}{V} \in \mathcal{S} \right\} < 0. \tag{3.3$'$}$$

Here, when p *is the square of an integer and* x, q_k *and* η *are matrices, then* $(X - \eta)'q_k$ *of* (3.3) *or* (3.3′) *should be considered as* $\mathrm{tr}((X - \eta)'q_k)$.

This lemma is an extension of Lemma 2.2 of Bai and Wu (1993).

PROOF Since $\{\boldsymbol{q}_k(.,.)\}$ is a sequence of uniformly bounded functions, the lemma can be proved by following the same lines as in the proof of Lemma 2.2 of Bai and Wu (1993). ////

Lemma 3.4 *If Conditions (E) and (U) hold, then*

$$\int \eta^{(j)} t^{(j)} u_1(\|\boldsymbol{t}\|_\Delta) f(\|\boldsymbol{t} + \boldsymbol{\eta}\|) d\boldsymbol{t} \le 0, . \tag{3.4}$$

and for any constant $\gamma > 0$, *there exists a constant* $\mu = \mu(\gamma) > 0$, *such that for all* $\|\boldsymbol{\eta}\| > \gamma$,

$$\int \boldsymbol{\eta}' \boldsymbol{t} u_1(\|\boldsymbol{t}\|_\Delta) f(\|\boldsymbol{t} + \boldsymbol{\eta}\|) d\boldsymbol{t} \le -\mu \|\boldsymbol{\eta}\|, \tag{3.5}$$

where $\eta^{(j)}$ *and* $t^{(j)}$ *are the* j *th components of* $\boldsymbol{\eta}$ *and* $\boldsymbol{t}$, *respectively and* Δ *is a diagonal matrix with positive diagonal elements.*

PROOF See Bai and Wu (1993). ////

4 Main Theorems

Consider the recursive procedure

$$\boldsymbol{\beta}_{n+1} = \boldsymbol{\beta}_n + S_{n+1}^{-1} a_n \boldsymbol{h}_1(\boldsymbol{\beta}_n, V_n, X_{n+1}, \boldsymbol{y}_{n+1}) \tag{4.1}$$

where $\boldsymbol{\beta}_0$ is arbitrary.

Theorem 4.1 *Suppose that the recursive estimates $\boldsymbol{\beta}_n$ are generated by (4.1) and the conditions (E), (X), (A1), (V) and (U) hold. Then for any $0 < \delta_1 < \delta_2 < \infty$, $\boldsymbol{\beta}_n \to \boldsymbol{\beta}_0$,* a.s. *as $n \to \infty$. When Condition (A1) is replaced by Condition (A2), the conclusion still holds.*

PROOF Without loss of generality, we can assume that $Q = I$, the true value of the coefficients $\boldsymbol{\beta} = \mathbf{0}$.

We have, for $n(k) < \ell \le n(k+1)$, by (4.1),

$$\begin{aligned}
\boldsymbol{\beta}_\ell &= \boldsymbol{\beta}_{n(k)} + \sum_{i=n(k)+1}^{\ell} S_i^{-1} a_{i-1} \boldsymbol{h}_1(\boldsymbol{\beta}_{i-1}, V_{i-1}, X_i, \boldsymbol{e}_i) \\
&= \boldsymbol{\beta}_{n(k)} + \sum_{i=n(k)+1}^{\ell} i^{-1} a_{i-1} \bar{\boldsymbol{h}}_1(\boldsymbol{\beta}_{n(k)}, V_{n(k)}, X_i) \qquad (4.2)\\
&\quad + R_{k,\ell} + T_{k,\ell} + W_{k,\ell}, \qquad (4.3)
\end{aligned}$$

where

$$R_{k,\ell} = \sum_{i=n(k)+1}^{\ell} i^{-1} a_{i-1} \left[\boldsymbol{h}_1(\boldsymbol{\beta}_{i-1}, V_{i-1}, X_i, \boldsymbol{e}_i) - \bar{\boldsymbol{h}}_1(\boldsymbol{\beta}_{i-1}, V_{i-1}, X_i)\right],$$

$$T_{k,\ell} = \sum_{i=n(k)+1}^{\ell} i^{-1} a_{i-1} \left[\bar{\boldsymbol{h}}_1(\boldsymbol{\beta}_{i-1}, V_{i-1}, X_i) - \bar{\boldsymbol{h}}_1(\boldsymbol{\beta}_{n(k)}, V_{n(k)}, X_i)\right],$$

$$W_{k,\ell} = \sum_{i=n(k)+1}^{\ell} i^{-1} \left[(i^{-1} S_i)^{-1} - I\right] a_{i-1} \boldsymbol{h}_1(\boldsymbol{\beta}_{i-1}, V_{i-1}, X_i, \boldsymbol{e}_i).$$

Since $\{a_i\}$ is bounded and $\|\boldsymbol{h}_1\| \le C\|X_i\|$ for some positive constant C, by Condition (X), (2.13) and Lemma 3.2, we obtain

$$\begin{aligned}
\max_{n(k)<\ell\le n(k+1)} |W_{k,\ell}| &\le \mathrm{o}(1) \left[\sum_{i=n(k)+1}^{n(k+1)} i^{-1} \|X_i\|\right] \\
&= \mathrm{o}(d_k).
\end{aligned}$$

Hereafter, the symbol o(.) is in the sense of "with probability one", unless specified. By Lemma 3.1, we get

$$\max_{n(k)<\ell\le n(k+1)} |R_{k,\ell}| = \mathrm{o}(d_k), \tag{4.4}$$

for either cases when Conditions (A1) or (A2) is true. By (2.17), (2.18) and Lemma 3.2, we have

$$\begin{aligned}
&\max_{n(k)<\ell\le n(k+1)} |T_{k,\ell}| \\
&\le C \sum_{i=n(k)+1}^{n(k+1)} i^{-1} \left(\|X_i\|^2 \|\boldsymbol{\beta}_{i-1} - \boldsymbol{\beta}_{n(k)}\| + \|X_i\| \|\tilde{V}_{i-1} - \tilde{V}_{n(k)}\| \right) \\
&\le C \sum_{i=n(k)+1}^{\ell} i^{-1} \left(\max_{n(k)<\ell\le n(k+1)} \|\boldsymbol{\beta}_\ell - \boldsymbol{\beta}_{n(k)}\| \|X_i\|^2 + \mathrm{o}(1)\|X_i\| \right) \\
&\le C \max_{n(k)<\ell\le n(k+1)} \|\boldsymbol{\beta}_\ell - \boldsymbol{\beta}_{n(k)}\| d_k + \mathrm{o}(d_k).
\end{aligned} \tag{4.5}$$

It is obvious that

$$\max_{n(k)<\ell\le n(k+1)} \|\boldsymbol{\beta}_\ell - \boldsymbol{\beta}_{n(k)}\| = \mathrm{O}(d_k), \tag{4.6}$$

follows from (4.3), Conditions (X) and (U), which, in turn, implies that

$$\max_{n(k)<\ell\le n(k+1)} |T_{k,\ell}| \le \mathrm{o}(d_k). \tag{4.7}$$

Therefore, if we can show that

$$\boldsymbol{\beta}_{n(k)} \to \mathbf{0}, \qquad \text{a.s. }, \tag{4.8}$$

Theorem 4.1 follows. Since $d_k = \mathrm{o}(1)$, from (4.3) - (4.6), we have

$$\boldsymbol{\beta}_{n(k+1)} = \boldsymbol{\beta}_{n(k)} + \sum_{i=n(k)+1}^{n(k+1)} i^{-1} a_{i-1} \bar{\boldsymbol{h}}_1(\boldsymbol{\beta}_{n(k)}, V_{n(k)}, X_i) + \mathrm{o}(d_k), \tag{4.9}$$

where the order of $\mathrm{o}(d_k)$ in (4.9) is uniform for $n(k) < \ell \le n(k+1)$.

Now we apply Lemma 3.3 to prove (4.8). Let $\boldsymbol{\eta}_k = \boldsymbol{\beta}_{n(k)}$ and $\boldsymbol{q}_k(\boldsymbol{\eta}, V) = d_k^{-1} \sum_{i=n(k)+1}^{n(k+1)} i^{-1} a_{i-1} \bar{\boldsymbol{h}}_1(\boldsymbol{\eta}, V, X_i)$. Define

$$\mathcal{S} = \{V : p \times p \text{ matrix with } 0 < \tilde{\delta}_1 \le \lambda_*(V) \le \lambda^*(V) \le \tilde{\delta}_2 < \infty\},$$

where $\tilde{\delta}_1 = \delta_1/\lambda^*(\Sigma) > 0$ and $\tilde{\delta}_2 = \delta_2/\lambda_*(\Sigma) < \infty$.

In order to complete the proof of Theorem 4.1, we only need to verify the uniform boundedness of $\boldsymbol{q}_k$ and the conditions (i) and (ii) in Lemma 3.3. Since $\|\boldsymbol{h}_1\| \le C\|X\|$, the uniformly boundedness of $\boldsymbol{q}_k$ follows from the boundedness of $\{a_i\}$ and Lemma 3.2. Thus, the condition (i) follows immediately from (2.10) and (2.15). What remains to be proved is to verify the condition (ii). By the transformation

$$\begin{cases} \boldsymbol{z} = O'\Sigma^{-\frac{1}{2}} \boldsymbol{e}, \\ W = O'\Sigma^{-\frac{1}{2}} X, \end{cases} \tag{4.10}$$

we obtain

$$\begin{aligned}\bar{\boldsymbol{h}}_1(\boldsymbol{\eta},\, V,\, X) &= |\Sigma|^{-\frac{1}{2}} \int X'\tilde{V}^{-1}(\boldsymbol{e} - X\boldsymbol{\eta})u_1(\|\boldsymbol{e} - X\boldsymbol{\eta}\|_{\tilde{V}})\, f\,(\|\boldsymbol{e}\|_{\Sigma})d\boldsymbol{e} \\ &= \int W'\Delta^{-1}(\boldsymbol{z} - W\boldsymbol{\eta})u_1(\|\boldsymbol{z} - W\boldsymbol{\eta}\|_{\Delta})f(\|\boldsymbol{z}\|)d\boldsymbol{z},\end{aligned}$$

where O is an $m \times m$ orthogonal matrix such that $\Delta = O'\Sigma^{-1/2}\tilde{V}\Sigma^{-1/2}O$ is a diagonal matrix. Therefore,

$$\begin{aligned}&\boldsymbol{\eta}'\boldsymbol{q}_k(\boldsymbol{\eta},\, V) = \\ &= d_k^{-1} \sum_{i=n(k)+1}^{n(k+1)} i^{-1}a_{i-1}\boldsymbol{\eta}'\bar{\boldsymbol{h}}_1(\boldsymbol{\eta},\, V,\, X_i) \\ &= d_k^{-1} \sum_{i=n(k)+1}^{n(k+1)} i^{-1}a_{i-1} \int \boldsymbol{\eta}'W_i'\Delta^{-1}(\boldsymbol{z} - W_i\boldsymbol{\eta})u_1(\|\boldsymbol{z} - W_i\boldsymbol{\eta}\|_{\Delta})f(\|\boldsymbol{z}\|)d\boldsymbol{z}.\end{aligned} \tag{4.11}$$

Let

$$\begin{cases} \boldsymbol{\gamma}_i = W_i\boldsymbol{\eta}, \\ \boldsymbol{t} = \boldsymbol{z} - \boldsymbol{\gamma}_i, \end{cases} \tag{4.12}$$

and let $\gamma > 0$ be a positive number and let $\|\boldsymbol{\eta}\| > \eta_0 > 0$. By (2.12), for all large k, we have

$$\frac{1}{n(k+1) - n(k)} \sum_{i=n(k)+1}^{n(k+1)} X_i'X_i I_{[\|X_i\| \le M]} \ge \frac{1}{2}I. \tag{4.13}$$

Hence,

$$\begin{aligned}&\frac{1}{n(k+1) - n(k)} \sum_{i=n(k)+1}^{n(k+1)} \|\boldsymbol{\gamma}_i\|^2 I_{[\|X_i\| \le M]} \\ = \;& \frac{1}{n(k+1) - n(k)} \sum_{i=n(k)+1}^{n(k+1)} \boldsymbol{\eta}'X_i'\Sigma^{-1}X_i I_{[\|X_i\| \le M]}\boldsymbol{\eta} \\ \ge \;& (\lambda^*(\Sigma))^{-1}\boldsymbol{\eta}' \left(\frac{1}{n(k+1) - n(k)} \sum_{i=n(k)+1}^{n(k+1)} X_i'X_i I_{[\|X_i\| \le M]} \right) \boldsymbol{\eta} \\ \ge \;& \frac{1}{2}(\lambda^*(\Sigma))^{-1}\|\boldsymbol{\eta}\|^2.\end{aligned} \tag{4.14}$$

Note that $\|X_i\| \le M$ implies $\|\boldsymbol{\gamma}_i\| \le (\lambda_*(\Sigma))^{-1/2}M\|\boldsymbol{\eta}\|$. Thus, by (4.14), we conclude that there are at least $b(n(k+1) - n(k))$ $\boldsymbol{\gamma}_i$'s satisfying $\|\boldsymbol{\gamma}_i\| \ge \gamma$, where

$$b = \frac{\lambda_*(\Sigma)}{4M^2\lambda^*(\Sigma)} > 0$$

and

$$\gamma = \frac{1}{2}(\lambda^*(\Sigma))^{-\frac{1}{2}}\|\boldsymbol{\eta}\| \geq \gamma_0 := \frac{1}{2}\eta_0(\lambda^*(\Sigma))^{-\frac{1}{2}}.$$

By Lemma 3.4, we have: For $\|\boldsymbol{\eta}\| > \eta_0$, there exists $\mu = \mu(\gamma_0) > 0$ such that

$$\begin{aligned}
\boldsymbol{\eta}'\boldsymbol{q}_k(\boldsymbol{\eta}, V) &= \qquad (4.15)\\
&= d_k^{-1}\sum_{i=n(k)+1}^{n(k+1)} i^{-1}a_{i-1}\int \boldsymbol{\gamma}_i'\Delta^{-1}\boldsymbol{t}u_1(\|\boldsymbol{t}\|_\Delta)f(\|\boldsymbol{t}+\boldsymbol{\gamma}_i\|)d\boldsymbol{t}\\
&\leq \nu_1(\lambda^*(\Delta)n(k+1)d_k)^{-1}\sum_{i=n(k)+1}^{n(k+1)}\int \boldsymbol{\gamma}_i'\boldsymbol{t}u_1(\|\boldsymbol{t}\|)f(\|\boldsymbol{t}+\boldsymbol{\gamma}_i\|)d\boldsymbol{t}\\
&\leq -\mu\nu_1(\lambda^*(\Delta)n(k+1)d_k)^{-1}\sum_{i=n(k)+1}^{n(k+1)}\|\boldsymbol{\gamma}_i\|I[\|\boldsymbol{\gamma}_i\| > \gamma_0]\\
&\leq -\frac{1}{2}\mu\nu_1 b(n(k+1)-n(k))(\lambda^*(\Delta)n(k+1)d_k)^{-1}(\lambda^*(\Sigma))^{-\frac{1}{2}}\|\boldsymbol{\eta}\|\\
&\leq -C\|\boldsymbol{\eta}\|. \qquad (4.16)
\end{aligned}$$

This proves Condition (ii) of Lemma 3.3 and consequently completes the proof of Theorem 4.1.

When Condition (A2) is true, checking the proof of Theorem 4.1, we only needs to split $R_{k,\ell}$ into $R^1_{k,\ell}$ and $R^2_{k,\ell}$, where

$$R^1_{k,\ell} = \sum_{i=n(k)+1}^{\ell} i^{-1}a_{n(k)}\left[\boldsymbol{h}_1(\boldsymbol{\beta}_{i-1}, V_{i-1}, X_i, \boldsymbol{e}_i) - \bar{\boldsymbol{h}}_1(\boldsymbol{\beta}_{i-1}, V_{i-1}, X_i)\right]$$

and

$$R^2_{k,\ell} = \sum_{i=n(k)+1}^{\ell} i^{-1}(a_{i-1}-a_{n(k)})\left[\boldsymbol{h}_1(\boldsymbol{\beta}_{i-1}, V_{i-1}, X_i, \boldsymbol{e}_i) - \bar{\boldsymbol{h}}_1(\boldsymbol{\beta}_{i-1}, V_{i-1}, X_i)\right].$$

It is easy to see that $R^1_{k,\ell}$ satisfies a similar estimate as (4.4). By (2.16) and strong limit theorems for independent random variables, one can show that

$$\max_{n(k)<\ell\leq n(k+1)}|R^2_{k,\ell}| = o(1)\left[\sum_{i=n(k)+1}^{n(k+1)} i^{-1}\|X_i\|\right] = o(d_k).$$

Then the second assertion of Theorem 4.1 follows. ////

Let Ω be the limit of V_n, By (2.2), Ω satisfies

$$\Omega = \mathrm{E}\,\boldsymbol{e}\,\boldsymbol{e}'u_2(\|\boldsymbol{e}\|^2_\Omega) = \Sigma^{\frac{1}{2}}\int \boldsymbol{z}\,\boldsymbol{z}'\Sigma^{\frac{1}{2}}u_2(\|\boldsymbol{z}\|^2_\Omega)f(\|\boldsymbol{z}\|^2)d\boldsymbol{z}$$

or

$$\tilde{\Omega} = \int z\, z' u_2(\| z \|_{\tilde{\Omega}}^2) f(\| z \|^2) d\, z\,,$$

where $\tilde{\Omega} = \Sigma^{-\frac{1}{2}} \Omega \Sigma^{-\frac{1}{2}}$, which implies that $\tilde{\Omega} = \omega^{-1} I$ or $\Sigma = \omega \Omega$ for some $\omega > 0$. Therefore, ω satisfies

$$m = \int \omega\, z' z\, u_2(\omega \| z \|^2) f(\| z \|^2) d\, z\,. \tag{4.17}$$

Note that the right hand side of (4.17) is a strictly increasing function of ω. Hence, by Condition (U), the equation (4.17) has a unique solution of ω.

Define $\zeta = \min(\zeta^*,\ 0.1)$ where ζ^* is defined by

$$\zeta^* = \frac{1}{4m} \int [\omega \| z \|^2 u_2(\omega \| z \|^2) - \frac{\omega}{1.5} \| z \|^2 u_2(\frac{\omega}{1.5} \| z \|^2)] f(\| z \|^2) d\, z\,.$$

By the condition (U), we know that ζ^* and hence ζ are positive constants. Define

$$\begin{cases} \boldsymbol{\beta}_{n+1} = \boldsymbol{\beta}_n + S_{n+1}^{-1}\, a_n \boldsymbol{h}_1(\boldsymbol{\beta}_n,\ V_n,\ X_{n+1},\ \boldsymbol{y}_{n+1}) \\ V_{n+1} = V_n + (n+1)^{-1} H_2(\boldsymbol{\beta}_n,\ V_n,\ X_{n+1},\ \boldsymbol{y}_{n+1}), \end{cases} . \tag{4.18}$$

Then we have

Theorem 4.2 *Suppose the constants* δ_1, δ_2 *in* (2.5) *satisfy* $\delta_1 < \zeta \lambda_*(\Omega)$ *and* $\delta_2 > 3\lambda^*(\Omega)$. *Let* $\boldsymbol{\beta}_n$ *and* V_n *be defined by the recursive procedure* (4.18). *Then, if Conditions (E), (X), (A1) or (A2) and (U) hold, as* $n \to \infty$,

$$(\boldsymbol{\beta}_n,\ V_n) \longrightarrow (\boldsymbol{\beta},\ \Omega) \quad \text{a.s.}\ .$$

PROOF As done in Bai and Wu (1993), one may easily see that the sequence of matrices V_n satisfies Condition (V). Applying Theorem 4.1, we have proved the strong consistency of the recursive estimators $\boldsymbol{\beta}_n$. Then, following the same steps as in the proof of Theorem 3.2 of Bai and Wu (1993), we can prove the theorem and the details of the proof are omitted. ////

Acknowledgements

The author was supported in part by Natural Sciences and Engineering Research Council of Canada and Faculty of Arts, York University.

References

[1] Bai, Z.D., Rao, C.R. and Wu, Y. (1991): Recent contributions to robust estimation in linear models. In *Probability, Statistics and Design Experiments—R.C. Bose Symposium Volume* (R.R. Bahadur, ed.), 33–50. Wiley, Eaton.

[2] Bai, Z. D. and Wu, Y. (1993): Recursive algorithm for M-estimates of regression coefficients and scatter parameters in linear models. *Sankhyā B* **55**(2) 199–218.

[3] Dodge, Y. (1987): *Statistical Data Analysis Based on the L_1-Norm and Related Methods*. North-Holland.

[4] Englund, J.-E, Holst, U. and Ruppert, D. (1988): Recursive M-estimators of location and scale for dependent sequences. *Scand. J. Statist.* **15** 147–159.

[5] Englund, J.-E. (1989): Multivariate recursive M-estimators of location for dependent sequences. *Sequential Anal.* **8** 293–315.

[6] Englund, J.-E. (1993): Multivariate recursive M-estimators of location and scatter for dependent sequences. *J. Multivariate Analysis* **45** 257–273.

[7] Huber, P.J. (1964): Robust estimation of a location parameter. *Ann. Math. Statist.* **35** 73–101.

[8] Huber, P.J. (1973): Robust regression: Asymptotics, conjectures and Monte Carlo. *Ann. Statist.* **1** 799–821.

[9] Maronna, R.A. (1976): Robust estimators of multivariate location and scatter. *Ann. Statist.* **4** 51–67.

[10] Ruppert, D. (1983): Convergence of stochastic approximation algorithms with non-additive dependent disturbances and applications. *Lecture Notes in Statistics* #20, 182–190. Springer-Verlag, New York.

List of Participants

Schloß Thurnau, June 10–12, 1994

Augustin, Thomas ◇ Seminar f. Spezialgebiete der Statistik, Universität München, Ludwigstr. 33, D-80539 München, Germany

Beran, Rudolf J. ◇ Department of Statistics, University of California, Evans Hall, Berkeley CA 94720, USA

Boltze, Lothar ◇ FB Mathematik und Informatik, Martin–Luther–Universität, Weinbergweg 17, D-06099 Halle, Germany

Buja, Andreas ◇ AT&T Bell Laboratories, Room 2C-261, P.O. Box 636, 600 Mountain Avenue, Murray Hill, NJ 07974-0636, USA

Christmann, Andreas ◇ Fachbereich Statistik, Universität Dortmund, Vogelpothsweg 87, D-44221 Dortmund, Germany

Croux, Christophe ◇ Department of Mathematics, University of Antwerp (U.I.A.), Universiteitsplein 1, B-2610 Wilrijk, Belgium

Davis, Laurie ◇ Fachbereich Mathematik, Universität Essen, Universitätsstraße 2, D-45141 Essen, Germany

Dinges, Hermann ◇ Fachbereich Mathematik, Universität Frankfurt, Robert–Mayer–Straße 6–10, D-60325 Frankfurt, Germany

Dodge, Yadolah ◇ Groupe de Statistique, Université de Neuchâtel, Pierre–à–Mazel 7, CH-2000 Neuchâtel, Switzerland

Donoho, David L. ◇ Department of Statistics, Stanford University, Sequoia Hall, Stanford CA 94305, USA

Dutter, Rudolf ◇ Institut für Statistik und Wahrscheinlichkeitsrechnung, TU Wien, Wiedner Hauptstraße 8–10/107, A-1040 Wien, Austria

Eicker, Friedhelm ◇ Math. Statistik und Anwendungen II, Universität Dortmund, Vogelpothsweg 87, D-44227 Dortmund, Germany

Gasser, Theo ◇ Abteilung Bio-Statistik, ISPM, Universität Zürich, Sumatrastraße 30, CH-8006 Zürich, Switzerland

Genton, Marc ◇ Departement de Mathematiques, Ecole Polytechnique Fédérale, MA (Ecublens), CH-1015 Lausanne, Switzerland

Gneiting, Tilmann ◇ Mathematik VII, Universität Bayreuth, Universitätsstraße 30, D-95440 Bayreuth, Germany

Haller, Gerd ◇ Mathematik III, Ruhr–Universität–Bochum, Universitätsstraße 150, D-44780 Bochum, Germany

Hampel, Frank ◇ Seminar für Statistik, ETH Zürich, CH-8092 Zürich, Switzerland

Heiler, Siegfried ◇ Fachbereich Statistik, Universität Konstanz, Universtitätsstr. 10, D-78434 Konstanz, Germany

Huber, Peter J. ◇ LSt. Mathematik VII, Universität Bayreuth, Universitätsstraße 30, D-95440 Bayreuth, Germany

Huber, Thomas ◇ Datavision AG, P.O.Box 471, CH-7250 Klosters, Switzerland

Huber–Carol, Catherine ◇ Universitè de Paris, 71 Blvd. des Etats Unis, F-78110 Le Vesiret, France

Hubert, Mia ◇ Department of Mathematics and Computing, Universitaire Instelling Antwerpen, Universiteitsplein 1, B-2610 Wilrijk, Belgium

Hummel, Tassilo ◇ Münchener Rückversicherung, Jella–Lepmann–Str. 13, D-81673 München, Germany

Kharin, Yurij ◇ Dept. of Mathematical Modelling and Data Analysis, Belarussian State University, 4 F. Skaryna Prospect, GUS-220080 Minsk, Belarus

Knüsel, Leo ◇ Institut für Statistik, Universität München, Geschw.–Scholl–Platz 1, D-80539 München, Germany

Koltchinskii, Vladimir I. ◇ Department of Mathematics and Statistics, University of New Mexico, Albuquerque, New Mexico, 87131-1141, USA

Lischer, Peter ◇ Eidg. Forschungsanstalt f. Agrikultur u. Umwelthygiene, Schwarzenburgstraße 155, CH-3097 Liebefeld, Switzerland

Maechler, Martin ◇ Seminar für Statistik, ETH Zürich, ETH Zentrum, SOL, CH-8092 Zürich, Switzerland

Marazzi, Alfio ◇ Institut Universitaire de Médecine Sociale et Preventive, Rue du Bugnon 17, CH-1005 Lausanne, Switzerland

Molinari, Luciano ◇ Kinderhospital Zürich, Universität Zürich, Steinwiesstraße 75, CH-8032 Zürich, Switzerland

Morgenthaler, Stephan ◇ Dept. de Mathematiques, Ecole Polytechnique Fédérale, MA (Ecublens), CH-1015 Lausanne, Switzerland

Müller, Christine ◇ Mathematisches Institut, Freie Universität Berlin, Arnimallee 2–6, D-14195 Berlin, Germany

Nagel, Matthias ◇ ZEG Berlin GmbH, Außenstelle Zwickau, Marchlewskistraße 2, D-08062 Zwickau, Germany

Nuesch, Peter E. ◇ Departement de Mathematiques, Ecole Polytechnique Fédérale, MA (Ecublens), CH-1015 Lausanne, Switzerland

Olbricht, Walter ◇ Mathematik VII, Universität Bayreuth, Universitätsstraße 30, D-95440 Bayreuth, Germany

Rieder, Helmut ◇ LSt. Mathematik VII, Universität Bayreuth, Universitätsstraße 30, D-95440 Bayreuth, Germany

Rousseeuw, Peter J. ◇ Dept. of Mathematics and Computing, Universitaire Instelling Antwerpen, Universiteitsplein 1, B-2610 Wilrijk, Belgium

Ruts, Ida ◇ Faculty of Applied Economic Sciences, UFSIA, Prinsstraat 13, B-2000 Antwerp, Belgium

Samarov, Alex ◇ CCREMS–Sloan School of Management, M.I.T., 1 Amherst Street, E40-131, Cambridge, MA 02139, USA

Sawitzki, Günther ◇ Institut für Angewandte Mathematik, Universität Heidelberg, Im Neuenheimer Feld 294, D-69120 Heidelberg, Germany

Schlather, Martin ◇ Mathematik VII, Universität Bayreuth, Universitätsstraße 30, D-95440 Bayreuth, Germany

Schönholzer, Hans-Peter ◇ Interstaatliche Ingenieurschule, Neu-Technikum Buchs, CH-9470 Buchs, Switzerland

Staudte, Robert G. ◇ Department of Statistics, La Trobe University, Bundoora Campus, Melbourne, Victoria 3083, Australia

Stützle, Werner ◇ Department of Statistics, GN22, University of Washington, Seattle, WA 98195, USA

Tabakis, Evangelos ◇ Mathematik VII, Universität Bayreuth, Universitätsstraße 30, D-95440 Bayreuth, Germany

Tyler, David E. Department of Statistics, Rutgers University, Hill Center, Busch Campus, New Brunswick, NJ 08903, USA

Unwin, Antony ◇ Universität Augsburg, Institut für Mathematik, Universitätsstr. 8, D-86159 Augsburg, Germany

Weichselberger, Kurt ◇ Seminar f. Spezialgebiete der Statistik, Universität München, Ludwigstr. 33, D-80539 München, Germany

Wolf–Ostermann, Karin ◇ Fachbereich Statistik, Universität Dortmund, 7-Eichen 3b, D-57271 Hilchenbach, Germany

Wu, Yuehua ◇ Department of Mathematics and Statistics, York University, 4700 Keele St., North York, Ontario M3J 1P3, Canada

Yohai, Victor J. ◇ Department of Statistics, University of Buenos Aires, Av. Liberator 5256, piso 7, RA-1426 Buenos Aires, Argentina

Zamar, Ruben H. ◇ Department of Statistics, University of British Columbia, 2021 West Mall, Vancouver, BC V6T 1W5, Canada

Lecture Notes in Statistics

For information about Volumes 1 to 22 please contact Springer-Verlag

Vol. 23: D.G. Saphire, Estimation of Victimization Prevalence Using Data from the National Crime Survey. v, 165 pages, 1984.

Vol. 24: T.S. Rao, M.M. Gabr, An Introduction to Bispectral Analysis and Bilinear Time Series Models. viii, 280 pages, 1984.

Vol. 25: E. Parzen (Editor), Time Series Analysis of Irregularly Observed Data. Proceedings, 1983. vii, 363 pages, 1984.

Vol. 26: J. Franke, W. Härdle and D. Martin (Editors), Robust and Nonlinear Time Series Analysis. Proceedings, 1983. ix, 286 pages, 1984.

Vol. 27: A. Janssen, H. Milbrodt, H. Strasser, Infinitely Divisible Statistical Experiments. vi, 163 pages, 1985.

Vol. 28: S. Amari, Differential-Geometrical Methods in Statistics. v, 290 pages, 1985.

Vol. 29: B.J.T. Morgan and P.M. North (Editors), Statistics in Ornithology. xxv, 418 pages, 1985.

Vol 30: J. Grandell, Stochastic Models of Air Pollutant Concentration. v, 110 pages, 1985.

Vol. 31: J. Pfanzagl, Asymptotic Expansions for General Statistical Models. vii, 505 pages, 1985.

Vol. 32: R. Gilchrist, B. Francis and J. Whittaker (Editors), Generalized Linear Models. Proceedings, 1985. vi, 178 pages, 1985

Vol. 33: M. Csörgo, S. Csörgo, L. Horváth, An Asymptotic Theory for Empirical Reliability and Concentration Processes. v, 171 pages, 1986.

Vol. 34: D.E. Critchlow, Metric Methods for Analyzing Partially Ranked Data. x, 216 pages, 1985.

Vol. 35: T. Calinski and W. Klonecki (Editors), Linear Statistical Inference. Proceedings, 1984. vi, 318 pages, 1985.

Vol. 36: B. Matérn, Spatial Variation. Second Edition. 151 pages, 1986.

Vol. 37: R. Dykstra, T. Robertson and F.T. Wright (Editors), Advances in Order Restricted Statistical Inference. Proceedings, 1985. viii, 295 pages, 1986.

Vol. 38: R.W. Pearson and R.F. Boruch (Editors), Survey Research Designs: Towards a Better Understanding of Their Costs and Benefits. v, 129 pages, 1986.

Vol. 39: J.D. Malley, Optimal Unbiased Estimation of Variance Components. ix, 146 pages, 1986.

Vol. 40: H.R. Lerche, Boundary Crossing of Brownian Motion. v, 142 pages, 1986.

Vol. 41: F. Baccelli, P. Brémaud, Palm Probabilities and Stationary Queues. vii, 106 pages, 1987.

Vol. 42: S. Kullback, J.C. Keegel, J.H. Kullback, Topics in Statistical Information Theory. ix, 158 pages, 1987.

Vol. 43: B.C. Arnold, Majorization and the Lorenz Order: A Brief Introduction. vi, 122 pages, 1987.

Vol. 44: D.L. McLeish, Christopher G. Small, The Theory and Applications of Statistical Inference Functions. vi, 124 pages, 1987.

Vol. 45: J.K. Ghosh (Editor), Statistical Information and Likelihood. 384 pages, 1988.

Vol. 46: H.-G. Müller, Nonparametric Regression Analysis of Longitudinal Data. vi, 199 pages, 1988.

Vol. 47: A.J. Getson, F.C. Hsuan, {2}-Inverses and Their Statistical Application. viii, 110 pages, 1988.

Vol. 48: G.L. Bretthorst, Bayesian Spectrum Analysis and Parameter Estimation. xii, 209 pages, 1988.

Vol. 49: S.L. Lauritzen, Extremal Families and Systems of Sufficient Statistics. xv, 268 pages, 1988.

Vol. 50: O.E. Barndorff-Nielsen, Parametric Statistical Models and Likelihood. vii, 276 pages, 1988.

Vol. 51: J. Hüsler, R.-D. Reiss (Editors). Extreme Value Theory, Proceedings, 1987. x, 279 pages, 1989.

Vol. 52: P.K. Goel, T. Ramalingam, The Matching Methodology: Some Statistical Properties. viii, 152 pages, 1989.

Vol. 53: B.C. Arnold, N. Balakrishnan, Relations, Bounds and Approximations for Order Statistics. ix, 173 pages, 1989.

Vol. 54: K.R. Shah, B.K. Sinha, Theory of Optimal Designs. viii, 171 pages, 1989.

Vol. 55: L. McDonald, B. Manly, J. Lockwood, J. Logan (Editors), Estimation and Analysis of Insect Populations. Proceedings, 1988. xiv, 492 pages, 1989.

Vol. 56: J.K. Lindsey, The Analysis of Categorical Data Using GLIM. v, 168 pages, 1989.

Vol. 57: A. Decarli, B.J. Francis, R. Gilchrist, G.U.H. Seeber (Editors), Statistical Modelling. Proceedings, 1989. ix, 343 pages, 1989.

Vol. 58: O.E. Barndorff-Nielsen, P. Blæsild, P.S. Eriksen, Decomposition and Invariance of Measures, and Statistical Transformation Models. v, 147 pages, 1989.

Vol. 59: S. Gupta, R. Mukerjee, A Calculus for Factorial Arrangements. vi, 126 pages, 1989.

Vol. 60: L. Györfi, W. Härdle, P. Sarda, Ph. Vieu, Nonparametric Curve Estimation from Time Series. viii, 153 pages, 1989.

Vol. 61: J. Breckling, The Analysis of Directional Time Series: Applications to Wind Speed and Direction. viii, 238 pages, 1989.

Vol. 62: J.C. Akkerboom, Testing Problems with Linear or Angular Inequality Constraints. xii, 291 pages, 1990.

Vol. 63: J. Pfanzagl, Estimation in Semiparametric Models: Some Recent Developments. iii, 112 pages, 1990.

Vol. 64: S. Gabler, Minimax Solutions in Sampling from Finite Populations. v, 132 pages, 1990.

Vol. 65: A. Janssen, D.M. Mason, Non-Standard Rank Tests. vi, 252 pages, 1990.

Vol 66: T. Wright, Exact Confidence Bounds when Sampling from Small Finite Universes. xvi, 431 pages, 1991.

Vol. 67: M.A. Tanner, Tools for Statistical Inference: Observed Data and Data Augmentation Methods. vi, 110 pages, 1991.

Vol. 68: M. Taniguchi, Higher Order Asymptotic Theory for Time Series Analysis. viii, 160 pages, 1991.

Vol. 69: N.J.D. Nagelkerke, Maximum Likelihood Estimation of Functional Relationships. V, 110 pages, 1992.

Vol. 70: K. Iida, Studies on the Optimal Search Plan. viii, 130 pages, 1992.

Vol. 71: E.M.R.A. Engel, A Road to Randomness in Physical Systems. ix, 155 pages, 1992.

Vol. 72: J.K. Lindsey, The Analysis of Stochastic Processes using GLIM. vi, 294 pages, 1992.

Vol. 73: B.C. Arnold, E. Castillo, J.-M. Sarabia, Conditionally Specified Distributions. xiii, 151 pages, 1992.

Vol. 74: P. Barone, A. Frigessi, M. Piccioni, Stochastic Models, Statistical Methods, and Algorithms in Image Analysis. vi, 258 pages, 1992.

Vol. 75: P.K. Goel, N.S. Iyengar (Eds.), Bayesian Analysis in Statistics and Econometrics. xi, 410 pages, 1992.

Vol. 76: L. Bondesson, Generalized Gamma Convolutions and Related Classes of Distributions and Densities. viii, 173 pages, 1992.

Vol. 77: E. Mammen, When Does Bootstrap Work? Asymptotic Results and Simulations. vi, 196 pages, 1992.

Vol. 78: L. Fahrmeir, B. Francis, R. Gilchrist, G. Tutz (Eds.), Advances in GLIM and Statistical Modelling: Proceedings of the GLIM92 Conference and the 7th International Workshop on Statistical Modelling, Munich, 13-17 July 1992. ix, 225 pages, 1992.

Vol. 79: N. Schmitz, Optimal Sequentially Planned Decision Procedures. xii, 209 pages, 1992.

Vol. 80: M. Fligner, J. Verducci (Eds.), Probability Models and Statistical Analyses for Ranking Data. xxii, 306 pages, 1992.

Vol. 81: P. Spirtes, C. Glymour, R. Scheines, Causation, Prediction, and Search. xxiii, 526 pages, 1993.

Vol. 82: A. Korostelev and A. Tsybakov, Minimax Theory of Image Reconstruction. xii, 268 pages, 1993.

Vol. 83: C. Gatsonis, J. Hodges, R. Kass, N. Singpurwalla (Editors), Case Studies in Bayesian Statistics. xii, 437 pages, 1993.

Vol. 84: S. Yamada, Pivotal Measures in Statistical Experiments and Sufficiency. vii, 129 pages, 1994.

Vol. 85: P. Doukhan, Mixing: Properties and Examples. xi, 142 pages, 1994.

Vol. 86: W. Vach, Logistic Regression with Missing Values in the Covariates. xi, 139 pages, 1994.

Vol. 87: J. Møller, Lectures on Random Voronoi Tessellations. vii, 134 pages, 1994.

Vol. 88: J. E. Kolassa, Series Approximation Methods in Statistics. viii, 150 pages, 1994.

Vol. 89: P. Cheeseman, R.W. Oldford (Editors), Selecting Models From Data: AI and Statistics IV. xii, 487 pages, 1994.

Vol. 90: A. Csenki, Dependability for Systems with a Partitioned State Space: Markov and Semi-Markov Theory and Computational Implementation. x, 241 pages, 1994.

Vol. 91: J.D. Malley, Statistical Applications of Jordan Algebras. viii, 101 pages, 1994.

Vol. 92: M. Eerola, Probabilistic Causality in Longitudinal Studies. vii, 133 pages, 1994.

Vol. 93: Bernard Van Cutsem (Editor), Classification and Dissimilarity Analysis. xiv, 238 pages, 1994.

Vol. 94: Jane F. Gentleman and G.A. Whitmore (Editors), Case Studies in Data Analysis. viii, 262 pages, 1994.

Vol. 95: Shelemyahu Zacks, Stochastic Visibility in Random Fields. x, 175 pages, 1994.

Vol. 96: Ibrahim Rahimov, Random Sums and Branching Stochastic Processes. viii, 195 pages, 1995.

Vol. 97: R. Szekli, Stochastic Ordering and Dependence in Applied Probablility. viii, 194 pages, 1995.

Vol. 98: Philippe Barbe and Patrice Bertail, The Weighted Bootstrap. viii, 230 pages, 1995.

Vol. 99: C.C. Heyde (Editor), Branching Processes: Proceedings of the First World Congress. viii, 185 pages, 1995.

Vol. 100: Wlodzimierz Bryc, The Normal Distribution: Characterizations with Applications. viii, 139 pages, 1995.

Vol. 101: H.H. Andersen, M.Højbjerre, D. Sørensen, P.S.Eriksen, Linear and Graphical Models: for the Multivariate Complex Normal Distribution. x, 184 pages, 1995.

Vol. 102: A.M. Mathai, Serge B. Provost, Takesi Hayakawa, Bilinear Forms and Zonal Polynomials. x, 378 pages, 1995.

Vol. 103: Anestis Antoniadis and Georges Oppenheim (Editors), Wavelets and Statistics. vi, 411 pages, 1995.

Vol. 104: Gilg U.H. Seeber, Brian J. Francis, Reinhold Hatzinger, Gabriele Steckel-Berger (Editors), Statistical Modelling: 10th International Workshop, Innsbruck, July 10-14th 1995. x, 327 pages, 1995.

Vol. 105: Constantine Gatsonis, James S. Hodges, Robert E. Kass, Nozer D. Singpurwalla(Editors), Case Studies in Bayesian Statistics, Volume II. x, 354 pages, 1995.

Vol. 106: Harald Niederreiter, Peter Jau-Shyong Shiue (Editors), Monte Carlo and Quasi-Monte Carlo Methods in Scientific Computing. xiv, 372 pages, 1995.

Vol. 107: Masafumi Akahira, Kei Takeuchi, Non-Regular Statistical Estimation. vii, 183 pages, 1995.

Vol. 108: Wesley L. Schaible (Editor), Indirect Estimators in U.S. Federal Programs. iix, 195 pages, 1995.

Vol. 109: Helmut Rieder (Editor), Robust Statistics, Data Analysis, and Computer Intensive Methods. xiv, 427 pages, 1996.